T4-ALG-799

Laboratory Safety Second Edition

PRINCIPLES AND PRACTICES

Laboratory Safety
Second Edition

PRINCIPLES AND PRACTICES

EDITORS

Diane O. Fleming, Ph.D.
Biosafety Consultant
Bowie, Maryland
Environmental Health Engineering
The Johns Hopkins University
School of Hygiene and Public Health
Baltimore, Maryland

John H. Richardson, D.V.M., M.P.H.
Biological Safety Consultant
Atlanta, Georgia

Jerry J. Tulis, Ph.D.
Biohazard Science Program
Division of Occupational and
Environmental Medicine
Duke University Medical Center
Durham, North Carolina

Donald Vesley, Ph.D.
Environmental and Occupational Health
School of Public Health
University of Minnesota
Minneapolis, Minnesota

ASM Press
Washington, D.C.

1325 Massachusetts Avenue, N.W.
Washington, D.C.

Library of Congress Cataloging-in-Publication Data

Laboratory safety: principles and practices / editors, Diane O. Fleming . . . [et al.].—2nd ed.
p. cm.
Includes bibliographical references and index.
ISBN 1-55581-047-0
1. Microbiological laboratories—Safety measures. I. Fleming, Diane O. II. American Society for Microbiology.
[DNLM: 1. Accidents, Occupational—prevention & control—United States.
2. Laboratories—standards—United States. WA 485 L123 1995]
QR64.7.L33 1995
576'.0289—dc20
DNLM/DLC
for Library of Congress 94-21372
CIP

Contents

Appendixes

Contributors

DONALD C. BLENDEN, D.V.M. (retired)
1506 North Circle Drive, Columbia, Missouri 65203

WALTER W. BOND, M.S.
Hospital Infections Program, Nosocomial Infections Laboratory Branch, National Center for Infectious Diseases, Centers for Disease Control and Prevention, Atlanta, Georgia 30333

JOHN M. BOYCE, M.D.
Division of Infectious Diseases, The Miriam Hospital, 164 Summit Avenue, Providence, Rhode Island 02096

MARK A. CHATIGNY
University of California, San Francisco, 507 Santa Ynez, San Leandro, California 94575

JONATHAN T. CRANE, A.I.A.
Lord, Aeck & Sargent, Inc., 400 Colony Square, Suite 300, 1201 Peachtree Street N.E., Atlanta, Georgia 30361

RICHARD L. EHRENBERG, M.D.
Office of the Deputy Director, National Institute for Occupational Safety and Health, Centers for Disease Control and Prevention, Building 1-1043, Mailstop D-26, Atlanta, Georgia 30333

MARTIN S. FAVERO, Ph.D.
Hospital Infections Program, Nosocomial Infections Laboratory Branch, National Center for Infectious Diseases, Centers for Disease Control and Prevention, Atlanta, Georgia 30333

DIANE O. FLEMING, Ph.D.
Biosafety Consultant, 15611 Plumwood Court, Bowie, Maryland 20716
Environmental Health Engineering, The Johns Hopkins University, School of Hygiene and Public Health, 615 North Wolfe Street, Baltimore, Maryland 21205

HOWARD FRUMKIN, M.D., M.P.H.
Division of Environmental and Occupational Health, Emory University School of Public Health, Atlanta, Georgia 30322

ROBYN R. M. GERSHON, M.T., M.H.S., Dr.P.H.
Department of Environmental Health Sciences, School of Hygiene and Public Health, The Johns Hopkins University, 615 North Wolfe Street, Baltimore, Maryland 21205

MARY J. R. GILCHRIST, Ph.D.
VA Medical Center, MDP 113, 3200 Vine Street, Cincinnati, Ohio 45220

LYNN HARDING, M.S.
Biosafety Consultant, 2846 Crestwood Avenue, Chattanooga, Tennessee 37415

BARBARA L. HERWALDT, M.D., M.P.H.
Parasitic Diseases Branch, National Center for Infectious Diseases, Centers for Disease Control and Prevention, 4770 Buford Highway N.E., Mailstop F-22, Atlanta, Georgia 30341

CAROL LAX HOERNER, Ph.D.
Carol Lax Hoerner, Ph.D., Inc., 19 Cazneau Avenue, Sausalito, California 94965

DEBRA L. HUNT, Dr.P.H.
Biological Safety, Duke University Medical Center, Box 3149, Durham, North Carolina 27710

DENNIS D. JURANEK, D.V.M., M.Sc.
Parasitic Diseases Branch, National Center for Infectious Diseases, Centers for Disease Control and Prevention, 4770 Buford Highway N.E., Atlanta, Georgia 30341

ARNOLD F. KAUFMANN, D.V.M.
Division of Bacterial and Mycotic Diseases, National Center for Infectious Diseases, Centers for Disease Control and Prevention, Atlanta, Georgia 30333

RALPH W. KUEHNE
Biosafety Consultant, 5596 Jollie Drive, Frederick, Maryland 21702

JAMES L. LAUER, Ph.D.
Department of Environmental Health and Safety, University of Minnesota, Minneapolis, Minnesota 55455

DANIEL F. LIBERMAN, Ph.D.
Boston University Medical Center, 80 East Concord Street, V-324, Boston, Massachusetts 02118

FRANK S. LISELLA, Ph.D.
Environmental Health and Safety, Emory University, P.O. Box AC, Atlanta, Georgia 30322

GERARD J. McGARRITY, Ph.D.
Genetic Therapy, Inc., 938 Clopper Road, Gaithersburg, Maryland 20878

SUSAN E. McNULTY, M.Sc.
Department of Environmental Health Sciences, SL-29, School of Public Health and Tropical Medicine, Tulane University, 1430 Tulane Avenue, New Orleans, Louisiana 70112

JOHN W. McVICAR, D.V.M.
Biosafety Consultant, 3905 McTyres Cove Court, Midlothian, Virginia 23112

JOHN H. RICHARDSON, M.P.H., D.V.M.
Biosafety Consultant, 2982 Danbyshire Court, Atlanta, Georgia 30345

JONATHAN Y. RICHMOND, Ph.D.
Office of Health and Safety, Centers for Disease Control and Prevention, Building 14, Mail Stop F05, 1500 Clifton Road, Atlanta, Georgia 30333

ROBERT S. RUNKLE
EA Engineering, Science, and Technology, Inc., Deerfield, Illinois 60015

JOSEPH R. SONGER, D.V.M.
Biosafety Consultant, 419 Ninth Avenue, Ames, Iowa 50010

BRUCE W. STAINBROOK
Law Environmental, Inc., Kennesaw, Georgia 30144

DAVID G. STUART, Ph.D.
The Baker Company, Inc., P.O. Drawer E, Sanford, Maine 04073

JANE SUEN, Ph.D.
Division of Public Health Systems, Centers for Disease Control and Prevention, Atlanta, Georgia 30333

SCOTT W. THOMASTON, M.S.
Environmental Health and Safety, Emory University, P.O. Box AC, Atlanta, Georgia 30322

WILLIAM A. TOSCANO, JR., Ph.D.
Department of Environmental Health Sciences, SL-29, School of Public Health and Tropical Medicine, Tulane University, 1430 Tulane Avenue, New Orleans, Louisiana 70112

DONALD VESLEY, Ph.D.
School of Public Health, University of Minnesota, Mayo Memorial Building, Box 807, 420 Delaware Street S.E., Minneapolis, Minnesota 55455

WILLIAM G. WINKLER, D.V.M. (retired)
5472 Hugh Howell Road, Stone Mountain, Georgia 30387

BARBARA G. ZIRKIN, Ed.D.
Department of Environmental Health Sciences, School of Hygiene and Public Health, The Johns Hopkins University, 615 North Wolfe Street, Baltimore, Maryland 21205

Foreword

THE MECHANICS OF CONDUCTING CLINICAL AND RESEARCH experiments change as biomedical science continues its explosive growth. New technologies emerge overnight to meet ever-increasing demands for more rapid techniques (often miniaturized, amplified, and automated) for laboratory activities, increasingly focused on marketable end products. Associated with these new technologies have been the rapid crossing and subsequent erosion of traditional boundaries between scientific disciplines, e.g., microbiologists using hazardous chemicals and radioactive materials, or chemists working with biohazardous microorganisms. DNA sequencers and synthesizers come to mind.

The potential for increased illnesses and injuries has gone up concomitantly, and new regulations promulgated to protect laboratory workers and the environment have proliferated. Costs for compliance (renovation of space, new construction, disposable equipment, waste management) take bigger chunks out of budgets. Meanwhile, the emergence of new diseases—HIV, multi-drug-resistant tuberculosis, and other nosocomial infections, and the new hantaviruses—continues to stimulate biosafety concerns among laboratorians.

This second edition of *Laboratory Safety: Principles and Practices* is a major contribution to the field of laboratory administration and management, and it will also serve as a valuable resource document for laboratorians and engineers.

This edition begins with a powerful statement on the epidemiology of laboratory-acquired infections during the past decade and includes, in subsequent chapters, documentation of additional previously unreported laboratory infections. The accompanying hazard assessments provide a structured means for evaluating the risks while keeping us mindful that occupationally acquired infections (occasionally resulting in death) still occur.

When the first edition was published, little attention was paid to infectious waste management or the field of biotechnology; in the intervening years, public and scientific interest has changed, and this edition provides excellent discussions on both subjects. These subject areas should be of particular interest in countries that have only recently begun to experience the initial growth phase of the new sciences.

Finally (it is interesting that this subject is often addressed last), the arguments for effective training, which also increasingly consumes considerable resources, are well presented. Focused training is often preceded by the development of safety manuals and associated with audits of laboratory practices and procedures. Increased worker awareness of the hazards and the methods for risk reduction brings increasing demands for better safety programs, a noble vision.

Jonathan Y. Richmond
Director, Office of Health and Safety
Centers for Disease Control and Prevention
Atlanta, Georgia

Introduction

I

Introduction

JOHN H. RICHARDSON

1

INTRODUCTION

The second edition of *Laboratory Safety: Principles and Practices* is written in recognition of the ongoing hazards of working with a variety of infectious agents and hazardous substances in the laboratory. The objectives of this book are to assist in the identification and the assessment of work-related risks; to provide an historical review of previously reported laboratory infections, illnesses, and deaths; to describe the hierarchical control methods for working with hazardous agents or materials; and to provide information on administrative controls available and proposed for use in safety management.

Important considerations facing laboratory management include the infectious agent(s) knowingly or unknowingly handled by laboratory workers; the level of awareness of infection risks to the employee; the appropriateness of laboratory practices used; the economics of laboratory safety programs, and the prioritization of organizational objectives to include safety management as a first-line goal.

INFECTIOUS AGENTS

Many published studies and surveys provide reasonable documentation of the hazards and risks of occupational infections in persons working in clinical and other categories of biomedical laboratories (5, 12, 13, 15, 16). Many of these studies are described in Chapter 2. Historically, bacterial diseases including typhoid, shigellosis, brucellosis, and tuberculosis represented the main occupational hazards of clinical laboratory workers during the first 60 years of this century. As procedures for the isolation and identification of viral pathogens came into general use, the incidence of occupational infections with various viral agents showed a corresponding increase (3, 4, 7).

In the preamble to a *Notice of Proposed Rule Making on Blood-borne Pathogens,* the Occupational Safety and Health Administration (OSHA) concluded that hepatitis B virus (HBV) represented the current and preeminent occupational infection hazard to the nation's estimated 5.3 million health care workers, including more than 500,000 workers in clinical and other biomedical laboratory activities (10). OSHA estimated that as many as 12,000 cases of occupational HBV infections and 200 hepatitis-associated deaths occur annually in health care workers. Although laboratory workers are not differentiated from other categories of health care workers in this OSHA document, many studies indicate that laboratory workers may experience incidence rates of HBV infection several times greater than those observed in the general population in respective geographic areas (3, 5, 7). Overall, OSHA estimated the incidence rate for HBV infection in susceptible health care workers (i.e., those workers without protective antibody from immunization or natural disease) ranges from 489 to 663 cases per 100,000 compared with 176 per 100,000 in overall adult populations. About 300,000 new cases of HBV are estimated to occur annually in the United States.

The availability of safe and efficacious vaccines against HBV infection has seemingly made only a modest impact on reducing the incidence of HBV in the workplace and in the community to date. Immunization of at-risk workers, mandated by the blood-borne pathogen standard, will cumulatively and positively have an effect on reducing the occupational risk of this viral disease. The Advisory Committee on Immunization Practices has now recommended HBV vaccine for all infants. This will provide a larger impact by herd immunity to HBV; however, this effect will not show up in the work force for at least 15 to 20 years. Further information about HBV and other hepatitis viruses can be found in Chapter 3.

There is no indication that another infectious agent is likely to replace HBV as the primary occupational infection hazard in the foreseeable future. Although there has been a recent upswing in reported cases of tuberculosis (including multiple drug-resistant strains) and systemic fungal and parasitic diseases in immunocompromised hosts, community and occupational infections with these agents are not likely to reach levels currently observed for HBV. Chapter 5 has been written to consider some of these special pathogens.

Although the consequences of infection with exotic hemorrhagic fever agents (e.g., Ebola, Marburg, and Lassa fever viruses) are grave, the risk of infection to health care and laboratory workers is low. Transmission patterns in community and occupational settings seem to parallel those seen with blood-borne pathogens. The recent history of these hemorrhagic fever agents suggests that the likelihood of imported cases is low. The morbid fear that erupted in the laboratory community after reports of laboratory-acquired human immunodeficiency virus (HIV) infection in 1987 was not followed by an epidemic of similar laboratory-related infections. A detailed description of HIV and work-related disease is found in Chapter 4.

In the 20 years since the Asilomar Conference, the rapidly growing field of genetic engineering has failed to produce the dreaded Andromeda strains predicted by some of the participants in that meeting. Oversight by the Recombinant Advisory Committee and voluntary guidelines, which are based on practices developed for conventional microorganisms, have proved satisfactory for minimizing risks to personnel and the environment. Further discussion of safety in biotechnology is found in Chapter 9.

LABORATORY PRACTICES

Safe practices are not new. In 1913, Eyre's *Bacteriologic Technique* (2) proposed regulations to be routinely followed in bacteriology laboratories. These recommended practices, which still apply, include

- Wearing of protective gloves
- Disinfection of hands after the handling of infectious materials
- Disinfection of instruments immediately after use
- Moistening of labels for specimen and culture containers with water (rather than by mouth)
- Disinfection of clinical specimens, cultures, and other contaminated wastes before discard
- Mandatory reporting of accidents and exposures to infectious materials

These basic good laboratory practices have subsequently been incorporated, in principle, in most current national and international guidelines, professional standards, safety manuals, and regulations pertaining to activities involving the use of infectious materials (1, 6, 9, 10, 14–17). Laboratory practices are described in more detail in Chapter 13.

In 1977, Pike (13) stated, "It appears that the knowledge, the technique and the equipment necessary to prevent most laboratory infections are available and that diligent application of preventive measures is necessary if a substantial decrease in the incidence of such infections is to be maintained." Behavioral aspects of safety that have an effect on the application of preventive measures are discussed in Chapter 18.

There can be no plausible argument against immediately implementing and diligently applying basic good laboratory practices in all laboratories where work with known or potentially infectious materials is conducted. The barriers to implementation continue to include the lack of true administrative support for the program and the economic considerations that do not include a true cost-benefit analysis.

ECONOMICS

Safety programs are the responsibility of institutional and organizational management; safety organization is the subject of Chapter 17. Provision and use of appropriate practices and barriers (Chapter 13), personal protective equipment (Chapter 11), suitable facilities (Chapter 12), occupational health services (Chapter 19), and other preventive measures may represent significant initial program expenditures as well as ongoing annual costs. The estimated minimum cost per employee is $466 per year for providing simple, basic personal protective equipment and barrier precautions in a typical clinical laboratory where manipulations of blood and other potentially infectious materials are handled on a regular basis. Additional per-employee costs may reasonably include charges for engineering and work practice controls (e.g., sharps disposal containers), development and implementation of a written exposure control plan, initial and annual training of employees covered by the exposure control plan, record keeping, and postexposure medical costs.

The costs for not implementing safety programs may predictably result in a significant budget impact in terms of lost time, decreased morale, and health care costs for infected workers, as well as penalties for failure to comply with mandated standards. Effective safety programs may, however, prove ultimately cost-effective. The costs for laboratory workplace safety will represent an ongoing and significant budget line item. Under mandated federal regulations, safety programs and practices are no longer voluntary.

AWARENESS OF RISKS

Few events, perhaps no event in this century, have created a greater level of awareness of occupational infection risk than the continuing pandemic of HIV. Although the demonstrated risk of occupational infection with HIV in health care workers has been consistently low (worst case, three to five cases per 1,000 after accidental needlesticks), the consequences of infection continue to be grave irrespective of the source of infection. For this reason, HIV is covered in detail in Chapter 4.

One overall positive aspect of the pandemic has, however, been to focus attention on both specific and generic occupational infection hazards. This awareness has served to create an overdue respect for the ongoing epidemic of HBV in health care workers and to stimulate the general use of common sense practices, barriers, and immunizations to prevent occupational infections.

Crude attack rates of viral hepatitis (HBV and non-A, non-B hepatitis) in British clinical pathologists declined from 123 cases per 100,000 for the period 1970 to 1974 to 27 cases per 100,000 for the period 1975 to 1979 (3). This dramatic reduction in the incidence of occupational infection occurred during a decade in which two highly publicized outbreaks of smallpox of laboratory origin occurred in Britain. During this same decade, studies indicated widespread contamination of equipment, specimens, specimen containers, and work surfaces in laboratories handling blood and other body fluids (3, 8). At that time, in the absence of a vaccine against HBV or national regulations governing laboratory activities, Grist (3) concluded that the decline in the incidence of viral hepatitis during the latter 5-year period was most plausibly explained by an increased awareness of occupational infection risk and a more consistent use of appropriate practices and barriers to minimize overt as well as inapparent exposures.

Risk awareness is an integral component of training mandated by the blood-borne pathogen standard. Optimistically, as the level of awareness of risks becomes universal, the incidence of occupational infection will show a corresponding decline. Section II of this book provides information on the risks associated with work using a variety of pathogenic microorganisms (Chapters 3 through 10).

OTHER MANAGEMENT CONSIDERATIONS

Modern clinical laboratories are complex entities that are expected to provide a variety of comprehensive services with a high degree of proficiency. Such entities typically require competent and skilled scientists, sophisticated and expensive equipment, and special facility features to accomplish daily tasks. Furthermore, clinical laboratories are subject to regulatory requirements for federal and state authorities as well as standards set by national accrediting organizations.

In this climate of intense competition for staff time and effort, safety programs, established under voluntary guidelines, are often relegated by management to a secondary or tertiary priority status. Safety became a first-level priority for clinical laboratory management in 1992, with the implementation of the OSHA blood-borne pathogen standard (11).

Knowledge of the agent(s) used, inherent risks, and appropriate preventive practices and measures as well as applicable regulatory standards are essential components of a laboratory safety program. Effective safety programs depend on a balanced application of appropriate preventive practices based on the assessed risk of the agent(s) or material(s) used and the activity conducted. The success of safety programs ultimately depends on the adoption and diligent, uniform application of appropriate safety practices and preventive measures by each institution and by each worker to perform tasks as well and as safely as laboratory scientists know how to perform them.

COURSES OF ACTION

The series of federal regulations applicable to private-sector laboratories (hazard communication, chemical hygiene, and blood-borne pathogen standards) implemented since 1987 establishes mandatory performance standards for facilities where work with hazardous chemicals and potentially infectious materials is conducted. These mandatory standards were developed in large part because voluntary standards were not achieving acceptable safety goals. Additional federal regulations governing private-sector laboratory operations can be expected based on OSHA's recent actions to protect health care workers and emergency response personnel.

In the future, it will continue to be incumbent on safety managers and laboratory scientists to become familiar with the regulatory process, to provide input in the rule-making procedure, and to work with regulatory agencies to ensure that current and future standards are scientifically based and realistically applied.

References

1. **Collins, C. H., E. G. Hartley, and R. Pilsworth.** 1974. The prevention of laboratory-acquired infection. Public Health Laboratory Service Monogr. Ser. No. 6. Her Majesty's Stationery Office, London.
2. **Eyre, J. W. H.** 1913. *Bacteriologic technique.* W. B. Saunders Co., Philadelphia.
3. **Grist, N. R.** 1982. Epidemiology and control of viral infections in the laboratory. *Yale J. Biol. Med.* **55:**213–218.
4. **Hanson, R. P., S. E. Sulkin, E. L. Buescher, W. M. Hammond, R. W. McKinney, and T. H. Work.** 1976. Arbovirus infection in laboratory workers. *Science* **1958:**1283–1286.
5. **Harrington, J. M., and H. S. Shannon.** 1976. Incidence of tuberculoses, hepatitis, brucellosis and shigelloses in British medical laboratory workers. *Br. Med. J.* **1:**759–762.
6. **Health and Welfare Canada and Medical Research Counsel of Canada.** 1990. *Laboratory biosafety guidelines.* Ministry of Supply and Services, Ottawa.
7. **Hicks, C. G., C. O. Hargiss, and J. R. Harris.** 1985. Prevalence survey for hepatitis B in high-risk university employees. *Am. J. Infect. Control* **12:**1–6.
8. **Lauer, J. L., N. A. Van Drumen, J. W. Washburn, and H. H. Balfour.** 1979. Transmission of hepatitis B virus in clinical laboratory areas. *J. Infect. Dis.* **140:** 513–516.
9. **National Committee for Clinical Laboratory Standards.** 1987. Protection of laboratory workers from infectious disease transmitted by blood and tissue. Proposed guidelines M29-P. National Committee for Clinical Laboratory Standards, Villanova, Pa.
10. **Occupational Safety and Health Administration.** 1989. Occupational exposure to blood-borne pathogens, proposed rule. 29CFR Part 1910.1030. *Fed. Regist.* **54** (102):23042–23139.
11. **Occupational Safety and Health Administration.** 1991. Occupational exposure to blood-borne pathogens, final rule. 29CFR Part 1910.1030. *Fed. Regist.* **56** (235):64175–64182.
12. **Pike, R. M.** 1976. Laboratory-associated infections: summary and analyses of 3,921 cases. *Health Lab. Sci.* **12**(2):105–114.
13. **Pike, R. M.** 1979. Laboratory-associated infections: incidence, fatalities, causes, and prevention. *Am. Rev. Microbiol.* **13:**41–66.
14. **Richmond, J. Y., and R. W. McKinney (ed.).** 1993. *Biosafety in microbiological and biomedical laboratories.* HHS Publication No. (CDC) 93-8395. U.S. Government Printing Office, Washington, D.C.
15. **Subcommittee on Arbovirus Laboratory Safety.** 1980. Laboratory safety for arboviruses and certain other viruses of vertebrates. *Am. J. Trop. Med. Hyg.* **29**(6):1359–1381.
16. **Wedum, A. G., W. E. Barkley, and A. Hellman.** 1972. Handling of infectious agents. *J. Am. Vet. Med. Assoc.* **161**(11):1547–1567.
17. **World Health Organization.** 1993. *Laboratory biosafety manual,* 2nd ed. World Health Organization, Geneva.

Epidemiology of Laboratory-Associated Infections

LYNN HARDING AND DANIEL F. LIBERMAN

2

INTRODUCTION

One of the unfortunate consequences of working with infectious materials is the potential for acquiring a laboratory-associated infection (LAI). History has shown that such infections occur and that laboratory workers are clearly at higher risk for infection with certain agents (i.e., hepatitis B virus [HBV], *Brucella*) than the general population. Our ability to quantify LAIs accurately is hampered by an indifference to and, frequently, an unwillingness to report these incidents.

In the absence of precise data on LAIs, epidemiologic methods provide the necessary tools to evaluate the extent and nature of personnel exposures. Ultimately, through the use of this information, potential exposures may be prevented by implementation of appropriate work practices and safety equipment, improvement of facilities, and more important, rigorous training of technical and support personnel.

In this chapter, we examine the extent of known LAIs. The discussion includes a review of the incidence of these infections, the means by which workers are exposed, the contributions to exposure made by human and environmental factors, and host susceptibility. The reader is encouraged to refer to other chapters in this book for detailed information on hazard assessment and prevention strategies.

EPIDEMIOLOGIC SEQUENCE

There is an orderly sequence that describes the actions to be taken to evaluate the causality between events and effects. First, the observer, an epidemiologist, defines the events or illnesses to be studied and establishes a "case" definition. Next, the cases or events that fit the case definition are identified, counted, and oriented to time, place, and person. From this information, the population at risk is identified and further studied. Using simple quantitative methods, rates of illness in this population are determined. The number of cases (numerator) is determined as a proportion or rate of the total population at risk (denominator). The calculated rates of the cases in this population (attack rates) can then be compared with the rates of occurrence in other populations. For example, attack rates can be compared in hematologists, nurses, immunologists, research investigators, janitors, maintenance personnel, or glassware washers. Using this information, the observer can make inferences about events related to a given health problem. The type of work can be correlated with exposure to certain agents (bacteria, fungi, or viruses) that cause health effects. The health status of the worker (susceptibility) might be related to relevant environmental factors such as the primary and secondary containment being used to protect the worker or perhaps the route of entry. Finally, interventions can be defined. Control measures, procedures and practices that may prevent the occurrence or recurrence of the event, can be implemented.

This chapter focuses on the application of epidemiologic methods to LAI to provide the tools for identifying and preventing health problems and disease processes that may emerge. Our goal is to tie historical and current facts together, pointing out observations that can provide the tools to control and prevent these LAIs.

HISTORY OF LAI (OBSERVATION)

During the past 40 years, efforts have been made to determine the significance of LAIs. Studies during the 1930s and 1940s demonstrated that bacterial, fungal, and rickettsial agents were potentially hazardous to individuals both within and in the vicinity of laboratories (53, 60).

Estimates of the extent of LAI are imprecise. Information available from publications, questionnaires, and personal communication usually summarizes acute symptomatic infections with minimal data on seroconversion or nonsymptomatic response to laboratory-associated microorganisms. Rarely was the host immune competence even considered. In the absence of any centralized reporting of infections and routine assessment of worker exposure, it is impossible to assess the true incidence of LAIs.

The potential for laboratory-acquired viral infections was addressed by Sulkin and Pike (61) in 1951. These authors also published cumulative surveys describing LAI during 1924 to 1977 (50–54, 60). In 1978, Pike (51) reported on 4,079 documented LAIs, 168 of which were fatal. As an apparent reflection of the work being performed in the responding laboratories at that time, bacteria accounted for 1,704 of the infections, viruses for 1,179, rickettsia for 598, fungi for 354, chlamydia for 128, and parasites for 116. Bacteria or viruses were associated with more than two-thirds of the lethal and nonlethal infections. Brucellosis, Q fever, hepatitis, typhoid, tularemia, and tuberculosis were the diseases most frequently reported (52). Although the risk of infection with these agents is great, Pike noted that most (96%) *Brucella* and typhoid fever cases and 60% of hepatitis cases were reported before 1955. Hepatitis infections continue to be of occupational significance for laboratory and health care workers (28).

In an attempt to update earlier LAI information, 58 publications between 1980 and 1991 were reviewed. These publications document symptomatic LAIs or seroconversions from a single agent or genus. Although the survey is incomplete, several interesting observations can be made. Three hundred seventy-five infections or seroconversions were reported, with five deaths.

Bacterial Infections

Most of the 65 bacterial infections were caused by *Salmonella typhi* (8), *Brucella melitensis* (2, 19, 26, 58, 59) and *Chlamydia* spp. (7, 40). The reappearance of *Salmonella* and *Brucella* infections in the modern laboratory, with increasingly sophisticated safety equipment, is of particular interest. These illnesses reinforce the need for worker training and adherence to good microbiological work practices. The typhoid infections were traced to educational and proficiency exercises used to assess a laboratory's ability to identify various microorganisms. *Neisseria meningitidis* was the microorganism responsible for the two bacterial deaths (14).

Viral Infections

Of 119 viral infections reported, 3 were fatal. Seventeen different viruses were identified. Deaths were attributed to herpes B virus (10, 11) and Creutzfeldt-Jakob disease virus (44). The availability of antiviral therapy is attributed to saving the lives of several herpes B-infected persons. HBV infection has been and continues to be among the most commonly reported LAIs (21). HBV, because of its significance in the workplace, is covered in detail in Chapter 3.

Approximately three-quarters of the 119 viral infections or seroconversions were caused by miscellaneous arboviruses (33, 45, 62) and hantaviruses (17, 37, 39, 69). Hantaan virus, a hantavirus (family *Bunyaviridae*), is the etiologic agent of Korean hemorrhagic fever. LAI related to hantavirus as well as to the human immunodeficiency virus (HIV) was first reported in the 1980s. Because of the projected high morbidity rate and long latency period associated with HIV, fear of HIV infection is foremost in the minds of laboratory and health care personnel. To date, the Centers for Disease Control and Prevention (CDC) has documented the occupational exposure and subsequent seroconversion of 39 health care workers in the United States (see also Chapter 4). Current trends indicate that the number of HIV LAIs continues to increase.

In 1989, Ebola-related filovirus seroconversions were documented among several asymptomatic animal handlers in U.S. primate handling facilities (13). Although infection with this virus did not produce the fatal hemorrhagic disease associated with other filovirus infections in Europe and Africa (Marburg virus and Ebola virus), it prompted closer scrutiny of nonhuman primate colonies and the development of guidelines for handling nonhuman primates during transit and quarantine (12). Other viruses associated with infection or seroconversion during this period included parvovirus B19 (16), echovirus 11 (43), orf virus (46), and the first report of infection (seroconversion) caused by a recombinant construct, a vaccinia virus expressing a vesicular stomatitis virus gene (35).

Protozoal Infections

Among the 58 current references reviewed for this survey, only 13 LAIs were caused by protozoa. They included *Leishmania* (22, 23, 57), *Trypanosoma* (18, 20, 32), *Plasmodium* (6, 67), *Cryptosporidium* (55), and *Toxoplasma* species (5). Most infections were related to exposure by needle puncture.

Rickettsial Infections

The number of reported rickettsial infections appears to have increased. One hundred fifty-three of 162 rickettsial infections were attributed to *Coxiella burnetii* (27, 42, 47). Research involving the use of sheep in laboratory and hospital facilities has exposed researchers and noninvolved personnel to *C. burnetii*, the etiologic agent of Q fever.

Fungal Infections

Those fungal LAIs reported were caused by saprophytic organisms or opportunistic pathogens (70).

MEANS OF EXPOSURE

There is no question that laboratory personnel continue to be at risk for occupational exposure to infectious agents. The most common routes of exposure are percutaneous inoculation (needle and syringe, cuts or abrasions from contaminated items, and animal bites), inhalation of aerosols generated by accident or by work practices and procedures, contact between mucous membranes (9) and contaminated material (hands or surfaces), and ingestion (52). Pike (51) indicated that approximately 18% of these episodes could be attributed to known accidents that were caused by either carelessness or human error. Twenty-five percent of those accidents involved hypodermic needles and syringes. The other main categories were spills and sprays (27%); injury with broken glass or other sharp objects (16%); aspiration through a pipette (13%); and animal bites, scratches, and contact with ectoparasites (13.5%).

Although accidents provide an easily identified, readily recognizable event, Pike noted that the source of most LAIs is less readily identified. The source of exposure for most laboratory-related illness (82%) is unknown. Although some could be ascribed to discarded glassware, handling infected animals, clinical specimens, or aerosols, all that was known with any certainty was that the individual had worked with or was in the vicinity of the agent.

Similar observations were found in the literature of the 1980s. Only a small proportion of the LAIs appears to be associated with a specific accident or event such as a needlestick, spill, or spray. The nonspecific source of LAI for most reported infections was working with the agent, being in the laboratory, or handling infected animals.

Bacterial Infections

Of the *S. typhi* infections reported by Blaser et al. (8), only one was related to a specific accident, several were caused by breaks in techniques such as eating and smoking in the laboratory, mouth pipetting, and working with high concentrations of organisms on the open bench, and the remainder were associated with working with the agent or being in the laboratory.

Rickettsial Infections

The same trend is observed among the reported rickettsial infections (27, 29, 42, 47, 48). Two exposures resulted from accidental inoculation and the remaining 160 infections were associated with working with the agent and/or animals and aerosols. All the *C. burnetii* infections were associated with infected sheep (27, 42, 47). Sheep are often naturally infected with *C. burnetii* and asymptomatic. They may carry the organism in blood, urine, feces, tissue, and milk. It is estimated that the placenta of infected sheep may contain 10^9 organisms per gram of tissue and 10^5 organisms per gram of milk. Wedum et al. (66) noted that the infectious dose for 25 to 50 volunteers by inhalation for *C. burnetii* is only 10 organisms (see also Chapter 7 for Q fever).

Viral Infections

Only 14 of 119 viral infections reported during the 1980s resulted from accidents. The remaining 105 infections were associated with working with the agent, working with animals, or being in the laboratory. The presence of animals or insects was associated with more than three-quarters of the viral infections. Hantavirus (17, 37, 39, 69), responsible for almost two-thirds of the viral infections, produces an asymptomatic infection among wild rodents. Inhalation of aerosols produced by chronically infected rodents appears to be the likely mode of transmission in the laboratory personnel studying the Korean hemorrhagic fever virus. A new strain of hantavirus associated with acute fatal pulmonary disease in healthy adults was recently identified in the Four Corners region of New Mexico (15). This outbreak underscores the infectivity and transmissibility of this virus and emphasizes the need for strict adherence to containment practices and procedures when handling viruses and infected animals in laboratories and animal facilities (15).

AEROSOLS

Laboratory studies of potential sources of infection have focused on hazards associated with routine microbiological techniques. Aerosols can be produced by almost all routine bacteriologic and virologic procedures. Table 1 lists data from several studies on the number of viable particles recovered within 2 ft of a work area. These data are based on an extensive series of air sampling determinations. Aerosols present two means of potential worker exposure: through minute respirable airborne particles and by the disposition of larger heavy droplets onto surfaces, equipment, and personnel. (See also Chapter 5.)

The mere presence of organisms in the air is insufficient to cause disease. The data in Table 1 indicate that standard laboratory procedures can generate aerosol particles that are respirable and, therefore, potentially hazardous to the laboratorian and to others in the vicinity (31).

With the current increase in diagnostic and research virology, an increase in viral LAIs can be expected among personnel who handle these agents. The large doses of bacterial agents required to produce disease through ingestion (Table 2) help to explain why all microbiologists do not become clinically ill from their cultures. Also, the very low inhalation dose for certain agents (e.g., *Francisella tularensis, C. burnetii,* measles virus, and coxsackie A21 virus) points out the importance of preventing aerosol formation in the laboratory environment and providing the appropriate containment equipment.

INCIDENCE OF LAI

The incidence of LAI and the distribution of causative agents is not known. When the known episodes were classified by Pike and Sulkin (53) according to the primary purpose of the work, research activity accounted for a majority (59%) of the infectious episodes, whereas diagnostic work (17%), biological production (3.4%), and teaching (2.7%)

TABLE 1 Concentration and particle size of aerosols created during representative laboratory techniques[a]

Operation	No. of viable colonies[b]	Particle size[c] (μm)
Mixing culture with:		
Pipette	6.0	3.5
Vortex mixer	0.0	0.0
Mixer overflow	9.4	4.8
Use of Waring blender:		
Top on	119.0	1.9
Top off	1500.0	1.7
Use of sonicator	6.0	4.8
Lyophilized cultures:		
Opened carefully	134.0	10.0
Dropped and broken	4838.0	10.0

[a]Adapted from references 36 and 56, with permission.
[b]Mean number of viable colonies per cubic foot of air sampled.
[c]Mean diameter of the particle.

TABLE 2 Infectious dose for humans[a]

Disease or agent	Dose[b]	Route of inoculation
Scrub typhus	3	Intradermal
Q fever	10	Inhalation
Tularemia	10	Inhalation
Malaria	10	Intravenous
Syphilis	57	Intradermal
Typhoid fever	10	Ingestion
Cholera	10	Ingestion
Escherichia coli	10	Ingestion
Shigellosis	10	Ingestion
Measles	0.2[c]	Inhalation
Venezuelan encephalitis	1[d]	Subcutaneous
Poliovirus 1	2[e]	Ingestion
Coxsackie A21	18	Inhalation
Influenza A2	790	Inhalation

[a]Adapted from reference 66, with permission.
[b]Dose in number of organisms.
[c]Median infectious dose in children.
[d]Guinea pig infective dose.
[e]Median infectious dose.

activities accounted for less. These data indicate that working with pathogenic microorganisms places a person at risk of infection. The data also show that the higher concentration of microorganisms associated with research, as opposed to a clinical laboratory, may also increase the risk of infection. Pike (51) and Phillips (49) cautioned that such comparisons may be misleading because they do not take attack rate into consideration. The attack rate is the number of cases divided by the population at risk. Accurate information on the number of technical people at risk in each activity during the period under study, as well as more precise information of the number of LAIs, is not available. The estimated attack rates suggest that the risk for researchers may be seven to eight times greater than for public health and hospital laboratory workers (60).

LAIs are not confined to scientific personnel. Although trained investigators, technical assistants, animal caretakers, and graduate students experienced more than three-quarters of research laboratory illness, the remainder occurred among clerical staff, dishwashers, janitors, and maintenance personnel. Of 369 illnesses evaluated at a military biological facility in 1964, scientific personnel accounted for 82.3%; janitors, dishwashers, and maintenance personnel accounted for 13.7%; and clerical personnel accounted for 3.7% of the illnesses reported (65). These results are consistent with the results reported by Pike and Sulkin (53) in the United States between 1930 and 1950. The Pike and Sulkin study noted that students, a category not represented at a military facility, sustained 4.9% of the infections.

In 1965, Phillips (49) estimated that the frequency of LAI (using available U.S. and European data) resulted in an expected number of one to five infections per million working hours. Assuming that the average laboratory person works 8 h per day for approximately 225 days per year, the probability of acquiring a work-related infection is on the order of 10^{-2} to 10^{-3} per year. The probability for those research personnel who handle more hazardous agents as well as new or rare agents would be expected to be higher. In 1973, Grist (28) reported that diagnostic laboratory workers who handle clinical blood specimens had a higher infection risk than other hospital workers. In 1985, Jacobson et al. (34) surveyed clinical laboratories in Utah and estimated an annual incidence of 3 LAIs per 1,000 employees, with the incidence among microbiologists being 9.4 infections per 1,000 employees. In 1988, Vesley and Hartmann (64) surveyed LAIs and injuries among 4,202 public health and 2,290 hospital clinical laboratory employees. The annual incidence rate for all full-time employees was calculated at 1.4 infections per 1,000 for public health laboratories and 3.5 infections per 1,000 for hospital laboratories. Persons who worked directly with infectious agents did have higher incident rates of 2.7 per 1,000 and 4.0 per 1,000 for public health and hospital laboratories, respectively. Vesley and Hartmann attributed the relatively low incidence of LAI to safety awareness and improvement in safety-related devices.

Human Factor

After reviewing LAI data, it is clear that the main cause of laboratory exposures is the worker (30, 68). The worker handles the agent, performs experiments, operates equipment, handles animals, and when needed, cleans up spills. Many, if not most, LAIs could have been prevented if proper procedures and practices had been followed. Even exposures in pharmaceutical production are worker-related, as supported by studies evaluating the human factor in laboratory accidents. In one such study (41), two groups of 33 people were selected according to the presence or absence of involvement in accidents. Each participant was asked to fill out questionnaires concerning habits and attitudes. There were several differences between groups. The accident-free group showed greater concern about infectious agents, were aware of the hazards associated with their work, and realized the importance of proper attitudes in safety endeavors. This group placed prime importance on understanding and following safety regulations. However, the accident-involved group were characterized as risk-takers. There was a presumption that they were less likely to follow known safety regulations. This behavior was regarded as an extension of their personal behavior in private life. There were more nonsmokers and nondrinkers in the accident-free group.

These findings underscore the role behavioral factors can play in accidents and imply some associated limitations of risk assessment. One way to minimize risk is to modify unsafe behavior patterns. The subject of behavior patterns has been reviewed comprehensively by Martin (41), who reported the principles of behavioral analysis as well as step-by-step procedures for behavior reinforcement.

Behavioral patterns of interest included the finding that men were involved in seven times more accidents than women (normalized to working hours). There was a difference in accident involvement between age groups as well. The rate in the 17- to 24-year-old group was about twice that of the 45-

to 64-year-old group. Unfortunately, the biomedical work force is usually young and innovative and falls into the higher accident group. If previous studies provide guidance, the behavior patterns in clinical and research laboratories will be worth close observation. We must also develop a better understanding as to types or kinds of work activities that are taking place. (Further information on behavioral factors may be found in Chapter 18.)

HOST SUSCEPTIBILITY

Most discussions of the typical laboratory worker assume a normal healthy individual who is young and reasonably fit. What the history of LAI tells us is that although we are all at risk, some of us, because of our underlying health status, may be at greater risk. There are medical conditions that will result in an increase in risk of an adverse health outcome. The most significant are diseases or drugs that alter host defense, allergic hypersensitivity, inability to receive a specific vaccination, and reproductive issues. It is important to recognize these risk factors and address them before initiating work with infectious agents (24).

Some workers may face increased risks for certain infections because they have conditions that alter or impair normal host defense mechanisms. For example, host defenses provided by healthy intact skin can be disrupted by diseases such as chronic dermatitis, eczema, and psoriasis, thus providing a portal of entry in the absence of personal protective clothing.

Immunodeficiency also might arise in workers with diabetes mellitus, certain connective tissues diseases, workers receiving cancer chemotherapy, or those with HIV infection (25). Some causes of immunodeficiency can be traced to steroid treatment for medical conditions such as asthma, inflammatory bowel disease, and acute viral infection. Pregnancy also brings with it the potential for some immunodeficiency. In more serious cases of immune deficiency, employees may need to be removed from working with any potentially infective organisms (63).

Concerns about occupational risks to the reproductive system may involve questions about infertility in either sex; exposures during pregnancy that might result in adverse outcomes such as spontaneous abortion or birth defects; male exposures that could result in adverse pregnancy outcomes through mechanisms such as damage to the sperm, transmission of toxic agents in seminal fluid, or contact of the pregnant woman with her partner's contaminated clothing; or breast feeding as a source of toxic exposure to the child after birth (4).

More commonly, concerns are directed to the potential congenital infection of a child as a result of a pregnant employee acquiring a work-related infection. Infectious agents can be described according to the level of risk of congenital transmission. Clinical laboratory workers with no direct patient contact run the risk of exposure to such agents as cytomegalovirus, rubella, HBV, and toxoplasma from handling human specimens. Because HIV has been associated with congenital infections in the community, concerns could be raised about a potential occupational exposure in a pregnant employee working with HIV. Concerns may also be raised when a pregnant employee is handling a pathogenic virus and the reproductive hazards of the virus are not specifically known (46).

In the biomedical laboratory setting, workers can develop allergies to proteins (biological products derived from raw materials, fermentation products, or enzymes), chemicals, and the dander or aerosolized urine proteins of animals (1, 3, 4), but these exposures are outside the scope of this chapter.

ENVIRONMENTAL FACTORS OR MEASURES

The most effective strategy for the prevention or minimization of LAI is to make certain that only approved procedures are consistently carried out. Practical planning for safety is hampered by the fact that safety cannot be measured directly. We plan for safety by evaluating its opposite (i.e., risk or relative degree of harm). We achieve safety in the research laboratory by recognizing what the risks are and then introducing procedures, practices, equipment, and facilities to control or reduce risk to acceptable levels (38).

A series of preventive measures have been developed to provide protection to workers. These measures are frequently referred to as containment. Containment represents the combination of personnel practices, procedures, and safety equipment, laboratory design, and engineering features to minimize the exposure of workers to hazards or potentially hazardous agents. These are reviewed in detail elsewhere in this book. Many investigators have demonstrated that by using increasingly stringent procedures and more sophisticated facilities, work can be performed with agents of increased hazard or risk (63).

It is not possible to describe all the practices and procedures that can be used to control LAI. The actual practices used are dependent on such factors as the agent, type of activity, equipment, facilities, and proficiency of personnel. The reader is encouraged to read Chapters 11 and 13 in this book, which describe these in greater detail.

The risks associated with work involving infectious agents can be minimized if appropriate attention is given to biological safety. Workers must be responsible for reviewing the laboratory procedures and protocols to assess whether the containment conditions (practices and procedures, requirements for protective clothing, decontamination and waste management practice, types of equipment, and design criteria of the laboratory) are adequate to prevent exposure of themselves, their colleagues, and the environment.

References

1. **Agrup, G., L. Belin, L. Sjostedt, and S. Skerfving.** 1986. Allergy to laboratory animals in laboratory technicians and animal keepers. *Br. J. Med.* **43:**192–198.
2. **Al-Aska, A. K., and A. H. Chagla.** 1989. Laboratory-associated brucellosis. *J. Hosp. Infect.* **14:**69–71.
3. **American Industrial Hygiene Association Biohazards Committee.** 1985. *Biohazards reference manual: allergies,* p. 13–19. American Industrial Hygiene Association, Akron, Ohio.
4. **Ammann, A. S.** 1987. Immunodeficiency diseases, p. 317–355. *In* D. P. Stites, J. D. Stobo, and J. U. Wells (ed.), *Basic and clinical immunology.* Appleton & Lange, Norwalk, Conn.
5. **Baker, C. C., C. P. Farthing, and P. Ratnesar.** 1984. Toxoplasmosis, an innocuous disease? *J. Infect.* **8:**67–69.
6. **Bending, M. R., and P. D. L. Maurice.** 1980. Malaria: a laboratory risk. *Postgrad. Med. J.* **56:**344–345.
7. **Bernstein, D. I., T. Hubbard, W. M. Wenman, B. L. Johnson, K. K. Holmes, H. Liebhaber, J. Schachter, R. Barnes, and M. A. Lovett.** 1984. Mediastinal and supraclavicular lymphadenitis and pneumonitis due to Chlamydia trachomatis serovars L_1 and L_2. *N. Engl. J. Med.* **311:**1543–1546.
8. **Blaser, M. J., F. W. Hickman, J. J. Farmer III, D. J. Brenner, A. Balows, and R. A. Feldman.** 1980. *Salmonella typhi:* the laboratory as a reservoir of infection. *J. Infect. Dis.* **142:**934–938.
9. **Bruins, S. C., and R. R. Tight.** 1979. Laboratory acquired gonococcal conjunctivitis. *JAMA* **41:**274–278.
10. **Centers for Disease Control.** 1987. B virus infection in humans—Pensacola, Florida. *Morbid. Mortal. Weekly Rep.* **36:**289–290, 295–296.
11. **Centers for Disease Control.** 1989. B virus infection in humans—Michigan. *Morbid. Mortal. Weekly Rep.* **38:**453–454.
12. **Centers for Disease Control.** 1990. Update: Ebola-related filovirus infection in non-human primates and interim guidelines for handling non-human primates during transit and quarantine. *Morbid. Mortal. Weekly Rep.* **39:**22–24, 29–30.
13. **Centers for Disease Control.** 1990. Update: filovirus infection in animal handlers. *Morbid. Mortal. Weekly Rep.* **39:**221.
14. **Centers for Disease Control.** 1991. Laboratory-acquired meningococcemia—California and Massachusetts. *Morbid. Mortal. Weekly Rep.* **40:**46–47, 55.
15. **Centers for Disease Control and Prevention.** 1993. Update: hantavirus pulmonary syndrome—United States, 1993. *Morbid. Mortal. Weekly Rep.* **42:**816–820.
16. **Cohen, B. J., A. M. Courouce, T. F. Schwarz, K. Okochi, and G. J. Kurtzman.** 1988. Laboratory infection with parvovirus B19. *J. Clin. Pathol.* **41:**1027–1028.
17. **Desmyter, J., K. M. Johnson, C. Deckers, J. W. LeDuc, F. Brasseur, and C. Van Ypersele de Strihou.** 1983. Laboratory rat associated outbreak of haemorrhagic fever with renal syndrome due to hantaan-like virus in Belgium. *Lancet* **ii:**1445–1448.
18. **De Titto, E. H., and F. G. Araujo.** 1988. Serum neuraminidase activity and hematological alterations in acute human chagas disease. *Clin. Immunol. Immunopathol.* **46:**157–161.
19. **Elidan, J., J. Michel, I. Gay, and H. Springer.** 1985. Ear involvement with human brucellosis. *J. Laryngol. Otol.* **99:**289–291.
20. **Emeribe, A. O.** 1988. Gambiense trypanosomiasis acquired from needle scratch. *Lancet* **i:**470–471.
21. **Evans, M. R., D. K. Henderson, and J. E. Bennett.** 1990. Potential for laboratory exposures to biohazardous agents found in blood. *Am. J. Public. Health* **80:**423–427.
22. **Evans, T. G., and R. D. Pearson.** 1988. Clinical and immunological responses following accidental inoculation of Leishmania donovani. *Trans. R. Soc. Trop. Med. Hyg.* **82:**854–856.
23. **Freedman, D. O., J. D. MacLean, and J. B. Viloria.** 1987. A case of laboratory acquired Leishmania donovani infection; evidence for primary lymphatic dissemination. *Trans. R. Soc. Trop. Med. Hyg.* **81:**118–119.
24. **Goldman, R. H.** 1989. Medical surveillance program, p. 305–322. *In* D. F. Liberman and J. G. Gordon (ed.), *Biohazards managements handbook.* Marcel Dekker, Inc., New York.
25. **Goldman, R. H.** 1991. Medical surveillance in the biotechnology industry, p. 209–225. *In* A. M. Ducatman and D. F. Liberman (ed.), *State of the art reviews in occupational medicine: the biotechnology industry.* Hanley and Belfus, Philadelphia.
26. **Goossens, H., L. Marcelis, P. Dekeyser, and J. P. Butzler.** 1983. Brucella melitensis: person-to-person transmission? *Lancet* **ii:**773. (Letter.)
27. **Graham, C. J., T. Yamauchi, and P. Rountree.** 1989. Q fever in animal laboratory workers: an outbreak and its investigation. *Am. J. Infect. Control* **17:**345–348.
28. **Grist, N. R.** 1973. Hazards in the clinical pathology laboratory. *Proc. R. Soc. Med.* **66:**795–796.
29. **Halle, S., and G. A. Dasch.** 1980. Use of a sensitive microplate enzyme-linked immunosorbent assay in a retrospective serological analysis of a laboratory population at risk to infection with typhus group Rickettsiae. *J. Clin. Microbiol.* **12:**343–350.
30. **Harrington, J. M., and H. S. Shannon.** 1976. Incidence of tuberculosis, hepatitis, brucellosis, and shigellosis in British medical laboratory workers. *Br. Med. J.* **1:**759–762.

31. **Hatch, T. F.** 1961. Distribution and deposition of inhaled particles in respiratory tract. **Bacteriol. Rep. 25:**237–240.
32. **Hofflin, J. M., R. H. Sadler, F. G. Araujo, W. E. Page, and J. S. Remington.** 1987. Laboratory-acquired chagas disease. *Trans. R. Soc. Trop. Med. Hyg.* **81:**437–440.
33. **Ilkal, M. A., V. Dhanda, J. J. Rodrigues, C. V. R. Mohan Rao, and D. T. Mourya.** 1984. Xenodiagnosis of laboratory acquired dengue infection by mosquito inoculation and immunofluorescence. *Indian J. Med. Res.* **79:**587–590.
34. **Jacobson, J. T., R. B. Orlob, and J. L. Clayton.** 1985. Infections acquired in clinical laboratories in Utah. *J. Clin. Microbiol.* **21:**486–489.
35. **Jones, L., S. Ristow, T. Yilma, and B. Moss.** 1986. Accidental human immunization with vaccinia virus expressing nucleoprotein gene. *Nature* **319:**543. (Letter.)
36. **Kenny, M. T., and F. L. Sabel.** 1968. Particle size distribution of *Serratia marcescens* aerosols created during common laboratory procedures and simulated laboratory accidents. *Appl. Microbiol.* **16:**1146–1150.
37. **Lee, H. W., and K. M. Johnson.** 1982. Laboratory-acquired infections with hantaan virus, the etiologic agent of Korean hemorrhagic fever. *J. Infect. Dis.* **146:**645–651.
38. **Liberman, D. F., and L. Harding.** 1989. Biosafety: the research/diagnostic laboratory perspective, p. 156–158. *In* D. F. Liberman and J. G. Gordon (ed.), *Biohazards management handbook.* Marcel Dekker, Inc., New York.
39. **Lloyd, G., and N. Jones.** 1986. Infection of laboratory workers with hantavirus acquired from immunocytomas propagated in laboratory rats. *J. Infect.* **12:**117–125.
40. **Marr, J. J.** 1983. The professor and the pigeon: psittacosis in the groves of academe. *Mo. Med.* **80:**135–136.
41. **Martin, J. C.** 1980. Behavior factors in laboratory safety: personnel characteristics and modification of unsafe acts, p. 321–342. *In* A. A. Fuscaldo, B. J. Erlick, and B. Hindman (ed.), *Laboratory safety: theory and practice.* Academic Press, New York.
42. **Meiklejohn, G., L. G. Reimer, P. S. Graves, and C. Helmick.** 1981. Cryptic epidemic of Q-fever in a medical school. *J. Infect. Dis.* **144:**107–113.
43. **Mertens, T., H. Hager, and H. J. Eggers.** 1982. Epidemiology of an outbreak in a maternity unit of infections with an antigenic variant of echovirus 11. *J. Med. Virol.* **9:**81–91.
44. **Miller, D. C.** 1988. Creutzfeldt-Jacob disease in histopathology technicians. *N. Engl. J. Med.* **318:**853–854.
45. **Mohan Rao, C. V. R., C. N. Dandawate, J. J. Rodrigues, G. L. N. Prasada Rao, V. B. Mandke, G. R. Ghalsasi, and B. D. Pinto.** 1981. Laboratory infections with Ganjam virus. *Indian J. Med. Res.* **74:**319–324.
46. **Moore, D. M., W. F. MacKenzie, F. Doepel, and T. N. Hansen.** 1983. Contagious ecthyma in lambs and laboratory personnel. *Lab. Anim. Sci.* **33:**473–475.
47. **Ossewaarde, J. M., and A. C. Hekker.** 1984. Q-fever infection probably caused by a human placenta. *Ned Tijdschr Geneeskd* **128:**2258–2260.
48. **Perma, A., S. Di Rosa, V. Intonazzo, A. Sgerlazzo, G. Tringali, and G. LaRosa.** 1990. Epidemiology of boutonneuse fever in western Sicily: accidental laboratory infection with a rickettsial agent isolated from a tick. *Microbiologica* **13:**253–256.
49. **Phillips, G. B.** 1965. Microbiological hazards in the laboratory, part I control. *J. Chem. Educ.* **42:**A43–A48.
50. **Pike, R. M.** 1976. Laboratory-associated infections: summary and analysis of 3921 cases. *Health Lab. Sci.* **13:**105–114.
51. **Pike, R. M.** 1978. Past and present hazards of working with infectious agents. *Arch. Pathol. Lab. Med.* **102:**333–336.
52. **Pike, R. M.** 1979. Laboratory-associated infections: incidence, fatalities, causes, and prevention. *Annu. Rev. Microbiol.* **33:**41–66.
53. **Pike, R. M., and S. E. Sulkin.** 1952. Occupational hazards in microbiology. *Sci. Monthly* **75:**222– 228.
54. **Pike, R. M., S. E. Sulkin, and M. L. Schulze.** 1965. Continuing importance of laboratory-acquired infections. *Am. J. Public Health* **55:**190–199.
55. **Reif, J. S., L. Wimmer, J. A. Smith, D. A. Dargatz, and J. M. Cheney.** 1989. Human cryptosporidiosis associated with an epizootic in calves. *Am. J. Public Health* **79:**1528–1530.
56. **Reitman, M., and A. G. Wedum.** 1956. Microbiological safety. *Public. Health Rep. (U.S.)* **71:**659–665.
57. **Sampaio, R. N., L. M. P. De Lima, A. Vexenat, C. C. Cuba, A. C. Barreto, and P. D. Marsden.** 1983. A laboratory infection with Leishmania braziliensis braziliensis. *Trans. R. Soc. Trop. Med. Hyg.* **77:**274. (Letter.)
58. **Smith, J. A., and A. G. Skidmore.** 1980. Brucellosis in a laboratory technician. *CMA J.* **122:**1231–1232. (Letter.)
59. **Staszkiewicz, J., C. M. Lewis, J. Colville, M. Zervos, and J. Band.** 1991. Outbreak of Brucella melitensis among microbiology laboratory workers in a community hospital. *J. Clin. Microbiol.* **29:**287–290.
60. **Sulkin, S. E., and R. M. Pike.** 1951. Laboratory-acquired infections. *JAMA* **147:**1740–1745.
61. **Sulkin, S. E., and R. M. Pike.** 1951. Survey of laboratory-acquired infections. *Am. J. Public Health* **41:**769–781.
62. **Tomori, O., T. P. Monath, E. H. O'Connor, V. H. Lee, and C. B. Cropp.** 1981. Arbovirus infections among laboratory personnel in Ibadan, Nigeria. *Am. J. Trop. Med. Hyg.* **30:**855–861.
63. **U.S. Department of Health and Human Services.** 1993. *Biosafety in microbiological and biomedical laboratories.* HHS Publication No. (NIH) 93-8395. U.S. Government Printing Office, Washington, D.C.
64. **Vesley, D., and H. M. Hartmann.** 1988. Laboratory-acquired infections and injuries in clinical laboratories: a 1986 survey. *Am. J. Public Health* **78:**1213–1215.
65. **Wedum, A. G.** 1964. Laboratory safety in research with infectious aerosols. *Public Health Rep. (U.S.)* **79:**619–633.
66. **Wedum, A. G., W. E. Barkley, and A. Hellman.** 1972. Handling infectious agents. *J. Am. Vet. Med. Assoc.* **161:**1557–1567.
67. **Williams, J. L., B. T. Innes, T. R. Burkot, D. E. Hayes, and I. Schneider.** 1983. Falciparum malaria: accidental transmission to man by mosquitoes after infection with culture-derived gametocytes. *Am. J. Trop. Med.*

Hyg. **32:**657–659.

68. **Williams, R. E. O.** 1981. In pursuit of safety. *J. Clin. Pathol.* **34:**232–239.
69. **Wong, T. W., Y. C. Chan, E. H. Yap, Y. G. Joo, H. W. Lee, P.-W. Lee, R. Yanagihara, C. J. Gibbs, Jr., and C. Gajdusek.** 1988. Serological evidence of hantavirus infection in laboratory rats and personnel. *Int. J. Epidemiol.* **17:**887–890.
70. **Zhicheng, S., and L. Pangcheng.** 1986. Occupational mycoses. *Br. J. Ind. Med.* **43:**500–501.

Hazard Assessment in the Laboratory

Transmission and Control of Laboratory-Acquired Hepatitis Infection

MARTIN S. FAVERO AND WALTER W. BOND

3

INTRODUCTION

This chapter describes the routes and mechanisms of transmission of hepatitis viruses in laboratory settings, as well as the basic strategies for prevention and control of infections caused by them.

Historically, viral hepatitis has been one of the most frequently reported infections acquired by laboratorians employed in clinical, research, and public health laboratories (14, 24, 26, 32). Currently, there are at least five viruses recognized that can cause hepatitis. These viruses are distinct and show no structural homology, taxonomic relationship, or similarity in the means by which they replicate in the human host. The current terminology used to describe these five viral agents is listed in Table 1, and the general morphologic, clinical, and epidemiologic features are in Table 2.

Although these five viral agents are unique viral entities, the clinical diseases they cause are very similar and cannot be distinguished by clinical signs and symptoms, routine laboratory liver enzyme tests, or histopathologic results from biopsies. The primary criteria for specific diagnosis are based on serologic tests (Table 2) developed over the past 25 years (36).

The following are brief descriptions of each of the five hepatitis viruses, specifically in the context of laboratory-acquired infections. Readers are referred to two review articles for detailed and broad descriptions of these viral agents, their disease epidemiologies, and descriptions of the variety of serologic tests used for their diagnosis (21, 36).

Hepatitis A

The hepatitis A virus (HAV) is transmitted by the fecal-oral route and occurs worldwide. In underdeveloped countries, virtually all children have been infected, usually without signs and symptoms of clinical hepatitis. In the United States and other industrialized nations, HAV infection occurs sporadically as well as in epidemics. HAV does not appear to be an occupational risk for laboratorians but has been shown to occur with relatively high frequency among animal handlers working in areas housing chimpanzees, especially in those laboratories that conduct HAV infectivity studies (32).

Hepatitis B

The hepatitis B virus (HBV) can cause acute hepatitis followed by disease resolution, but a certain percentage of these infections can result in chronic active hepatitis. HBV is responsible for most morbidity and mortality of acute and chronic liver dis-

TABLE 1 Terminology of viral hepatitis

Agent	Preferred terminology	Other terminology
Hepatitis A virus	Hepatitis A	Infectious hepatitis
Hepatitis B virus	Hepatitis B	Serum hepatitis
Hepatitis C virus	Hepatitis C	Parenterally transmitted non-A, non-B hepatitis
Hepatitis D virus	Hepatitis D	Delta hepatitis
Hepatitis E virus	Hepatitis E	Enterically transmitted non-A, non-B hepatitis

ease worldwide. HBV is spread by the blood-borne route, and the efficiency of transmission is directly related to extremely high numbers of circulating viruses during active infections. Hepatitis B is the most often reported laboratory-acquired infection and occurs at a high rate among laboratorians who frequently handle blood or blood-contaminated fluids or items (20, 24). Also, animal handlers who work in areas housing chimpanzees have been reported to be at relatively high risk of HBV infection. Fortunately, there is an effective hepatitis B vaccine available, and as discussed later, laboratorians and other health care workers who are at risk for HBV infection should be vaccinated.

The basic reservoir of HBV is the human chronic or transient carrier of hepatitis B surface antigen (HBsAg), and the principal source of infectious virions is plasma or serum. There is general agreement that a fecal-oral route of hepatitis B transmission does not exist (16, 25). Many other body fluids, secretions, and excretions, including urine, feces, saliva, nasopharyngeal washings, breast milk, bile, semen, synovial fluid, sweat, tears, and cerebrospinal fluid, have been hypothesized as being infective because they contain detectable HBsAg. Among these, however, only saliva and semen have been shown to be infective (1). Blood is acknowledged to be the most consistently effective vehicle of HBV transmission. The efficiency of the various mechanisms of hepatitis B transmission within the hospital and clinical laboratory environments is related significantly, if not solely, to the extraordinary amounts of circulating HBV in the blood of infected individuals who are either in the acute phase of infection or chronic HBsAg carriers and positive for hepatitis B e antigen (HBeAg). It has been shown, for example, that human serum positive for HBsAg and HBeAg can be diluted to 10^{-8} and still produce hepatitis B infection when injected into susceptible chimpanzees (35).

Because of this high viral titer, blood can be diluted to such an extent that even in the absence of visible or chemically detectable blood, HBV can be present in relatively small inocula or on laboratory environmental surfaces. Further, HBV can be present in a variety of secretions and excretions and other body fluids that contain small amounts of serum under normal conditions or under conditions of trauma. With the variables of exposure and host susceptibility being constant, the efficiency and

TABLE 2 Hepatitis

	Virus	Family	Size (nm)	Genome	Acute mortality (%)	Chronic infection	Spread	Antigens	Antibodies
A	HAV	*Picornaviridae*	27	ssRNA	0.2–0.4	No	Fecal-oral	HAV Ag	Anti-HAV
B	HBV	*Hepadnaviridae*	42	dsDNA	0.2–1	Yes	Blood Sexual Perinatal	HBsAg HBcAg[a] HBeAg	Anti-HBs Anti-HBc Anti-HBe
C	HCV	*Flaviviridae*	30–60	ssRNA	1–2	Yes	Blood Sexual Perinatal	HCV Ag	Anti-HCV
D	HDV	Satellite	40	ssRNA	2–20	Yes	Blood Sexual	HDV Ag	Anti-HDV
E	HEV	Calicivirus-like	32	ssRNA	0.2–3	No	Fecal-oral	HEV Ag	Anti-HEV

[a]HBc, core antigen; HBe, e antigen; HBs, surface antigen.

probability of disease transmission are directly related to the quantity of HBV in a particular body fluid or on a fomite.

Hepatitis C

The hepatitis C virus (HCV) is a recently described agent responsible for most cases of parenterally transmitted non-A, non-B hepatitis (2, 3). Before the serologic tests that can be used to detect antibodies to antigens of HCV in serum of infected individuals, the diagnosis of non-A, non-B hepatitis was made by exclusion by using serologic tests that can specifically diagnose acute or resolved infections with HAV or HBV (17). The epidemiology of hepatitis C is just beginning to be defined, but it occurs worldwide and is commonly found in underdeveloped countries as well as industrialized nations. HCV is transmitted by the blood-borne route, but the efficiency of transmission appears to be less than that of HBV, probably because of a much lower titer of circulating virus in individuals with hepatitis C. Laboratorians and other health care workers who frequently handle blood may be at increased risk to HCV infection but apparently not to the same degree as they are to HBV infection (22). As described later, the primary strategy for infection control with HCV is based on control strategies developed for HBV safety in laboratory and health care settings.

Hepatitis D

The hepatitis delta virus (HDV) is a relatively small virus with defective RNA and causes hepatitis only in individuals with active HBV infection. Hepatitis D is a severe disease with a relatively high mortality rate as a result of acute infection and a high likelihood of the development of chronic disease. Hepatitis D is endemic in certain areas of the Middle East and the Amazon Basin, but worldwide, it is usually associated with certain high-risk groups including parenteral drug abusers and recipients of multiple blood transfusions or plasma products. HDV can infect an individual chronically infected with HBV (superinfection), or it can be transmitted at the same time with HBV (coinfection). HDV may occur in 2 to 3 logs higher concentration in the blood of infected individuals than HBV and, therefore, can be transmitted more efficiently than HBV in certain circumstances. Although this situation could theoretically provide for occupational transmission of HDV among laboratorians and health care workers who handle blood and blood products, there have not been documented reports of laboratory-acquired hepatitis D infections. Strategies for control of hepatitis D are basically the same as those developed for the prevention of hepatitis B, obviously including HBV vaccination.

Hepatitis E

The hepatitis E virus (HEV) causes acute infections and is commonly referred to as "enterically transmitted non-A, non-B hepatitis." HEV infection does not lead to chronic hepatitis nor has a carrier state been described. HEV is transmitted by the fecal-oral route by the same modes that have been described for HAV and is a common cause of sporadic and epidemic hepatitis in underdeveloped areas of the world, such as the Indian subcontinent and in Asia. In industrialized nations including the United States, cases of hepatitis E have only been detected in patients who have been infected in other parts of the world. There have been no reports of HEV infections associated with health care or laboratory settings, but presumably, infection risk would be similar to that of HAV. The same precautions and strategies appropriate for preventing hepatitis A among laboratorians would be effective in preventing hepatitis E in these same settings.

In summary, hepatitis B, C, and D are blood-borne infections, and hepatitis A and E are enterically transmitted infections. The overriding strategy for the prevention and control of viral hepatitis in laboratories necessarily must be considered from the standpoint of viral hepatitis type B and blood precautions or what are now referred to as universal precautions (9, 10). This is not to imply that one should not use care in handling fecal specimens or contaminated materials, but rather that standard hygienic and environmental precautions described elsewhere in this book are sufficient to prevent transmission of hepatitis A and E.

ROUTES AND MECHANISMS OF HEPATITIS TRANSMISSION IN LABORATORIES

The following are modes of hepatitis (primarily type B) transmission that can occur in a variety of settings including laboratories, and they are listed in the probable order of efficiency of disease transmission.

Direct Contact

Since direct physical contact with infective material confers the highest risk of disease transmission, most guidelines and recommendations for laboratory safety emphasize a variety of strategies to prevent the following types of laboratory worker exposures:

1. Direct percutaneous inoculation by needle of contaminated serum or plasma such as occurs by injection of contaminated blood or blood products and by accidental needlesticks.

2. Percutaneous transfer of infective serum or plasma in the absence of overt needle puncture, such as by contamination of minute cutaneous scratches, abrasions, or burns during laboratory manipulations.

3. Contamination of mucosal surfaces by infective serum or plasma or infective secretions or excretions such as occurs with accidents associated with mouth pipetting, label licking, splashes or spatterings, or other means of skin-to-mouth or skin-to-eye contact.

Indirect Contact

Hepatitis transmission, especially type B hepatitis, can occur by indirect means via common environmental surfaces in a laboratory, such as test tubes, laboratory benches, laboratory accessories, and other surfaces contaminated with infective blood, serum, secretions, or excretions that can be transferred to the skin or mucous membranes. The probability of disease transmission with a single exposure of this type may be remote, but the frequency of such exposures makes this mechanism of transmission potentially an efficient one over a long period of time. Activities in laboratories such as nail biting, smoking, eating, and a variety of hand-to-nose, -mouth, and -eye actions contribute to indirect transmission.

Fecal-Oral Transmission

The fecal-oral route does not appear to be a mode of HBV transmission nor has it been reported for HCV or HDV. Therefore, handling feces does not constitute the same degree of hazard as handling blood specimens. HBV-contaminated materials can initiate infection when they enter the mouth, but the precise route of infection is associated with entering the host's vascular system via mucosal surfaces rather than via an intestinal route. This type of transmission is not enteric but rather peroral transmission.

Patients infected with HAV and probably HEV do not have a viremic stage that is extensive enough to pose a problem of disease transmission through direct contact with blood. Transmission of hepatitis A can occur by direct or indirect contact via the fecal-oral route. The degree of conservatism practiced in strategies for handling feces or fecally contaminated specimens is controlled by several factors. First, the carrier state of HAV virus in feces has never been demonstrated (13). Second, highest mean levels of HAV in feces of infected patients occur during the incubation phase of the disease, and by the time a patient is hospitalized, the level of HAV in feces has been significantly reduced. Because HAV may be present in low numbers, fecal specimens being examined in laboratories, as well as contaminated objects, should be considered potentially infective. However, the efficiency of HAV infection is considered to be positively correlated to high levels of HAV in feces. Mosley (28), for example, pointed out that maximum (high titer) HAV excretion is equivalent, epidemiologically, to optimum disease communicability. Third, patients' fecal specimens, except in highly specialized research laboratories, are not routinely examined for the presence of HAV, because clinical diagnosis of the disease is based on serologic tests for antibodies to HAV. Consequently, a practical risk of disease transmission would be associated with those fecal specimens being examined microbiologically that are from patients hospitalized for other reasons but who also may be in the incubation phase of HAV infection. This situation would be more likely for pediatric patients (34).

Precautions used in laboratories for handling fecal specimens and contaminated items should be sufficient to prevent acquisition of hepatitis A by laboratorians. Laboratory personnel conducting infectivity studies with HAV using chimpanzees or marmosets or caring for these animals should be aware that handling of feces and cage-cleaning activities can create situations that favor fecal-oral transmission. These situations can occur readily by direct contact with infective feces and by generation of droplets that ultimately can enter the mouth and be ingested.

Airborne Transmission

Although hepatitis B transmission by means of the airborne route has been hypothesized, it has not been documented. We have performed two studies that show true airborne transmission of hepatitis B from infective blood or saliva is not likely. In one study (31), a filter-rinse technique capable of detecting low levels of aerosolized airborne HBsAg was devised and evaluated. Laboratory tests showed the procedure had an efficiency of 22% and was capable of detecting as little as 5×10^{-5} ml of aerosolized HBsAg-positive serum in a single air sample. The technique was used in a hemodialysis center serving a patient population with a high prevalence of HBsAg seropositivity and one in which conditions favored the production of aerosols. Samples were collected from the patient area during treatment, from the laboratory area while blood samples were being processed, and from the area where reusable dialyzers were disassembled and cleaned. HBsAg was not detected in any of 60 air samples collected. However, random swab-rinse samples of surfaces in and around hemodialysis centers showed that 15% of them, particularly ones frequently touched, were positive for HBsAg (4).

In another study (30), air samples were collected from the dental operatory of an institution for the mentally handicapped, where residents had a high incidence of HBsAg seropositivity. Although gingival swab samples from nearly all the patients showed the presence of HBsAg, air samples collected during procedures of scaling, extraction, high-speed drilling, and other procedures favoring aerosol production were uniformly negative for HBsAg. True aerosols (i.e., inspirable particles less than 10 μm in diameter) of blood, saliva, or mouth washings did not appear to be efficiently produced by these procedures.

However, events such as splashing, centrifuge accidents, or removal of rubber stoppers from tubes can account for disease transmission by means of large droplet transfer into the mouth, eyes, or minor cuts or scratches or onto abraded skin. This is not airborne transmission but rather transmission by direct droplet contact.

SPECIFIC PRECAUTIONS FOR PREVENTING LABORATORY-ACQUIRED VIRAL HEPATITIS

The primary strategy for preventing laboratory-acquired viral hepatitis is the recognition of the concept and consistent practice of universal precautions (9, 10). In this context, "universal precautions" can be considered as the practice of blood precautions universally with all blood and blood-derived or blood-contaminated specimens. This is true not only in laboratories where blood, serum, and other specimens from patients who are known to be infected with HBV are processed but in all clinical laboratories involved in the chemical, hematologic, and microbiological assay of blood and blood products. The prevalence of HBsAg in serum of patients whose blood is being assayed varies with the population being served by the laboratory. This prevalence can be as high as 1% of hospital admissions, and with specific high-risk groups such as parenteral drug abusers or male homosexuals, the HBsAg carriage rate may be significantly higher. Laboratories that process hundreds to thousands of blood samples per day will no doubt handle a certain number of blood specimens that contain HBV but are not labeled as such. Until the mid-1980s, it was common practice of many hospital laboratories to prominently identify specimens and specimen containers with a "hepatitis" label when patients were known to be hospitalized with hepatitis. Unfortunately, two methods of handling blood were sometimes practiced: very careful ones with those tubes with a special label and fairly lax ones with unlabeled specimens from other patients. Universal precautions should be used at all times when handling blood and other blood-derived or blood-contaminated body secretions or excretions.

The following guidelines for the prevention of laboratory-acquired viral hepatitis are also applicable in routine laboratory practices to limit the acquisition of other infectious diseases (12, 19).

1. *Safety officer.* The responsibility for laboratory safety resides ultimately with the director of the laboratory; however, from an operational standpoint a safety officer who is familiar with laboratory practices and biohazards should be appointed from among the laboratory staff. The safety officer should be responsible for giving advice and consultation to the laboratory staff on matters of biohazards; instructing new members in safety procedures; procuring protective equipment, disposal bags, and disinfectants; developing and maintaining a laboratory accident reporting system; periodically reviewing and updating safety procedures; and monitoring serologic surveillance data for the laboratory staff if such data are collected. In hospital laboratories, the laboratory safety officer should consult with and report new information to the hospital's infection control practitioner. Further, when appropriate, reports should be made to the hospital's infection control committee, as well as to local and state health departments.

2. *Reporting of accidents.* Accidents such as cuts, needlesticks, and skin abrasions with instruments possibly contaminated with blood, soiling of broken skin, or contamination of the eyes or mouth must be reported promptly to the safety officer, who should maintain records and make sure that proper medical consultation and treatment, if necessary, are available. Prompt reporting is imperative because the efficacy of prophylactic postexposure immunoglobulins (IG) is greatest when administered immediately after accidental exposure occurs. Significant spills of blood, even if not associated with personnel contamination, should also be reported to the safety officer.

3. *Handwashing.* Frequent handwashing by laboratorians is an important safety precaution that should be practiced after contact with specimens and laboratory procedures, especially those associ-

ated with blood or body fluids. Hands should always be washed before entering and leaving the laboratory, before eating, drinking, or smoking, and after completing analytical work. Frequent handwashing should be performed even when gloves are used and, in particular, each time gloves are removed. Handwashing facilities should be conveniently located to encourage frequent use. A handwashing product that is widely acceptable to personnel and not irritating to skin is desirable. Liquid soap, granule soap, or soap impregnated tissues can be used. Ideally, sinks should be equipped with foot- or knee-operated faucets.

4. *Gloves.* All laboratorians who have direct or indirect contact with blood or articles contaminated with blood should wear gloves. Wearing of gloves is obviously for the protection of the laboratorian. It should be realized that gloves themselves can become contaminated, so surfaces such as telephones, door knobs, and laboratory equipment frequently touched by other laboratorians should not be touched with contaminated gloves. Disposable gloves are preferred because they can be changed frequently. It is not necessary for laboratorians to wear sterile gloves when performing laboratory activities. Gloves made of vinyl, latex, or other materials can be used. Gloves should be compatible with being used for long periods of time without causing skin reactions and should have no visible tears or defects. Accidentally torn or punctured gloves should be removed immediately, and hands should be washed before donning fresh gloves.

5. *Protective clothing.* Staff members should wear a gown with a closed front or coat with an overlapping front when they are in the laboratory. A disposable plastic apron also may be desirable. Gowns, aprons, and gloves must be removed and hands washed before staff members leave the laboratory for any purpose. Disposable gloves and aprons should be worn only once. Gowns and coats should be put in a laundry bag at the end of each appropriate period of work. If a gown or coat is accidentally contaminated, it should be discarded in the laundry bag at once and a fresh one obtained. A face shield or safety glasses and mask should be worn when it is anticipated that there is a potential for blood and other types of specimens to be spattered (7).

6. *Personal hygiene.* Smoking, eating, or drinking must not be done in the laboratory. Employees working in technical and analytical areas should be given an opportunity for regular breaks in accordance with laboratory policy, and there should be an area where employees can have these breaks and eat lunch. Care should be taken not to put fingers, pencils, or other objects into the mouth, specimen tube labels must not be licked, and hands should be washed after every procedure in which they may have become contaminated. Cosmetic application and contact lens maintenance should not be performed in laboratory areas.

7. *Pipetting.* Mouth pipetting should be prohibited; automatic pipettes with disposable plastic tips are recommended. If disposable pipette tips are used, they should be discarded after pipetting each specimen (e.g., they should not be rinsed several times in water between each specimen). Other pipettes should be used with rubber bulbs or an automatic suction device, and fluids should never be drawn up to the top of the pipette. The contents of the pipette should be expelled gently down the wall of the receptacle to avoid splashing, and pipettes should be held in a near-vertical position while in use. Contaminated pipettes must not be placed on the laboratory bench; they should be placed gently in a flat discard pan. Any rubber bulb that may have become contaminated internally during use should be placed immediately into the discard pan.

8. *Receipt of specimens.* Incoming blood or serum specimens should be received in a designated area of the laboratory and examined closely to be sure they have been properly closed or packed. Soiled or leaking containers should be brought to the attention of the safety officer to decide whether they should be decontaminated and discarded without being unpacked. When handling blood specimens, receiving technicians should wear disposable gloves and should open the specimen container slowly to reduce the possibility of spatter.

9. *Labeling, processing, and storage of blood tubes.* Blood, serum, and biological specimens from patients who are known to be infected with hepatitis or for whom hepatitis-specific serologic tests or serum enzyme tests are being performed do not need to be identified with special labels. Precautions used by laboratorians when handling these types of specimens should be no different from those used for all specimens (i.e., universal precautions are the appropriate strategy). Because blood tubes can be contaminated on the outside (8) as well as contain infective blood, they must be handled and stored with care. If blood tubes must be refrigerated, they should be capped and placed in a designated refrigerator or in a designated portion of a refrigerator. Blood tubes should never be kept in a refrigerator that contains food or beverages. If blood or serum is held in frozen storage, it should be placed in a container designed for low-temperature storage. Organized storage of these containers is desirable to prevent unnecessary handling of specimens.

10. *Needles and syringes.* Particular caution should be taken with needles and syringes whether or not they are contaminated with blood. Disposable needles and syringes should be used, and they should be discarded after a single use. Used needles should not be recapped by hand; they should be placed in permanently labeled, leak- and puncture-resistant containers designed for this purpose. Needle nippers should not be used, and needles should not be purposely bent or broken by hand because accidental needle punctures can occur. Used syringes should be placed in leak- and puncture-resistant containers. If reusable syringes are used, they should be rinsed thoroughly with cold water in a designated sink by a person wearing gloves. Syringes and associated components should then be placed in a leakproof container before decontamination or further reprocessing.

11. *Disposal of waste specimens and contaminated material.* As mentioned previously, all blood and most biological specimens from humans and from experimental animals must be viewed as potentially infective. Accordingly, when small volumes of fluid or items contaminated with these specimens become waste, they must be disposed of in a safe manner. Each laboratory should have special receptacles for these wastes. Preferably, the wastes in these receptacles should be decontaminated by autoclaving or other means in the laboratory or transported in impervious bags to an autoclave or approved incinerator in another area. When it is not feasible to autoclave or decontaminate larger volumes of blood or other patient fluids, they may be poured down a single designated sink drain. Gloves should be worn during this procedure, and care should be taken to prevent splashing onto the walls of the sink. After a sink drain is used for fluid disposal, it should be thoroughly flushed with water (minimizing splashing) and the sink and drain should be flushed with an appropriate amount of liquid disinfectant such as sodium hypochlorite.

12. *Centrifuging.* Blood specimens should be centrifuged with tubes tightly capped. If a tube breaks in the centrifuge, the bucket containing the spilled blood and broken glass should be placed carefully in a pan of disinfectant. The surfaces of the centrifuge head, bowl, trunions, and remaining buckets should be swabbed with an appropriate disinfectant; alternately, the trunions and buckets can be autoclaved. Microhematocrit centrifuges and blood bank serofuges should be cleaned and then disinfected daily. The top of the centrifuge should always be closed when the unit is in operation, and the top should not be opened until the unit has come to a complete standstill. Centrifuge tubes to be used in angle-head centrifuge must never be filled to the point that liquid is in contact with the lip of the tube when it is placed in the rotor. When the tube lip becomes wet, liquid will be forced past the cap seal and over the outside of the tube.

13. *Automated bioassay equipment.* Automated equipment capable of performing several biochemical assays simultaneously is becoming commonplace in clinical laboratories. Although there is a variety of configurations of these types of equipment, we have not observed any that, because of their external and internal design, constitute significant risks of hepatitis transmission. Rather, the potential risks of hepatitis transmission are associated with procedures of handling, preparation, and delivery of specimens to the automated equipment. Tubes must not be packed too tightly; they must be capped up until the point of insertion into the automated system. Blood-contaminated plastic tubing should be cleaned periodically during a workday. Gloves should be worn at all times by operators of this equipment. If blood or serum ultimately is collected in a reservoir that is not piped into the sewer system, contents of the reservoir should be decontaminated before disposal. If this is not feasible, the contents should be poured into a designated sink drain as described previously for fluid wastes.

14. *Care of laboratory bench tops.* Each working area should be supplied with a wash bottle containing an appropriate intermediate-level disinfectant. The disinfectant solution should be mixed and renewed according to the directions on the manufacturer's label, and the bench surface must be cleaned and then wiped with disinfectant at the beginning and end of each day or more frequently as spills or contamination occur. Because accidents and errors are most likely to happen when the laboratory work area is crowded with equipment and materials, care should be taken to keep the laboratory work area tidy. Tubes and other containers should be placed only in the appropriate rack or tray, never directly on the bench. Disposable, absorbent, plastic-backed pads can be used to protect laboratory bench tops if spattering or spills are common or are anticipated.

15. *Purification or concentration of infectious material.* In some laboratories actively engaged in hepatitis research, biophysical and biochemical procedures are often used to concentrate or purify infective materials. These procedures increase the biohazard associated with these materials, and the safety practices and precautions that have been mentioned may not provide adequate protection. For example, the hepatitis viruses collectively require biosafety level 2 (12) for routine diagnostic procedures. However, biosafety level 3 is needed

when they are purified, concentrated, or prepared in large volumes, and all laboratory procedures should then be performed in a class II biological safety cabinet (29) (see Chapter 11, Primary Barriers). Also, it would be prudent to conduct potentially hazardous procedures such as homogenization of infectious stool and liver preparations in a biological safety cabinet even though purification and concentrations had not yet occurred. Ultracentrifuges used for procedures involving hepatitis viruses should be installed in a containment system that provides for the adequate single-pass filtered exhaust of any material that might be released in the event of an accident.

16. *Postmortem examinations.* As elsewhere in the hospital environment, an autopsy schedule will include several individuals infected with HBV or other infectious agents that have not been identified in advance by clinical serologic testing. Consequently, all autopsies should be conducted with the same precautions as would be routinely done (i.e., universal precautions). Specifically, universal precautions would include the wearing of gloves and appropriate protective clothing at all times. If procedures used can cause spattering of blood, feces, or other material, the use of face shields or eye protection and masks should be considered. Disinfectants of at least intermediate-level activity should be used routinely after each postmortem examination to disinfect those items or surfaces that become contaminated.

17. *Animal areas.* In laboratory settings where chimpanzees, marmosets, or other animals are housed or are being used for hepatitis infectivity studies, both blood (universal precautions) and enteric precautions should be practiced. Personnel, when performing ordinary types of animal manipulation such as feeding and injections, should wear protective clothing including gowns, gloves, caps, masks, and shoe covers. For cage cleaning, face shields or masks and eye protection are recommended. Areas should be cleaned daily with a detergent-disinfectant. An appropriate system should be designed for the disposal of feces, urine, and other wastes. In the event that soiling by animal waste occurs outside the immediate vicinity of the cages, the contaminated area should be cleaned and decontaminated immediately with an appropriate disinfectant.

IMMUNE PROPHYLAXIS

The Advisory Committee on Immunization Practices of the Public Health Service periodically publishes recommendations and strategies for use of the hepatitis B vaccine as well as IGs for protection against several types of viral hepatitis (11). These recommendations are updated periodically in the Centers for Disease Control's *Morbidity and Mortality Weekly Report.*

This section is based on recent recommendations of the Advisory Committee on Immunization Practices for protection against viral hepatitis. Briefly, they are summarized as follows:

1. Hepatitis A

Pre-exposure prophylaxis: not necessary for laboratorians; animal handlers working with primates who are experimentally or naturally infected with hepatitis A probably should receive IG (or, preferably, the hepatitis A vaccine; see reference 12 and below).

Postexposure prophylaxis: IG.

Vaccination: a commercially available effective hepatitis A vaccine may be available by mid- to late 1995. Although the Advisory Committee on Immunization Practices has not issued recommendations for its use, it may be prudent to vaccinate laboratorians who work in laboratories doing research with HAV, as well as animal handlers involved in the care and feeding of primates experimentally or naturally infected with HAV.

2. Hepatitis B

Pre-exposure prophylaxis: hepatitis B vaccine; hepatitis B immunoglobulin (HBIG).

Postexposure prophylaxis: hepatitis B vaccine and/or HBIG.

Vaccination: hepatitis B vaccine. The various options for the use of hepatitis B vaccine and HBIG are discussed in detail below.

3. Hepatitis C

Pre-exposure prophylaxis: none.

Postexposure prophylaxis: none.

Vaccination: none.

4. Delta hepatitis

Because HDV is dependent on an HBV-infected host for replication, prevention of hepatitis B infection either pre-exposure or postexposure will suffice to prevent HDV infection for persons susceptible to hepatitis B. Percutaneous exposure to blood or serum known to be positive for both HBV and HDV should be treated exactly as such exposures to HBV alone. For persons who are HBsAg carriers and are at risk of HDV infection, no products are available to be used for pre-exposure or postexposure prophylaxis.

Vaccination: hepatitis B vaccine.

5. Hepatitis E

Pre-exposure prophylaxis: none.

Postexposure prophylaxis: none.

Vaccination: none.

The hepatitis B vaccine is currently available for both pre-exposure and postexposure prevention of hepatitis B. The recommended doses and schedules of currently licensed vaccines for adults are as follows:

Recombivax HB—Three doses of 10 μg (1.0 ml) each at 0, 1, and 6 months.

Engerix-B—Four doses of 20 μg (1.0 ml) each at 0, 1, 2, and 12 months.

The vaccine is 80 to 95% effective in preventing HBV infection, has minimal side effects, and is recommended for pre-exposure prevention of hepatitis B. Laboratorians, especially those who frequently handle human or primate blood, are at high-to-moderate risk of hepatitis B infection and should be vaccinated.

IGs are sterile solutions of antibodies prepared from large pools of human plasma. IG (formerly referred to as "immune serum globulin" or "gamma globulin") produced in the United States contains antibodies specific for HAV and HBV. Tests of IG lots since 1977 indicate that anti-HAV and anti-HBs have uniformly been present at stable titers. For example, all lots tested since that time have contained an anti-HBs titer of at least 1:100 by radioimmunoassay (RIA). HBIG is prepared from plasma preselected for high levels of specific antibody and, in the United States, has an anti-HBs titer greater than 1:100,000 by RIA. The price of a dose of HBIG is approximately 20 times that of IG.

In the context of the hospital and laboratory environment, immune prophylaxis for protection against HAV infection usually is not warranted. In the event of a known source of exposure to HAV such as accidental ingestion of fecal material known to contain HAV, a single intramuscular (i.m.) dose of 5 ml of IG can be given. Pre-exposure prophylaxis is not recommended, but certain institutions that manage primates used in experimental HAV infection studies administer IG periodically to animal handlers who are negative for anti-HAV. In this instance, 5 ml i.m. should be given every 5 to 6 months.

Recommendations for Postexposure Prophylaxis Caused by Percutaneous or Mucous Membrane Exposure to Blood

For accidental percutaneous (needlestick, laceration, or bite) or permucosal (ocular or mucous membrane) exposure to blood, the decision to provide prophylaxis must include consideration of several factors: (i) whether the source of the blood is known, (ii) the HBsAg status of the source, and (iii) the hepatitis B vaccination and vaccine-response status of the exposed person (11). Such exposures usually affect persons for whom hepatitis B vaccine is recommended. For any exposure of a person not previously vaccinated, hepatitis B vaccination is recommended.

Following any such exposure, a blood sample from the person who was the source of the exposure (index patient) should be tested for HBsAg. The hepatitis B vaccination status and anti-HBs response status (if known) of the exposed person should be reviewed. The outline below and Table 3 summarize prophylaxis for percutaneous or permucosal exposure to blood according to the HBsAg status of the index source and the vaccination status and vaccine response of the exposed person.

For greatest effectiveness, passive prophylaxis with HBIG, when indicated, should be given as soon as possible after exposure (its value beyond 7 days after exposure is unclear).

1. Source of exposure HBsAg-positive

a. Exposed person has not been vaccinated or has not completed vaccination. Hepatitis B vaccinations should be initiated. A single dose of HBIG (0.06 ml/kg) should be given as soon as possible after exposure and within 24 h if possible. The first dose of hepatitis B vaccine should be given i.m. at a separate site (deltoid for adults) and can be given simultaneously with HBIG or within 7 days of exposure. Subsequent doses should be given as recommended for the specific vaccine. If the exposed person has begun but not completed the vaccination regimen, one dose of HBIG should be given immediately, and vaccination should be completed as scheduled.

b. Exposed person has already been vaccinated against hepatitis B, and anti-HBs response status is known.

(1) If the exposed person is known to have had adequate response in the past, the anti-HBs level should be tested unless an adequate level has been demon-

TABLE 3 Recommendations for hepatitis B prophylaxis after percutaneous or permucosal exposure

Exposed person	Treatment when source is found to be:		
	HBsAg-positive	**HBsAg-negative**	**Source not tested or unknown**
Unvaccinated	HBIG × 1[a] and initiate HB vaccine[b]	Initiate HB vaccine[b]	Initiate HB vaccine[b]
Previously vaccinated	Test exposed for anti-HBs	No treatment	No treatment
Known responder	1. If adequate,[c] no treatment 2. If inadequate, HB vaccine booster dose		
Known nonresponder	HBIG × 2 or HBIG × 1 plus 1 dose HB vaccine	No treatment	If known high-risk source, may treat as if source were HBsAg-positive
Response unknown	Test exposed for anti-HBs 1. If inadequate,[c] HBIG × 1 plus HB vaccine booster dose 2. If adequate, no treatment	No treatment	Test exposed for anti-HBs 1. If inadequate,[c] HB vaccine booster dose 2. If adequate, no treatment

[a]HBIG dose, 0.06 ml/kg i.m.
[b]Hepatitis B (HB) vaccine dose—see text.
[c]Adequate anti-HBs is ≥10 sample ratio units by RIA or positive enzyme immunoassay.

strated within the last 24 months. Although current data show that vaccine-induced protection does not decrease as antibody level wanes, most experts consider the following approach to be prudent.

(a) If anti-HBs level is adequate, no treatment is necessary.

(b) If anti-HBs level is inadequate (an adequate antibody level is ≥10 mIU/ml, approximately equivalent to 10 sample ratio units by RIA or positive by enzyme immunoassay), a booster dose of hepatitis B vaccine should be given.

(2) If the exposed person is known not to have responded to the primary vaccine series, the exposed person should be given either a single dose of HBIG and a dose of hepatitis B vaccine as soon as possible after exposure, or two doses of HBIG (0.06 ml/kg), one given as soon as possible after exposure and the second 1 month later. The latter treatment is preferred for those who have failed to respond to at least four doses of vaccine.

c. Exposed person has already been vaccinated against hepatitis B, and the anti-HBs response is unknown. The exposed person should be tested for anti-HBs.

(1) If the exposed person has adequate antibody, no additional treatment is necessary.

(2) If the exposed person has inadequate antibody on testing, one dose of HBIG (0.06 ml/kg) should be given immediately and a standard booster dose of vaccine given at a different site.

2. Source of exposure known and HBsAg-negative

a. Exposed person has not been vaccinated or has not completed vaccination. If unvaccinated, the exposed person should be given the first dose of hepatitis B vaccine within 7 days of exposure, and vaccination should be completed as recommended. If the exposed person has not completed vaccination, vaccinations should be completed as scheduled.

b. Exposed person has already been vaccinated against hepatitis B. No treatment is necessary.

3. Source of exposure unknown or not available for testing

a. Exposed person has not been vaccinated or has not completed vaccination. If unvaccinated, the exposed person should be given the first dose of hepatitis B vaccine within 7 days of exposure and vaccination completed as recommended. If the exposed person has not completed vaccination, vaccination should be completed as scheduled.

b. Exposed person has already been vaccinated against hepatitis B, and anti-HBs response status is known.

(1) If the exposed person is known to have had adequate response in the past, no treatment is necessary.

(2) If the exposed person is known not to have responded to the vaccine, prophylaxis as described earlier in section 1.b.(2) under "Source of exposure HBsAg-positive" may be considered if the source of the exposure is known to be at high risk of HBV infection.

c. Exposed person has already been vaccinated against hepatitis B, and the anti-HBs response is unknown. The exposed person should be tested for anti-HBs.
   (1) If the exposed person has adequate anti-HBs, no treatment is necessary.
   (2) If the exposed person has inadequate anti-HBs, a standard booster dose of vaccine should be given.

DISINFECTION, STERILIZATION, AND DECONTAMINATION

General concepts and terminology in the application of disinfection, sterilization, and decontamination procedures are presented in detail elsewhere (15, 18).

The human hepatitis viruses, however, have been difficult to study because most (HBV, HCV, HDV, and HEV) have not yet been grown in tissue culture. With the exception of HAV, comparative virucidal testing for the most part has not been performed as it has for other types of viruses that can be conveniently cultured and tested in the laboratory. To date, HAV, a nonlipid enteric virus, dried in feces, is known to survive both storage at 25°C and 42% relative humidity for at least 30 days (27) as well as 1-min exposures to all but a limited number of the most potent germicidal chemicals (37), none of which are suitable for routine decontamination of environmental surfaces. When suspended in the absence of feces, however, the susceptibility of HAV to germicidal chemicals is not unusual and is similar to that of other enteric viruses such as poliovirus (37). Similarly, although HBV is capable of surviving drying and storage at 25°C and 42% relative humidity for at least 7 days (6), it has been shown to be inactivated by several intermediate- to high-level disinfectants including glutaraldehyde, 500 ppm chlorine from sodium hypochlorite, an iodophor hard-surface disinfectant, and isopropyl and ethyl alcohols (5, 23). Recent studies have confirmed that HBV is relatively susceptible to disinfection, being inactivated by proprietary phenolic or quaternary ammonium environmental germicides (33).

We are aware of no evidence to suggest that any of the human hepatitis viruses are intrinsically more resistant to physical or chemical agents than most viruses or that the general resistance levels can even approach that of bacterial endospores. We therefore propose, as have others (37), that the resistance levels of the human hepatitis viruses that have not been studied in great detail (e.g., HCV, HDV, HEV) be considered near that of *Mycobacterium tuberculosis* subsp. *bovis* (a comparatively resistant standard test bacterium) and nonlipid viruses (e.g., poliovirus) but much less than that of bacterial endospores.

Conventional sterilization processing such as steam autoclaving or ethylene oxide can be relied on to inactivate hepatitis viruses effectively. This is also true for liquid chemical disinfectants that have been registered by the Environmental Protection Agency (EPA) as sterilant-disinfectants (sporocides). These sterilization procedures, however, are primarily used for medical devices that are reprocessed for use on patients in health care facilities. Liquid chemical agents capable of being used as sterilants are not appropriate for use on environmental surfaces.

In the context of laboratory procedures in which a liquid chemical germicide may be used to disinfect laboratory work tops or laboratory instruments that have been directly exposed to cultured or concentrated hepatitis viruses or human- or animal-source specimens containing these agents, we recommend chemical disinfectants and their appropriate concentrations and contact times that are capable of producing at least an intermediate level of disinfection activity (i.e., those germicides having label claims for or are known to inactivate *M. tuberculosis* subsp. *bovis* and a spectrum of viruses including nonlipid ones such as poliovirus) (Table 4).

For general housekeeping purposes such as cleaning floors, walls, and other environmental surfaces in the laboratory area, any disinfectant-detergent registered by the EPA for such purposes would be suitable.

DECONTAMINATION

In some high-risk areas such as laboratories and hemodialysis units, one is confronted with the problem of decontaminating large and small blood spills, patient care equipment that becomes contaminated with blood, and frequently touched instrument surfaces such as control knobs, levers, or buttons, which may play a role in environmentally transmitted hepatitis. The strategies for applying the principles of HBV inactivation vary according to the item or surface being considered, its potential role in the risk of hepatitis virus transmission, and to a certain extent, the thermal and chemical lability of the surface or instrument. For example, if a significant spill of blood occurred on the floor or on a countertop in a laboratory, the objective of the procedure to inactivate HBV or other blood-borne hepatitis viruses would be one of decontamination or disinfection

TABLE 4 Some physical and chemical methods for inactivating hepatitis viruses[a]

Method	Class, concentration, or level	Activity
Sterilization		
Heat		
Moist heat (steam under pressure)	250°F (121°C), 15 min Prevacuum cycle 270°F (132°C), 5 min	
Dry heat	170°C, 1 h 160°C, 2 h 121°C, 16 h or longer	
Gas		
Ethylene oxide	450–500 mg/liter, 55–60°C	
Liquid[b]		
Glutaraldehyde, aqueous	Variable, usually 1–2%	
Hydrogen peroxide, stabilized	6–10%	
Formaldehyde, aqueous	8–12%	
Disinfection		
Heat		
Moist heat	75–100°C	High
Liquid		
Glutaraldehyde, aqueous[c]	Variable	High to intermediate
Hydrogen peroxide, stabilized	6–10%	High
Formaldehyde, aqueous[d]	3–8%	High to intermediate
Iodophors[e]	40–50 mg/liter free iodine; up to 10,000 mg/liter available iodine	Intermediate
Chlorine compounds[f]	500–5,000 mg/liter free available chlorine	Intermediate
Phenolic compounds[g]	0.5–3%	Intermediate

[a] *Comment:* Adequate precleaning of surfaces is the first prerequisite for any disinfecting or sterilizing procedure. The longer the exposure to a physical or chemical agent, the more likely it is that all pertinent microorganisms will be eliminated. Ten minutes' exposure may not be adequate to disinfect many objects, especially those that are difficult to clean because of narrow channels or other areas that can harbor organic material as well as elevated numbers of microorganisms; thus, longer exposure times (i.e., 20–30 min) may be necessary. This is especially true when high-level disinfection is to be achieved. Although alcohols (e.g., isopropanol, ethanol) have been shown to be effective in killing HBV, we do not recommend that they be used generally for this purpose due to rapid evaporation and consequent difficulty in maintaining reasonable contact times. Also, isopropanol exhibits an irregular pattern of activity against nonlipid viruses (37). Immersion of small precleaned items in alcohols could be considered. Sterilization, disinfection, and hepatitis viruses are discussed in greater detail elsewhere (15, 18, 33, 37).

[b] This list of liquid chemical germicides contains generic formulations. Other commercially available formulations based on the listed active ingredients can also be considered for use. Users should ensure that the proprietary formulations are registered with the EPA. Information in the scientific literature or presented at symposia or scientific meetings can also be considered in determining the suitability of certain formulations.

[c] Several glutaraldehyde-based proprietary formulations are on the U.S. market. Manufacturer's instructions regarding use as a sterilant or disinfectant or anticipated dilution during use should be closely followed.

[d] Because of the ongoing controversy of the role of formaldehyde as a potential occupational carcinogen, the use of formaldehyde is recommended only in limited circumstances under carefully controlled conditions of ventilation or vapor containment (e.g., disinfection of certain hemodialysis equipment).

[e] Only those iodophors registered with the EPA as hard-surface disinfectants should be used, and manufacturer's instructions regarding proper use dilution and product stability should be closely followed. Check product label claims for demonstrated activity against *Mycobacterium* spp. (tuberculocidal activity) as well as a spectrum of lipid and nonlipid viruses. Iodophors formulated as antiseptics are not suitable for use as disinfectants.

[f] See text.

[g] Check product label claims for demonstrated activity against *Mycobacterium* spp. (tuberculocidal activity) as well as a spectrum of lipid and nonlipid viruses.

and not sterilization. Consequently, in such a situation we would recommend that gloves be worn and the blood spill be absorbed with disposable towels. The spill site should be cleaned of all visible blood, and then the area should be wiped down with clean towels soaked in an appropriate intermediate-level disinfectant such as a 1:100 dilution of commercially available household bleach (0.05% sodium hypochlorite in final dilution). All soiled towels should be put in a plastic bag or other leakproof container that can be placed in the medical waste of that particular department. The concentration of disinfectant used depends primarily on the type of surface that is involved. For example, in the case of a direct spill on a porous surface that cannot be physically cleaned before disinfection, 0.5% sodium hypochlorite (5,000 mg available chlorine per liter) should be used. However, if the surface is hard and smooth and has been cleaned appropriately, 0.05% sodium hypochlorite (500 mg available chlorine per liter) is sufficient. For commercially available chemical disinfectants, the use concentrations specified by the manufacturer should be followed.

Other types of environmental surfaces of concern include surfaces that are touched frequently, such as control knobs or panels on laboratory instruments. Ideally, gloves should be worn and manipulated in a manner appropriate not only to avoid skin contact with patient materials but to avoid "finger painting" of this contamination to a variety of other frequently touched surfaces. Because this ideal is seldom fully realized in a busy laboratory setting, laboratory instrument and equipment surfaces (including fixtures such as light switches and door pulls or push plates) should be routinely cleaned and disinfected. The objective here would be to reduce the level of possible contamination to such an extent that disease transmission is remote. In a practical sense, this could mean that a cloth soaked in either 0.05% sodium hypochlorite or a suitable proprietary disinfectant or disinfectant-detergent could be used. In this context, the element of physical cleaning is as important as, if not more important than, the choice of the disinfectant. It is not necessary, cost-effective, or in many cases, even feasible to attempt more powerful germicidal procedures with these types of items or surfaces.

GENERAL HOUSEKEEPING PROCEDURES

As a rule, routine daily cleaning procedures used for general microbiological laboratories can be used for laboratories in which blood specimens are processed. Obviously, special attention should be given to areas or items visibly contaminated with blood or feces. Furthermore, cleaning personnel must be alerted to the potential hazards associated with blood, serum, and fecal contamination. Floors and other housekeeping surfaces contaminated in this manner should be thoroughly cleaned of gross material and then treated with a detergent-disinfectant. Gloves should be worn by cleaning personnel doing these duties. However, in the case of large blood spills as mentioned above, this type of procedure may have to be augmented by specific site decontamination using a more potent chemical agent such as an intermediate-level disinfectant.

SUMMARY

Viral hepatitis is one of the most frequently reported laboratory-associated infections. Hepatitis B and perhaps hepatitis C constitute the most significant risks to laboratorians as compared with hepatitis A, D, and E. Unvaccinated clinical laboratorians involved with handling blood, serum, and other body fluids are at increased risk of acquiring hepatitis B. The primary mode of transmission is by accidental needlesticks. The presence of blood or serum on hands, whether from direct or indirect sources (such as contaminated environmental or laboratory instrument surfaces), can result in the HBV gaining access to the vascular system subcutaneously by cuts or needlesticks, by contamination of pre-existing skin lesions, or by nasal, oral, or ocular exposure (7). Although many biological and biochemical assays performed on blood are currently done by automated instrumentation, the element of automation in itself has not been recognized as presenting an additional environmental hazard. Rather, activities associated with sample reception, opening, transfer, and inappropriate disposal of those specimens are most likely to be associated with potential HBV exposure, especially when poor techniques are used. Infection control strategies should stress proper techniques for handling needles, other sharps, blood and specimen containers, consistent use of gloves, protective clothing, and face-eye barriers, frequent handwashing, good awareness of personal hygienic practices, and the appropriate use of cleaning, disinfection, and sterilization techniques. Infection control activities, including disease surveillance, should be performed by either a designated safety officer who is on the staff of the laboratory or a hospital infection control practitioner.

References

1. **Alter, H. J., R. H. Purcell, J. L. Gerin, W. T. London, P. M. Kaplan, V. J. McAuliffe, J. Wagner, and P. V. Holland.** 1977. Transmission of hepatitis B to chimpanzees by hepatitis B surface antigen-positive saliva and semen. *Infect. Immun.* **16:**928–933.
2. **Alter, M. J.** 1991. Inapparent transmission of hepatitis C: footprints in the sand. *Hepatology* **14:**389–391.
3. **Alter, M. J.** 1993. The detection, transmission, and outcome of hepatitis C virus infection. *Infect. Agent Dis.* **2:**155–166.
4. **Bond, W. W., N. J. Petersen, and M. S. Favero.** 1977. Viral hepatitis B: aspects of environmental control. *Health Lab. Sci.* **14:**235–252.
5. **Bond, W. W., M. S. Favero, N. J. Petersen, and J. W. Ebert.** 1983. Inactivation of hepatitis B virus by inter-

mediate-to-high level disinfectant chemicals. *J. Clin. Microbiol.* **18:**535–538.

6. **Bond, W. W., M. S. Favero, N. J. Petersen, C. R. Gravelle, J. W. Ebert, and J. E. Maynard.** 1981. Survival of hepatitis B virus after drying and storage for one week. *Lancet* **i:**550–551.
7. **Bond, W. W., N. J. Petersen, M. S. Favero, J. W. Ebert, and J. E. Maynard.** 1982. Transmission of type B viral hepatitis via eye inoculation of a chimpanzee. *J. Clin. Microbiol.* **15:**533–534.
8. **Centers for Disease Control.** 1980. Hepatitis B contamination in a clinical laboratory—Colorado. *Morbid. Mortal. Weekly Rep.* **29:**459–460.
9. **Centers for Disease Control.** 1987. Recommendations for prevention of HIV transmission in health care settings. *Morbid. Mortal. Weekly Rep.* **36**(No. 2S)**:**1–18.
10. **Centers for Disease Control.** 1988. Update: universal precautions for prevention of transmission of human immunodeficiency virus, hepatitis B virus, and other bloodborne pathogens in health-care settings. *Morbid. Mortal. Weekly Rep.* **37:**377–388.
11. **Centers for Disease Control.** 1991. Hepatitis B virus: a comprehensive strategy for eliminating transmission in the United States through universal childhood vaccination: recommendations of the Immunization Practices Advisory Committee (ACIP). *Morbid. Mortal. Weekly. Rep.* **40**(No. RR-13)**:**1–25.
12. **Centers for Disease Control/National Institutes of Health.** 1993. *Biosafety in microbiological and biomedical laboratories,* 3rd ed. HHS Publication No. (CDC)93-8395. U.S. Government Printing Office, Washington, D.C.
13. **Deinhardt, F.** 1980. Predictive value of markers of hepatitis virus infection. *J. Infect. Dis.* **141:**229–305.
14. **Dienstag, J. L., and D. M. Ryan.** 1982. Occupational exposure to hepatitis B virus in hospital personnel: infection or immunization? *Am. J. Epidemiol.* **115:**26–39.
15. **Favero, M. S., and W. W. Bond.** 1991. Chemical disinfection of medical and surgical materials, p. 617–641. *In* S. S. Block (ed.), *Disinfection, sterilization and preservation,* 4th ed. Lea & Febiger, Philadelphia.
16. **Favero, M. S., J. E. Maynard, R. T. Leger, D. R. Graham, and R. E. Dixon.** 1979. Guidelines for the care of patients hospitalized with viral hepatitis. *Ann. Intern. Med.* **91:**872–876.
17. **Francis, D. P., and J. E. Maynard.** 1979. The transmission and outcome of hepatitis A, B, and non-A, non-B: a review. *Epidemiol. Rev.* **1:**17–31.
18. **Garner, J. S., and M. S. Favero.** 1985. *Guidelines for handwashing and hospital environmental control.* HHS Publication No. 99-1117. Centers for Disease Control, Atlanta.
19. **Groeschel, D. H. M., and B. A. Strain.** 1991. Laboratory safety in clinical microbiology, p. 49–58. *In* A. Balows, W. J. Hausler, K. L. Hermann, H. D. Isenberg, and H. J. Shadomy (ed.), *Manual of clinical microbiology,* 5th ed. American Society for Microbiology, Washington, D.C.
20. **Hadler, S. C., I. L. Doto, J. E. Maynard, J. Smith, B. Clark, J. Mosley, T. Eickhoff, C. K. Himmelsbach, and W. R. Cole.** 1985. Occupational risk of hepatitis B infection in hospital workers. *Infect. Control* **6:**24–31.
21. **Hoofnagle, J. H., and A. M. DiBisceglie.** 1991. Serologic diagnosis of acute and chronic viral hepatitis. *Semin. Liver Dis.* **11:**73–83.
22. **Kiyosawa, K., T. Sodeyama, E. Tanaka, Y. Nakano, S. Furuta, K. Nishioka, R. H. Purcell, and H. J. Alter.** 1991. Hepatitis C in hospital employee with needlestick injuries. *Ann. Intern. Med.* **115:**367–369.
23. **Kobayashi, H., M. Tsuzuki, K. Koshimizu, M. Toyama, N. Yoshihara, T. Shikata, K. Abe, K. Mizuno, M. Otomo, and T. Oda.** 1984. Susceptibility of hepatitis B virus to disinfectants and heat. *J. Clin. Microbiol.* **20:**214–216.
24. **Levy, B. S., J. C. Harris, J. L. Smith, J. W. Washburn, J. Mature, A. Davis, J. T. Crosson, H. Polesky, and M. Hanson.** 1977. Hepatitis B in ward and clinical laboratory employees of a general hospital. *Am. J. Epidemiol.* **106:**330–335.
25. **Maynard, J. E.** 1978. Modes of hepatitis B virus transmission, p. 125–137. *In* Japan Medical Research Foundation (ed.), *Hepatitis viruses.* University of Tokyo Press, Tokyo.
26. **Maynard, J. E.** 1978. Viral hepatitis as an occupational hazard in the health care profession, p. 293–305. *In* Japan Medical Research Foundation (ed.), *Hepatitis viruses.* University of Tokyo Press, Tokyo.
27. **McCaustland, K. A., W. W. Bond, D. W. Bradley, J. W. Ebert, and J. E. Maynard.** 1982. Survival of hepatitis A virus in feces after drying and storage for one month. *J. Clin. Microbiol.* **16:**957–958.
28. **Mosley, J. W.** 1978. Epidemiology of HAV infection, p. 85–104. *In* G. N. Vyas, S. N. Cohen, and R. Schmid (ed.), *Viral hepatitis: a contemporary assessment of epidemiology, pathogenesis and prevention.* The Franklin Institute Press, Philadelphia.
29. **National Sanitation Foundation Standard 49.** 1976. *Class II (laminar flow) biohazard cabinetry.* Ann Arbor, Mich.
30. **Petersen, N. J., W. W. Bond, and M. S. Favero.** 1979. Air sampling for hepatitis B surface antigen in a dental operatory. *J. Am. Dent. Assoc.* **99:**465–467.
31. **Petersen, N. J., W. W. Bond, J. H. Marshall, M. S. Favero, and L. Raij.** 1976. An air sampling technique for hepatitis B surface antigen. *Health Lab. Sci.* **13:**233–237.
32. **Pike, R. M.** 1979. Laboratory-associated infections: incidence, fatalities, causes, and prevention. *Annu. Rev. Microbiol.* **33:**41–46.
33. **Prince, D. L., H. N. Prince, O. Thraenhart, E. Muchmore, E. Bonder, and J. Pugh.** 1993. Methodological approaches to disinfection of human hepatitis B virus. *J. Clin. Microbiol.* **31:**3296–3304.
34. **Rosenblum, L. S., M. E. Villarino, O. V. Nainan, M. E. Melish, S. C. Hadler, P. P. Pinsky, W. R. Jarvis, C. E. Ott, and H. S. Margolis.** 1991. Hepatitis A outbreak in a neonatal intensive care unit: risk factors for transmission and evidence of prolonged viral excretion among preterm infants. *J. Infect. Dis.* **164:**476–482.
35. **Shikata, T., T. Karasawa, K. Abe, T. Uzawa, H. Suzuki, T. Oda, M. Imai, M. Mayumi, and Y. Moritsugu.** 1977. Hepatitis B antigen and infectivity of hepatitis B virus. *J. Infect. Dis.* **136:**571–576.
36. **Swenson, P. D.** 1991. Hepatitis viruses, p. 959–983. *In* A. Balows, W. J. Hausler, K. L. Hermann, H. D. Isenberg, and H. J. Shadomy (ed.), *Manual of clinical microbiology,* 5th ed. American Society for Microbiology, Washington, D.C.
37. **Thraenhart, O.** 1991. Measures for disinfection and control of viral hepatitis, p. 445–471. *In* S. S. Block (ed.), *Disinfection, sterilization and preservation,* 4th ed. Lea & Febiger, Philadelphia.

Human Immunodeficiency Virus Type 1 and Other Blood-Borne Pathogens

DEBRA L. HUNT

4

INTRODUCTION

Laboratory-acquired infections from blood-borne pathogens have been recognized since 1949 when a laboratory worker was reported to have been infected with "serum hepatitis" in a blood bank (161). Skinhoj (242, 243) reported subsequent increases in occurrences in laboratory-acquired hepatitis and found a sevenfold higher rate of hepatitis in laboratory workers when compared with the general population.

The potential occult infectivity of blood has been emphasized since the documentation of occupationally transmitted human immunodeficiency virus (HIV) infection. As of October 1993, 39 documented occupationally acquired infections with HIV-1 have been reported, with an additional 81 possible work-related infections (134). Since the first occupational transmission was reported in 1984 (4), health care and laboratory administrators, as well as those in the public sector, have re-examined the infection control aspects of their work practices and have begun to develop equipment and procedures to minimize exposures. Because infection with HIV is not always clinically apparent and the infectious potential of blood and other body fluids is not always known, the Centers for Disease Control (CDC) recommended "universal blood and body fluid precautions" in 1987 (42). This approach emphasizes the consistent use of blood and body fluid precautions for *all* patients and their clinical specimens and tissues.

The "universal precautions" strategy has formed the foundation for federal guidelines through the CDC and regulations from the Occupational Safety and Health Administration (OSHA) (see OSHA standard, Appendix II). Both organizations recognize that this practical approach to safety will not only minimize the risk of occupationally acquired HIV-1 infection but will also serve to protect against occupational infection with other blood-borne pathogens such as hepatitis B, hepatitis C, human T-cell leukemia viruses I and II, or HIV-2.

This chapter provides an overview of the epidemiology, risk of transmission, and the recommended or regulated strategies to prevent occupational transmission of HIV and other blood-borne pathogens (the hepatitis viruses were discussed in detail in Chapter 3). All laboratory workers are encouraged to keep abreast of applicable rules or recommendations from federal, state, or professional agencies.

HUMAN IMMUNODEFICIENCY VIRUS TYPE 1

On June 5, 1981, the CDC reported several cases of *Pneumocystis carinii* pneumonia in young male homosexuals (31). Several weeks later, Kaposi's sarcoma was reported in 26 male homosexuals, some of whom also were diagnosed with *P. carinii* pneumonia (32). These reports represented the first recognized cases of what is now defined as AIDS. Twelve years later, in September 1993, more than 339,000 AIDS cases had been reported to the CDC among persons of all ages in the United States (61), with an estimated 8 to 10 million infected adults and 1 million infected children throughout the world (57).

Within 3 years of the recognized syndrome, the virus causing AIDS was isolated and found to be a new human retrovirus (12, 96, 212). Retroviruses had been studied primarily in animal diseases but were found to be a cause of a human disease in 1980 by Poiesz et al. (209). Although the virus that causes AIDS was originally called human T-cell lymphotropic retrovirus III (HTLV-III) by Gallo and lymphadenopathy-associated virus by Montagnier, the virus has subsequently been termed *HIV-1* by committee (67).

Biological Characteristics

Certain biological characteristics of HIV-1 are important to the epidemiologic and clinical aspects of the disease and contribute to the risk involved with viral transmission. These include (16)

1. HIV-1 belongs to a group of RNA viruses known as human retroviruses, named for the novel reverse transcriptase (RT) enzyme. The RT enzyme allows a DNA chain to be copied from the viral RNA. The double-stranded DNA material from the viral template is then incorporated into the host cell genome. This step is important epidemiologically because the retroviruses are able to exist in a latent phase for prolonged periods before disease develops. More specifically, HIV-1 belongs to the lentivirus group (lenti- = "slow"). It is estimated that the average incubation period between HIV-1 infection and the development of the disease AIDS for both homosexual men and adults with transfusion-associated HIV infection via transfusion is 8 years (171). For infants infected with HIV via transfusion, the estimated incubation period is approximately 2 years (172).
2. The RT enzyme is a natural target for antiviral agents. For example, zidovudine (AZT) inhibits HIV-1 replication by blocking RT activity.
3. Because RT activity is specific for replication of retroviruses, its detection provides an excellent indicator of retroviral activity in laboratory tests.
4. The polymerase enzyme of HIV that is involved in transcription is error-prone, contributing to the antigenic hypervariability on the viral envelope. This complicates the development of a universal vaccine and perhaps influences the virulence of the different strains of virus.
5. Surrounding the RNA viral core and viral enzymes, the lipoprotein envelope contains several important glycoproteins that help bind the virus to host cell receptors and are the focus of several vaccine studies. Evidence of antibodies to glycoproteins GP120, GP160, and GP41 is essential to laboratory testing for infection with HIV-1 (268).
6. Retroviruses, like other enveloped viruses, are rapidly destroyed by common laboratory disinfectants and detergents (151).
7. The HIV-1 viruses replicate intracellularly in the host. The main target cells are those that possess the CD4 protein receptor, primarily the T4 lymphocytes. The T4 cell is lysed or severely limited in function by viral replication, leading to eventual depletion of immunologic capabilities.

   The monocyte-macrophage cell, another target cell for HIV-1, harbors the virus but is more resistant to the cytopathic effects. The monocyte-macrophage cell serves as a reservoir for the latent viral state, a "Trojan horse" that transports the virus throughout the body, protecting it from host defenses. Evidence also indicates the Langerhans cells of the skin may also harbor the virus (21).
8. The HIV-1 virus is found in body fluids as cell-associated as well as cell-free. The numbers of virally infected cells and infectious viruses in plasma vary with the stage of HIV-1 infection. For example, HIV p24 antigen, a marker of HIV-1 replication, has been demonstrated during the acute stage of HIV infection and at the late stages of infection when CD4 lymphocytes decrease in number (232). Also, increased HIV-1 plasma titers are associated with the later stages of the disease (130). This higher "dose" of virus may be an important determinant of an increased risk of viral transmission.
9. The HIV-1 virus has been cultured from blood (96), semen (274), vaginal and cervical secretions (210, 264), saliva (112), breast milk (258), tears (93), urine (165), cerebrospinal fluid (131), alveolar fluid (275), and amniotic fluid (194); however, proven human transmission of the virus has only occurred via blood, bloody body fluids, semen, vaginal-cervical secretions, breast milk, or concentrated viral material.

Inactivation Studies

Retroviruses are classified by Klein and Deforest (151) as protein and lipid viruses and, as such, are susceptible to many common disinfectants found in the laboratory. Since 1984, several studies have eval-

TABLE 1 Environmental survival of HIV-1

Condition	Temp. (°C)	Parameters	D_{10}[a]	Comments	References
Heat	60	2 min		Virus in factor VIII preparations	McDougal et al. (185)
	60	2 h		ophilized virus	McDougal et al. (185)
	56	10 min			Martin et al. (177)
	56	20 min		50% serum	Spire et al. (246)
	56	10 min	2 min		McDougal et al. (185)
	56	5 h	20 min	50% plasma	Resnick et al. (224)
	50		24 min		McDougal et al. (185)
	45		3.3 h		McDougal et al. (185)
	37		4.8 days		McDougal et al. (185)
	37	11 days		50% plasma	Resnick et al. (224)
	37	6 days		Dried virus	Prince et al. (216)
Aqueous solution	RT[b]	15 days			Resnick et al. (224)
	RT	>7 days			Barre-Sinoussi et al. (13)
Dried virus	RT	>3 days	9 h	50% plasma in petri dish	Resnick et al. (224)
	RT	>7 days			Barre-Sinoussi et al. (13)
	RT	>7 days	8 h	5% serum on glass	Prince et al. (216)

[a]D_{10}, amount of time required to reduce viral infectivity by 1 log or 90%.

[b]RT, room temperature (20 to 27°C).

uated the inactivation of HIV-1 by a variety of physical and chemical means. The methods of testing for viability of HIV-1 after exposure to disinfectants or physical methods include the determination of RT activity and the ability of the treated virus to infect T-cell lines in tissue culture. The tissue culture assay appears to be more sensitive for small amounts of virus than the RT assay (224). It is yet unknown if the tissue culture assay is able to measure the critical human infectious dose. Therefore, the observed log reductions in virus titer and extrapolated decay rates using high concentrations of virus allow for inferences about effectiveness of the disinfectant or method of inactivation.

Environmental Stability

Under experimental laboratory conditions and grown in high concentrations of 7 to 10 logs tissue culture infectious dose ($TCID_{50}$), HIV-1 demonstrates stability at room temperature both in the dry or liquid form. (One $TCID_{50}$ is the amount of virus required to infect half the cells in tissue culture.) In aqueous suspensions of tissue culture fluid, the virus has remained viable after 1 to 2 weeks (224). Several authors have demonstrated the recovery of viable HIV-1 after 3 to 7 days in the dried state as a viral film on glass or a petri dish (13, 216, 224). Resnick et al. (224) and Prince et al. (216) calculated the amount of time required to reduce viral infectivity by 1 log or 90% (the D_{10} value) in a dried state at room temperature to be 8 to 9 h. It follows that a blood spill containing 1 to 3 logs of virus per milliliter (130) in a clinical or laboratory setting could potentially contain viable HIV-1 for more than 1 day if allowed to dry. Prompt cleaning with appropriate disinfectants should be initiated to remedy this situation.

Heat Inactivation

Although HIV-1 appears stable at room temperature, it is very heat-labile. McDougal et al. (185) found that the virus follows first-order kinetics and calculated the D_{10} values for a series of temperatures (Table 1). In a liquid suspension, they found little difference in the thermal decay rate when the virus was suspended in culture medium, serum, or liquid factor VIII but found that the virus in the lyophilized state was somewhat resistant to heat.

Martin, as well as other authors (177, 246), reported inactivation of HIV-1 suspensions at 56°C within 10 to 20 min (D_{10} value = 2 min). Resnick et al. (224), however, found that heating at 56°C for 5 h was necessary, calculating a D_{10} value of 20 min. The reason for discrepancies in these studies is undetermined.

A 1988 survey of laboratories evaluating HIV-1 tests for the CDC (52) reported that 3.9% of the laboratories heat-inactivated serum specimens at 56°C as a safety measure before testing. However, the heating process can cause false-positive results for enzyme immunoassay and Western blot tests (108, 180), changes in laboratory enzyme levels, and turbidity problems with plasma (156). The CDC recommended that heat inactivation of serum does not preclude the use of standard precautions and should *not* be used as a routine means of protection of laboratory workers (52).

The heating process has better applicability in the preparation of safe therapeutic blood products. Piszkiewicz et al. (207) found that pasteurization of antithrombin III concentrate at 60°C for 7 min reduced HIV-1 to below detectable levels. Others have found alternative methods of inactivation of blood products, including exposure to tri-(*n*-butyl)-phosphate and sodium cholate for 20 min at room temperature (214). A promising method that destroys HIV-1 but does not effect changes in hematologic parameters is the "photodynamic method," a hematoporphyrin photosensitizer (179).

Physical Methods of Inactivation

Spire et al. (246) demonstrated that HIV-1 is fairly radio-resistant, requiring higher doses of both gamma and UV irradiation for inactivation of the virus than the doses routinely used for food irradiation or in laminar flow safety cabinets. Martin et al. (176) found that sonication does not destroy HIV-1, but both high pH and low pH (<1 or >13) will inactivate the virus. Kempf et al. (148) also found that the virus is inactivated at a pH less than 4 when suspended in immunoglobulin.

Chemical Inactivation

Although the virus survives in liquid cultures for more than 2 weeks (224) and in the dried state for more than 3 days (13, 216, 224), results have confirmed that, in most cases, disinfectants at concentrations below commonly used levels are sufficient to inactivate high titers of HIV-1. Table 2 lists several chemical disinfectants and the parameters tested. Most conditions of testing in these studies have represented in-use situations such as room temperature (21 to 25°C) or less than 10-min exposure times.

Resnick et al. (224) found that common chemicals frequently used in laboratory procedures inactivated the virus. For example, nonionic detergent (Nonidet P-40) inactivated HIV-1 at 0.5% and is used in the preparation of disrupted virus at a concentration of 1%. The laboratory fixative, acetone:alcohol, also was found effective against HIV-1 after 20 min of exposure. Martin et al. (176, 177) tested paraformaldehyde (used as a laboratory fixative at 1%) and found that infected cells treated with >0.1% paraformaldehyde could no longer transmit the virus to susceptible cells.

Disinfectants that are effective against aqueous HIV-1 suspensions cannot be assumed to be equally effective against dried HIV-1. Data exist that show other viruses are more resistant to disinfectants in a dried state than suspended viruses (151, 167, 237). Several studies found alcohols to be effective against HIV-1 in culture suspensions (177, 224, 247); however, Hanson et al. (116) failed to inactivate dried, cell-free HIV within 10 min of exposure to 70% ethanol. The use of 70% ethanol as a laboratory surface disinfectant should be carefully reconsidered.

The standard antiseptics povidine-iodine and chlorhexidine gluconate, as well as common disinfectants used in the clinical setting (bleach, quaternary ammonium chlorides, phenolics, and glutaraldehyde), quickly inactivate the virus in culture (see Table 2). Hanson et al. (116) found 1% glutaraldehyde effective against dried HIV-1 at 1-min exposure but found it failed to inactivate the dried virus even after 15 min when serum was added. Other data have also shown that lipophilic viruses, including HIV, become appreciably resistant (or the disinfectant less active) in the presence of high organic load such as in blood, semen, and feces (216). Increasing the concentration of disinfectant, increasing exposure time, or simply removing as much of the organic load as possible by thorough cleaning before disinfection is recommended to alleviate this problem (see also Chapter 3).

Discrepancies in the studies to date may be explained by the variety of methodologies used to determine virucidal activity such as testing culture suspensions versus dried virus on carriers, end point determinations, organic load, contact times, neutralizers, and composition of cell culture media. Perhaps variation in HIV-1 strains may also account for differences in resistance to germicides. The need for an internationally accepted standard for virucidal assays was expressed by the American Society for Microbiology Working Group for Viruses at a symposium evaluating chemical germicides (215). Because most of the studies on inactivation of HIV-1 have tested suspension cultures of extracellular HIV-1, the Working Group called for additional data on dried viruses with an organic load challenge.

The Environmental Protection Agency has developed guidelines for the testing of virucides that involve the use of dried viral films on carriers such as glass slides or petri dishes (83) and are the basis for virucide registration in the United States. Recently,

TABLE 2 Chemical and physical methods of inactivating HIV[a]

Method	Concentration tested with RTA[b] reduction	Concentration tested reducing $>10^5$ TCID[c] HIV	Comments	References
Chemical				
Disinfectants				
Sodium hypochlorite	—[d]	0.1%		Martin et al. (177)
	—	0.5%		Resnick et al. (224)
	0.2%	—		Spire et al. (235)
Chlorine dioxide (LD)	1:200	1:200		Sarin et al. (235)
Alcohol	—	70%		Resnick et al. (224)
Ethanol	25%	—		Spire et al. (247)
	—	50%		Martin et al. (177)
	—	70%	Dried, cell-free virus >10 min required	Hanson (116)
Isopropyl	—	35%		Martin et al. (177)
Methylalcohol: acetone	—	1:1	20 min required	Resnick et al. (224)
Quaternary ammonium chloride	—	0.08%		Resnick et al. (224)
Hydrogen peroxide	—	0.3%		Martin et al. (177)
Phenolic	—	0.5%		Martin et al. (177)
Paraformaldehyde	—	0.5%		Martin et al. (177)
	—	0.1%	2 h	Martin et al. (176)
Neutral buffered formalin	—	1%		Martin et al. (176)
Nonidet P-40	—	0.5%	In 50% human plasma	Resnick et al. (224)
	—	1.0%		Martin et al. (177)
Glutaraldehyde	1%	—		Spire et al. (247)
	—	1%	Dried, cell-free	Hanson et al. (116)
	—	2%	Dried, in serum	Hanson et al. (116)
	—	1%	Dried, in serum Requires >15 min	Hanson et al. (116)
	—	2%	Dried, cell-associated	Hanson et al. (116)
Sodium hydroxide	40 mmol/liter	—		Spire et al. (247)
Antiseptics				
Povidone-iodine	—	0.25%	37°C	Kaplan et al. (145)
	—	0.5%	37°C	Harbison and Hammer (118)
Chlorhexidine gluconate	—	0.2% (1:20)	37°C	Harbison and Hammer (118)
Physical				
Gamma irradiation	2.5×10^5 rad			Spire et al. (246)
UV radiation	5×10^3 J/m^2			Spire et al. (246)
pH	—	<1, >13		Martin et al. (177)
	—	<4	Virus in IgG	Kempf et al. (148)

[a]All tests conducted at room temperature and <10 min contact time unless otherwise noted.

[b]RTA, reverse transcriptase activity.

[c]TCID, tissue culture infectious dose.

[d]Not tested by the given method.

the Environmental Protection Agency TSS-7 guidelines have been used for approved testing protocols for HIV inactivation (215). In general, HIV-1 is inactivated by chemical germicides that are effective against lipophilic viruses. Labels of Environmental Protection Agency-registered disinfectants should be scrutinized for virucidal claims and directions for in-use concentrations should be followed.

The HIV-1 virus has not been shown to be transmitted to date through environmental exposures, although the potential exists for inadvertent contamination of hands by touching soiled surfaces and subsequent inoculation of mucous membranes. The risk for this mode of exposure increases in a research laboratory situation in which high titers of virus may be manipulated. Routine cleaning and prompt decontamination of spills are the best methods to minimize this risk. In fact, the CDC does not recommend any changes in the standard guidelines for sterilization, disinfection, or housekeeping practices to handle HIV-1 (42).

Epidemiology

Since the recognition and reporting of AIDS in 1981, more than 339,000 persons with AIDS have been reported to public health departments in the United States for a nationwide rate of 17.2 per 100,000 (61). More than 204,000 (60%) of these have died. In 1992, AIDS became the leading cause of death among U.S. men 25 to 44 years of age and the fourth leading cause of death among U.S. women 15 to 44 years of age (60).

Epidemiologic information gathered since the early 1980s indicates that the modes of transmission of HIV-1 have remained the same. HIV-1 is transmitted through sexual contact, percutaneous or mucous membrane exposure to blood, birth or breast feeding from an infected mother, or transfusion of HIV-contaminated blood. Homosexual-bisexual men and intravenous (IV) drug users have represented more than three-fourths of all AIDS cases reported from 1981 to 1990 (56). However, the largest recent proportionate increases have occurred among women, blacks, Hispanics, persons living in the South, and persons exposed through heterosexual contact. Cases in children associated with perinatal HIV transmission have also continued to rise.

The annual incidence of AIDS cases associated with blood transfusions and therapeutics for hemophilia has stabilized since the serologic screening of blood donations and heat treatment of clotting factors was initiated in 1985. Currently, it is estimated that the risk of any unit of blood being contaminated with HIV after the screening process is 1:150,000 (69). Immune globulin preparations (40), recent therapeutic products for hemophilia patients (49), and hepatitis B vaccines prepared from pooled human sera (35) have been shown to be free of the virus.

In 1993, 4.7% of all AIDS cases were assigned to a "no identified risk" (NIR) category, representing a large caseload of those recently diagnosed (61). For many of these cases, follow-up investigation is incomplete or the patient died before an investigation could be performed. Historically, on investigation, 83% of the NIRs are classified into an identified risk category. Overall, the NIR category represents about 3% of all reported AIDS cases. Generally, 10% of the NIR cases are health care workers, and this percentage has remained stable over time (57).

Occupational HIV-1 Transmission

AIDS in Health Care Workers

National surveillance data of health care workers demonstrate that there is no higher risk for developing AIDS in those working in the health care or laboratory setting than for those in the general public. As of 1988, approximately 5.3% of reported AIDS cases with a known work history had related a history of working in a health care or laboratory setting since 1978 (63). This percentage is comparable to the proportion of the U.S. population working in health care (5.7%) (25). Moreover, 95% of this group have recognized nonoccupational risk factors. Health care workers with AIDS are more likely to be homosexual or bisexual than other persons with AIDS. After surveillance investigations are completed, only 1.4% of health care workers with AIDS are classified with NIR other than employment in a health care setting (63). Further examination of the remaining NIR cases in health care workers shows that they are demographically more similar to other AIDS cases than to health care workers in general. For example, 73% of the NIR health care workers versus 23% of all U.S. health care workers are male (25). Also, the only occupation that is overrepresented among the NIR cases is that of maintenance workers (20% for NIR cases versus 6% for cases with identified risk, $P<0.004$) (63) and not the occupations that are at risk for percutaneous blood exposures such as surgeons or laboratory technologists.

More indirect evidence that the risk of transmission of HIV-1 in the health care setting is small is found in 13 HIV prevalence studies conducted on cohorts of health care workers around the country, many of whom work in areas of high community seroprevalence (18, 79, 100, 106, 119, 129, 153, 170,

175, 239, 267–269). These prevalence studies have examined 6,619 U.S. health care workers with 1,208 reported HIV exposures and found 8 seropositive (0.12%) health care workers with no identified community risk. The prevalence of infection in health care workers does not appear to be any higher than that of the comparable population-at-large.

The lack of association of HIV transmission in the health care setting has also been demonstrated in prevalence studies from Kinshasa, Zaire (173, 199), where community prevalence of HIV is high (8.4% in women attending an antenatal clinic and 6.5% among men donating blood). No higher rates of seropositivity were found in the hospital staff, nor were there any significant differences among the medical, administrative, and manual workers (6.5, 6.4, and 6%, respectively). Of note is a lack of seroconversion in laboratory workers between 1984 and 1986 (199). This finding is important when differences in infection control practices in the developing countries are compared with those in the United States and Europe. For example, resources such as gowns, gloves, and disinfectants are not routinely available, and needles and syringes are nearly always washed by hand, then sterilized for reuse. These findings reaffirm the apparent low risk for occupational transmission of HIV.

Documented Case Studies

Occupational HIV infection after a specific exposure is the best indicator of the mode of HIV transmission in the health care setting. Although the risk of occupational HIV transmission appears to be low, a few case reports of health care workers infected with the virus through occupational exposure have been reported. Since 1981, 39 health care workers have experienced clearly defined seroconversions after exposures (4, 5, 39, 42, 43, 46, 47, 101, 107, 114, 123, 134, 136, 168, 169, 188, 198, 202, 218, 254, 256, 260, 265, 267). The modes of transmission for these infections appear to be 34 (87%) percutaneous (i.e., needlestick, laceration, blood in nonintact skin), 4 (10%) mucocutaneous, and 1 (3%) both percutaneous and mucocutaneous. Thirty-six (92%) of the exposures that resulted in infection were to blood; one exposure was to visibly bloody fluid, one to an unspecified fluid, and one to concentrated virus in the laboratory. Also, 81 possible occupational transmissions have been reported from health care workers who have been investigated and are without identifiable behavioral or transfusion risks but who have experienced nondocumented percutaneous or mucocutaneous exposures or contact with laboratory levels of HIV (6, 9, 14, 27, 28, 106, 111, 115, 134, 153, 166, 211, 236, 267, 269).

Sixteen laboratory workers per se have documented seroconversions, representing the largest health care occupation group among the 39 infected workers (41%) (134). Examples of specific exposures that resulted in infections include cuts with contaminated objects such as a broken Vacutainer tube (123) or a blunt stainless steel cannula used to clean equipment contaminated with concentrated virus (267). Before "universal precautions" were recommended in 1987, a laboratory worker's bare hands and arms were contaminated from a blood spill from an apheresis machine. The worker also had dermatitis and, subsequently, seroconverted (43). Fifteen additional laboratory workers are among those with possible occupational infections with no documented specific exposure that resulted in seroconversion (134).

Occupational Risk Assessment

Prevalence and epidemiologic studies indicate that occupational HIV infection does not occur frequently. Documented HIV seroconversions caused by exposures demonstrate that an occupational risk of HIV transmission does exist. Factors that may contribute to the magnitude of that risk have been addressed by Henderson (121) and include the type or extent of injury, the body fluid involved, the "dose" of inoculum, environmental factors, and recipient susceptibility. The interactions and additive effect of these factors on the individual laboratory worker are complex and unknown. However, some data are available that can help further define risks associated with several procedures or circumstances.

Route or Extent of Exposure

Parenteral. Of the 39 occupationally acquired HIV infections reported, 34 (87%) have been associated with parenteral exposure (needlesticks, cut with contaminated objects, or nonintact skin exposure to blood). Many of these have been associated with an extensive injury (4, 42, 47, 122, 182, 254, 267) involving partial injection of blood or deep intramuscular injections or cuts. However, at least three have described only superficial injuries (28, 198, 202).

The best direct measure of risk of HIV transmission by a single exposure is accomplished through prospective cohort studies that document an HIV exposure event with follow-up serologic monitoring of the exposed health care worker (Table 3). In 15 prospective studies to date that have reported 3,579 percutaneous exposures in health care workers, 8 instances of seroconversion have been documented, for an overall risk of transmission per

TABLE 3 Summary of published prospective studies of the risk for occupational HIV-1 transmission in the clinical setting

Exposure type (references)	No. of studies	No. of HCWs[a]	No. of exposures	No. infected	Infection rate (%)
Percutaneous (80, 90, 100, 123, 125, 136, 142, 154, 187, 208, 218, 219, 259, 273)	15	3,328	3,579	8	0.25
Mucous membrane (80, 90, 100, 123, 136, 142, 154, 187, 218, 219, 259, 273)	13	906	1,364	1	0.07
Cutaneous (123)	1	149	5,568	0	0
Routine patient care activities (no exposures) (100, 122, 154)	3	929	NA[b]	0	0

[a]HCWs, health care workers.
[b]NA, not applicable.

percutaneous injury from an HIV-infected source of 0.25% (80, 90, 100, 123, 125, 136, 142, 154, 187, 208, 218, 219, 259, 273).

Mucous membrane. Four mucous membrane exposures resulting in HIV infection have been reported in health care workers (43, 107, 134), although in one instance, nonintact skin contact with blood could not be ruled out as a route of exposure. In this case, a laboratory worker's face was splattered with blood when a Vacutainer top flew off the tube while collecting blood from a patient. She also reported having acne (43).

Thirteen prospective studies have included mucous membrane exposures in their risk evaluations and have reported only one seroconversion from 1,364 exposures (80, 90, 100, 123, 136, 142, 154, 187, 218, 219, 259, 273). Therefore, the risk of transmission of HIV via mucous membrane exposure is extremely low (0.07%), much lower than that of a percutaneous injury (i.e., <0.3% per exposure).

Cutaneous. The identification of the Langerhans cell in the subepithelial tissue as a target cell for the HIV (21) has caused concern among some health care workers that cutaneous exposure to HIV may result in transmission of the virus via these cells into the body (73). Infection of the Langerhans cells is usually a consequence of septic HIV infection (21). Enormous protection against all pathogens is provided by intact skin; however, penetration of the skin into the subepithelial tissues and subsequent inoculation of the Langerhans cells might occur during a needlestick or cut injury or other breaks in the skin (99).

One reported case of HIV transmission via skin contact has been reported in a mother caring for her HIV-infected child (39). Although no specific cuts, punctures, or splashes were noted, the mother reported that she used no barrier precautions such as gloves or gowns and did not always wash her hands after caring for her child. She frequently handled blood and bloody body fluids. It is likely that tiny cuts on the skin may have actually been the route of transmission of the virus.

A report of a laboratory worker infected with a laboratory strain of HIV (267) considered the source of that exposure to be "contact of the individual's gloved hand with H9/HTLV-III_B culture supernatant with inapparent and undetected exposure to skin." The subject worked with concentrated HIV and reported wearing gowns and gloves routinely. The subject admitted episodes when pinholes or tears in gloves required that they be changed. The subject also related accounts of leakage of virus-positive culture fluid from equipment and the subsequent decontamination efforts with a hand brush. The subject also recalled an episode of nonspecific dermatitis on the arm that was always covered by a gown.

A subgroup of 98 other laboratory workers who also worked with concentrated HIV were seronegative. An incidence rate of 0.48 per 100 person-years exposure has been calculated for prolonged laboratory exposure to *concentrated* virus, approximately the same magnitude of risk of infection as health care workers who experience a needlestick HIV exposure (267). Over a 45-year career, this rate would lead to a risk of 195 infections per 1,000 exposed workers in research and production facilities.

Three prospective studies have reviewed the risk of HIV transmission as a result of routine patient care activities without a known percutaneous or mucous membrane exposure (100, 122, 154). None of the 929 health care workers studied have been infected. Recently, Henderson et al. (123) summarized a 6-year ongoing study of the risk of HIV transmission to health care workers sustaining a variety of occupational exposures, including cutaneous exposure. Responding to a questionnaire, 149 National

Institutes of Health (NIH) workers reported 5,568 cutaneous exposures to blood or other bodily fluids from HIV-infected patients or their specimens and more than 10,000 cutaneous exposures to blood from all patients. No seroconversion occurred from these exposures despite the high frequency of occurrence, confirming the lack of evidence of any measurable risk of transmission of HIV by cutaneous exposure in a clinical setting.

Other routes of exposure. There have been no documented cases of HIV transmission through the respiratory, ingestion, or vector route of exposure. Some have questioned the possibility of respiratory transmission of HIV (139), specifically with the research laboratory-acquired infection with no documented percutaneous exposure (267). It is well known that common laboratory procedures using blenders and centrifuges have been evaluated and shown to produce infectious aerosols (3, 147, 205, 222, 249, 250). Before the CDC and NIH recommendations for biological containment in laboratories, agents such as rabies (271) or arboviruses (117) that are not transmitted by aerosols in the community or clinical setting were documented to cause infection under laboratory conditions when *concentrated* agents were aerosolized by blending or purification procedures. The reported laboratory worker infected with the laboratory strain of HIV may have been exposed to aerosols released during reported rotor-seal failures involving the continuous-flow zonal centrifuge.

However, an expert safety review team convened by the director of the NIH addressed this issue and agreed that the potential for direct contact transmission was much greater than for aerosol transmission (46). Procedures that generated aerosols were carried out in biological safety cabinets (BSCs). They cite other instances involving overt aerosol exposure in laboratory and production facilities involving concentrated HIV that have not resulted in seroconversions in exposed workers (41). Nevertheless, the occurrence of infection through an unknown exposure emphasizes the need for laboratory workers, particularly in research or production facilities, to adhere strictly to published safety guidelines.

The potential for respiratory transmission of HIV in individuals performing aerosol-producing procedures in a clinical setting (i.e., surgery, dentistry) has also been raised (73, 139). No epidemiologic information supports this theory. In fact, several studies have shown a low prevalence of HIV infection in dentists who are routinely exposed to aerosolized body fluids (88, 104, 119, 153, 170). Likewise, surgeons are not overrepresented in the reported AIDS cases compared with other health care workers (63).

Johnson and Robinson (139) demonstrated that HIV can remain viable in cool vapors and aerosols generated by common surgical power instruments but not in the heated vapors produced from electrocautery. In a companion study, Heinsohn et al. (120) demonstrated that aerosols of sub- and micron particle sizes are produced by the instruments. Questions remain whether any respirable size particles generated contain viable HIV and whether there exists an infectious dose required for aerosol transmission of HIV.

Other Factors

Viral concentration. The transmission of HIV and subsequent infection may also depend on the "dose" of the virus present at time of exposure. The dose is determined by the size of the inoculum or the concentration of virus in the inoculum. The dose of HIV required to infect humans is unknown. Fultz et al. (95) studied the infection of chimpanzees with HIV-1 and found that those receiving >1 $TCID_{50}$ by IV injection were persistently infected for up to 18 months. Chimpanzees inoculated with low doses (0.1 $TCID_{50}$) did not become infected, suggesting that immune systems can manage to contain small inocula of virus.

A large inoculum of HIV-infected blood such as a unit of transfused blood carries a higher likelihood of virus transmission. Donegan et al. (76) examined recipients of infected blood units with no other risk factors for HIV infection and found that 89.5% were seropositive. Ho et al. (130) estimated that 250 ml of HIV-contaminated blood contains 10^4 to 10^6 $TCID_{50}$ of HIV. In contrast, a much lower risk is associated with occupational exposures (0.25%) in which the amount of blood involved is unknown but calculated by Ho to contain 0.06 to 7 $TCID_{50}$ of HIV.

The concentration of virus in blood or bodily fluid is dependent on the stage of the patient's illness and the antiviral treatment of the patient (130, 232). Since 1983, the CDC Cooperative Needlestick Surveillance Project (260) has evaluated 1,103 workers with percutaneous injuries that resulted in four seroconversions, all of whom sustained exposures to blood from source patients with CDC-defined AIDS. Saag et al. (232) evaluated the plasma viremia levels in patients infected with HIV and found none of the asymptomatic adults, 12% of adults with AIDS-related complex, and 93% of AIDS patients had cell-free infectious virus in their plasma. Titers of the virus ranged from 10 to 10^8 TCID/ml of plasma, with a mean of $10^{2.8}$ TCID/ml in patients with AIDS. Patients with acute HIV infection had viral titers of 10 to 10^3 TCID/ml.

Saag et al. (232) also found that therapy with AZT led to a significant decline in titer. Ho et al. (130) found a 25-fold lower titer mean in AIDS patients treated with AZT versus untreated AIDS patients.

Research or production laboratory workers, by the nature of the work performed, are placed at greater risk because of the high viral concentrations in culture (>10^8 TCID/ml). Published recommended barrier protection and precautions developed by the NIH and the CDC reduce worker exposure to high-risk operations (33, 34, 36, 42, 46).

Specimen age. The length of time the blood has been removed from the source before exposure may also influence the number of infectious viruses present in the inoculum. Although most occupational infections have occurred after exposure to "fresh" blood, HIV has demonstrated stability in the environment in both liquid and dry states (224) and may survive for hours to days at room temperature.

Other. Other factors contributing to the overall risk of HIV transmission may include the virulence of the viral strain (10), postexposure first-aid or prophylactic practices, or health care worker-related factors such as skin integrity, immunologic status (100), or inflammation around the exposure site (numbers of $CD4^+$ cells available) (123).

Relative Risk for Occupational Infection

The anxiety surrounding HIV in the laboratory setting has been partially caused by the historical problem of occupational hepatitis B virus (HBV) infection and its designation as a model for transmission of a blood-borne pathogen. The CDC estimates that 12,000 health care workers will become occupationally infected with HBV each year, resulting in more than 250 deaths (51). In contrast, the total number of occupationally acquired HIV infections in 12 years is estimated to be 120 (39 documented, 81 possible). The risk of HBV infection after a parenteral exposure to hepatitis B surface antigen positive blood (6 to 30%) (36) is much greater than the risk of HIV infection from a similar exposure to HIV-infected blood (0.25%). The difference in rate of transmission and numbers of deaths is probably a result of the lower concentrations of virus in the blood of HIV-infected persons compared with that of HBV-infected persons (57). If the incidence of fatality is estimated for hepatitis B as 1 to 2% and for HIV infection as ultimately 100%, the risk of mortality from parenteral exposure to both viruses is essentially the same (0.06 to 0.3%). Repeated exposures will increase the overall risk to the laboratory worker for occupational infection. The hepatitis B vaccine is available and effective if used and is a preventive measure currently unavailable for HIV.

The risk of transmission associated with occupational exposures to other blood-borne pathogens has yet to be determined. Nevertheless, standard precautions taken to prevent exposures to hepatitis B will prevent occupational transmission of HIV and other blood-borne pathogens. A thorough discussion of hepatitis can be found in Chapter 3.

OTHER BLOOD-BORNE PATHOGENS

Retroviruses

Since 1980, five types of human retrovirus have been isolated: HTLV-I, HTLV-II, HIV-1 (formerly HTLV-III), HIV-2 (formerly HTLV-IV), and HTLV-V. Based on morphologic features and molecular hybridization studies, HTLV-I, -II, and -V are classified as oncornaviruses and are associated primarily with malignancies such as leukemia and lymphoma. HIV-1 and HIV-2 have been classified together as lentiviruses (16) and cause cell lysis and death.

Human T-Lymphotropic Virus Type I

HTLV-I, the first human retrovirus to be isolated (209), has been associated with adult T-cell leukemia-lymphoma and with a chronic neurologic disease called tropical spastic paraparesis (105). Compared with HIV-1, HTLV-I infection is a more chronic, endemic infection, largely confined to populations defined by geography, race, and age. Endemic areas appear to be concentrated in southwestern Japan, Africa, the Caribbean basin, and the southeastern United States (17, 127, 272). Although seroprevalence can be high in the endemic areas of Japan (20% of adults) or the Caribbean (2 to 5% of black adults), HTLV-I seroprevalence is low in the general population in the United States. Blood donations screened by the American Red Cross in 1989 found a seropositive HTLV-I/II rate of 1.4 per 10,000 (54), similar to the rate of HIV-1 (1.72 per 10,000) (57).

HTLV-I, like HIV-1, is transmitted by sexual contact, perinatally, and through contaminated blood. HTLV-I is, however, considerably less infectious than HIV-1, and its transmission is closely cell-associated. For example, HTLV-1 has been transmitted through whole blood, packed cells, and platelets but not through fresh-frozen plasma (201) nor through pooled clotting factor concentrates from seropositive donors in the United States (65). Parenteral exposure to contaminated needles is a documented risk accounting for high rates of seropositivity (18 to 49%) among drug abusers (158, 159, 228). One seroconversion in a health care worker who unintentionally inoculated himself with blood from an infected patient with adult T-cell leukemia-lymphoma has been reported in Japan (146).

Human T-Lymphotropic Virus Type II

A related virus, HTLV-II, was isolated in 1982 from a patient with a T-cell variant of hairy cell leukemia (143). Other cases of infection associated with hematologic abnormalities have been identified, but the HTLV-II involvement in disease is unknown (251). The epidemiology of HTLV-II is unknown because of the paucity of cases and the lack of a specific screening test for HTLV-II (HTLV-I and -II cross-react in the HTLV-I screen). The virus is presumably transmitted via the same mechanisms as HTLV-I. The main known association of HTLV-II seropositivity is IV drug abuse. In one study, a surprising 52% of the HTLV-I/II seropositive U.S. blood donors analyzed by a specific DNA amplification test were infected with HTLV-II and were more associated with a risk factor of IV drug use than the HTLV-I seropositives (160).

Human Immunodeficiency Virus Type 2

In 1986, a second human retrovirus capable of causing AIDS was isolated from patients of West African origin and was named HTLV-IV (144). The virus, later renamed HIV-2, was very closely related to simian immunodeficiency viruses, suggesting a recent divergence from a common ancestor (223).

HIV-2 is endemic to western Africa where it is the dominant HIV. Although HIV-2 seroprevalence rates are high in this region (8.9% of adults in Guinea-Bissau), the rate of AIDS is low (204), suggesting that the ability to cause disease is less efficient in HIV-2 than in HIV-1. HIV-2 cases have also been reported with more frequency in Europe and Canada, and seven cases have been reported in the United States (53). All U.S. subjects had West African origins. The incidence in nonendemic populations is difficult to estimate at this time until more specific laboratory tests become available for widespread screening. The current enzyme immunoassay laboratory tests will only detect 46 to 96% of HIV-2-positive sera (74). A positive enzyme immunoassay test for HIV-1 with an indeterminate Western blot or clinical AIDS with a weak or negative HIV-1 test should raise the possibility of HIV-2 infection (135).

HIV-2 seems to be transmitted in the same way as HIV-1. To date, no occupational infections have been documented, although there is documentation of parenteral transmission through IV drugs and blood transfusions (77).

Human T-Lymphotropic Virus Type V

HTLV-V is the designation given to an apparently new retrovirus isolated from a cluster of patients in southern Italy with a clinical syndrome resembling mycosis fungoides. This virus is significantly cross-reactive with and genetically related to HTLV-I (16). As with HIV-2 and HTLV-II, there is a lack of epidemiologic data regarding HTLV-V.

Future retrovirologic research will no doubt reveal other retroviral agents responsible for other diseases. For example, RT activity has been detected in the cells of Kawasaki disease and may indicate a retroviral etiologic agent (16).

Other Pathogens

Concern over laboratory-acquired infections has focused on HIV and the hepatitis viruses in recent years. Additional infectious diseases may include a septic phase in which agents are found in blood, many for prolonged lengths of time. Table 4 categorizes some of the documented occupationally or nosocomially transmitted blood-borne agents into three transmission groupings:

1. As with HIV and hepatitis viruses, transmission of the agent and subsequent infection of the health care worker might occur after percutaneous or mucous membrane exposure to infected blood.
2. Some infectious agents have been transmitted nosocomially through blood transfusions and pose a *potential* occupational hazard via blood exposure.
3. Other blood-borne agents also infect tissues, and laboratory-acquired infections have occurred via contact with concentrated infectious material or with experimentally infected animals, blood, or excreta.

No published information is available regarding the risk of occupational transmission of any of these agents (Table 4). Most of the agents are rarely found in the United States and may not pose a significant risk at this time. For example, Chagas' disease is endemic in Latin America, where blood transfusions frequently transmit the trypanosomes. However, Kerndt et al. (150) discovered a 2.4% seropositivity rate in Los Angeles after testing more than 1,000 blood donations.

Syphilis cases in the United States, however, are increasing in number with an incidence of 14.6 cases per 100,000 persons, a rate similar to that of HIV and HBV (261). *Treponema pallidum,* the causative agent of syphilis, is found in highest numbers during the

TABLE 4 Nosocomial transmission of blood-borne pathogens other than HIV or hepatitis viruses

1. Agents known to cause occupational infections in health care workers from percutaneous or mucous membrane exposure to blood
   - HTLV-I (146)
   - *Treponema pallidum* (syphilis) (252)
   - *Plasmodium* (malaria) (15, 20, 29)
   - *Borrelia* (261)
   - *Rickettsia rickettsii* (Rocky Mountain spotted fever) (238)
   - *Mycobacterium leprae* (leprosy) (174)
   - Viral hemorrhagic fever viruses:
     - Lassa (89)
     - Marburg (244)
     - Ebola (126)
     - Crimean-Congo (26)
2. Agents known to cause nosocomial transmission of infection through blood transfusions or tattoos (potential occupational hazards)
   - *Treponema pallidum* (syphilis) (252)
   - *Plasmodium* sp. (malaria) (97, 113)
   - *Babesia microti* (110, 245)
   - *Brucella* (261)
   - Colorado tick fever virus (190)
   - Cytomegalovirus (8)
   - *Trypanosoma brucei gambiense* (African trypanosomiasis) (128)
   - *Trypanosoma cruzi* (Chagas' disease) (24, 150)
   - *Leishmania* sp. (24)
   - *Mycobacterium leprae* (leprosy) (203)
   - Parvovirus B19 (191)
3. Blood-borne agents associated with laboratory-acquired infection via highly concentrated material or infected animals
   - *Plasmodium* sp. (malaria) (137)
   - *Leptospira* sp. (leptospirosis) (206)
   - Arboviruses (117)
   - Colorado tick fever virus (248)
   - Ebola virus (81)
   - *Trypanosoma cruzi* (22, 30, 132)
   - *Leishmania* sp. (92, 234)
   - *Toxoplasma gondii* (85, 220)
   - *Rickettsia ricketsii* (Rocky Mountain spotted fever) (140)
   - Parvovirus B19 (68)
   - *Brucella* (133, 206)
   - *Treponema pallidum* (syphilis) (206)
   - Lassa fever virus (162)
   - *Trypanosoma brucei gambiense* (African trypanosomiasis) (229)
   - *Borrelia* sp. (84)

secondary hematogenous stage but is also found intermittently in blood if syphilis is left untreated. Nosocomial transmission of syphilis has occurred by needlesticks, blood transfusions, and tattooing (252), as well as in laboratory environments (206).

Most of these agents have not been implicated in documented occupational infections with clinical exposures, but the amount of infectious agents present during septic phases of infection indicate the real potential for percutaneous transmission. For example, *Babesia microti* is present in 30 to 85% of peripheral red blood cells during the parasitemia stage of infection (255) and has been transmitted through blood transfusions (110, 245). In acute *Brucella melitensis* infections, 70 to 90% of blood cultures will grow the organism (7, 109). Human parvovirus B19 can demonstrate a high viral titer (10^{10} virions/ml) during a brief viremic stage (8) and has been transmitted through blood transfusions. In fact, Barbara and Contreras (8) estimated that up to 90% of recipients of factor VIII are likely to be seropositive for parvovirus B19, the causative agent of erythema infectiosum, also known as fifth disease.

Amplification of some blood-borne agents in the laboratory environment has resulted in laboratory-acquired infections due to contact with higher doses of agent than is found in clinical situations. In many of the reported laboratory-acquired cases listed in category 3 of Table 4, no specific incident for exposure could be recalled. Rather, the infected employee simply "worked with the agent" (117, 206), implying either aerosolization of high titers of organisms or inadvertent inoculation of mucous membranes or nonintact skin. Most of the agents listed in category 3 of Table 4 caused an occupational infection before the publication of the standard laboratory containment guidelines (225) designed to protect laboratory workers from aerosols, splashes, and other hazardous exposures.

Table 4 is not a hierarchy of risk of transmission nor is it all inclusive for blood-borne agents. The agents listed in the table with potential risk of transmission and the increasing prevalence of recognized and new retroviruses should be a reminder that laboratory safety policies should not focus on the risk of transmission of one or two agents. Emphasis should be placed on the development of standard laboratory practices for handling blood and other potentially infectious materials from all human sources in the clinical setting and the need to comply with published safety guidelines based on potential routes of transmission and procedures performed in the clinical and research laboratory environment.

STRATEGIES FOR INFECTION PREVENTION

Within 1 year of the first recognized cases of the newly defined disease AIDS, the CDC issued guidelines (33) for clinical and laboratory staff regarding appropriate precautions for handling specimens collected from AIDS patients. Updates from the CDC in 1983 (34) and 1985 (37, 38) re-emphasized precautions that had been recommended previously for handling specimens from patients known to be infected with hepatitis B (i.e., minimizing the risk for transmission by the percutaneous, mucous membrane, and cutaneous routes of infection). After anecdotal laboratory-associated infections with HIV were reported, the CDC issued its first agent summary statement for work with HTLV-III/LAV in 1986 (41). The statement included a summary of laboratory-associated infections with HTLV-III (HIV), the hazards that might be encountered in the laboratory, and advice on the safety precautions that should be taken by laboratories. Biosafety level (BSL) 2 precautions were recommended for work with clinical specimens, body fluids, or tissues from humans or laboratory animals *known or suspected to contain* HTLV-III/LAV (HIV). BSL 3 additional practices and containment equipment were recommended for activities involved with culturing research laboratory-scale amounts of the virus. A BSL 3 facility and BSL 3 practices and procedures were recommended for all work involving industrial-scale, large-volume concentrations of the virus (Table 5).

Reports issued by the CDC (43) in May 1987 documented that laboratory workers and other clinical staff were occupationally infected with HIV via nonintact skin and mucous membrane exposures. Because the HIV serostatus of the patient sources was unknown at the time of exposure and the exposures were nonparenteral, the CDC issued the "universal blood and body fluid precautions" recommendations in August 1987 (42). The main premise involves the *careful* handling of all blood and body fluids as if *all* were contaminated with HIV, HBV, or other blood-borne pathogens. This "universal precautions" concept formed the basis for all subsequent recommendations from the CDC (48, 51) and other professional organizations such as the National Committee for Clinical Laboratory Standards (NCCLS) (196). A summary of the universal precautions recommendations for clinical laboratories appears in Table 6.

The counterpart of universal precautions in a laboratory situation involves the *consistent* use of BSL 2 facilities and practices as outlined in the CDC/NIH manual *Biosafety in Microbiological and Biomedical Laboratories* (225–227; Appendix I). The BSL 2 precautions are most appropriate for clinical settings or when exposure to human blood, primary human tissue, or cell cultures is anticipated. Standard microbiological practices form the basis for BSL 2, with additional protection available from personal protective equipment (PPE) and BSCs when appropriate.

In 1988, two reports of research laboratory workers with documented occupational HIV infection prompted an investigation by an expert team to review possible sources of exposure and any need to revise current practices to reduce hazards in the research laboratory (46). Subsequently, an agent summary update was issued and included in the 1988 edition of *Biosafety in Microbiological and Biomedical Laboratories* (226). The expert team did not advise alteration of the CDC/NIH biosafety recommendations for laboratories but stressed the need for reinforcement of safety practices through proficiency and administrative discipline.

In addition to the advisory nature of the CDC/NIH guidelines, OSHA issued a final standard to regulate occupational exposure to blood-borne pathogens (262) (see Appendix II). The standard builds on the implementation of "universal precautions," specifying the need for control methods, training, compliance, and record keeping. The components of the OSHA standard are outlined in Table 7.

OSHA also recognizes that employees in HIV/HBV research laboratories and production facilities may be placed at a higher risk of infection after an exposure because of the concentrated preparations of viruses. Requirements for practices and special

TABLE 5 CDC/NIH recommended precautions for laboratory work with HIV-1[a]

Facility	Practices and procedures	Activities involving:
BSL 2	BSL 2	Clinical specimens Body fluids Human or animal tissues infected with HIV
BSL 2	BSL 3	Growing HIV at research laboratory scale Growing HIV-producing cell lines Working with concentrated HIV preparations Droplet or aerosol production
BSL 3	BSL 3	HIV at industrial-scale levels Large volume or high concentration Production and manipulation

[a]Adapted from reference 197.

TABLE 6 Summary of universal precautions for laboratories—CDC[a]

1. Universal precautions should apply to blood and all body fluids containing visible blood, semen, vaginal secretions, tissues, cerebrospinal fluid, synovial fluid, pleural fluid, peritoneal fluid, pericardial fluid, and amniotic fluid.
2. Hands should be washed immediately when contaminated with blood or other bodily fluids, after removing gloves, and after completing laboratory activities.
3. Use of needles and syringes should be limited to situations in which there is no alternative. If used, needles should not be recapped, purposely bent or broken by hand, removed from disposable syringes, or otherwise manipulated by hand. After use, disposable syringes and needles, scalpel blades, or other sharp items should be placed in puncture-resistant containers for disposal; these containers should be located as close as practical to the use area. Reusable sharps should be placed in a puncture-resistant container for safe transport to the processing area.
4. Laboratory workers should use protective barriers appropriate for the laboratory procedure and the type and extent of exposure anticipated. For example:
   All persons processing blood specimens should wear gloves.
   Phlebotomists should wear gloves when they have cuts, scratches, or other breaks in the skin, when hand contamination is predictable (i.e., uncooperative patient, or heel or finger sticks on infants and children), and when receiving training in phlebotomy.
   Surgical or examination gloves should not be washed or disinfected for reuse.
   General-purpose utility gloves should be used for housekeeping, instrument cleaning, and decontamination procedures and can be decontaminated and reused as long as they remain intact.
   Masks and protective eyewear or face shields should be worn if mucous membrane contact with blood or bodily fluids is anticipated (i.e., removing tops from vacuum tubes).
   Gowns, laboratory coats, or aprons should be worn during procedures that are likely to generate splashes of blood or bodily fluids and should be removed before leaving the laboratory.
   Routine procedures, such as histologic and pathologic studies or microbiologic culturing, do not require a BSC. BSCs (class I or II) should be used whenever procedures are conducted that have a high potential for generating droplets (i.e., blending, sonicating, and vigorous mixing).
5. All specimens of blood should be put in a well-constructed container with a secure lid to prevent leakage during transport.
6. Mechanical pipetting devices should be used in the laboratory. Mouth-pipetting must not be performed.
7. Laboratory work surfaces should be cleaned of visible material and then decontaminated with an appropriate chemical germicide after a blood or bodily fluid spill and when work activities are completed.
8. Contaminated materials used in laboratory tests should be decontaminated before reprocessing or be placed in bags and disposed of in accordance with institutional and regulatory policies for disposal of infective waste.
9. Contaminated scientific equipment should be clean and then decontaminated before repair in the laboratory or transport to the manufacturer.
10. Area posting of warning signs should be considered to remind employees of continuing hazards of infectious disease transmission in the laboratory.

[a]Modified from references 42, 48, and 51.

procedures, facility design, and additional training for these workplace situations are included in the OSHA standard and are consistent with the CDC/NIH laboratory biosafety guidelines for BSL 2 and 3.

Specific Precautions

OSHA has issued the blood-borne pathogen standard as a "performance" standard. That is, the employer has a mandate to develop an exposure control plan to provide a safe work environment but is allowed some flexibility in accomplishing this goal. The OSHA standard includes the basic philosophy of the CDC "universal precautions," along with combinations of engineering controls, work practices, and PPE to accomplish the intent of the standard (Appendix II). Exposure control plans for laboratories must adhere to the rules of the OSHA standard but can also benefit from safety recommendations from other professional organizations such as the CDC, NIH, or NCCLS. The following recommendations may be used to augment a laboratory safety plan .

Sharps Precautions

Because injuries from contaminated sharps represent the highest risk for HIV transmission, clinical and research laboratory safety plans should restrict the use of needles and other sharp instruments in the laboratory for use only when there is no alternative, such as performing phlebotomy. For many laboratory procedures, blunt cannulas or small-bore tubing can be substituted. If needles must be used,

TABLE 7 Summary of OSHA blood-borne pathogen standard requirements[a]

I. Exposure control plan (ECP): the establishment's written or oral policy for implementation of procedures relating to control of infectious disease hazards

II. Components of the ECP include:
   A. Exposure risk determination for all employees
   B. Control methods:
      1. Universal precautions: a method of infection control in which all human-derived blood and potentially infectious materials are treated as if known to be infectious for HIV or HBV
      2. Engineering controls: use of available technology and devices to isolate or remove hazards from the worker (biosafety cabinets [BSCs], puncture-resistant sharps containers, mechanical pipetting devices, and covered centrifuge canisters)
      3. Work practice controls: alterations in the manner in which a task is performed to reduce the likelihood of exposure to the worker (standard microbiologic practices in laboratories [e.g., Biosafety Level 1 practices], disposal of needles without recapping or breaking)
      4. Personal protective equipment (PPE): specialized clothing or equipment used by workers to protect themselves from exposures (gloves, gowns, laboratory coats, fluid-resistant aprons, face shields, masks, eye protection, and head and foot coverings)
      5. Additional requirements for HIV and HBV research laboratories and production facilities:
         Biosafety level 3 (BSL 3) special practices and containment equipment
         Research facility meets BSL 2 design criteria
         Production facility meets BSL 3 design criteria
   C. Housekeeping practices
   D. Laundry practices
   E. Regulated waste disposal
   F. Tags, labels, and bags
   G. Training and education programs: additional training for employees in HIV or HBV research laboratories and production facilities *before* work with HIV or HBV
   H. Hepatitis B vaccination
   I. Postexposure evaluation and follow-up
   J. Record keeping: includes medical records, training records, and maintaining availability of records

III. Administrative controls: to develop the ECP; to provide support of the ECP and provide accessibility of control methods, monitor compliance, and survey for effectiveness; and to investigate exposures for prevention of future occurrences

[a]From reference 262.

the use of "self-sheathing" needles designed to prevent needlesticks should be considered. Used needles should never be bent, broken, recapped, removed from disposable syringes, or otherwise manipulated by hand before disposal; rather, they should be carefully placed in conveniently located puncture-resistant containers (227). Removal of needles from nondisposable Vacutainer sleeves or syringes should be accomplished with a mechanical device such as forceps or hemostats or by using notched slots designed into needle boxes for safe removal of the needle. The OSHA standard allows a "one-handed" recapping technique only if there is no alternative feasible. All disposable sharps encountered in the laboratory, including pipettes, microtome blades, micropipette tips, capillary tubes, and slides, should also be carefully placed in conveniently located puncture-resistant containers for disposal. Nondisposable sharps should be placed in a hard-walled container for transport to a processing area.

Plasticware should be substituted for glassware whenever possible. Broken glassware should never be handled directly by hand but must be removed by mechanical means such as a brush and dustpan, tongs, or forceps. Cotton swabs can be used to retrieve small slivers of glass.

Engineering Controls

Recognizing that human behavior is inherently less reliable than mechanical controls, OSHA advocates the use of available technology and devices to isolate or remove hazards from the worker. The use of self-sheathing needles is an example of an engineering control to help isolate the worker from the hazard of needlestick exposure.

Another engineering control in the laboratory is the use of a properly maintained BSC to enclose work with a high potential for creating aerosols or droplets (i.e., blending, sonication, necropsy of infected animals, intranasal inoculation of animals, or opening lyophilized vials under pressure). *All* work with infectious material in an HIV research laboratory should be performed in a BSC or other physical containment device. For example, high-energy activities such as centrifugation that are performed outside a BSC should be designed for aerosol containment. Sealed safety cups or rotors should be used for centrifugation and changed out in a BSC. Before centrifugation tubes should be examined for cracks, and any glass fragments in the centrifuge cups should be carefully removed with forceps or hemostats. Microwell plate lids can be sealed with tape or replaced with adhesive-backed mylar film before centrifugation.

Plastic shielding can be used to reduce the exposure to splatter or droplets from fluorescent activated cell sorters or other automated laboratory equipment that might generate droplets of infectious material. Likewise, the Plexiglas radiation shield used in RT assays offers protection from splatter. However, if used in a BSC, the sloped top may divert airflow in the cabinet and must be removed to provide optimal protection by the BSC.

High-speed blenders and grinders are available that contain aerosols of infectious material but need to be opened in a BSC after processing. Enclosed electrical incinerators are preferable to open Bunsen burner flames for decontaminating bacteriologic loops to prevent splatter and may be used within or outside of a BSC.

Work Practice Controls

The manner in which a task is performed can minimize the likelihood of exposures in the laboratory. For example, careful disposal of used needles without recapping or otherwise manipulating by hand can reduce the likelihood of needlesticks.

Standard microbiological practices have been recommended by CDC and NIH guidelines (226, 227) for all laboratory containment levels (see Appendix I). Most of the practices are designed to prevent indirect transmission of infectious material from environmental surfaces to the hands and from hands to the mouth or mucous membranes. Such practices include prohibition of mouth pipetting, eating, drinking, smoking, applying cosmetics, or handling contact lenses in the laboratory and attention to environmental decontamination.

One of the best work practices for any laboratory setting is that of frequent and adequate handwashing when hands are visibly contaminated, after completion of work, before leaving the laboratory, after removing gloves, and before eating, drinking, smoking, or changing contact lenses. Any standard handwashing product is adequate, but products should be avoided that disrupt skin integrity. When knee- or foot pedal-controlled faucets are not available, faucets should be turned off by using the same paper towels used to dry hands to prevent recontamination of hands. Proper attention to handwashing will prevent inadvertent transfer of infectious material from hands to mucous membranes (see also Chapter 17).

In most clinical settings, skin lesions may be covered by occlusive dressings and, if lesions are on the hands, gloves worn over the dressings to prevent contamination of nonintact skin. However, workers with skin lesions or dermatitis on hands or wrists should not perform procedures with concentrated HIV material even if wearing gloves. Other work practices can reduce the amount of splatter from laboratory procedures. Covering pressurized vials with plastic-backed or alcohol-soaked gauze when removing needles or when removing tops of pressurized Vacutainer tubes will minimize the exposure to splatter. To prevent popping stoppers on evacuated tubes or vials, blood should never be forced into the tube by exerting pressure on the syringe plunger; rather, tubes and vials should be filled by internal vacuum only. Extreme caution should be used when handling pressurized systems such as continuous-flow centrifuges, apheresis, or dialysis equipment. Use of imperviously backed absorbent material ("lab diapers") can reduce the amount of splatter on laboratory work surfaces when liquids accidentally leak or fall during laboratory procedures and can aid in laboratory cleanup. The air intake and exhaust grills in BSCs must be kept clear of any surface covers or equipment.

Safe transport of specimens or infectious material within the laboratory or to other areas can minimize the potential for accidental spills or injuries. Specimens should be placed in a closed leakproof primary container and covered with a secondary container (i.e., a plastic bag of appropriate size and strength) to contain any leaks during transport. The OSHA regulations do not mandate labeling or color-coding specimens if the specimens are handled only within the facility, a policy implementing "universal precautions" is in effect, and the containers are recognizable as human specimens. Bulk samples may be safely transported in a rack within a sealable plastic container such as a modified "tackle box." The box should be labeled with a biohazard symbol or be color-coded if the contents are not clearly visible as specimens.

Luer caps should be used to transport syringes (after needles are removed with forceps or hemostats and properly disposed) or needles carefully recapped using a one-handed technique. Capillary tubes should be transported in a solid-walled secondary container such as a screw-top test tube. Transport of cultures or hemocytometers from the BSC within the laboratory may be facilitated by placing them on a tray to limit the number of trips and opportunities for spillage.

Designation of "clean" versus "dirty" areas of the laboratory or within BSCs can help prevent inadvertent contamination. Work should be planned to move from clean areas to dirty areas. Routine cleaning of work surfaces must be performed after procedures are completed and at the end of each work shift, with additional decontamination as needed for spills. Routine cleaning can be accomplished using a variety of disinfectants including iodophors registered as hard-surface disinfectants, phenolics, and 70% ethanol (with consideration given to the need for longer contact time when decontaminating dried viral cultures [116]). Diluted bleach has been most widely used for routine disinfection (10% bleach [0.5% sodium hypochlorite] for porous surfaces and 1% bleach [0.05% sodium hypochlorite] for cleaned, hard, and smooth surfaces). Aldehydes are not recommended for surfaces because of their potential toxicity. Further information on disinfection may be found in Chapter 17.

Prompt decontamination is important after spills of infectious materials, because HIV is able to survive for several hours in the environment (Table 1). Appropriate spill clean-up in a clinical setting should be carried out by a trained employee using appropriate PPE and the following methods:

1. Absorb the spill with towels or "lab diapers" to remove the extraneous organic material.
2. Clean with soap and water.
3. Decontaminate with an appropriate disinfectant. (The CDC recommends an EPA-registered "hospital disinfectant" that is also "tuberculocidal"; or a 1 to 10% bleach solution [one part of household bleach to nine parts of water] is sufficient [42].)

Large spills of cultured or concentrated agents may be safely handled with an extra step:

1. Flood the spill with an appropriate disinfectant *or* absorb the spill with granular material impregnated with disinfectant.
2. Carefully soak up the liquid material with absorbent material (paper towels) or wipe or scrape up the granular absorbent material and dispose of according to the waste disposal policy.
3. Clean the area with soap and water.
4. Decontaminate with fresh disinfectant.

Laboratory equipment (analyzers, centrifuges, pipettors) should be checked routinely for contamination and appropriately decontaminated after use with potentially infectious materials. Equipment sent for repair must be decontaminated before leaving the laboratory or labeled as to the biohazard involved and packaged to prevent the exposure of transport and repair personnel.

Because the intent of the OSHA blood-borne pathogen standard is worker protection, the rules for appropriate waste disposal emphasize adequate packaging. Sharps disposal containers must be puncture- and leakproof as well as easily accessible. Other regulated ("infectious" or "medical") waste must be placed in leakproof containers or bags that are color-coded red or orange or labeled with the word *biohazard* or the universal biohazard symbol. All disposal containers should be replaced before they are full.

Blood or body fluids may be disposed of by carefully pouring down the sanitary sewer if local health codes permit but preferably not poured into a sink used mainly for handwashing. Liquid and solid culture materials, however, *must* be decontaminated before disposal, most commonly by steam sterilization (autoclaving). Tissues, body parts, and infected animal carcasses are generally incinerated. *All* contaminated laboratory waste from HIV research-scale laboratories or production facilities and animal rooms must be decontaminated before disposal (BSL 3 practices). Additional regulated waste definitions and requirements may exist locally and must be consulted for proper disposal policies.

Personal Protective Equipment

Another strategy to minimize worker exposure to infectious material is the use of PPEs that are appropriate for the laboratory procedure and the type and extent of exposure anticipated. Examples include a variety of gloves, gowns, aprons, and face, shoe, and head protection. Appropriate selection and use of PPEs are discussed in more detail in Chapters 11 and 13. PPE may be used in combination with engineering controls or work practices for maximum worker protection.

Gloves are required by OSHA when hand contact with blood, other potentially infectious materials, mucous membranes, or nonintact skin is reasonably anticipated. The federal regulations also require gloves when handling or touching contaminated items or surfaces and for performing vascular access procedures. (The exception for trained phlebotomists [Table 6] is only allowed in volunteer blood donation centers.) Gloves are appropriate in the laboratory when handling clinical specimens, infected animals, or soiled equipment, when performing all laboratory procedures with potentially infectious materials in clinical or research laboratories, when cleaning spills, and when handling waste.

For routine procedures, vinyl or latex gloves are effective when appropriately used for prevention of skin exposure to infectious materials. These gloves are not sufficient protection against puncture wounds from needles or sharps. However, there is evidence of a "wiping" function that may reduce the amount of blood or infectious material brought through from the outside of the needle as it penetrates a glove or combination of gloves. Johnson et al. (138) found that two or three layers of latex gloves appeared to reduce the frequency of HIV-1 transfer by surgical needles to cell cultures. They also found that Kevlar gloves (untreated), Kevlar gloves (treated with the virucidal compound nonoxynol-9), and nonoxynol-9-treated cotton gloves used as intermediate layers between two layers of latex gloves significantly reduced the amount of HIV-1 transfer when compared with a single latex glove barrier. Gerberding et al. (103) reported that when surgeons wear double gloves, the rate of puncture of the inner glove is three times less than the rate of puncture of a single glove.

Other gloves are available that provide puncture "resistance" such as stainless steel mesh (chain mail) gloves to protect against injury from large sharp edges such as knife blades. Nitrile gloves (synthetic rubber) have some degree of puncture resistance that may eliminate problems of tears from rings or fingernails, yet retain the necessary dexterity required for performing laboratory procedures. A thin leather glove has been developed that can be worn under latex gloves for an additional barrier against needlesticks or animal bites. Even heavyweight utility gloves (dishwashing gloves) provide extra protection and should be worn when the procedure permits, such as cleaning contaminated equipment or spills.

Undetected holes and leaks require that gloves be inspected by the user and changed when necessary. The Food and Drug Administration has issued acceptable quality limits for defects at 2.5% defective for surgeons' gloves and 4.0% for latex examination gloves (263), although the acceptable quality limit varies widely among manufacturers. The reported percentage of defects caused by holes for nonsterile latex gloves ranges from 0 to 32%, and for nonsterile vinyl gloves, from 0 to 42% (196). Clearly, for high-risk situations such as gross contamination of gloves with blood, bloody body fluid, or high concentrations of HIV-1, the use of double gloves will lower the risk of hand contamination from seepage through undetected glove defects. Although they are more puncture-resistant, nitrile gloves are designed to tear apart when any pressure is applied to a hole in the glove, so that any defect in the glove can be detected.

Disposable latex and vinyl gloves must not be washed or disinfected for reuse. Detergents may cause enhanced penetration of liquids through undetected holes, causing a "wicking" effect (48, 196). Disinfectants, such as 70% ethanol, can also enhance the penetration of the glove barrier and facilitate deterioration (152, 196).

Gloves must be changed when visibly contaminated, torn, or defective or when tasks are completed. Because hands may be inadvertently contaminated from laboratory surfaces, gloves should be removed before handling telephones, doorknobs, or "clean" equipment. Alternatively, "dirty" equipment may be designated and marked to be handled only with gloved hands. Laboratory workers should practice the aseptic technique for glove removal (i.e., the contaminated outside of the gloves is turned inside as gloves are removed to protect the worker from skin contamination). Hands should always be washed after glove removal.

When soiling of clothing is anticipated, laboratory coats, gowns, or aprons are recommended. However, when a potential for splashing or spraying exists, solid-front, appropriately fluid-resistant gowns should be selected. If the anticipated exposure involves soaking, solid-front fluidproof gowns are required, as well as hoods or caps, facial protection, and shoe covers. BSL 3 practices advise a solid-front or wrap-around, long-sleeved gown or coveralls for adequate protection in research laboratories or production facilities.

Gowns with tightly fitting wrists or elasticized sleeves should be worn for work in BSCs. Alternatively, water-resistant "gauntlets" that provide a barrier between the glove and the laboratory coat are available to reduce skin exposure of the wrist and arm.

Laboratory coats or gowns should not be worn outside the laboratory. In HIV-1 research laboratories or production facilities, the gowns or other protective clothing must be decontaminated before laundering or disposal (BSL 3 practices).

When splashing of blood or infectious material into the mucous membranes of the face is anticipated, a mask and goggles or face shield must be used. Most laboratory procedures involving this degree of potential exposure should be conducted within containment equipment such as a BSC. Face protection might be needed for activities conducted outside a BSC, such as performing an arterial puncture, removing cryogenic samples from liquid nitrogen, or in some animal care areas. Masks and eye goggles or face shields also serve a passive function as a means of preventing accidental contact of contaminated gloved hands with the eyes, nose, and mouth during the course of work activities.

Whatever the PPE needs of any particular laboratory, OSHA requires that the employer provide an adequate supply of PPEs in the appropriate sizes. Hypoallergenic gloves must be available for employees who develop allergies to glove material or the powder inside gloves. Any defective PPE must be replaced, and reusable protective clothing must be laundered and maintained by the institution. Finally, all laboratory workers must be instructed in the proper use of PPEs and their location.

Animal Research

Nonhuman Primates

Until recently, HIV studies in animals were restricted to nonhuman primates. Chimpanzees can be successfully infected with HIV-1, although they do not demonstrate disease (91). A potentially new animal model of HIV infection has been reported by Agy et al. (2). This group has successfully infected *Macaca nemestrina* with HIV-1. Like chimpanzees, this animal demonstrates infection but not disease. The advantages are the lower cost and greater availability of the animals.

Simian immunodeficiency virus (SIV), a retrovirus closely related to HIV-1 and HIV-2, has been isolated from several species of nonhuman primates (African green monkeys [200], sooty mangabeys [94], and macaques [71]) and is used as a model for HIV infection. Two laboratory workers have sustained exposures (one needlestick, one open skin lesion) and subsequently seroconverted with antibodies to SIV (59). Both remain well, with no clinical or laboratory evidence of immunodeficiency to date. These two laboratory-acquired infections emphasize the need to adhere to the specific guidelines that have been published for research with nonhuman primates to minimize the risk of HIV-1 and SIV transmission to laboratory workers and animal handlers (23, 45, 50). BSL 2 practices, containment equipment, and facilities for animals (see Appendix I) are recommended in both instances, with an additional note to use face shields or surgical masks and eye shields as appropriate in animal rooms to protect the mucous membranes of the eyes, nose, and mouth from excreta that may be thrown by animals. To avoid accidental injuries during inoculations or other procedures on animals, both chemical and physical restraints may be used (270).

SCID Mice

Two mouse models were developed in 1988 that allow in vivo studies of HIV in human cells. The models were constructed by engrafting either human peripheral blood leukocytes (192) or human thymus, lymph node, and liver tissue (SCID-hu) (183) into severe combined immunodeficient (SCID) mice. The human peripheral blood leukocyte SCID mice and SCID-hu mice can then be infected with HIV-1 and demonstrate syndromes that closely resemble human infection (195).

To minimize the potential risk to laboratory workers and the community from research with HIV-infected mice, all studies were initially conducted in containment facilities following practices that met or exceeded BSL 3 standards. In 1990, a working group sponsored by the National Institute of Allergy and Infectious Diseases reviewed studies conducted on the C.B17-*scid/scid* mouse (189). The participants concluded that the level of viremia in these infected mice does not exceed that of humans, and the likelihood of HIV interaction with endogenous viruses to produce a pseudotype virus with an altered route of transmission is low. It was agreed that research on the HIV-infected SCID mouse could be conducted safely in BSL 2 facilities using BSL 3 practices. However, BSL 2 facilities should incorporate design features that prevent escape of HIV-infected animals. The workshop participants also agreed that animal studies likely to generate HIV pseudotypes should be conducted under BSL 3 conditions until further evaluation of this potential. Also, any research in which HIV is purposely coinfected with mouse amphotropic or xenotropic retroviruses (e.g., murine leukemia virus) or human lymphocytotrophic viruses (e.g., Epstein-Barr virus or cytomegalovirus) should be conducted at BSL 3. Appropriate precautions should also be taken for work with SCID mice other than type C.B17. Information on the risk of exposure to replication-competent retroviruses can be found in Chapter 9.

Transgenic Mice

Transgenic mice have been developed as another animal model that contain inserted copies of HIV proviral DNA. Leonard et al. (163) demonstrated that some transgenic mice develop a fatal disease that mimics AIDS in humans. Although no biosafety recommendations have been published specifically addressing this animal model, concerns over animals escaping, mating in the wild, and possibly extending the host range of the virus have prompted researchers to use BSL 3 facilities primarily for animal containment (270).

Other

Another small animal model, the New Zealand white rabbit, is used for HIV in vivo research (155), constructed by injection of an HIV-infected human T-cell line. Infected rabbits have also been studied in BSL 2 facilities with manipulations carried out using BSL 3 practices by the laboratory workers.

Other Research Concerns

Vaccinia Vectors

Advances in recombinant DNA research techniques and peptide synthesis have enabled researchers to identify and clone proviral DNA forms of HIV, with expression of viral proteins. The viral proteins can then be used for drug and vaccine development studies (240). The vaccinia virus has been genetically engineered to contain foreign DNA from one or more infectious agents, including HIV. The recombinant vaccinia virus can then express the protein antigens of HIV (62, 193). Because laboratory-acquired infections with vaccinia and recombinant vaccinia have been reported (141, 206), the CDC recommends that the vaccinia vaccine be given to laboratory workers who directly handle vaccinia cultures or who handle animals contaminated or infected with vaccinia or recombinant vaccinia (58).

HIV Proteins

Because some of the expressed proteins may be immunogenic, concern has been expressed that laboratory workers exposed to the *proteins* may develop HIV-specific antibodies and demonstrate a false-positive Western blot test for determination of HIV infection (221). Although the initial concern is a psychosocial one, the impact on the workers' immune response to any subsequent HIV vaccination attempt is unknown.

HIV Proviral DNA

Letvin et al. (164) infected macaques by inoculation with SIV proviral DNA. They speculate that HIV viral DNA may be infectious for laboratory workers and call for researchers to evaluate DNA handling techniques carefully.

The fast pace of HIV research developments and techniques requires careful planning to prevent exposures to HIV, HIV proteins, or HIV proviral DNA. Attention to the BSL 2 or 3 recommendations, as well as training and monitoring laboratory workers for safe work practices, will minimize this risk.

Employee Training and Monitoring

One of the most important components of an exposure control plan for the laboratory is a formal training program. Mere "on-the-job" training is not acceptable as adequate safety training in the laboratory. Recommendations from the CDC (46) and the NCCLS (196) that emphasize education of laboratory workers have been incorporated into the OSHA blood-borne pathogen standard (262; Appendix II).

Interactive training sessions must be conducted on initial hire and with annual updates by a person knowledgeable about the standard. Employees must be educated regarding their risks and the institution's plan to control these risks. Specific required components of the training program may be found in Appendix II.

Recognizing the increased risk of working with concentrated viral preparations, OSHA requires that employees in HIV research laboratories and production facilities receive *additional* initial training. Employees in these situations must demonstrate proficiency in standard microbiological practices as well as practices and techniques specific to the facility before work with HIV. This might include prior experience in handling human pathogens or tissue cultures or participation in a training program with a progression of work activities to develop proficiency before pathogens are handled.

Employers must ensure compliance with the OSHA standard. The CDC (46) and the NCCLS (196) recommend that workplace practices be monitored at regular intervals by a biosafety expert. The NCCLS suggests that audits be conducted to evaluate the existence and effectiveness of training programs and the appropriate training of the safety instructors. The audit should also examine the adequacy of the laboratory facilities and equipment, the standard operating practices, and the written safety protocols. Corrective measures should be implemented if needed. If breaches in protocol are detected, employees should be re-educated and, if necessary, disciplinary action taken.

OCCUPATIONAL HEALTH ISSUES

Vaccination

Historically, vaccines have been the most effective and safest prevention strategy in a biosafety program. One of the best examples of this approach for the laboratory is the hepatitis B vaccine, now mandated by OSHA as an important component of a workplace exposure control program (262). However, the unique properties of the HIV virus and the complex social issues involved with the disease have made efforts to design and test a safe effective vaccine very difficult. Obstacles to the development of a vaccine include the many antigenic variants of the virus, the fact that HIV is not only transmitted as free virus but integrated into the DNA of the host cells, the risk of immunopathology by viral proteins, no known animal model that consistently demonstrates infection and disease, the need for systemic *and* mucosal immunity, the unknown long-term protective response of antibodies, and the complex issues associated with strategies for large-scale testing of a candidate vaccine (19, 66).

Vaccine Strategies

The most common approach to HIV vaccine development to date is the use of protein subunits of the virus, via recombinant DNA and genetic engineering (231), purified native proteins from infected cells (11), or synthetic peptides from the HIV envelope or core (149). Seven clinical trials are in progress using HIV-1 recombinant envelope vaccines and two trials using core peptides (98). Multiple antigens may be required, however, for effective protection. Soluble subunit vaccines rarely induce cytotoxic responses unless presented in a special adjuvant or delivery system.

Such an approach is that of live virus recombinants (non-HIV viruses such as vaccinia) used as vectors combined with HIV envelope genes (62, 193) with a follow-up booster with recombinant envelope protein (186). The "prime-boost" combination produces a strong cellular immune response. However, large efficacy trials using vaccinia as a vector will not be entirely safe unless the vaccinia virus is further attenuated (98). Other attenuated live vectors in development include recombinant adenovirus (64), poliovirus (82), *Salmonella* (1), and Calmette-Guerin bacillus (253). Again, safety for immunocompromised recipients is a concern when inoculating with live vectors. Other vaccine strategies proposed have been the use of a killed HIV-1 virus (178), a nonpathogenic, genetically altered HIV variant (75, 87), or an anti-idiotype vaccine to mimic CD4 receptors to "neutralize" the virus (70).

Problems with the experimental vaccines to date include the requirement for multiple injections over several months, the short duration of immunity provided by the vaccine, and the lack of protection when animals are challenged by IV or genital infection with high doses of virus (98). Also, immunity has not been demonstrated against the wide variety of HIV strains found in nature, nor has the necessary induction of $CD8^+$ cytotoxic lymphocyte activity been demonstrated.

Postinfection Vaccination

Although vaccines to prevent infection may not be currently attainable, vaccines are being developed that may prevent or delay the onset of the disease AIDS (233). This "postinfection" or "therapeutic" strategy relies on the immunization of asymptomatic HIV-infected individuals to boost their protective response, reduce the viral burden, and prevent the clinical progression to AIDS. Therapeutic trials with HIV gp160 vaccine and killed HIV (the Salk vaccine) are under way in humans.

Despite the failure to date to develop an effective, practical vaccine for humans, the commitment of the federal government, private industry, and academia is encouraging for the future. There is no doubt that when such a vaccine is approved, OSHA will require its widespread use in the health care and laboratory settings as an integral component of an exposure control program.

Postexposure Evaluation Program

The implementation of a laboratory exposure control program that includes universal precautions and the recommendations from the CDC and the NIH may reduce the incidence of occupational exposure to HIV; however, use of prevention strategies will not entirely eliminate the risk of accidental exposures to HIV and subsequent occupational infection. OSHA estimates that full compliance with the blood-borne pathogen standard would reduce the risk of mucous membrane and skin exposure by 90% and the risk of parenteral exposure by 50% (261). Therefore, a postexposure evaluation program is a necessary and mandated component of a laboratory safety program.

Laboratories handling blood, other potentially infectious materials, or concentrated HIV viral material must adhere to the OSHA postexposure protocol that requires confidential medical evaluation, follow-up, and documentation of an exposure incident. OSHA defines an "exposure incident" as a "specific eye, mouth, other mucous membrane, non-intact skin, or parenteral contact with blood or other potentially infectious materials that results from the performance of an employee's duties" (262). The NCCLS and the CDC have also published guidelines for HIV-exposure evaluation that can serve to augment and complement the OSHA standard (42, 43, 196). The recommendations from these three organizations are summarized in Table 8.

TABLE 8 Postexposure evaluation and management for HIV-1[a]

	OSHA	NCCLS	CDC
	Document route and circumstances of exposure		Document route and circumstances of exposure
Source	Identify source and test after consent within legal restrictions and as soon as possible	Test source with consent or source material in an anonymous manner	Test source with consent
Exposed worker (general)	Counsel employee about: Source serostatus Applicable laws Further needed evaluations		
	Employee's blood collected and tested after consent		
Source is HIV(+) or refuses test	Not addressed	Counsel employee about: Risk of infection Reporting any acute illness Implications of HIV testing AZT prophylaxis Following CDC recommendations for prevention of HIV transmission during the first 6–12 weeks after exposure	Counsel employee about: Risk of infection Reporting any acute illness Following CDC recommendations for prevention of HIV transmission during first 6–12 weeks after exposure
		Test exposed employee after consent within 48 h	Medical evaluation of exposed employee
		Medical evaluation of exposed employee	Test exposed employee after consent as soon as possible
Follow-up	Evaluate any illness	Evaluate any illness	Evaluate any illness
		If baseline negative, test employee for HIV infection at 6 and 12 weeks and 6, 9, 12, and 24 months after exposure	If baseline negative, test employee for HIV infection at 6 and 12 weeks and 6 months after exposure (minimum)
Treatment	As recommended by Public Health Service	If worker requests and is counseled, AZT should be given, case should be reported to CDC, and follow-up period should be extended	No recommendations for or against the use of AZT prophylaxis If AZT given, counseling and informed consent of employee obtained
Source is HIV(–)	Not addressed	Test exposed employee with consent at 3 and 6 months	Optional baseline test of exposed employee and at 12 weeks postexposure (particularly if source is in high-risk group)
Other	Employee receives a health care professional's written opinion of evaluation Medical records maintained for term of employment plus 30 years	Serologic testing available for employees who think they may have been occupationally exposed	Serologic testing available for employees who think they may have been occupationally exposed

[a]Modified from references 42, 43, 196, and 262.

All three organizations suggest that the exposure source material be evaluated for evidence of HIV-1 virus that might include HIV antibody testing or nontraditional tests such as HIV antigen or HIV DNA in peripheral blood mononuclear cells using the polymerase chain reaction. Because HIV antigens and viral DNA may be present in serum before antibodies are detectable, the nontraditional tests may be valuable when testing a source individual who tests HIV antibody negative but who is a member of a high-risk group for HIV infection. The NCCLS also suggests that seronegative sources who are members of high-risk groups be retested at 3 to 6 months if possible (196). The CDC recommends an optional baseline test of the exposed employee and a follow-up test at 12 weeks if the seronegative source is in a high-risk group (42).

The three organizations also require counseling of the exposed employees regarding the source serostatus, the associated risk of HIV-1 transmission, any applicable legal consequences, and the need for medical evaluation of any acute febrile illness the employee may experience postexposure. Such an illness, characterized by fever, rash, or lymphadenopathy, developing within 12 weeks of exposure has been reported in documented occupationally acquired HIV infections (175) and may indicate an acute HIV infection. During the first 6 to 12 weeks after the exposure, when infected employees may not have yet seroconverted, exposed workers should also be informed to follow public health service recommendations for preventing transmission of HIV (44). These recommendations include refraining from blood donations, abstaining from sex or use of safe sex measures, and no sharing of personal items such as razors or toothbrushes.

Although OSHA requires serologic testing of employees (with consent) after any occupational exposure incident, the CDC and the NCCLS recommend voluntary baseline testing only after exposures to an HIV-positive source or a source who refuses testing. The CDC and the NCCLS both recommend sequential serologic HIV testing of the exposed employee if the initial baseline serum is negative. The CDC recommends retesting periodically for a minimum of 6 months postexposure (e.g., 6 weeks, 12 weeks, and 6 months). Delayed seroconversion among health care workers has not been documented, so routine testing beyond 6 months is probably not indicated (102). The NCCLS, however, recommends additional testing at 9, 12, and 24 months because of the *possibility* of late seroconversion. The extension of postexposure serologic testing should also be considered if the employee has been given prophylactic AZT, because the drug has been shown to alter the course of HIV infection in animals (184), resulting in later detection of the virus than in animals not given AZT. Certainly, the employee should be tested if any signs or symptoms of HIV infection should occur after the routine testing period is over.

In addition to traditional HIV antibody testing, Henderson et al. (123) used antigen capture assays and polymerase chain reaction evaluations for employees with known exposure to HIV at the NIH. These tests failed to provide an earlier diagnosis of occupationally transmitted infection than could be detected by the traditional HIV antibody test. However, such nontraditional testing might be useful for evaluation of an exposed employee experiencing an acute illness before antibody conversion.

Both the CDC and the NCCLS recommend serologic testing be made available to employees who think they have been occupationally exposed to HIV-1. This might occur after an exposure to a source that cannot be identified or tested. No organization recommends routine serologic testing of clinical health care workers for HIV-1; however, the CDC/NIH Biosafety Guidelines call for "medical surveillance programs" in "all laboratories that test specimens, do research, or produce reagents involving HIV" (45).

The CDC and the NIH recommend that baseline serum samples for all HIV-1 research laboratory personnel be collected and stored, according to BSL 3 practices (45). Also, the expert team convened by the director of the NIH in 1988 to review safety practices in laboratories producing highly concentrated HIV recommended a medical surveillance serology program for these facilities (46). They recommended that serum samples be obtained at least once a year from the laboratory workers and analyzed for seroconversion. Results should be reported to individual workers in a timely and confidential manner, with counseling services available for any occupationally infected workers.

Postexposure Treatment

Azidothymidine

Because there is no approved effective vaccine for HIV-1, prophylactic use of AZT (azidothymidine or zidovudine) has been suggested and implemented in several medical centers in an attempt to prevent infection with the HIV virus (124). Fischl et al. (86) provided the first evidence of the efficacy of AZT in prolonging the life of HIV-1-infected patients. The basis for the efficacy of AZT is its inhibition of the viral RT activity that is necessary for viral replication. If AZT is to be used prophylactically, it is essential that it be given as soon as possible after exposure.

Attempts have been made to assess the efficacy of AZT prophylaxis after retroviral exposure in animals (184, 230, 241, 257) and humans (55) with mixed results. Animal studies have demonstrated prevention or alteration in the course of infection. AZT given between 1 and 4 h after inoculation of animal retroviruses seems to provide protection in mice and cats (230, 257). However, a study of macaque monkeys failed to demonstrate prevention of infection with SIV in two of three monkeys treated with AZT within 1 h of viral inoculation (181).

Recently, the SCID-hu mouse model has been used to evaluate AZT prophylactic activity on human hematolymphoid systems exposed to HIV (184, 241). McCune et al. (184) demonstrated that, even when AZT coverage is started 24 h before inoculation of HIV, some cells in the human thymus of SCID-hu mice show signs of infection at 2 weeks postinjection with HIV. The same research group (241) found that AZT suppressed HIV infection in a time-dependent manner (<2 h) after HIV inoculation of SCID-hu mice. The relevance of these animal studies to HIV infection is unknown.

Insufficient data exist to assess the efficacy of prophylactic AZT in humans after occupational exposure. Because of the small risk of occupational HIV transmission after exposure, it is estimated that a minimum of 3,000 enrollees in a double-blind, placebo-controlled trial would be needed to demonstrate significance of AZT prophylactic efficacy at a 5% level (121). Two attempts at such a study have been initiated. One study was sponsored by Burroughs-Wellcome in which a 6-week course of AZT (200 mg every 4 h) was administered to 49 clinically exposed health care workers (45 received placebo) (55). None was infected in either group. Because of the difficulty encountered in subject accrual, the study was discontinued in 1989 with no conclusive information. Likewise, the National Institute of Allergy and Infectious Diseases sponsored an open study for AZT prophylaxis after massive exposure to HIV-infected material (55) such as HIV-contaminated blood transfusion or exposure to high concentrations of HIV. Of three persons enrolled in this study and reported, two remained seronegative 3 and 11 months after exposure and AZT prophylaxis. This study was also discontinued in February 1991 with no conclusive information.

In 1991, investigators at the Clinical Center of the NIH and San Francisco General Hospital, with support from Burroughs-Wellcome, initiated a multicenter study of AZT toxicity that involves 14 hospitals. It is hoped that this large study will provide much-needed data regarding the toxicity and, perhaps, efficacy of prophylactic AZT (102).

The CDC has not made a recommendation for or against the use of AZT for prophylaxis because of the lack of relevant data. However, in 1988, the CDC expanded its ongoing prospective, voluntary surveillance of workers with occupational exposure to HIV to include information on postexposure chemoprophylaxis. Between October 1988 and June 1992, 444 workers reported percutaneous injuries and were followed for toxicity and efficacy of AZT prophylaxis. No seroconversions occurred in the workers who were not given AZT, but 1 seroconversion occurred among the 143 workers who were given AZT prophylactically, even though AZT was taken within 2 h after a needlestick (260).

Other reports of AZT prophylaxis failure have been reported in seven health care workers after percutaneous exposure to HIV-infected blood (5, 168, 169, 256, 260). Anecdotal reports of AZT failure have also been reported after blood transfusion from an HIV-infected donor (55), an accidental IV injection of 100 to 200 μl of HIV-contaminated blood (157), and self-inoculation of HIV-contaminated blood (78).

Decisions regarding AZT prophylaxis should be made on an individual basis in accordance with a defined institutional policy. Considerations for AZT administration should include

- The unknown efficacy of prophylactic AZT
- The possible side effects and toxicity
- The unknown teratogenicity and mutagenicity
- The unknown optimal dose and duration of dose
- The unknown "grace period" during which AZT should be given

Employees must be counseled regarding these issues and the diversity of opinions among physicians regarding this experimental use of AZT. Informed consent of the employee should be obtained and blood counts and chemistries monitored every 2 weeks. The CDC and the NCCLS also recommend an extended period of follow-up serologies if AZT is given, because the animal studies suggest an alteration in the infection process. AZT should be used only as an experimental option for informed employees when resources and expertise to monitor therapy are available (102).

Passive Immunotherapy

Another approach to postexposure treatment for prevention of HIV-1 infection is that of passive immunity with high concentrations of HIV-1 antibody, similar to the hepatitis B immunoglobulin strategy, for immediate temporary protection. Prince et al. (213) found that anti-HIV-1 globulin extracted from HIV-infected persons failed to provide protection in chimpanzees when challenged with a large dose of HIV (400 $TCID_{50}$). A later study demonstrated that a single high dose of anti-HIV-1 globulin or high titers of soluble CD4-IgG given before inoculation of a smaller challenge of HIV-1 (120 $TCID_{50}$) protected chimpanzees from infection (266). Likewise, passive immunization with high antibody titers from immune macaques prevented infection of five of seven cynomolgous monkeys with HIV-2 and SIV (217). Further studies on postexposure prophylaxis are in progress and may provide additional strategies for prevention of HIV infection in infants born to HIV-infected mothers or in employees after occupational exposures.

References

1. **Aggarwai, A., J. Sadoff, P. Markham, R. Gallo, and G. Franchini.** 1991. Induction of cytotoxic T lymphocytes against HIV-2 in mice immunized with recombinant *Salmonella typhimurium,* abstr. MA1036. 7th Int. Conf. AIDS/III STD World Congr., Florence, Italy.
2. **Agy, M. B., L. R. Frumkin, L. Corey, R. W. Coombs, S. W. Wolinsky, J. Koehler, W. R. Morton, and M. G. Katze.** 1992. Infection of *Macaca nemestrina* by human immunodeficiency virus type 1. *Science* **257:** 103–106.
3. **Anderson, R. E., L. Stein, M. L. Moss, and N. H. Gross.** 1952. Potential infectious hazards of bacteriological techniques. *J. Bacteriol.* **64:**473–481.
4. **Anonymous.** 1984. Needlestick transmission of HTLV-III from a patient infected in Africa. *Lancet* **ii:**1376–1377.
5. **Anonymous.** 1993. HIV seroconversion after occupational exposure despite early prophylactic zidovudine therapy. *Lancet* **341:**1077–1078.
6. **Aoun, H.** 1989. When a house officer gets AIDS. *N. Engl. J. Med.* **321:**693–696.
7. **Arnow, P. M., M. Smaron, and V. Ormiste.** 1984. Brucellosis in a group of travelers to Spain. *JAMA* **251:**505–507.
8. **Barbara, J. A. J., and M. Contreras.** 1990. Infectious complications of blood transfusion: viruses. *Br. Med. J.* **300:**450–453.
9. **Barker, T.** 1987. Physician sues Johns Hopkins after contracting AIDS. *Am. Med. Assoc. News.* **19:**8.
10. **Barnes, D. M.** 1988. Research news: AIDS virus creates lab risk. *Science* **239:**348–349.
11. **Barnes, D. M.** 1988. Obstacles to an AIDS vaccine. *Science* **240:**719–721.
12. **Barre-Sinoussi, F., J. C. Chermann, F. Rey, M. T. Nugeyre, S. Chamaret, J. Gruest, C. Dauguet, C. Axler-Blin, F. Vezinet-Brun, C. Rouzioux, W. Rozenbaum, and L. Montagnier.** 1983. Isolation of a T-lymphotrophic retrovirus from a patient at risk for acquired immune deficiency syndrome (AIDS). *Science* **220:**868–871.
13. **Barre-Sinoussi, F., M. T. Nugeyre, and J. C. Chermann.** 1985. Resistance of AIDS virus at room temperature. *Lancet* **ii:**721–722.
14. **Belani, A., D. Dutta, S. Rosen, R. Dunning, V. Jiji, M. L. Levin, D. Glasser, S. Sigelman, and S. Baker.** 1984. AIDS in a hospital worker. *Lancet* **i:**676. (Letter.)
15. **Bending, M. R., and P. D. L. Maurice.** 1980. Malaria: a laboratory risk. *Postgrad. Med. J.* **56:**344–345.
16. **Blattner, W. A.** 1989. Retroviruses. p. 545–592. In A. S. Evans (ed.), *Viral infections of humans: epidemiology and control,* 3rd ed. Plenum Press, New York.
17. **Blayney, D. W., W. A. Blattner, M. Robert-Guroff, E. S. Jaffe, R. I. Fisher, P. A. Bunn, Jr., M. G. Patton, R. C. Gallo, and H. R. Rarick.** 1983. The human T-cell leukemia-lymphoma virus in the southeastern United States. *JAMA* **250:**1048–1052.
18. **Boland, M., J. Keresztes, P. Evans, J. Oleske, and E. Connor.** 1986. HIV seroprevalence among nurses caring for children with AIDS/ARC, abstr. THP.212. 3rd Int. Conf. AIDS, Washington, D.C.
19. **Bolognesi, D. P.** 1991. AIDS vaccines: progress and unmet challenges. *Ann. Intern. Med.* **114:**161–162.
20. **Borsch, G., J. Odendahl, G. Sabin, and D. Ricken.** 1982. Malaria transmission from patient to nurse. *Lancet* **ii:**1212.
21. **Braathen, L. R., G. Ramirez, R. O. F. Kunze, and H. Gelderblom.** 1987. Langerhans cells as primary target cells for HIV infection. *Lancet* **ii:**1094.
22. **Brener, Z.** 1987. Laboratory-acquired Chagas' disease: comment. *Trans. R. Soc. Trop. Med. Hyg.* **81:**527.
23. **Broderson, J. R.** 1987. Biosafety in acquired immunodeficiency syndrome (AIDS) studies using nonhuman primates. *J. Med. Primatol.* **16:**131–138.
24. **Bruce-Chwatt, I. J.** 1972. Blood transfusion and tropical disease. *Trop. Dis. Bull.* **69:**825–862.
25. **Bureau of Labor Statistics.** 1988. *Employment and earnings. U.S. Department of Labor, Bureau of Labor Statistics,* Washington, D.C.
26. **Burney, M. I., A. Ghafoor, M. Saleen, P. A. Webb, and J. Casals.** 1980. Nosocomial outbreak of viral hemorrhagic fever caused by Crimean hemorrhagic fever-Congo virus in Pakistan, January, 1976. *Am. J. Trop. Med. Hyg.* **29:**941–947.
27. **Bygbjerg, J. C.** 1984. AIDS in a Danish surgeon (Zaire, 1976). *Lancet* **i:**676. (Letter.)
28. **Cadman, E. C.** 1990. Physicians in training and AIDS. *N. Engl. J. Med.* **322:**1392–1393.
29. **Cannon, N. J., S. P. Walker, and W. E. Dismukes.** 1972. Malaria acquired by accidental needle puncture. *JAMA* **222:**1425.
30. **Centers for Disease Control.** 1980. Chagas' disease—Michigan. *Morbid. Mortal. Weekly Rep.* **29:**147–148.
31. **Centers for Disease Control.** 1981. Pneumocystis pneumonia—Los Angeles. *Morbid. Mortal. Weekly Rep.* **30:**250–252.
32. **Centers for Disease Control.** 1981. Kaposi's sarcoma and pneumocystis pneumonia among homosexual men—New York and California. *Morbid. Mortal. Weekly Rep.* **30:**305–308.
33. **Centers for Disease Control.** 1982. Acquired immune deficiency syndrome: precautions for clinical and laboratory staff. *Morbid. Mortal. Weekly Rep.* **31:**577–580.

34. **Centers for Disease Control.** 1983. A.I.D.S.: precautions for health-care workers and allied professionals. *Morbid. Mortal. Weekly Rep.* **32:**450–451.
35. **Centers for Disease Control.** 1984. Hepatitis B vaccine: evidence confirming lack of AIDS transmission. *Morbid. Mortal. Weekly Rep.* **33:**685–687.
36. **Centers for Disease Control.** 1985. Recommendations for protection against viral hepatitis. *Morbid. Mortal. Weekly Rep.* **34:**313–335.
37. **Centers for Disease Control.** 1985. Update: evaluation of human T-lymphotropic virus type III/lymphadenopathy-associated virus infection in health care personnel—United States. *Morbid. Mortal. Weekly Rep.* **34:**575–578.
38. **Centers for Disease Control.** 1985. Summary: recommendations for preventing transmission of infection with human T-lymphotropic virus III/LAV in the workplace. *Morbid. Mortal. Weekly Rep.* **34:**681–695.
39. **Centers for Disease Control.** 1986. Apparent transmission of human T-lymphotropic virus type III/lymphadenopathy-associated virus from a child to a mother providing health care. *Morbid. Mortal. Weekly Rep.* **35:**75–79.
40. **Centers for Disease Control.** 1986. Safety of therapeutic immune globulin preparations with respect to transmission of human T-lymphotropic virus type III/lymphadenopathy-associated virus infection. *Morbid. Mortal. Weekly Rep.* **35:**231–233.
41. **Centers for Disease Control.** 1986. HTLV III/LAV: agent summary statement. *Morbid. Mortal. Weekly Rep.* **35:**540–549.
42. **Centers for Disease Control.** 1987. Recommendations for prevention of HIV transmission in health care settings. *Morbid. Mortal. Weekly Rep.* **36**(Suppl. 2):3S–18S.
43. **Centers for Disease Control.** 1987. Update: human immunodeficiency virus infections in health care workers exposed to blood of infected patients. *Morbid. Mortal. Weekly Rep.* **36:**285–289.
44. **Centers for Disease Control.** 1987. Public health service guidelines for counseling and antibody testing to prevent HIV infection and AIDS. *Morbid. Mortal. Weekly Rep.* **36:**509–515.
45. **Centers for Disease Control.** 1988. Agent summary statement for human immunodeficiency viruses (HIVs); including HTLV-III, LAV, HIV-1, and HIV-2, and report on laboratory-acquired infection with human immunodeficiency virus. *Morbid. Mortal. Weekly Rep.* **37**(Suppl. 4):1–17.
46. **Centers for Disease Control.** 1988. Occupationally acquired human immunodeficiency virus infections in laboratories producing virus concentrates in large quantities: conclusions and recommendations of an expert team convened by the director of the National Institutes of Health (NIH). *Morbid. Mortal. Weekly Rep.* **37**(Suppl. 4):19–22.
47. **Centers for Disease Control.** 1988. Update: acquired immunodeficiency syndrome and human immunodeficiency virus infection among health care workers. *Morbid. Mortal. Weekly Rep.* **37:**229–235.
48. **Centers for Disease Control.** 1988. Update: universal precautions for prevention of transmission of human immunodeficiency virus, hepatitis B virus, and other bloodborne pathogens in health-care settings. *Morbid. Mortal. Weekly Rep.* **37:**377–387.
49. **Centers for Disease Control.** 1988. Safety of therapeutic products used for hemophilia patients. *Morbid. Mortal. Weekly Rep.* **37:**441–450.
50. **Centers for Disease Control.** 1988. Guidelines to prevent simian immunodeficiency virus infection in laboratory workers and animal handlers. *Morbid. Mortal. Weekly Rep.* **37:**696–703.
51. **Centers for Disease Control.** 1989. Guidelines for prevention of transmission of human immunodeficiency virus and hepatitis B virus to health-care and public-safety workers. *Morbid. Mortal. Weekly Rep.* **38**(Suppl. 6):3–31.
52. **Centers for Disease Control.** 1989. Problems created by heat-inactivation of serum specimens before HIV-1 antibody testing. *Morbid. Mortal. Weekly Rep.* **38:**407–413.
53. **Centers for Disease Control.** 1989. Update: HIV-2 infection—United States. *Morbid. Mortal. Weekly Rep.* **38:**572–580.
54. **Centers for Disease Control.** 1990. Human T-lymphotropic virus type 1 screening in volunteer blood donors—United States, 1989. *Morbid. Mortal. Weekly Rep.* **39:**915–924.
55. **Centers for Disease Control.** 1990. Public health service statement on management of occupational exposure to human immunodeficiency virus, including considerations regarding zidovudine postexposure use. *Morbid. Mortal. Weekly Rep.* **39**(RR-1):1–14.
56. **Centers for Disease Control.** 1991. Mortality attributable to HIV infection/AIDS—United States, 1981–1990. *Morbid. Mortal. Weekly Rep.* **40:**41–44.
57. **Centers for Disease Control.** 1991. The HIV/AIDS epidemic: the first ten years. *Morbid. Mortal. Weekly Rep.* **40:**357–369.
58. **Centers for Disease Control.** 1991. Vaccinia (smallpox) vaccine: recommendations of the immunization practices advisory committee (ACIP). *Morbid. Mortal. Weekly Rep.* **40**(RR-14):1–10.
59. **Centers for Disease Control.** 1992. Seroconversion to simian immunodeficiency virus in two laboratory workers. *Morbid. Mortal. Weekly Rep.* **41:**678–681.
60. **Centers for Disease Control and Prevention.** 1993. Update: mortality attributable to HIV infection among persons aged 25–44 years—United States, 1991 and 1992. *Morbid. Mortal. Weekly Rep.* **42:**869–872.
61. **Centers for Disease Control and Prevention.** 1993. *AIDS statistical information, quarterly HIV/AIDS surveillance report.* Centers for Disease Control, Atlanta.
62. **Chakrabarti, S., M. Robert-Guroff, F. Wong-Staal, and R. C. Gallo.** 1986. Expression of the HTLV-III envelope gene by a recombinant vaccinia virus. *Nature* **320:**535–537.
63. **Chamberland, M., L. Conley, A. Lifson, C. White, K. Castro, and T. Dondero.** 1988. AIDS in health care workers: a surveillance report, abstr. 9020. 4th Int. Conf. AIDS, Stockholm, Sweden.
64. **Chanda, P. K., R. J. Natuk, B. B. Mason, B. M. Bhat, L. Greenberg, S. K. Dheer, K. L. Molnar-Kimber, S. Mizutani, M. D. Lubeck, and A. R. Davis.** 1990. High-level expression of the envelope glycoprotein of the human immunodeficiency virus type 1 in presence of *rev* gene using helper-independent adenovirus type 7 recombinants. *Virology* **175:**535–574.
65. **Chorba, T. L., J. M. Jason, R. B. Ramsey, K. Lechner, I. Pabinger-Fasching, V. S. Kalyanaraman, J. S.**

McDougal, C. D. Cabradilla, L. C. Tregillus, and D. N. Lawrence. 1985. HTLV-I antibody status in hemophilia patients treated with factor concentrates prepared from US plasma sources, and in hemophilia patients with AIDS. *Thromb. Haemost.* **53:**180–182.

66. **Clements, M. L.** 1990. AIDS vaccines, p. 1112–1119. *In* G. L. Mandell, R. G. Douglas, and J. E. Bennett (ed.), *Principles and practice of infectious diseases,* 3rd ed. Churchill Livingstone, New York.
67. **Coffin, J., A. Haase, J. Levy, L. Montagnier, S. Oroszian, W. Teich, H. Temin, K. Tayoshima, H. Varmus, P. Vogt, and R. Weiss.** 1986. Human immunodeficiency virus. *Science* **232:**697. (Letter.)
68. **Cohen, B. J., A. M. Courouce, T. F. Schwarz, K. Okochi, and G. J. Kurtzman.** 1988. Laboratory infection with parvovirus B19. *J. Clin. Pathol.* **41:**1027–1028. (Letter.)
69. **Cumming, P. D., E. L. Wallace, J. B. Schorr, and R. Y. Dodd.** 1989. Exposure of patients to human immunodeficiency virus through the transfusion of blood components that test antibody-negative. *N. Engl. J. Med.* **321:**941–946.
70. **Dalgleish, A. G., B. J. Thomson, J. C. Chanh, M. Malkovsky, and R. Kennedy.** 1987. Neutralisation of HIV isolates by anti-idiotypic antibodies which mimic the T4 (CD4) epitope: a potential AIDS vaccine. *Lancet* **ii:**1047–1050.
71. **Daniel, M. D., N. L. Letvin, N. W. King, M. Kannagi, P. K. Sehgal, R. D. Hunt, P. J. Kanki, M. Essex, and R. C. Desrosiers.** 1986. Isolation of T-cell tropic HTLV-III-like retrovirus from macaques. *Science* **228:**1201–1204.
72. **Davis, A. R., B. Kostek, B. B. Mason, C. L. Hsiao, J. Morin, S. K. Dheer, and P. P. Hung.** 1985. Expression of hepatitis B surface antigen with a recombinant adenovirus. *Proc. Natl. Acad. Sci. USA* **82:**7560–7564.
73. **Day, L. J.** 1989. AIDS: an occupational hazard for orthopaedic surgeons? *Orthop. Rev.* **18:**493–497.
74. **Denis, F., G. Leonard, M. Mounier, A. Sangare, G. Gershy-Damet, and J. L. Rey.** 1987. Efficacy of five enzyme immunoassays for antibody to HIV in detecting antibody to HTLV-IV. *Lancet* **i:**324–325.
75. **Desrosiers, R. C.** 1992. HIV with multiple gene deletions as a live attenuated vaccine for AIDS. *AIDS Res. Hum. Retroviruses* **8:**411–421.
76. **Donegan, E., M. Stuart, J. C. Niland, H. S. Sacks, S. P. Azen, S. L. Dietrich, C. Faucett, M. A. Fletcher, S. H. Kleinman, E. A. Operskalski, H. A. Perkins, J. Pindyck, E. R. Schiff, D. P. Stiles, P. A. Tomasulo, J. W. Mosley, and the Transfusion Safety Group.** 1990. Infection with human immunodeficiency virus type 1 (HIV1) among recipients of antibody-positive donations. *Ann. Intern. Med.* **113:**733–739.
77. **Dufoort, G., A.-M. Courouce, R. Ancelle-Park, and O. Bletry.** 1988. No clinical signs 14 years after HIV-2 transmission after blood transfusion. *Lancet* **ii:**510. (Letter.)
78. **Durand, E., C. Le Jeunne, and F.-C. Hughes.** 1991. Failure of prophylactic zidovudine after suicidal self-inoculation of HIV-infected blood. *N. Engl. J. Med.* **324:**1062. (Letter.)
79. **Ebbesen, P., M. Melbye, F. Scheutz, A. J. Bodner, and R. J. Biggar.** 1986. Lack of antibodies to HTLV-III/LAV in Danish dentists. *JAMA* **256:**2199. (Letter.)
80. **Elmslie, K., and J. V. O'Shaughnessy.** 1987. National surveillance program on occupational exposure to HIV among health care workers in Canada. *Can. Dis. Week. Rep.* 13:163–166.
81. **Emond, R. T. D., B. Evans, E. T. W. Bowen, and G. Lloyd.** 1977. A case of Ebola virus infection. *Br. Med. J.* **2:**541–544.
82. **Evans, D. J., J. McKeating, J. M. Meredith, and K. I. Burke.** 1989. An engineered poliovirus chimaera elicits broadly reactive HIV-1 neutralizing antibodies. *Nature* **339:**385–388.
83. **Federal Register.** 1975. *EPA virucide disinfectant efficacy testing. Fed. Regist.* **40:**11625 (June 25, 1975).
84. **Felsenfeld, O.** 1971. *Borrelia: strains, vectors, human and animal borreliosis.* Warren H. Green, Inc., St. Louis.
85. **Field, P. R., G. G. Moyle, and P. M. Parnell.** 1972. The accidental infection of a laboratory worker with *Toxoplasma gondii. Med. J. Aust.* **2:**196–198.
86. **Fischl, M. A., D. D. Richman, M. H. Greico, M. S. Gottleib, P. A. Volberding, O. L. Laskin, J. M. Leedom, J. E. Groopman, D. Mildvan, R. T. Schooley, G. G. Jackson, D. T. Durack, D. King, and the AZT Collaborative Working Group.** 1987. The efficacy of azidothymidine (AZT) in the treatment of patients with AIDS and AIDS-related complex: a double-blind, placebo-controlled trial. *N. Engl. J. Med.* **317:**185–191.
87. **Fisher, A. G., L. Ratner, H. Mitsuya, L. M. Marselle, M. E. Harper, S. Broder, R. C. Gallo, and F. Wong-Staal.** 1986. Infectious mutants of HTLV-III with changes in the 3′ region and markedly reduced cytopathic effects. *Science* **233:**655–659.
88. **Flynn, N. M., S. M. Pollet, J. R. Van Horne, R. Elvebakk, S. D. Harper, and J. R. Carlson.** 1987. Absence of HIV antibody among dental professionals exposed to infected patients. *West. J. Med.* **146:**439–442.
89. **Frame, J. D., J. M. Baldwin, Jr., D. J. Gocke, and J. M. Troup.** 1970. Lassa fever, a new virus disease of man from West Africa: I. Clinical description and pathological findings. *Am. J. Trop. Med. Hyg.* **19:**670–676.
90. **Francavilla, E., P. Cadrobbi, P. Scaggiante, G. DiSilvestro, F. Bortolotti, and A. Bertaggia.** 1989. Surveillance on occupational exposure to HIV among health care workers in Italy, abstr. D623. Vth Int. Conf. AIDS, Montreal, Quebec, Canada.
91. **Francis, D. P., P. M. Feorino, J. R. Broderson, H. M. McClure, J. P. Getchell, C. R. McGrath, B. Swenson, J. S. McDougal, E. L. Palmer, A. K. Harrison, F. Barre-Sinoussi, J. C. Chermann, L. Montagnier, J. W. Curran, C. D. Cabradilla, and V. S. Kalyanaraman.** 1984. Infection of chimpanzees with lymphadenopathy-associated virus. *Lancet* **ii:**1276–1277.
92. **Freedman, D. O., J. D. MacLean, and J. B. Viloria.** 1987. A case of laboratory acquired *Leishmania donovani* infection: evidence for primary lymphatic dissemination. *Trans. R. Soc. Trop. Med. Hyg.* **81:**118–119.
93. **Fujikawa, L. S., S. Z. Salahuddin, A. G. Palestine, H. Masur, R. B. Nussenbatt, and R. C. Gallo.** 1985. Isolation of human T-lymphotropic virus type III from the tears of a patient with acquired immune deficiency syndrome. *Lancet* **ii:**529–530.

94. **Fultz, P. N., H. M. McClure, D. C. Anderson, R. B. Swenson, R. Anand, and A. Srinivasan.** 1986. Isolation of a T-lymphotropic retrovirus from naturally infected sooty mangabey monkeys (*Cercocebus atys*). *Proc. Natl. Acad. Sci. USA* **83:**5286–5290.
95. **Fultz, P. N., H. M. McClure, R. B. Swenson, C. R. McGrath, A. Brodie, J. P. Getchell, F. C. Jensen, D. C. Anderson, J. R. Broderson, and D. P. Francis.** 1986. Persistent infection of chimpanzees with human T-lymphotropic virus type III/lymphadenopathy-associated viruses: a potential model for acquired immunodeficiency syndrome. *J. Virol.* **58:**116–124.
96. **Gallo, R. C., S. Z. Salahuddin, M. Popovic, G. M. Shearer, M. Kaplan, B. F. Haynes, T. J. Palker, R. Redfield, J. Oleske, and B. Safai.** 1984. Frequent detection and isolation of cytopathic retroviruses (HTLV-III) from patients with AIDS and at risk for AIDS. *Science* **224:**500–503.
97. **Garcia, J. J. G., F. Arnalich, J. M. Pena, J. J. Garcia-Alegria, F. F. Garcia Fernandez, C. Jimenez Herraez, and J. J. Vazquez.** 1986. An outbreak of *Plasmodium vivax* malaria among heroin users in Spain. *Trans. R. Soc. Trop. Med. Hyg.* **80:**549–552.
98. **Gardner, M. B.** 1993. Acquired immunodeficiency syndrome vaccines. *West. J. Med.* **158:**296–297.
99. **Gerberding, J. L.** 1989. Risks to health care workers from occupational exposures to hepatitis B virus, HIV, and CMV. *Infect. Dis. Clin. North Am.* **3:**735–745.
100. **Gerberding, J. L., C. E. Bryant-LeBlanc, K. Nelson, A. R. Moss, D. Osman, H. F. Chambers, J. R. Carlson, W. L. Drew, J. A. Levy, and M. A. Sande.** 1987. Risk of transmitting the human immunodeficiency virus, cytomegalovirus, and hepatitis B virus to health care workers exposed to patients with AIDS and AIDS-related conditions. *J. Infect. Dis.* **156:**1–8.
101. **Gerberding, J. L., and D. K. Henderson.** 1987. Design of national infection control policies for human immunodeficiency virus infections. *J. Infect. Dis.* **156:**861–864.
102. **Gerberding, J. L., and D. K. Henderson.** 1992. Management of occupational exposures to bloodborne pathogens: hepatitis B virus, hepatitis C virus, and human immunodeficiency virus. *Clin. Infect. Dis.* **14:**1179–1185.
103. **Gerberding, J. L., C. Littell, A. Tarkington, A. Brown, and W. P. Schecter.** 1990. Risk of exposure of surgical personnel to patients' blood during surgery at San Francisco General Hospital. *N. Engl. J. Med.* **322:**1788–1793.
104. **Gerberding, J. L., K. Nelson, D. Greenspan, J. Greenspan, J. Greene, and M. A. Sande.** 1987. Risk to dental professionals from occupational exposure to human immunodeficiency virus (HIV): followup, abstr. 27th Intersci. Conf. Antimicrobial Agents Chemother., New York.
105. **Gessain, A., F. Barin, J. C. Vernant, O. Gout, L. Maurs, A. Calender, and G. de The.** 1985. Antibodies to human T-lymphotropic virus type-1 in patients with tropical spastic paralysis. *Lancet* **ii:** 407–409.
106. **Gilmore, N., M. L. Ballachey, and M. O'Shaughnessy.** 1986. HTLV-III/LAV serologic survey of health care workers in a Canadian teaching hospital, abstr. 200. 2nd Int. Conf. AIDS, Paris, France.
107. **Gioananni, P., A. Sinicco, G. Cariti, A. Lucchini, G. Paggi, and O. Gichino.** 1988. HIV infection acquired by a nurse. *Eur. J. Epidemiol.* **4:**119–120.
108. **Goldfarb, M. F.** 1988. Effect of heat-inactivation on results of HIV-antibody detection by Western blot assay. *Clin. Chem.* **34:**1661–1662.
109. **Gotuzzo, E., C. Carrillo, J. Guerra, and L. Llosa.** 1986. An evaluation of diagnostic methods for brucellosis—the value of bone marrow culture. *J. Infect. Dis.* **153:**122–125.
110. **Grabowski, E. F., P. J. V. Giardina, D. Goldberg, H. Masur, S. E. Read, R. L. Hirsch, and J. L. Benach.** 1982. Babesiosis transmitted by a transfusion of frozen-thawed blood. *Ann. Intern. Med.* **96:**466–467.
111. **Grint, P., and M. McEvoy.** 1985. Two associated cases of the acquired immune deficiency syndrome (AIDS). *PHLS Commun. Dis. Rep.* **42:**4.
112. **Groopman, J. E., S. Z. Salahuddin, M. G. Sarngadharan, P. D. Markham, M. Gnda, A. Sliski, and R. C. Gallo.** 1984. HTLV-III in saliva of people with AIDS related complex and healthy homosexual men at risk for AIDS. *Science* **226:**447–449.
113. **Guerrero, I., B. Weniger, and M. G. Schultz.** 1983. Transfusion malaria in the United States, 1972–1981. *Ann. Intern. Med.* **99:**221–226.
114. **Gurtler, L. G., J. Eberle, and L. Bader.** 1993. HIV transmission by needlestick and eczematous lesion—three cases from Germany. *Infection* **21:**40–41.
115. **Haley, C., V. J. Reff, and F. K. Murphy.** 1989. Report of a possible laboratory-acquired HIV infection, abstr. 5th Int. Conf. AIDS, Montreal.
116. **Hanson, P. J. V., D. Gor, D. J. Jeffries, and J. V. Collins.** 1989. Chemical inactivation of HIV on surfaces. *Br. Med. J.* **298:**862–864.
117. **Hanson, R. P., S. E. Sulkin, E. L. Buescher, W. M. Hammon, R. W. McKinney, and T. H. Work.** 1967. Arbovirus infections of laboratory workers. *Science* **158:**1283–1286.
118. **Harbison, M. A., and S. M. Hammer.** 1989. Inactivation of human immunodeficiency virus by betadine products and chlorhexidine. *J. Acquired Immune Defic. Syndr.* **2:**16–20.
119. **Harper, S., N. Flynn, J. VanHorne, S. Jain, J. Carlson, and S. Pollet.** 1986. Absence of HIV antibody among dental professionals, surgeons, and household contacts exposed to persons with HIV infection., abstr. THP215. 3rd Int. Conf. AIDS, Washington, D.C.
120. **Heinsohn, P. A., D. L. Jewett, C. H. Bennet, and A. Rosen.** 1989. Characterization of blood aerosols created by some surgical power tools, abstr. 32nd Annu. Meeting Am. Biol. Safety Assoc.
121. **Henderson, D. K.** 1990. HIV-1 in the health care setting, p. 2221–2236. In G. L. Mandell, R. G. Douglas, and J. E. Bennett (ed.), *Principles and practice of infectious diseases,* 3rd ed. Churchill Livingstone, New York.
122. **Henderson, D. K., B. J. Fahey, J. Saah, J. Schmitt, and H. C. Lane.** 1988. Longitudinal assessment of the risk for occupational nosocomial transmission of human immunodeficiency virus-1 in health care workers, abstr. 28th Intersci. Conf. Antimicrobial Agents Chemother., Los Angeles.
123. **Henderson, D. K., B. J. Fahey, M. Willy, J. M. Schmitt, K. Carey, D. E. Koziol, H. C. Lane, J. Fedio, and A. J. Saah.** 1990. Risk for occupational transmis-

sion of human immunodeficiency virus type I (HIV-1) associated with clinical exposures. *Ann. Intern. Med.* **113:**740–746.

124. **Henderson, D. K., and J. L. Gerberding.** 1989. Prophylactic zidovudine after occupational exposure to the human immunodeficiency virus: an interim analysis. *J. Infect. Dis.* **160:**321–327.
125. **Hernandez, E., J. M. Gatell, T. Puyuelo, D. Mariscal, J. M. Barrera, and C. Sanchez.** 1988. Risk of transmitting the human immunodeficiency virus to health care workers exposed to HIV infected body fluids, abstr. 9003. 4th Int. Conf. AIDS, Stockholm, Sweden.
126. **Heymann, D. L., J. S. Weisfeld, P. A. Webb, K. M. Johnson, T. Cairns, and H. Berquist.** 1980. Ebola hemorrhagic fever: Tandala, Zaire, 1977–1978. *J. Infect. Dis.* **142:**372–376.
127. **Hinuma, Y., H. Komoda, T. Chosa, T. Kondo, M. Kohakura, T. Takenaka, M. Kikuchi, M. Ichimaru, K. Yunoki, I. Sato, R. Matsuo, Y. Takiuchi, H. Uchino, and M. Hanaoka.** 1982. Antibodies to adult T-cell leukemia virus-associated antigens (ATLA) in sera from patients with ATL and controls in Japan: a nation-wide serepidemiologic study. *Int. J. Cancer* **29:**631–635.
128. **Hira, P. R., and S. F. Husein.** 1979. Some transfusion-induced parasitic infections in Zambia. *J. Hyg. Epidemiol. Microbiol. Immunol.* **4:**436–444.
129. **Hirsch, M. S., G. P. Wormser, R. T. Schooley, D. D. Ho, D. Felsenstein, C. C. Hopkins, C. Joline, F. Duncanson, M. G. Sarngadharan, and C. Saxinger.** 1985. Risk of nosocomial infection with human T-cell lymphotropic virus III (HTLV-III). *N. Engl. J. Med.* **312:**1–4.
130. **Ho, D. D., T. Moudgil, and M. Alam.** 1989. Quantitation of human immunodeficiency virus type 1 in the blood of infected persons. *N. Engl. J. Med.* **321:**1621–1625.
131. **Ho, D. D., T. R. Rota, R. T. Schooley, J. C. Kaplan, J. D. Allan, J. E. Groopman, L. Resnick, D. Felsenstein, C. A. Andrews, and M. S. Hirsch.** 1985. Isolation of HTLV-III from cerebrospinal fluid and neural tissue of patients with neurologic syndromes related to the acquired immune deficiency syndrome. *N. Engl. J. Med.* **313:**1493–1497.
132. **Hoflin, J. M., R. H. Sadler, F. G. Araujo, W. E. Page, and J. S. Remington.** 1987. Laboratory-acquired Chagas disease. *Trans. R. Soc. Trop. Med. Hyg.* **81:**437–440.
133. **Huddleson, I. F., and M. Munger.** 1940. A study of the epidemic of brucellosis due to *Brucella melitensis. Am. J. Public Health* **30:**944–954.
134. **Hudson, C. (National Institute for Occupational Safety and Health, Centers for Disease Control and Prevention).** 1993. Personal communication.
135. **Hughes, A., and T. Corrah.** 1990. Human immunodeficiency virus type 2 (HIV-2). *Blood Rev.* **4:**158–164.
136. **Ippolito, G., V. Puro, and G. DeCarli.** 1993. The risk of occupational human immunodeficiency virus infection in health care workers. Italian Multicenter Study. The Italian Study Group on Occupational Risk of HIV Infection. *Arch. Intern. Med.* **153**(12)**:**1451–1458.
137. **Jensen J. B., T. C. Capps, and J. M. Carlin.** 1981. Clinical drug-resistant falciparum malaria acquired from cultured parasites. *Am. J. Trop. Med. Hyg.* **30:**523–525.
138. **Johnson, G. K., T. Nolan, H. C. Wuh, and W. S. Robinson.** 1991. Efficacy of glove combinations in reducing cell culture infection after glove puncture with needles contaminated with human immunodeficiency virus type 1. *Infect. Control Hosp. Epidemiol.* **12:**435–438.
139. **Johnson, G. K., and W. S. Robinson.** 1991. Human immunodeficiency virus-1 (HIV-1) in the vapors of surgical power instruments. *J. Med. Virol.* **33:**47–50.
140. **Johnson, J. E., III, and P. J. Kadull.** 1967. Rocky Mountain spotted fever acquired in a laboratory. *N. Engl. J. Med.* **277:**842–847.
141. **Jones, L., S. Ristow, T. Yilma, and B. Moss.** 1986. Accidental human vaccination with vaccinia virus expressing nucleoprotein gene. *Nature* **319:**543. (Letter.)
142. **Jorbeck, H., M. Marland, and E. Steinkeller.** 1989. Accidental exposures to HIV-positive blood among health care workers in 2 Swedish hospitals, abstr. A517. 5th Int. Conf. AIDS, Montreal, Quebec, Canada.
143. **Kalyanaraman, V. S., M. G. Sarngadharan, M. Robert-Guroff, M. Miyoshi, D. Golde, and R. C. Gallo.** 1982. A new subtype of human T-cell leukemia virus (HTLV-II) associated with a T-cell variant of hairy cell leukemia. *Science* **218:**571–573.
144. **Kanki, P., F. Barin, S. M'Boup, F. Denis, J. S. Allan, T. H. Lee, and M. Essex.** 1986. New human T-lymphotropic retrovirus related to simian T-lymphotropic virus type III ($STLV\text{-}III_{AGM}$). *Science* **232:**238–243.
145. **Kaplan, J. C., D. C. Crawford, A. G. Durno, and R. T. Schooley.** 1987. Inactivation of human immunodeficiency virus by betadine. *Infect. Control* **8:**412–414.
146. **Kataoka, R., N. Takehara, and Y. Iwahara.** 1990. Transmission of HTLV-1 by blood transfusion and its prevention by passive immunization in rabbits. *Blood* **76:**1657–1661.
147. **Keeny, M. Y., and F. L. Sabel.** 1968. Particle distributions of *Serratia marcescens* aerosols created during common laboratory procedures and simulated laboratory accidents. *Appl. Microbiol.* **16:**1146–1150.
148. **Kempf, C., P. Jentsch, F. Barre-Sinoussi, B. Poirier, J.-J. Morgenthaler, A. Morell, and D. Germann.** 1991. Inactivation of human immunodeficiency virus (HIV) by low pH and pepsin. *J. Acquired Immune Defic. Syndr.* **4:**828–830.
149. **Kennedy, R. C., R. D. Henkel, D. Pauletti, J. S. Allan, T. H. Lee, M. Essex, and G. R. Dreesman.** 1986. Antiserum to a synthetic peptide recognizes the HTLV-III envelope glycoprotein. *Science* **231:**1556–1559.
150. **Kerndt, P., H. Waskin, F. Steurer, et al.** 1988. *Trypanosoma cruzi* antibody among blood donors in Los Angeles, abstr. 28th Intersci. Conf. Antimicrobial Agents Chemother., Los Angeles.
151. **Klein, M., and A. Deforest.** 1983. Principles of viral inactivation, p. 422–434. *In* S. S. Block (ed.), *Disinfection, sterilization, and preservation,* 3rd ed. Lea & Febiger, Philadelphia.
152. **Klein, R. C., E. Party, and E. L. Gershey.** 1989. Safety in the laboratory. *Nature* **341:**288. (Letter.)
153. **Klein, R. S., J. A. Phelan, K. Freeman, C. Schable, G. H. Friedland, N. Rieger, and N. H. Steigbigel.** 1988. Low occupational risk of human immuno-

deficiency virus infection among dental professionals. *N. Engl. J. Med.* **318:**86–90.

154. **Kuhls, T. L., S. Viker, N. B. Parris, A. Garakian, J. Sullivan-Bolyai, and J. Cherry.** 1987. Occupational risk of HIV, HBV, and HSV-2 infections in health care personnel caring for AIDS patients. *Am. J. Public Health* **77:**1306–1309.
155. **Kulaga, H., T. Folks, R. Rutledge, M. E. Truckenmiller, E. Gugel, and T. J. Kindt.** 1989. Infection of rabbits with human immunodeficiency virus 1. *J. Exp. Med.* **169:**321–326.
156. **Lai, L., G. Ball, J. Stevens, and D. Shanson.** 1985. Effect of heat treatment of plasma and serum on biochemical indices. *Lancet* **ii:**1457–1458.
157. **Lange, J. M., C. A. Boucher, C. E. Hollack, E. H. Wiltink, P. Reiss, E. A. VanRoyen, M. Roos, S. A. Danner, and J. Goudsmit.** 1990. Failure of zidovudine prophylaxis after accidental exposure to HIV-1. *N. Engl. J. Med.* **322:** 1375–1377.
158. **Lee, H., P. Swanson, V. S. Shorty, J. A. Zack, J. D. Rosenblatt, and I. S. Che.** 1989. High rate of HTLV-II infection in seropositive IV drug abusers from New Orleans. *Science* **244:**471–475.
159. **Lee, H., S. Weiss, L. Brown, D. Mildvan, V. Shorty, L. Saravolatz, A. Chu, H. M. Ginzburg, N. Markowitz, and D. C. DesJarlais.** 1990. Patterns of HIV-1 and HTLV-I/II in intravenous drug abusers from the middle Atlantic and central regions of the USA. *J. Infect. Dis.* **162:**347–352.
160. **Lee, H. H., P. Swanson, J. D. Rosenblatt, I. S. Y. Chen, W. C. Sherwood, D. E. Smith, G. E. Tegtmeier, L. P. Fernando, C. T. Fang, M. Osame, and S. H. Kleinman.** 1991. Relative prevalence and risk factors of HTLV-I and HTLV-II infection in US blood donors. *Lancet* **337:**1435–1439.
161. **Leibowitz, S., L. Greenwald, I. Cohen, and J. Litwins.** 1949. Serum hepatitis in a blood bank worker. *JAMA* **140:**1331–1333.
162. **Leifer, E., D. J. Gocke, and H. Bourne.** 1970. Lassa fever, a new virus disease of man from West Africa: II. Report of a laboratory-acquired infection treated with plasma from a person recently recovered from the disease. *Am. J. Trop. Med. Hyg.* **19:**677–679.
163. **Leonard, J. M., J. W. Abramczuk, D. S. Pezen, R. Rutledge, J. H. Belcher, F. Hakim, G. Shearer, L. Lamperth, W. Travis, T. Fredrickson, A. L. Notkins, and M. A. Martin.** 1988. Development of disease and virus recovery in transgenic mice containing HIV proviral DNA. *Science* **242:**1665–1670.
164. **Letvin, N. L., C. I. Lord, N. W. King, and M. S. Wyand.** 1991. Risks of handling HIV. *Nature* **349:**573. (Letter.)
165. **Levy, J. A., L. S. Kaminsky, W. J. W. Morrow, K. Steimer, P. Luciw, D. Dina, J. Hoxie, and L. Oshiro.** 1985. Infection by the retrovirus asssociated with AIDS—clinical, biological, and molecular features. *Ann. Intern. Med.* **103:**694–699.
166. **Lima, G., and C. Traina.** 1988. [Remarks on a case of AIDS-related syndrome (ARC/LAS) in a nurse.] *Minerva Med.* **79:**141–143. (In Italian.)
167. **Lloyd-Evans, N., V. S. Springthorpe, and S. A. Sattar.** 1986. Chemical disinfection of human rotavirus-contaminated surfaces. *J. Hyg. Camb.* **97:** 163–173.
168. **Looke, D. F. M., and D. I. Grove.** 1990. Failed prophylactic zidovudine after needlestick injury. *Lancet* **335:**1280. (Letter.)
169. **Lot, F., and D. Abiteboul.** 1992. Infections professionnelles par le V.I.H. en France: le point au 31 mars 1992. *Bull. Epidemiol. Hebdomadoire* (Paris) **26:**117–119.
170. **Lubick, H. A., and L. D. Schaeffer.** 1986. Occupational risk of dental personnel survey. *J. Am. Dent. Assoc.* **113:**10–12. (Letter.)
171. **Lui, K.-J., W. W. Darrow, and G. W. Rutherford III.** 1988. A model-based estimate of the mean incubation period for AIDS in homosexual men. *Science* **240:**1333–1335.
172. **Lui, K.-J., T. A. Peterman, D. N. Lawrence, and J. R. Allen.** 1988. A model-based approach to characterize the incubation period of pediatric transfusion-associated acquired immunodeficiency syndrome. *Stat. Med.* **7:**395–401.
173. **Mann, J. M., H. Francis, T. C. Quinn, K. Bila, P. K. Asila, N. Bosenge, N. Nzilambi, L. Jansegers, P. Piiot, K. Ruti, and J. W. Curran.** 1986. HIV serprevalence among hospital workers in Kinshasa, Zaire. *JAMA* **256:**3099–3102.
174. **Marchoux, P. E.** 1934. Un cas d'inoculation accidentelle du bacille de Hanson en pays non lepreux. *Int. J. Lepr.* **2:**1–7.
175. **Marcus, R.** 1988. The cooperative needlestick surveillance group: CDC's health-care workers surveillance project: an update, abstr. 9015. 4th Int. Conf. AIDS, Stockholm, Sweden.
176. **Martin, L. S., S. L. Loskoski, and J. S. McDougal.** 1987. Inactivation of human T-lymphotropic virus type III/lymphadenopathy-associated virus by formaldehyde-based reagents. *Appl. Environ. Microbiol.* **53:**708–709.
177. **Martin, L. S., J. S. McDougal, and S. L. Loskoski.** 1985. Disinfection and inactivation of the human T-lymphotropic virus type III/lymphadenopathy-associated virus. *J. Infect. Dis.* **152:**400–403.
178. **Marx, P. A., N. C. Pedersen, N. W. Lerche, K. G. Osborn, L. J. Lowenstine, A. A. Lackner, D. H. Maul, H. S. Kwang, J. D. Kluge, and C. P. Zaiss.** 1986. Prevention of simian acquired immune deficiency syndrome with a formalin-inactivated type D retrovirus vaccine. *J. Virol.* **60:**431–435.
179. **Matthews, J. L., J. T. Newman, F. Sogandares-Bernal, M. M. Judy, H. Skils, J. E. Leveson, A. J. Marengo-Rowe, and T. C. Chanh.** 1988. Photodynamic therapy of viral contaminants with potential blood banking applications. *Transfusion* **28:**81–83.
180. **McBride, J. H., P. J. Howanitz, D. O. Rodgerson, J. Miles, and J. B. Peter.** 1987. Influence of specimen treatment on nonreactive HTLV-III sera. *AIDS Res. Hum. Retroviruses* **3:**333–340.
181. **McClure, H. M., D. C. Anderson, P. Fultz, A. Ansari, A. Brodie, and A. Lehrman.** 1989. Prophylactic effects of AZT following exposure of macaques to an acutely lethal variant of SIV, abstr. 5th Int. Conf. AIDS, Montreal, Quebec, Canada.
182. **McCray, E.** 1986. The cooperative needlestick surveillance group. Occupational risk of the acquired immunodeficiency syndrome among health care workers. *N. Engl. J. Med.* **314:**1127–1132.
183. **McCune, J. M., R. Namikawa, H. Kaneshima, L. D. Shultz, M. Lieberman, and I. L. Weissman.** 1988. The SCID-hu mouse: murine model for the analysis of human hematolymphoid differentiation and function. *Science* **241:**1632–1639.

184. **McCune, J. M., R. Namikawa, C. C. Shih, L. Rabin, and H. Kaneshima.** 1990. 3′-Azido-3′-deoxythymidine suppresses HIV infection in the SCID-hu mouse. *Science* **247:**564–566.
185. **McDougal, J. S., L. S. Martin, S. P. Cort, M. Mozen, C. M. Heldebrant, and B. L. Evatt.** 1985. Thermal inactivation of the acquired immune deficiency virus, human T-lymphotropic virus-III/lymphadenopathy-associated virus, with special reference to antihemophilic factor. *J. Clin. Invest.* **76:**875–877.
186. **McElrath, J., E. Peterson, and J. Dragavon.** 1992. Combination prime-boost approach to HIV vaccination in seronegative individuals; enhanced immunity with additional subunit gp160 protein boosting, abstr. MoA0027. 8th Int. Conf. AIDS/III STD World Cong. Amsterdam.
187. **McEvoy, M., K. Porter, P. Mortimer, N. Simons, and D. Shanson.** 1987. Prospective study of clinical, laboratory, and ancillary staff with accidental exposures to blood or body fluids from patients infected with HIV. *Br. Med. J.* **294:**1595–1597.
188. **Michelet, C., F. Cartier, A. Ruffault, C. Camus, N. Genetet, and R. Thomas.** 1988. Needlestick HIV infection in a nurse, abstr. 9010. 4th Int. Conf. AIDS, Stockholm, Sweden.
189. **Milman, G., and P. D'Souza.** 1990. HIV infections in SCID mice: safety considerations. *ASM News* **56:**639–642.
190. **Monath, T. P.** 1990. Colorado tick fever, p. 1233–1240. *In* G. L. Mandell, R. G. Douglas, and J. E. Bennett (ed.), *Principles and practice of infectious diseases,* 3rd ed. Churchil Livingstone, New York.
191. **Mortimer, P. P., N. L. C. Luban, J. F. Kellener, and B. J. Cohen.** 1983. Transmission of serum parvovirus-like virus by clotting factor concentrates. *Lancet* **ii:**482–484.
192. **Mosier, D. E., R. J. Gulizia, S. M. Baird, and D. B. Wilson.** 1988. Transfer of a functional human immune system to mice with severe combined immunodeficiency. *Nature* **335:**256–259.
193. **Moss, B., and C. Flexner.** 1987. Vaccinia virus expression vectors. *Annu. Rev. Immunol.* **5:**305–324.
194. **Mundy, D. C., R. F. Schinazi, A. R. Gerber, A. J. Nahmias, and H. W. Randall, Jr.** 1987. Human immunodeficiency virus isolated from amniotic fluid. *Lancet* **ii:**459–460. (Letter.)
195. **Namikawa, R., H. Kaneshima, M. Lieberman, I. L. Weissman, and J. M. McCune.** 1988. Infection of the SCID-hu mouse by HIV-1. *Science* **242:**1684–1686.
196. **National Committee for Clinical Laboratory Standards.** 1991. *Protection of laboratory workers from infectious disease transmitted by blood, body fluids, and tissue.* Tentative guideline. M29-T2. National Committee for Clinical Laboratory Standards, Villanova, Pa.
197. **National Institutes of Health.** 1988. *Working safely with HIV in the research laboratory.* Guidance document. Division of Safety, Occupational Safety and Health Branch, Bethesda, Md.
198. **Neisson-Verant, C., S. Arfi, D. Mathez, J. Leibowitch, and N. Monplanir.** 1986. Needlestick HIV seroconversion in a nurse. *Lancet* **ii:**814. (Letter.)
199. **N'Galy, B., R. W. Ryder, K. Bila, K. Mwandagalirwa, R. Colebunders, H. Francis, J. M. Mann, and T. C. Quinn.** 1988. Human immunodeficiency virus infection among employees in an African hospital. *N. Engl. J. Med.* **319:**1123–1127.
200. **Ohta, Y., T. Masuda, H. Tsujimoto, K. Ishikawa, T. Kodama, S. Morikawa, M. Nakai, S. Honjo, and M. Hayami.** 1988. Isolation of simian immunodeficiency virus from African green monkeys and seroepidemiologic survey of the virus in various nonhuman primates. *Int. J. Cancer* **41:**115–122.
201. **Okochi, K., H. Sato, and Y. Hinuma.** 1984. A retrospective study on transmission of adult T-cell leukemia virus by blood transfusion: seroconversion in recipients. *Vox Sang.* **46:**245–253.
202. **Oksenhendler, E., M. Harzic, J. M. LeRoux, C. Rabian, and J. P. Clauvel.** 1986. HIV infection with seroconversion after a superficial needlestick injury to the finger. *N. Engl. J. Med.* **315:**582. (Letter.)
203. **Parritt, R. J., and R. E. Olsen.** 1947. Two simultaneous cases of leprosy developing in tattoos. *Am. J. Pathol.* **23:**805–817.
204. **Paulsen, A. G., B. Kvinesdal, P. Aarby, K. Molbak, K. Frederiksen, F. Dias, and E. Lauritzen.** 1989. Prevalence of and mortality from human immunodeficiency virus type 2 in Bissau, West Africa. *Lancet* **i:**827–831.
205. **Phillips, G. B.** 1961. *Microbiological safety in US and foreign laboratories.* Technical study 35.AD-268635. National Technical Information Service, Springfield, Va.
206. **Pike, R. M.** 1976. Laboratory-associated infections: summary and analysis of 3,921 cases. *Health Lab. Sci.* **13:**105–114.
207. **Piszkiewicz, D., R. Apfelzweig, L. Bourret, K. Hattley, C. D. Cabradila, J. S. McDougal, and D. Menache.** 1988. Inactivation of HIV in antithrombin III concentrate by pasteurization. *Transfusion* **28:**198–199.
208. **Pizzocolo, G., R. Stellini, G. P. Cadeo, S. Casari, and P. L. Zampini.** 1988. Risk of HIV and HBV infection after accidental needlestick, abstr. 9012. 4th Int. Conf. AIDS, Stockholm, Sweden.
209. **Poiesz, B. J., F. W. Ruscetti, A. F. Gazdar, P. A. Bunn, J. D. Minna, and R. C. Gallo.** 1980. Detection and isolation of type-C retrovirus particles from fresh and cultured lymphocytes of patients with cutaneous T-cell lymphoma. *Proc. Natl. Acad. Sci. USA* **77:**7415–7419.
210. **Pomerantz, R. J., S. M. de la Monte, S. P. Donegan, T. R. Rota, M. W. Vogt, D. E. Craven, and M. S. Hirsch.** 1988. Human immunodeficiency virus (HIV) infection of the uterine cervix. *Ann. Intern. Med.* **108:**321–327.
211. **Ponce de Leon, R. S., G. Sanchez-Mejorada, and M. Zaidi-Jacobsen.** 1988. AIDS in a blood bank technician. *Infect. Control Hosp. Epidemiol.* **318:**86–90.
212. **Popovic, M., M. G. Sarngadharan, E. Read, and R. C. Gallo.** 1984. Detection, isolation, and continuous production of cytopathic retrovirus (HTLV-III) from patients with AIDS and pre-AIDS. *Science* **224:**497–500.
213. **Prince, A. M., B. Horowitz, L. Baker, R. W. Shulman, H. Ralph, J. Valinsky, A. Cundell, B. Brotman, W. Boehle, F. Rey, M. Piet, H. Reesink, N. Lelie, M. Tersmette, F. Mieckma, L. Barbosa, G. Nemo, C. L. Nostala, A. J. Langlois, J. S. Allan, D. R. Lee, and J. W. Eichberg.** 1988. Failure of a human immunodeficiency virus (HIV) immune globulin to protect chimpanzees against experimental challenge with HIV. *Proc. Natl. Acad. Sci. USA* **85:**6944–6948.

214. **Prince, A. M., B. Horowitz, and B. Brotman.** 1986. Sterilization of hepatitis and HTLV-III viruses by exposure to tri (*n*-butyl) phosphate and sodium cholate. *Lancet* **i**:706–710.
215. **Prince, H. N.** 1987. Summary: working group IV, viruses, p. 139–142. *In* M. S. Favero and D. H. M. Groschel (ed.), *Chemical germicides in the health care field.* American Society for Microbiology, Arlington, Va.
216. **Prince, H. N., D. L. Prince, and R. N. Prince.** 1991. Principles of viral control and transmission, p. 411–444. *In* S. S. Block (ed.), *Disinfection, sterilization, and preservation,* 4th ed. Lea & Febiger, Philadelphia.
217. **Putkonen, P., R. Thorstensson, L. Ghavamzadeh, J. Albert, K. Hild, G. Biberfeld, and E. Norrby.** 1991. Prevention of HIV-2 and SIV_{sm} infection by passive immunization in cynomolgus monkeys. *Nature* **352**:436–438.
218. **Ramsey, K. M., E. N. Smith, and J. A. Reinarz.** 1988. Prospective evaluation of 44 health care workers exposed to human immunodeficiency virus-1, with one seroconversion. *Clin. Res.* **36**:1a. (Abstr.)
219. **Rastrelli, M., D. Ferrazzi, B. Vigo, and F. Giannelli.** 1989. Risk of HIV transmission to health care workers and comparison with the viral hepatitidies, abstr. A503. 5th Int. Conf. AIDS, Montreal, Quebec, Canada.
220. **Rawal, B. D.** 1959. Laboratory infection with toxoplasma. *J. Clin. Pathol.* **12**:59–61.
221. **Rebar, R., and T. J. Concannon.** 1989. Working safely with human immunodeficiency recombinant proteins, abstr. American Biological Safety Association, Biological Safety Conference, New Orleans.
222. **Reitman, M., and G. B. Phillips.** 1956. Biological hazards of common laboratory procedures. III. The centrifuge. *Am. J. Med. Technol.* **22**:14–16.
223. **Reitz, M. S., and R. C. Gallo.** 1990. Human immunodeficiency virus, p. 1344–1352. *In* G. L. Mandell, R. G. Douglas, and J. E. Bennett (ed.), *Principles and practice of infectious diseases,* 3rd ed. Churchill Livingstone, New York.
224. **Resnick, L., K. Veren, S. Z. Salahuddin, S. Tondreau, and P. D. Markham.** 1986. Stability and inactivation of HTLV-III/LAV under clinical and laboratory environments. *JAMA* **255**:1887–1891.
225. **Richardson, J. H., and W. E. Barkley (ed.).** 1984. *Biosafety in microbiological and biomedical laboratories.* HHS Publication No. (CDC) 84-8395. U.S. Department of Health and Human Services, Public Health Service, Washington, D.C.
226. **Richardson, J. H., and W. E. Barkley (ed.).** 1988. *Biosafety in microbiological and biomedical laboratories,* 2nd ed. HHS Publication No. (NIH) 88-8395. U.S. Department of Health and Human Services, Public Health Service, Washington, D.C.
227. **Richmond, J. Y., and R. W. McKinney (ed.).** 1993. *Biosafety in microbiological and biomedical laboratories,* 3rd ed. HHS Publication No. (CDC) 93-8395. U.S. Department of Health and Human Services, Public Health Service, Washington, D.C.
228. **Robert-Guroff, M., S. H. Weiss, J. A. Giron, A. M. Jennings, H. M. Ginzburg, I. B. Margolis, W. A. Blattner, and R. C. Gallo.** 1986. Prevalence of antibodies to HTLV-I, and -II in intravenous drug abusers from an AIDS endemic region. *JAMA* **255**:3133–3137.
229. **Robertson, D. H. H., S. Pickens, J. H. Lawson, and B. Lennox.** 1980. An accidental laboratory infection with African trypanosomes of a defined stock. I. The clinical course of infection. *J. Infect.* **2**:105–112.
230. **Ruprecht, R. M., L. G. O'Brien, and S. Nusinoff-Lehrman.** 1986. Suppression of mouse viremia and retroviral disease by 3′-azido-3′-deoxythymidine. *Nature* **323**:467–469.
231. **Rusche, J. R., D. L. Lynn, M. Robert-Guroff, A. J. Langlois, H. K. Lyerly, H. Carson, K. Krohn, A. Ranki, R. C. Gallo, and D. P. Bolegnesi.** 1987. Humoral immune response to the entire human immunodeficiency virus envelope glycoprotein made in insect cells. *Proc. Natl. Acad. Sci. USA* **84**:6924–6928.
232. **Saag, M. S., M. J. Crain, W. D. Decker, S. Campbell-Hill, S. Robinson, W. E. Brown, M. Leuther, R. J. Whitley, B. H. Hahn, and G. M. Shaw.** 1991. High-level viremia in adults and children infected with human immunodeficiency virus: relation to disease stage and CD4+ lymphocyte levels. *J. Infect. Dis.* **164**:72–80.
233. **Salk, J.** 1987. Prospects for the control of AIDS by immunizing seropositive individuals. *Nature* **327**:473–476.
234. **Sampaio, R. N., L. M. P. deLima, A. Vexenat, C. C. Cuba, A. C. Barreto, and P. D. Marsden.** 1983. A laboratory infection with *Leishmania braziliensis. Trans. R. Soc. Trop. Med. Hyg.* **77**:274.
235. **Sarin, P. S., D. I. Scheer, and R. D. Kross.** 1985. Inactivation of human T-cell lymphotropic retrovirus (HTLV-III) by LD (tm). *N. Engl. J. Med.* **313**:1416.
236. **Schmidt, C. A.** 1988. [HIV infection from a needle-stick injury.] *Dtsch. Med. Wochenschr.* **113**(2):76. (Letter in German.)
237. **Schurman, W., and H. J. Eggers.** 1983. Antiviral activity of an alcoholic hand disinfectant: comparison of the in vitro suspension test with in vivo experiments on hands and on individual fingertips. *Antiviral Res.* **3**:25–41.
238. **Sexton, D. J., H. A. Gallis, J. R. McRae, and T. R. Cate.** 1975. Possible needle-associated rocky mountain spotted fever. *N. Engl. J. Med.* **292**:645. (Letter.)
239. **Shanson, D. C., R. Evans, and L. Lai.** 1985. Incidence and risk of transmission of HTLV-III infection to a staff at a London hospital, 1982–85. *J. Hosp. Infect. Suppl.* **C**:15–22.
240. **Shaw, G. M., B. H. Hahn, S. K. Ayra, J. E. Groopman, R. C. Gallo, and F. Wong-Staal.** 1984. Molecular characterization of human T-cell leukemia (lymphotropic) virus type III in the acquired immune deficiency syndrome. *Science* **226**:1165–1170.
241. **Shih, C.-C., H. Kaneshima, L. Rabin, R. Namikawa, P. Sager, J. McGowan, and J. M. McCune.** 1991. Postexposure prophylaxis with zidovudine suppresses human immunodeficiency virus type 1 infection in SCID-hu mice in a time-dependent manner. *J. Infect. Dis.* **163**:625–627.
242. **Skinhoj, P.** 1974. Occupational risks in Danish clinical chemical laboratories. II. Infections. *Scand. J. Clin. Lab. Invest.* **33**:27–29.
243. **Skinhoj, P., and M. Soeby.** 1981. Viral hepatitis in Danish healthcare personnel, 1974–78. *J. Clin. Pathol.* **34**:408–410.
244. **Smith, D. H., B. K. Johnson, M. Isaacson, R. Swanapoel, K. M. Johnson, A. Bagshawe, T. Siongok, and**

W. K. Keruga. 1982. Marburg virus disease in Kenya. *Lancet* **i:**816–820.

245. **Smith, R. P., A. T. Evans, M. Popovsky, L. Mills, and A. Spielman.** 1986. Transfusion acquired babesiosis and failure of antibiotic treatment. *JAMA* **256:**2726–2727.
246. **Spire, B., F. Barre-Sinoussi, D. Dormont, L. Montagnier, and J. C. Chermann.** 1985. Inactivation of lymphadenopathy-associated virus by heat, gamma rays, and ultraviolet light. *Lancet* **i:**188–189.
247. **Spire, B., L. Montagnier, F. Barre-Sinoussi, and J. C. Chermann.** 1984. Inactivation of lymphadenopathy associated virus by chemical disinfectants. *Lancet* **ii:**899–901.
248. **Spruance, S. L., and A. Bailey.** 1973. Colorado tick fever. A review of 115 laboratory confirmed cases. *Arch. Intern. Med.* **131:**288–293.
249. **Stein, L., R. E. Anderson, and N. H. Gross.** 1949. Potential infectious hazards of common bacteriological techniques. *Bacteriol. Proc.*, 10–11.
250. **Stern, E. L., J. W. Johnson, D. Vesley, M. M. Halbert, L. E. Williams, and P. Blume.** 1974. Aerosol production associated with clinical laboratory procedures. *Am. J. Clin. Pathol.* **62:**591–600.
251. **Stoeckle, M.** 1990. Introduction-type C oncoviruses including human T-cell leukemia virus types I and II, p. 1336–1341. *In* G. L. Mandell, R. G. Douglas, and J. E. Bennett (ed.), *Principles and practice of infectious diseases,* 3rd ed. Churchill Livingstone, New York.
252. **Stokes, J. H., H. Beerman, and N. R. Ingraham.** 1945. *Modern clinical syphilology, diagnosis, treatment, case study,* 3rd ed. The W.B. Saunders Co., Philadelphia.
253. **Stover, C. K., V. F. DeLa Cruz, T. R. Fuerst, J. E. Burlein, L. A. Benson, L. T. Bennett, G. P. Bansal, J. F. Young, M. H. Lee, and G. F. Hatfull.** 1991. New use of BCG for recombinant vaccines. *Nature* **351:**456–460.
254. **Stricof, R. L., and D. L. Morse.** 1986. HTLV-III/LAV seroconversion following a deep intramuscular needlestick injury. *N. Engl. J. Med.* **314:**1115.
255. **Sun, T., M. J. Tenenbaum, J. Greenspan, S. Teichberg, R. T. Wang, T. Degnan, and M. H. Kaplan.** 1983. Morphologic and clinical observations in human infection with *Babesia microti*. *J. Infect. Dis.* **148:**239–248.
256. **Tait, D. R., D. J. Pudifin, V. Gathiram, and I. M. Windsor.** 1992. HIV seroconversions in health care workers, Natal, South Africa, abstr. no. PoC4141. 8th Int. Conf. AIDS, Amsterdam.
257. **Tavares, L., C. Roneker, K. Johnston, S. Nusinoff-Lehrman, and F. de Noronha.** 1987. 3′-azido-3′-deoxythymidine in feline leukemia virus-infected cats: a model for therapy and prophylaxis of AIDS. *Cancer Res.* **47:**3190–3194.
258. **Thiry, L., S. Sprecher-Goldberger, T. Jonckheer, J. Levy, P. VanPerre, P. Henrivaux, J. Cogniaux-LeClerc, and N. Clumeck.** 1985. Isolation of AIDS virus from cell-free breast milk of three healthy virus carriers. *Lancet* **ii:**891–892. (Letter.)
259. **Tokars, J. I., R. Marcus, and the Cooperative Needlestick Surveillance Group.** 1990. Surveillance of health care workers exposed to blood from patients infected with the human immunodeficiency virus, abstr. 490. 30th Intersci. Conf. Antimicrobial Agents Chemother. Atlanta.
260. **Tokars, J. I., R. Marcus, D. H. Culver, C. A. Schable, P. S. McKibben, C. I. Bandea, D. M. Bell, and the CDC Cooperative Needlestick Surveillance Group.** 1993. Surveillance of HIV infection and zidovudine use among health care workers after occupational exposure to HIV-infected blood. *Ann. Intern. Med.* **118:**913–919.
261. **U.S. Department of Labor/Occupational Safety and Health Administration.** 1989. Occupational exposure to bloodborne pathogens: proposed rules. *Fed. Regist.* **54:**23042–23139.
262. **U.S. Department of Labor/Occupational Safety and Health Administration.** 1991. Occupational exposure to bloodborne pathogens; final rule. *Fed. Regist.* **56:**64175–64182.
263. **U.S. Food and Drug Administration.** 1990. Medical devices; patient examination and surgeons' gloves; adulteration; final rule. 21 CFR Part 800. *Fed. Regist.* **55:**51254–51258.
264. **Vogt, M. W., D. J. Witt, D. E. Craven, R. Byington, D. F. Crawford, R. T. Schooley, and M. S. Hirsch.** 1986. Isolation of HTLV-III/LAV from cervical secretions of women at risk for AIDS. *Lancet* **i:**525–527.
265. **Wallace, M. R., and W. O. Harrison.** 1988. HIV seroconversion with progressive disease in a health care worker after needlestick injury. *Lancet* **i:**1454. (Letter.)
266. **Ward, R. H. R., D. J. Capon, C. M. Jett, K. K. Murthy, J. Mordenti, C. Lucas, S. W. Frie, A. M. Prince, J. D. Green, and J. W. Eichberg.** 1991. Prevention of HIV-1 IIIB infection in chimpanzees by CD4 immunoadhesion. *Nature* **352:**434–436.
267. **Weiss, S. H., J. J. Goedert, S. Gartner, M. Popovic, D. Waters, P. Markham, B. D. M. Veronese, M. H. Gail, W. E. Barkley, J. Gibbons, F. A. Gill, M. Leuther, G. M. Shaw, R. C. Gallo, and W. A. Blattner.** 1988. Risk of human immunodeficiency virus (HIV-1) infection among laboratory workers. *Science* **239:**68–71.
268. **Weiss, S. H., J. J. Goedert, M. G. Sarngadharan, A. J. Bodner, R. C. Gallo, and W. A. Blattner.** 1985. Screening test for HTLV-III (AIDS agent) antibodies: specificity, sensitivity, and applications. *JAMA* **253:**221–225.
269. **Weiss, S. H., W. C. Saxinger, D. Rechtman, M. H. Grieco, J. Nadler, S. Holman, H. M. Ginzburg, J. E. Groopman, J. J. Goedert, and P. D. Markham.** 1985. HTLV-III infection among health care workers: association with needlestick injuries. *JAMA* **254:** 2089–2093.
270. **Wilson, D. (Division of Safety, National Institutes of Health, Bethesda, Md.).** 1991. Personal communication.
271. **Winkler, W. G., T. R. Fashinell, L. Leffingwell, P. Howard, and J. P. Conomy.** 1973. Airborne rabies transmission in a laboratory worker. *JAMA* **226:** 1219–1222.
272. **Wong-Staal, F., and R. C. Gallo.** 1985. Human T-lymphotropic viruses. *Nature* **317:**395–403.
273. **Wormser, G. P., C. Joline, S. Sivak, and Z. A. Arlin.** 1988. Human immunodeficiency virus infections: considerations for health care workers. *Bull. NY Acad. Med.* **64:**203–215.
274. **Zagury, D., J. Bernard, J. Leibowitch, B. Safai, J. E. Groopman, M. Weldman, M. G. Sarngadharan,**

and R. C. Gallo. 1984. HTLV-III in cells cultured from semen of two patients with AIDS. *Science* **226:**449–451.

275. **Ziza, J. M., F. Brun-Vezinet, A. Venet, C. H. Rouzioux, J. Traversat, B. Israel-Biet, F. Barre-Sinoussi, J. C. Chermann, and P. Godeau.** 1985. Lymphadenopathy-associated virus isolated from bronchoalveolar lavage fluid in AIDS-related complex with lymphoid interstitial pneumonitis. *N. Engl. J. Med.* 313:183. (Letter.)

Biosafety Precautions for Airborne Pathogens

MARY J. R. GILCHRIST

5

INTRODUCTION

The previous two chapters dealt with safety with respect to organisms that are borne primarily in blood and body fluids. With such agents, contact of these fluids with mucous membranes or nonintact skin, as well as penetration of intact skin with sharp objects bearing such fluids, is the predominant means of transmission from human to human in the community, the hospital, and the laboratory. In contrast, this chapter deals with those organisms for which the airborne route is the predominant means of transmission to humans, from the habitat in nature for the fungi and from humans, animals, and water for the pathogenic mycobacteria. Although the fungi and mycobacteria share in common the airborne mode of transmission, they are very different in many substantive elements of the airborne route of transmission. It is only with a clear knowledge of these differences that the laboratorian can adequately assess hazards and implement safety protocols to abate those hazards.

The term *laboratory,* in its broadest definition, applies to many different situations, including the collection and evaluation of materials in their natural environment. Laboratorians who go on field trips to collect specimens or to study the ecology of organisms often consider themselves in a laboratory situation, although not strictly in a laboratory environment. In such instances, the laboratorian may be exposed to winds or air currents containing pathogens (e.g., fungal spores). Of particular concern are the spores of fungi known to naturally infect normal hosts, in particular, those fungi that produce systemic mycoses. In North American soils, these are primarily *Coccidioides immitis* in the desert southwest, and *Blastomyces dermatitidis* and *Histoplasma capsulatum,* both in the midwest and southeast. The human host who becomes infected with these agents is not normally a source of infection to social contacts or hospital employees. This is reflected in the lack of any recommendation for isolation precautions for such patients by the Centers for Disease Control and Prevention (CDC) (6). The absence of communicability is due to unique properties of the fungi. These organisms are dimorphic. The form in which the organisms are transmissible is the hyphal form that occurs in nature and, under certain circumstances, in the laboratory. The hyphal forms of these fungi produce conidia (spores) that are readily transmissible by the airborne route. The tissue forms of these fungi are yeasts, or spherules, and not readily transmissible to humans either by direct contact or by the airborne route. However, there are isolated case reports of percutaneous transmission in laboratorians or health care workers. Infection with these agents has occurred when infectious tissue of the patient was introduced via trauma (e.g., at autopsy) into the subcutaneous tissues of the laboratorian. These cases are exceptions. The primary source of infectious hazards caused by these fungi in the health care environment is in the laboratory via an airborne route. It is in the laboratory manipulation of these organisms that the infectious form of the agent, the conidia, is propagated. To isolate and identify the agents of the systemic mycoses accurately, the laboratorian incubates the cultures at a temperature that is permissive for the production of these more hazardous, easily airborne forms. This laboratory hazard is the focus of a section of this chapter on hazard assessment and containment of systemic dimorphic fungi.

The transmission of the agent *Mycobacterium tuberculosis* is the primary focus for laboratory safety in the mycobacteriology laboratory. With *M. tuberculosis,* the natural transmission to humans is almost always from another human by infectious droplet nuclei. The mechanism of transmission from human to human is largely analogous to the most common means of acquisition of the organism in the laboratory setting. In both cases, the primary means of transmission is via droplet nuclei. Efficient means of production of droplet nuclei in nature include sneezing, coughing, and vibration of the larynx, all of which introduce energy that subdivides fluids into tiny droplets. Analogous mechanisms in the laboratory include sonication, vortexing, blending, and all other manipulations of fluids. Thus, with tuberculosis, it is useful to understand natural transmission as a paradigm for accidental laboratory acquisition. Hazard assessment and abatement, relative to acquisition of disease caused by mycobacteria, are addressed in a section of this chapter.

Several other agents may represent a hazard in the laboratory via the airborne mechanism of transmission but may not also be hazardous by the same route in natural infections of humans. Several circumstances dictate that transmission in the artificial laboratory setting may be enhanced over that of the natural setting. The first of these involves the high concentration of microorganisms that is often manipulated in the laboratory. Culture amplification of microorganisms increases their number by many orders of magnitude over the numbers commonly found in human clinical material. With amplified culture fluids, the inoculum necessary to produce an infection by the airborne route might be easily achievable by modest manipulations that produce minor aerosols. The second of these involves the efficiency of production of aerosols that may be available in the laboratory. Unlike clinical disease affiliated with tuberculosis, not all other infectious processes involve such an efficient transmitting host response, the cough. Thus, it is not only in the laboratory where the organism may be manipulated with such an efficient means of aerosol production that the agent is transmissible by an airborne route. A third circumstance involves a combination of the first two circumstances. The laboratorian may work with large volumes of the fluids in which organisms are amplified and also may manipulate the fluids. Thus, the net amount of aerosol produced may be much larger because of the large volume of fluid manipulated, even with an inefficient means of aerosol production. In all three circumstances, organisms effectively become more hazardous than their counterparts in human fluids. The CDC has recognized this situation by recommending the use of biosafety cabinets (BSC) for all work with certain agents normally handled at biosafety level (BSL) 2 (3). Organisms such as these are the subject of the third section of this chapter.

The current laboratory safety-oriented biolevel classification of microorganisms into four BSLs reflects what is known regarding the tendency for an organism to be transmitted by an airborne route. The respective BSL practices in these four categories take into account the associated tendency for aerosol transmission. Organisms are assigned to BSL 1 because they are not known to cause disease in healthy adults; thus, the practices associated with BSL 1 do not include containment of aerosols. Organisms to be handled under BSL 2 are those primarily transmitted by autoinoculation, contact with mucous membranes, or ingestion. However, consideration has been given to the potential for such an organism to infect via an airborne route when present at high levels. For that reason, BSL 2 practices call for the use of a BSC or other containment equipment when aerosol-generating procedures are carried out with culture-amplified organisms. Organisms assigned BSL 3 containment are those that cause human disease and have a pronounced tendency to be transmitted in the laboratory by aerosols. Affiliated practices for BSL 3 call for containment in the BSC whenever culture-amplified materials are manipulated or when aerosol-generating activities are carried out on body substances thought likely to contain organisms that produce human infection when present in small numbers. BSL 4 is used for organisms that must be more strictly contained because they have a greater, or perhaps unknown, tendency to pose a high risk of a serious or fatal disease in humans. They may be transmitted by any of the routes mentioned above. The absolute containment used at BSL 4 may be a class III cabinet line or a ventilated suit to protect the employee while working in a class II BSC. The selection of appropriate biosafety practices may not always be clear-cut. For example, a good portion of the work with microorganisms in clinical laboratories involves the identification of unknown agents isolated from patients with diseases of unknown etiology. The last section of this chapter provides guidance on ways in which hybrid biosafety practices may be used when one takes into account the statistical probability of the agent being present and the hazard associated with the agent and activity.

HAZARD ASSESSMENT AND CONTAINMENT

Fungi

Certain microorganisms can readily be transported from the cell mass that is grown in the laboratory to suspension in the air with little apparent physical intervention to disperse them. In nature, most fungal hyphae develop structures intended for dispersal in air, either conidia on specialized aerial fruiting bodies or hyphal elements that mature into transmissible subsegments (arthroconidia). As these transmissible elements evolved specifically to disperse the organisms via air currents in nature, the dispersing bodies also evolved an ability to resist desiccation and the germicidal effects of UV lights. Moreover, the conidial forms are structured to be readily discharged into the air and to remain aloft for long periods. When inhaled by a susceptible host, the conidia develop into the yeast phase or spherule, the alternate forms of the fungus found in the tissue of the infected host. *Histoplasma capsulatum* is present in tissues as a yeast; with rare exceptions (e.g., endocarditis), it is detected in tissue in its hyphal form. *Coccidioides immitis* is found in tissues as an endosporulating spherule or as individual endospores from a ruptured spherule. *Blastomyces dermatitidis* occurs in tissues as a characteristically broad-based budding yeast. When cultivated in the laboratory, the systemic dimorphic fungi tend to form hyphal structures during growth on artificial media at lower temperatures, 25 to 30°C. The hyphal structures ultimately produce the specialized conidial forms. With *C. immitis* hyphal in the laboratory at both temperatures of incubation, the growth in vitro of fungal hyphae undergoes a transition within a few days to segments, known as arthroconidia, that are readily dispersed into the air. The hazard to the laboratorian comes when the culture plate of the unknown fungus is opened, easily dispersing these arthrocondidia for inhalation. The other two agents generally require a more prolonged incubation period before the infectious conidia are formed. Once conidial structures are present, the laboratory isolates of these fungi represent a hazard if the containers are opened and the conidia are lofted into the air. Of 4,000 laboratory-acquired infections, approximately 9% were caused by fungi (13). Although not a large percentage, it must certainly represent a larger percentage than that predicted by the relative frequency of detection of systemic dimorphic fungi in the laboratory. Although, undoubtedly, many laboratory-acquired fungal infections are unreported, mycotic morbidity is a well-recognized occupational risk for mycologists (5).

For work with *known* isolates of the systemic dimorphic fungi, hazard assessment is relatively simple and precautions are straightforward (3). As airborne hazards that cause relatively serious human diseases, these agents must be handled in the laboratory under BSL 3 containment practices. Because the infectious conidia are easily airborne, the mere lifting of a culture plate lid is sufficient to release an infectious dose. Thus, with all fungal cultures grown in the laboratory at temperatures of 25 to 30°C and with *Coccidioides* at both temperatures, containers of fungal agents that have been culture-amplified are sealed with tape and opened only in the BSC. It is prudent to handle all such materials, even those grown at elevated temperatures (35 to 37°C), under containment (e.g., the BSC), because these agents have a tendency to assume the infectious hyphal-spore form unless there is rigorous control of the conditions of incubation and evaluation. With *Coccidioides,* the form that grows at both temperatures is the one that develops arthroconidia.

Rationalized protocols for work with such organisms in culture call for the implementation of further safety considerations. There should be a protocol to follow when there is an accident involving a container that breaks outside of the BSC (11). There should be provisions in the procedure manual that prohibit the making of slide cultures with fungal agents in which conidial structures have not developed sufficiently to rule out the systemic dimorphic fungi (9). Until confirmation of an agent other than a systemic dimorphic fungus is made, the assessment of the hyphal structures should be performed with a tease-mount rather than a Scotch tape preparation (9). Moreover, there should be administrative controls in place to make certain that cultures of the organisms are not disseminated to untrained outsiders for use in science fairs or research under uncontrolled conditions. Finally, there should be strict adherence to the rules for shipping samples of the organisms (3).

For work with unknown fungal isolates, in particular in the clinical laboratory, hazard assessment and abatement must be anticipatory. For all clinical and research situations in which fungal cultures are requested, culture plates should be taped and culture tubes should have secure screw caps, both of which should be opened only in the BSC. In no case should early growth or pigment production be considered evidence against the prospect that a fungal isolate is one of the systemic dimorphic fungi. For example, *C. immitis* often grows within a few days and, even though commonly described as a colorless mycelial mass, may elaborate pink to green pigmentation. The infectious arthroconidia may be absent from the growth on certain media (e.g., cycloheximide media) but present on other media. The absence of characteristic macroconidia should not be considered as evidence to rule out *H. capsulatum;* microconidia are most likely to serve as a source of airborne infection. Moreover, it is common to encounter hyphal forms that are slow to produce the characteristic tuberculate macroconidia. Whenever sterile hyphae are encountered, particularly when they are small in cross section and able to grow on cycloheximide agar, the unknown should be considered a strong candidate for an infectious fungal hazard until proved otherwise. The most problematic setting for laboratory safety with the dimorphic fungi is not the mycology laboratory, where their growth is anticipated and safety procedures are generally fully implemented. However, in the bacteriology laboratory, it is not unusual for a laboratorian to open a plate and encounter "fuzzy" colony. There are two basic protocols that should be implemented in the bacteriology laboratory.

First, when culture plates inoculated with clinical specimens are saved for 3 or more days, the lids should be taped or sealed or the technologist should be forewarned not to open the plate without first examining the surface growth for evidence of hyphal forms. Second, when a plate with evidence of hyphal growth is accidentally opened before the growth is discovered, the technologist should quickly close the plate and perform all further workup of the organisms in the BSC. Moreover, the hyphal growth should be evaluated to rule out systemic dimorphic fungi, even if the protocol does not call for the full identification of the organism in question. If there are sterile hyphae present and the growth does not readily suggest the identity of the organism, the supervisor should be contacted for an assessment (i.e., whether to refer the isolate to the mycology laboratory for identification). The presence or absence of conidial structures at the time of exposure should be noted, along with the potential numbers of conidia present at the time of the exposure, so that a medical decision can be made regarding the incident, should the organism prove to be one of the hazardous fungal agents. Unknown fungal growth referred as a potential systemic dimorphic fungus should be shipped or transported according to packaging instructions for the dimorphic fungi, because the infectious conidia may develop in transit and place at risk those who transport or open the container.

There are instances in which organisms, although dimorphic, are not known to represent an infectious aerosol hazard in the laboratory. *Sporothrix schenckii,* although potentially pathogenic if inoculated subcutaneously or splashed into the eye (15), does not appear to constitute a laboratory hazard with regard to the respiratory route under usual clinical or investigative conditions and is handled at BSL 2. There is a suggestion that the dematiacious organisms might represent a hazard via the respiratory route when growing in their hyphal form, based on their infectiousness to immunocompetent hosts who encounter them growing in nature (3). It is prudent to manipulate the sporulating growth of *S. schenckii* in the BSC, minimizing respiratory exposure, assuming that the spores will become aerosolized. Although many of the opportunistic and saprophytic fungi do not represent a substantial infection risk to the immunocompetent laboratorian, the dispersal of their conidia in the laboratory is problematic because there may be allergic reactions as well as contamination of media. Therefore, to protect the worker and the work, gross exposures should be minimized by only opening the culture containers in the BSC.

Mycobacteria

Aerosol Transmission in Nature

Studies of the transmission of tuberculosis conducted during the first half of this century (18) led to the concept of the "droplet nucleus." Photographs of sneezes and coughs revealed the discharge of thousands of droplets ranging in size from a few micrometers to several hundred micrometers in diameter. Further studies with more sophisticated photographic techniques and physical modeling revealed that particles that are very small when they are discharged evaporate very readily (Table 1), transforming within hundredths of a second into a dehydrated mass containing the previously dissolved solutes of the discharged solution and any particulates that were carried with the droplet. The fineness of the division of the discharged particle determines its ultimate fate. The larger particles drop to the floor

TABLE 1 Evaporation time and falling distance of droplets based on size[a]

Diameter of droplet (μm)	Evaporation time (s)	Distance fallen (before evaporation) (ft)
200	5.2	21.7
100	1.3	1.4
50	0.31	0.085
25	0.08	0.0053

[a]Adapted from reference 18.

within seconds, where they most often form an aggregate of dust that is not readily redispersed into the air. For droplets discharged from a height of 6 ft (1.8 m), those larger than 140 μm tend to fall to the ground before they evaporate and those smaller than 140 μm would be more likely to evaporate before contacting the ground (18). The aerodynamic properties of the residues of these evaporated droplets are such that they remain aloft for very long periods. These particles have been designated "droplet nuclei."

An understanding of the formation and persistence of droplet nuclei allows us to devise methods to prevent the transmission of tuberculosis. A tissue held over the mouth of an individual who is coughing collects all of the tiny droplets as they are discharged, before they have time to evaporate. This practice minimizes the production of droplet nuclei because the tiniest droplets have no time to evaporate before they coalesce as a mass of fluid in the paper tissue. This mass will ultimately dry, but the coalesced material cannot be converted into droplet nuclei without the intervention of powerful physical forces. Thus, the contaminated tissue may be safely disposed of in the wastebasket without transmitting disease to others under normal circumstances. A patient who is compliant, using tissues when coughing, can reduce the number of microorganisms released into the room air. The laboratorian who actually has contact with a tuberculosis patient is at little excess risk if pathogenic microbes potentially released by patient coughing are contained by the patient's use of a simple mask or tissue. There is no reciprocity between the means of prevention of the actual formation of droplet nuclei (coughing into a tissue) and the means of prevention of exposure (barriers to breathing in the droplet nuclei). Once a droplet nucleus has been allowed to form, its small size can penetrate the fibers of a tissue or a surgical mask. Thus, these products do not represent adequate physical barriers to the aerosol transmission of droplet nuclei. The appropriate barrier is a well-fitted respirator that does not allow leakage of air around the edges and blocks passage of microorganisms in the filter media (fibers or pores) through which air is inspired. Although a simple surgical mask applied to a tuberculosis patient who must be transported outside the isolation room will prevent the dispersal of organisms as droplet nuclei, such a mask does not provide adequate protection to the individual who must breathe air containing droplet nuclei.

Aerosol Transmission of Mycobacteria in the Laboratory

The buttery or pastelike material that forms the mass of most colonies of bacteria and yeasts grown in the laboratory is cohesive in most cases and does not readily promote the distribution of the individual units of the mass into the air. Thus, with bacteria and yeasts, the mere opening of a petri dish lid is not thought to pose the hazard that has previously been described for the fungi that elaborate infectious conidia for natural dispersion. With mycobacteria, the infectious dose is extremely low (about 10 organisms) and colonies are dry; thus, the containers (e.g., test tubes and petri dishes) of growth on solid media are never intentionally opened outside of the BSC. Intentional manipulation of the colony mass increases the chance of dispersal of the organisms into the air. A particularly aerosol-prone procedure is the incineration of organisms from the bacteriologic needle or loop. Thus, it is prudent to substitute, for the conventional flame, an electric incinerator into which the loop is inserted or use a phenol sand trap (10, 14).

To ensure safety, it is necessary to plan for accidental breakage of containers outside the BSC. In this instance, the matrix containing the microorganism determines the hazard potential. Accidents involving breakage of containers of *M. tuberculosis* grown on solid media are rated much less aerosol-prone than those involving organisms suspended in fluid (10). In this reference, dropping a tube or plate of solid medium, or an acid-fast bacillus-positive sputum, is rated as producing a "minimal" aerosol, requiring disinfection. In contrast, dropping a container of culture-amplified *M. tuberculosis* suspended in a fluid is rated as a substantial aerosol hazard, requiring immediate evacuation and sealing of the room before decontamination. Organisms on solid medium cannot be readily dispersed as droplet nuclei. Thus, one can understand why their hazard potential is low, but one may wonder why the hazard potential of acid-fast bacillus-positive sputum is rated as minimal. Acid-fast bacilli in sputum present a lower risk than those in culture-amplified fluids because of their lower numbers and because the sputum is more tenacious than water-based suspensions and thus less readily subdivided into droplets during impact on a hard surface (10).

Bacteria suspended or grown in fluid are more easily distributed into air than are those in a cell mass grown as a colony. Mycobacteria suspended in dispersing agents (such as Tween 80) or subjected to physical forces (e.g., vortexing, sonicating, blending, homogenizing) to disperse the clumps into single organisms represent the greatest risk of aerosol production (4). The droplets aerosolized by the manipulation of fluid become droplet nuclei if they dry before they land on a horizontal surface. If the organisms in the droplet nuclei are not clumped, they are more likely to have effective diameters less than 5 μm and thus be respirable into the alveoli.

Hazard Assessment and Abatement for Mycobacteria in the Laboratory

Hazard assessment in the mycobacteriology laboratory should focus critical attention on the manipulation of fluids. All opening of tubes, pipetting, transfers, sonication, and vortexing of fluid materials should be carefully performed in the BSC or with appropriate containment equipment (e.g., HEPA-filtered or sealed rotor centrifuges). The greater emphasis should be placed on those fluids in which the organism has been amplified by culture. Many safety rules in laboratories are so poorly formulated that they fail in execution.

1. For example, the laboratorian may centrifuge mycobacteria in an aerosol-proof safety container but fail to carry through and open these containers in the BSC. During centrifugation, some tubes tend to leak, and thus there may be droplet nuclei in the safety container; these should be contained in a BSC.
2. For centrifugation, the sealed (aerosol-proof) safety container must be purchased and maintained as such. Many centrifuge containers may appear to provide safety, in that they provide a physical enclosure for the tubes, but they may not actually be rated as aerosol-proof. Furthermore, a good preventive maintenance program must be in place to ensure the continuing containment of aerosols in these safety containers. For example, the O-rings should be inspected regularly and replaced if they are cracked or dry.
3. Although all work may apparently be carefully conducted in the BSC, it will fail to protect the worker if the material is removed from the BSC and discarded in an unsafe fashion. For example,

Tubes that contain fluids that have been vortexed to disperse organisms most likely contain substantial numbers of droplet nuclei, even though the tubes may have been allowed to stand for some time after the vortexing process. Removing the tubes from the BSC when the lids are partially open may allow the release of these droplet nuclei by diffusion of air from the tube interior. These tubes should be carefully sealed and wiped off or submerged in disinfectant before they are removed from the BSC for handling, transport, or storage.

Containers into which disposable loops, swabs, or sticks are placed may also contain droplet nuclei. If small droplets of fluids are released from these materials, droplet nuclei can form. A plastic bag is not a sufficient container to remove such products from the BSC and transport them to the autoclave, because the plastic bag is compressible and air can be released by pressure against the bag during transport. Thus, all such materials should either be submerged in disinfectant or transported in a sealed (aerosol-proof) container.

Hazard assessment involves the critical evaluation of each step in the analytical process (2). It also involves the judgment of the workers who are involved. These workers must understand the rationale for the rules and help to develop them. For laboratorians requiring examples, a sample program is available in the literature (8). It would be inappropriate to adopt such a program without a thorough review and understanding of the hazards. The laboratory director is encouraged to spend time actually working in the mycobacteriology laboratory, at equal risk with the technologists, so that appropriate insight into the hazards and needs might be generated. The programs that are developed should be integrative, including consideration for blood-borne pathogens and aerosol-borne organisms whenever they are relevant. For example, in a laboratory where human immunodeficiency virus and tuberculosis were prominent in the blood of the patients, a barrier to splashing would protect against human immunodeficiency virus but not tuberculosis. In such cases, a gauze used to open a tube, instead of a barrier, would protect against droplets and droplet nuclei, much like the analogy of the handkerchief with the coughing patient.

Stratification of Laboratory Rules According to Extent of Services

The most recent CDC/NIH guidelines for safety (3) call for safety stratification according to the American Thoracic Society extent of services guidelines published more than 10 years ago (1). The laboratory that inoculates but does not work with propagated organisms is distinguished from those that isolate and then identify organisms. This classification of work is not relevant to many laboratories. For example, it is more common for a laboratory to incubate cultures and refer them to other laboratories once growth is observed. This will involve manipulation of culture-amplified organisms to the extent that they may be transferred and stained before referral. Moreover, for the more advanced laboratories, the type of work that is performed with mycobacteria has changed vastly over the decade. The concept of "usual laboratory procedures" is evolving and often involves a greater intensity of aerosol-prone procedures than in the past decades; this will intensify during the progression into the molecular era of tuberculosis diagnosis. Currently, many laboratories have adopted the use of the radiometric method for isolation, identification, and susceptibility testing. During the next several years, other nonradiometric methods of amplification are expected to proliferate. These methods necessitate the manipulation of substantial volumes of fluids in which the organisms have been propagated. Because the practices in laboratories do not conform to the 1983 American Thoracic Society classification, the safety practices cannot be strictly associated. The CDC guidelines are not standards of laboratory practice but are intended for interpretation and appropriate application. Thus, rather than associating the level of laboratory service with a level of safety based on an irrelevant 1983 classification, it is more useful to stratify safety according to types of procedures used and organisms encountered. This concept is provided below:

1. Work with mycobacterial organisms present in biological fluids should be conducted in the BSC but may be performed in a facility that is engineered at BSL 2.

2. Work with mycobacterial organisms grown on solid media and manipulated with procedures that are not aerosol-prone may be safely accomplished using BSL 3 practices in the BSL 2 facility. Solid media may involve the use of the thin plate method for rapid propagation of microorganisms (17) as well as the more classical tubed and plated media.

3. Work with organisms grown or suspended in fluids to high levels, particularly those dispersed by detergent or mechanical action, should be considered the greatest risk, and each element of the work should be individually evaluated to determine whether it is necessary. If deemed necessary, each step should be carefully assessed and containment protocols developed. When it is deemed probable that aerosols might be released in the BSC at substantial levels and frequencies, then full BSL 3 facilities, complemented with appropriate respiratory protection for all in the shared air space, are indicated.

4. Procedures that produce aerosols on a more substantial basis (blending) or with organisms of greater risk (multiple-drug-resistant organisms) might best be conducted in complete containment cabinets (class III) because it is often easier to engineer a smaller space (BSC) than it is to rely on the ongoing engineering of the room (and building) ventilation system, the physical environment, the respirator, and the class I or II BSC. The CDC/NIH guidelines offer the option of an airline suit when working at BSL 3 or 4 with a class I or II BSC, but this option may place outsiders at risk in facilities where expertise in managing negative air pressure and directional air flow is lacking.

The above suggestions are compiled in Table 2, which provides specific examples of tasks applicable to each category.

The section on mycobacteria should not conclude without reference to those mycobacteria that represent little or no aerosol threat to the laboratorian. Mycobacteria other than *M. tuberculosis* and *Mycobacterium bovis* are handled at a lower BSL because their transmissibility via aerosol is limited. A derivative of *M. bovis,* Bacillus Calmette-Guérin, that is used in the vaccine is also ranked as a BSL 2 organism as a result of its attenuated virulence. When the mycobacteria other than tuberculosis or bovis are manipulated in the clinical laboratory, it is common to manipulate them under conditions equivalent to that of *M. tuberculosis.* This is particularly true in two cases. In the first, they are being worked up as unknown agents that resemble *M. tuberculosis* and must be treated as such until identified. At the point of identification, work is usually discontinued by the laboratory, and thus the capability of using relaxed biosafety standards is never realized. In the second case, work with the organisms may be in a plant involved in the production of vaccine. In such cases, large quantities of organisms, which have been assessed for risk during the research and development stages, are manipulated, and the containment level is either increased one level or the appropriate large-scale guidelines are used (12).

Miscellaneous Bacteria

The many bacteria that have caused laboratory-acquired infections include *Brucella, Francisella,* and *Pseudomonas pseudomallei* (13). When infections

TABLE 2 Stratification of tasks according to risk of aerosol spread of tuberculosis

Task and risk assessment	Practice	Special instructions
Manipulation of body fluids potentially containing *M. tuberculosis* or *M. bovis*. BSL 2 task: agents are present in lower numbers in patient body fluids. Respirator program or mask not required.	All work with open vessels is conducted in a BSC. Centrifugation outside the BSC is conducted in sealed, break-proof containers.	1. Pour into splash-proof container and rinse funnel with disinfectant. 2. Pipette over disinfectant-soaked pad. Do not "blow out" pipette. 3. Immerse used pipettes in disinfectant or seal discard container before removing from BSC. 4. Vortex tightly sealed tubes and allow to stand for 30 min before opening. 5. Open safety centrifuge cups in BSC. Inspect surface of tubes for leakeage: disinfect cups, if contaminated, before reuse. 6. Safety centrifuge cups are part of a routine preventive maintenance program; O-rings are replaced to ensure adequate seal. 7. All empty vessels are submerged in disinfectant or tightly sealed before removal from BSC for transport to autoclave.
Manipulation of colonies of *M. tuberculosis* or *M. bovis*. BSL 3 task: organism numbers are amplified by culture: risk is less than with culture-amplified organisms suspended in fluids. Respirator program not necessary.	Plates are sealed with tape or Shrink-Seals and opened only in the BSC. All manipulations of opened tubes or plates are conducted only in the BSC.	1. Colonies are transferred from solid medium to solid medium, with streaking of plates restricted to those plates with a smooth surface. 2. Loops are disinfected in safety Bacticinerator or microorganisms are removed from the loop by a phenol sand trap before incineration in a flame. (See text for precautions on use of gas burners in BSC.) Disposable loops are immersed in disinfectant before removal from BSC. 3. Plates and tubes for discard are sealed with an aerosol-proof seal before removal from BSC for transport to autoclave.
Manipulation of fluids containing culture-amplified *M. tuberculosis* or *M. bovis*. BSL 3 task: volume and extent of potential aerosolization determines whether a BSL 3 facility is required or whether BSL 3 practices in a BSL 2 facility will suffice.	1. Pipetting or aspirating fluid from sealed bottles 2. Vortexing 3. Centrifuging 4. Sonicating 5. Blending	A respiratory protection program and negative room air are recommended, especially for 2–5 below, and particularly when organisms are well dispersed, for example, with Tween 80. 1. Work over disinfectant-soaked pad. Do not blow out pipette or syringe. Immerse used devices in disinfectant or seal discard container before removing from BSC. 2. Vortex tightly sealed tubes and allow to stand for 30 min before opening. 3. Centrifuge in sealed centifuge cups, open in BSC. If centrifuge is installed in BSC, evaluate for any interference with BSC operation. 4. Always sonicate in BSC, even if using a closed container, to protect from organisms introduced onto external surfaces or from accidentally opened tubes. 5. Use special containment blenders or total containment BSC (class III) when blending: this process is usually done on large volumes of organisms.

caused by these agents have occurred, it has often been because they were being manipulated as unknown organisms in a clinical bacteriology setting routinely operating as a BSL 2 laboratory. But, in fact, in many such laboratories, aerosol-prone procedures are not always performed in the BSC, as required by BSL 2 practices. Unfortunately, it is not uncommon for the laboratorian to dispense with safety concerns for a microorganism once it is suspended in fluid. The increasing use of automated instruments for identification and susceptibility testing further places the laboratorian at risk because most of the operations involve manipulation of organisms suspended in fluids. The laboratory director should evaluate each step of each procedure to detect aerosol-prone conditions that may require use of the BSC. However, rather than perform much of the work in the BSC, it is more likely that many of the aerosol-prone procedures can be better contained. For example, all bacteriology laboratories should use microincinerators for sterilization of bacteriologic loops. It is judged that when such safety measures are instituted in bacteriology laboratories, the occasional encounter with agents such as *Brucella,* although classified as BSL 3, will not result in infection.

Other candidate organisms include those such as *Legionella* and newer agents such as *Rochalimaea* (*Bartonella*) *henselae* or *Rochalimaea quintana.* These organisms involve the prolonged incubation of media, often inoculated with specimens from patients at risk for tuberculosis. It has been documented that *M. tuberculosis* will grow on the media and under the incubation conditions used for the isolation of these agents (7). In fact, routine blood and chocolate agar support the development of microscopic colonies of *M. tuberculosis* within 1 week of incubation. Moreover, Brucella agar and pertussis media are particularly good in supporting the growth of this agent; even the charcoal media used for isolation of *Legionella* yield very tiny colonies of *M. tuberculosis.* Surprisingly, microscopic appearance of colonies on these media precedes the microscopic appearance of *M. tuberculosis* colonies on Middlebrook (7H11) medium. In contrast, mature colonies of several weeks' duration are always smaller on routine media than they are on mycobacterial media. Because they may not progress to form macroscopically visible colonies, their presence will be unrecognized. They will be present in an amplified but undetected form. It is hoped that such information will discourage the laboratory practices of the past that have included "hot looping" agar (i.e., touching the surface of agar that has no apparent colonies on it to cool a loop might produce a substantial aerosol from a macroscopically invisible colony of *M. tuberculosis*). It is further hoped that microincinerator use will be adopted by all routine and special bacteriology laboratories working to isolate the agents that cause pertussis and legionellosis and all newer agents that require prolonged incubation of media and may be sought in specimens that might contain *M. tuberculosis.* A clinical laboratory that first was successful in the isolation of *R. henselae* reports that it is not infrequently receiving organisms referred for confirmation of *Rochalimaea* identity that prove to be *M. tuberculosis* that has been unsuspected by the referring laboratory (16). Thus, bacteriology laboratories are advised to review each procedure that is used in the laboratory and consider whether it might be necessary to carry out the examination of plates subjected to prolonged incubation in the BSC.

References

1. **American Thoracic Society.** 1983. Levels of laboratory services for mycobacterial diseases. *Am. Rev. Resp. Dis.* **128:** 213.
2. **Buesching, W. J., J. C. Neff, and H. M. Sharma.** 1989. Infection hazards in the clinical lab: a program to protect laboratory personnel. *Clin. Lab. Med.* **9:**351–361.
3. **Centers for Disease Control and National Institutes of Health.** 1993. *Biosafety in microbiological and biomedical laboratories,* 3rd ed. HHS Publication No. (CDC)93-8395. U.S. Department of Health and Human Services, U.S. Government Printing Office, Washington, D.C.
4. **Collins, C. H.** 1993. *Laboratory acquired infections,* 3rd ed. Butterworth/Heinemann, Oxford.
5. **DiSalvo, A. F.** 1987. Mycotic morbidity—an occupational risk for mycologists. *Mycopathologia* **99:**147–153.
6. **Garner, J. S., and B. P. Simmons.** 1983. Guideline for isolation precautions in hospitals. *Infect. Control* **4:**245–325.
7. **Gilchrist, M. J. R.** 1994. Unpublished data.
8. **Gilchrist, M. J. R., J. Hindler, and D. O. Fleming.** 1992. Laboratory safety managemen, p. xxix–xxxvii. *In* H. I. Isenberg (ed.), *Clinical microbiology procedures handbook.* American Society for Microbiology, Washington, D.C.
9. **Haley, L. D., and C. S. Callaway.** 1979. *Laboratory methods in medical mycology.* U.S. Department of Health Education and Welfare, U.S. Government Printing Office, Washington, D.C.
10. **Kent, P. T., and G. P. Kubica.** 1985. *Public health mycobacteriology. A guide for the level III laboratory.* U.S. Department of Health and Human Services, Public Health Service, U.S. Government Printing Office, Washington, D.C.
11. **McGinnis, M. R.** 1980. *Laboratory handbook of medical mycology.* Academic Press, Inc., New York.
12. **National Institutes of Health.** 1991. Action under the guidelines. NIH guidelines for research with recombinant DNA molecules. *Fed. Regist.* **56**(138):33174–33183.
13. **Pike, R. M.** 1976. Laboratory-associated infections: summary and analysis of 3921 cases. *Health Lab. Sci.* **13:**104–114.

14. **Strong, B. E., and G. P. Kubica.** 1981. *Isolation and identification of Mycobacterium tuberculosis—a guide for the level II laboratory.* U.S. Department of Health and Human Services, U.S. Government Printing Office, Washington, D.C.
15. **Thompson, D. W., and W. Kaplan.** 1977. Laboratory-acquired sporotrichosis. *Sabouraudia* **15:**167–170.
16. **Welch, D. F. (University of Oklahoma Health Sciences Center).** Personal communication.
17. **Welch, D. F., A. P. Guruswamy, S. J. Sides, C. J. Shaw, and M. J. R. Gilchrist.** 1993. Timely culture for mycobacteria which utilizes a microcolony method. *J. Clin. Microbiol.* **31:**2178–2184.
18. **Wells, W. F.** 1955. *Airborne contagion and air hygiene.* Harvard University Press, Cambridge, Mass.

Protozoa and Helminths

BARBARA L. HERWALDT AND DENNIS D. JURANEK

6

INTRODUCTION

Parasitic diseases are receiving increasing attention in developed countries because of the importance of these diseases in travelers, immigrants, and immunosuppressed persons. Renewed clinical interest in parasitic diseases has stimulated laboratory research. Many persons in research and clinical laboratories may be exposed to parasites and at risk for infection. The risks faced by laboratorians depend on such factors as which parasites, stages of their life cycles, and types of specimens are handled (e.g., clinical specimens, cultures, or infected animals or vectors) as well as the nature of the laboratory work. Because some parasitic diseases can be life-threatening even in immunocompetent persons, the first reaction to a laboratory accident typically is fear. Even persons knowledgeable about parasitic diseases may not know what clinical manifestations to expect when natural modes of transmission are bypassed and how to monitor and treat someone who has had a laboratory accident.

This chapter focuses on the hazards of handling specimens containing viable parasites and the diseases that can result. Table 1 provides information about parasitic diseases that can be laboratory-acquired, as well as about protective measures and diagnostic tests. The protozoa are discussed in more depth in the chapter than are the helminths because the former generally pose more risk to laboratorians. We review reported cases of laboratory-acquired infections and also describe previously unreported cases to illustrate the spectrum of laboratory accidents and resultant illnesses. Inoculum sizes are only occasionally known. The proportion of laboratorians that become infected when working with a given parasite is unknown, as are the risks of transmission after various exposures.

Presumably only a fraction of all laboratory-acquired parasitic infections are recognized and reported. The reported ones may represent the more interesting or unusual end of the spectrum of all infections. Of 4,079 laboratory-associated infections reported by Pike (56) in 1978, 116 (3%) were caused by 17 different parasites; 74 (64%) of the 116 cases had been published. We became aware of previously unreported cases through means such as requests for antiparasitic drugs available through the Centers for Disease Control and Prevention (CDC) Drug Service and telephone consultations provided by CDC personnel after laboratory accidents. Persons who provided information about these cases are acknowledged at the end of the chapter.

Basic principles of personal hygiene, as well as the CDC/National Institutes of Health assessment of laboratory risks, environmental-exposure precautions, and specimen-disposal techniques within

TABLE 1 Parasites to which laboratory workers may be exposed[a]

Organism	Routes of infection	Infectious stages	Protective measures	Diagnostic tests	Common signs and symptoms
Blood and tissue protozoa					
Acanthamoeba	Wound, eye (aerosol?)	Trophozoite, cyst	Gloves, mask, gown, negative-pressure hood	Brain biopsy, culture, corneal scraping (serology?)	Headache, neurologic impairment, skin abscess, pneumonitis, keratitis, conjunctivitis
Babesia	Needle, wound, vector	Intraerythrocytic stages, sporozoite	Gloves, wound and needle precautions	Blood smear, serology, animal inoculation	Fever, chills, fatigue, anemia
Leishmania	Needle, wound, transmucosal, vector	Amastigote, promastigote	Gloves; wound, mucous membrane,[b] and needle precautions	Cutaneous: lesion scraping, biopsy + impression smear, culture, animal inoculation Visceral: serology, biopsy, culture Mucosal: serology, biopsy, culture	Cutaneous: nodules/ulcers Visceral: fever (early), hepatosplenomegaly and pancytopenia (late) Mucosal: nasal symptoms
Naegleria	Transmucosal (nasopharynx), aerosol	Trophozoite (flagellate?) (cyst?)	Gloves, mask, gown, negative-pressure hood	CSF[c] exam and culture	Headache, obtundation, neurologic impairment
Plasmodium	Needle, wound, vector	Intraerythrocytic stages, sporozoite	Gloves, wound and needle precautions	Blood smear, culture, animal inoculation (serology?)	Fever, chills, fatigue, anemia
Sarcocystis	Oral	Tissue cyst	Gloves, handwashing	Stool exam	Gastrointestinal symptoms
Toxoplasma	Oral, needle, wound, transmucosal	Oocyst, tissue cyst, tachyzoite	Gloves; wound, mucous membrane,[b] and needle precautions	Serology (IgM), animal inoculation, tissue culture	Adenopathy, fever, malaise, rash
Trypanosoma (American)	Needle, wound, transmucosal, vector (aerosol?)	Trypomastigote, epimastigote, amastigote	Gloves; wound, mucous membrane,[b] and needle precautions	Blood smear, culture, biopsy, animal inoculation, xenodiagnosis, serology	Swelling/redness at inoculation site, fever, rash, adenopathy, ECG[d] changes
Trypanosoma (African)	Needle, wound, transmucosal, vector (aerosol?)	Trypomastigote	Gloves; wound, mucous membrane,[b] and needle precautions	Blood smear, culture, biopsy, animal inoculation, CSF[c] exam (serology?)	Swelling/redness at inoculation site, fever, rash, adenopathy, headache, fatigue, neurologic signs
Intestinal protozoa					
Cryptosporidium	Oral, transmucosal (aerosol?)	Oocyst, sporozoite	Gloves, handwashing, mucous membrane precautions[b]	Stool exams with concentration and special stains, immunodiagnostic test for antigen in stool	Diarrhea, abdominal pain
E. histolytica	Oral	Cyst	Gloves, mask, handwashing	Stool exams with concentration, serology (for invasive disease)	Diarrhea (may be bloody), abdominal pain

Giardia lamblia (*intestinalis*)	Oral	Cyst	Gloves, mask, handwashing	Stool exams with concentration, immunodiagnostic test for antigen in stool	Diarrhea, abdominal cramps, nausea, flatulence, bloating
Helminths					
Ascaris	Oral	Egg	Gloves, mask, handwashing	Stool exam	Cough, fever, pneumonia; abdominal cramps, diarrhea or constipation[e]
	Skin contact	Worm (allergen)	Gloves	History of exposure	Hypersensitivity reactions, especially pruritus
Enterobius	Oral	Egg	Gloves, mask, handwashing, nail cleaning	Scotch tape test	Perianal pruritus
Hookworm	Percutaneous[f]	Larva	Gloves, gown, handwashing	Stool exam	Animal species: creeping eruption or "ground itch" (skin) Human species: diarrhea, abdominal pain, anemia[e]
Hymenolepis nana	Oral	Egg	Gloves, mask, handwashing	Stool exam	Abdominal pain, diarrhea
Schistosoma	Percutaneous[f]	Cercaria	Gloves, gown	Stool exam, serology	Dermatitis, fever, hepatosplenomegaly, adenopathy
Strongyloides	Percutaneous[f]	Larva	Gloves, gown, caution when handling specimens	Stool exams; motile larvae may be seen in wet preparations	Cough and chest pain followed by abdominal pain, cramping[e]
Taenia solium	Oral	Egg, cysticercus	Gloves, hood, handwashing, specimen disposal[g]	Cysticercosis: serology, brain scan, soft tissue X ray Worm: stool exams	Cysticercosis: neurologic symptoms Worm: usually asymptomatic but may cause vague abdominal symptoms
Trichinella	Oral	Larva	Gloves, mask, handwashing	Serology, muscle biopsy	Abdominal and muscle pain[e]
Trichuris	Oral	Egg	Gloves, mask, handwashing	Stool exam	Abdominal pain, tenesmus[e]

[a]The parasites listed should be handled in accordance with biosafety level 2 standards.

[b]Use of a hood provides optimal protection against exposure of the mucous membranes of the eyes, nose, and mouth.

[c]CSF, cerebrospinal fluid.

[d]ECG, electrocardiogram.

[e]Symptoms are unusual unless infecting inoculum is heavy, which would be unlikely in most laboratory-acquired infections.

[f]Parasite capable of penetrating intact skin.

[g]*Taenia solium* eggs are highly resistant to chemical disinfectants and are not killed by routine sewage treatment.

microbiology laboratories, are outlined elsewhere in this book (Appendix I). The parasites discussed in this chapter should be handled under the containment conditions known as biosafety level 2, which are described elsewhere in this book. Laboratorians handling specimens or animals harboring parasites may also be exposed to nonparasitic hazards such as bacterial or viral agents or bites. Persons working with fresh (versus fixed) human stool specimens should be adequately immunized against polio, and persons working with clinical blood specimens should be immunized against hepatitis B. Immunizations for parasitic infections are not yet available.

Relevant topics that are discussed elsewhere include nosocomial transmission of parasitic diseases (41, 42) and the clinical management and diagnostic evaluation of persons suspected or confirmed to have parasitic infections (3, 46, 71, 77). Also, the Division of Parasitic Diseases at the CDC can provide consultation (404-488-7760). For questions about the availability of antiparasitic drugs, contact the CDC Drug Service (404-639-3670). For emergency questions during evenings, weekends, and holidays, call 404-639-2888.

PROTOZOAN INFECTIONS

Blood and Tissue Protozoa

Babesia species

In nature, babesiosis is transmitted from reservoir hosts to humans by the bite of infected *Ixodes* ticks (e.g., *Ixodes dammini*). *Babesia microti*, which is the most common etiologic agent of zoonotic babesiosis in the northeastern and midwestern United States, can infect and cause illness in both spleen-intact and asplenic persons. In contrast, other *Babesia* species (e.g., *B. divergens* in Europe) reportedly cause illness primarily in persons who are asplenic or otherwise immunocompromised. *B. microti* infections in young, otherwise healthy, spleen-intact persons may be asymptomatic or resemble a mild case of malaria. However, immunocompromised or asplenic persons infected with any *Babesia* species may become severely ill.

Although no cases of laboratory-acquired babesiosis have been reported, such infections could be acquired through contact with the tick vector or with blood from infected animals or humans. Because ticks are more easily controlled in the laboratory than are mosquitoes, the risk of becoming infected through contact with infected ticks is relatively low. Persons handling infected animals or blood specimens should guard against accidentally inoculating themselves with infected blood; cases of babesiosis acquired by blood transfusion have been reported (49, 69, 79).

The diagnosis of babesiosis generally is made by examining Giemsa-stained thick and thin blood smears. Because *Babesia* and malaria parasites can resemble each other, the slides may require examination by a reference laboratory. Pre-exposure and acute- and convalescent-phase serologic specimens may help confirm the diagnosis of acute *Babesia* infection; an indirect immunofluorescent antibody (IFA) test is available through the CDC. Hamster inoculations may be of diagnostic value for low-grade *B. microti* infections in which parasites are not demonstrable on blood smears. The infection may require from 1 to 8 weeks to become patent in hamsters. Persons with mild illness may not require therapy. Persons requiring therapy should be treated with clindamycin and quinine.

Leishmania species

Leishmaniasis, which is caused by various species of the genus *Leishmania*, is transmitted in nature from reservoir hosts to humans by the bite of infected female phlebotomine sand flies. The promastigote form of the parasite is found in sand fly vectors and cultures and the amastigote form in macrophages of mammalian hosts. The main clinical syndromes are visceral leishmaniasis, which affects internal organs; cutaneous leishmaniasis, which causes skin lesions; and mucosal leishmaniasis, which may affect nasal, oral, and pharyngeal mucosae.

Leishmaniasis can be acquired in the laboratory by several means. One is inadvertent contact with an infected sand fly; containment measures for infected flies should be strictly followed. Another means of infection is by handling cultured parasites or specimens from infected animals or persons. The parasite can infect persons who have needlestick injuries or pre-existing skin abrasions, which may be small and inapparent. Blood specimens from infected animals or persons should be handled with care (i.e., biosafety level 2 precautions, especially the use of gloves and mechanical pipettes) even though fewer parasites generally are found in the bloodstream than in infected tissues.

Laboratory-acquired infections with *Leishmania tropica*, *L. braziliensis*, and *L. donovani* have been reported (13, 18, 21, 66, 67, 72). A student who sustained a needlestick injury while passaging *L. tropica* amastigotes in mice developed an erythematous nodule at the inoculation site 4 weeks later; the nodule ulcerated in 2 weeks (66). The student developed cellular immunity to leishmanial antigens, but no organisms were demonstrated histologically or by culture of a biopsy specimen obtained at 12 weeks.

A student who passaged infective suspensions of *L. braziliensis* amastigotes in hamsters without wearing gloves developed a skin ulcer on a finger (67). The diagnosis of leishmaniasis was parasitologically confirmed by demonstrating amastigotes in an impression smear of a biopsy specimen from the lesion and by inoculating a hamster with biopsy material. Also, the student's Montenegro skin test result was positive, and anti-*Leishmania* antibody was detected with an IFA test.

Of the four laboratorians reportedly infected with *L. donovani* (13, 18, 21, 72), only one, a person in China who accidentally contaminated his fingers and oral mucosa (during mouth-pipetting) with blood from infected squirrels (13), developed symptoms suggestive of visceral involvement. One of the other three laboratorians punctured her hand with a needle containing amastigotes (5×10^8 amastigotes/ml, Humera strain) in a suspension of spleen tissue from a hamster (21). The strain had been passaged in hamsters for more than a decade. Three weeks after the accident, she developed swelling and intermittent erythema distal to the inoculation site. A nodule was noted at the inoculation site 4 weeks later, and regional adenopathy was detected the following week. Amastigotes were noted in a skin biopsy specimen obtained at this time, and a culture of the specimen was positive. Anti-*Leishmania* antibody was detected by the CDC's complement-fixation test. Results of bone marrow microscopy and culture were negative.

While recapping a needle, a physician accidentally inoculated himself with amastigotes from a hamster infected with *L. donovani;* the isolate (MHOM/SU/00/S3) had been maintained in laboratory animals for decades (18). He developed a nodule at the inoculation site 6 months later but did not develop adenopathy or systemic symptoms. Amastigotes were seen in a biopsy specimen from the nodule, a culture of the specimen was positive, and the patient's lymphocytes had an accentuated response to leishmanial antigen.

A technician who had been bitten periodically on his fingers by laboratory animals infected with *L. donovani* developed a swollen finger and regional adenopathy (72). Whether he became infected through the bites or through subsequent contamination of the wounds is not known. Culture of a biopsy specimen from a lymph node was positive, and amastigotes were noted on an impression smear of the specimen. No parasites were found in smears or in a culture of a bone marrow biopsy specimen.

We are aware of three unreported cases from the 1980s of laboratory-acquired leishmaniasis in the United States, one of which resulted in mucosal disease, and also of six laboratorians with needlestick injuries (1981 to 1993) who did not develop leishmaniasis. The person who ultimately developed mucosal disease had scratched herself with a needle containing amastigotes of *L. amazonensis* (Maria strain) and developed a local erythematous nodule within 3 months. Culture of a biopsy specimen from the lesion was positive. The patient was treated with what now would be considered an inadequate course of the pentavalent antimonial sodium stibogluconate (Pentostam). The lesion regressed but recurred, and the patient was treated again with the drug and also with heat. Although the local lesion healed, she developed mucosal leishmaniasis several years later. The second of the three unreported cases resulted from accidental inoculation with *L. guyanensis* by a laboratorian injecting mice. The person noted itching at the inoculation site 3 months later, a skin ulcer developed during the next 2 months, and a culture of a biopsy specimen was positive. The third case occurred in a person who became infected with *L. tropica* after accidentally inoculating himself while injecting an animal. A leishmanial nodule developed at the inoculation site 3 weeks later.

Our approach to managing the care of persons who have had laboratory accidents is to monitor for clinical and laboratory evidence of leishmaniasis rather than to treat presumptively. Skin lesions that develop at or near the site of exposure should be evaluated. Depending on the nature of the lesion, all or some of the following may be appropriate: lesion scrapings, aspirates, and biopsy with impression smear (51). The specimens should be stained with Giemsa and examined for amastigotes. Aspirates and biopsy specimens are useful for in vitro culturing; culture medium is available commercially and through the Division of Parasitic Diseases at the CDC. Animal inoculation, DNA probe analysis, and monoclonal antibody analysis can also be used to confirm the diagnosis. Persons who may have been infected with *L. donovani, L. chagasi,* or *L. infantum* should be monitored serologically every month for 6 to 12 months; IFA and complement-fixation assays are performed at the CDC. If seroconversion is noted or clinical illness (e.g., unexplained fever) develops, further evaluation (e.g., bone marrow aspiration) may be indicated.

Persons requiring therapy for leishmaniasis generally are treated with a pentavalent antimonial compound, although other antileishmanial agents may have merit in some situations. Sodium stibogluconate, the antimonial available in the United States, can be obtained through the CDC

Drug Service; it is administered parenterally, generally for at least several weeks (30).

Naegleria fowleri and *Acanthamoeba* species

Naegleria fowleri and *Acanthamoeba* species are free-living amebae that cause life-threatening infection of the central nervous system (CNS) (44, 74). Infection with *Naegleria fowleri* is typically acquired by swimming in fresh water. The parasite invades the CNS through the nasal mucosa and the cribriform plate and causes primary amebic meningoencephalitis, a disease that generally is rapidly fatal. *Acanthamoeba* species cause more subacute or chronic infection, either granulomatous amebic encephalitis (which may result from hematogenous dissemination in the context of pulmonary or skin lesions) or keratitis (in contact lens wearers or persons with corneal abrasions). Although no cases of laboratory-acquired infections with these organisms have been reported, the possibility of becoming infected from inhaling infectious aerosols should be considered. Immunologically impaired persons should not be permitted in areas where these amebae are cultured. Infections with these parasites are difficult to treat regardless of the host's immune status.

Plasmodium species

Malaria is transmitted in nature by the bite of infected female anopheline mosquitoes. The four species of this intraerythrocytic parasite that most commonly are associated with human infection are *Plasmodium falciparum, P. vivax, P. ovale,* and *P. malariae.*

Malaria can be acquired in the laboratory by several mechanisms. One of the most common is inadvertent contact with an infected vector from a mosquito colony, but such exposure is often unrecognized. Strict containment measures should be followed for infected mosquitoes. Mosquito-borne (sporozoite-induced) infections with *P. cynomolgi* (14, 15, 19, 23, 50) and *P. falciparum* (12, 76) have been reported. Also, we are aware of unreported infections with *P. cynomolgi* (one), *P. falciparum* (one), and *P. vivax* (three). Some infections with *P. cynomolgi* occurred before it was recognized that this species, which naturally infects Asian monkeys, could be transmitted to humans by mosquitoes. Persons infected with *P. cynomolgi* typically have low-level or submicroscopic parasitemias; the diagnosis can be confirmed by injecting the patient's blood into a monkey and monitoring the animal for parasitemia.

Malaria can also be acquired through accidental contact (through needlestick injuries or open wounds) with infected blood from human or animal hosts or with cultured parasites, thus bypassing the hepatic stage of the parasite's life cycle. Four cases of laboratory-acquired *P. falciparum* infections acquired through accidental contact with infected blood have been reported and had incubation periods ranging from 4 to 17 days: a student with skin excoriations became ill 4 days after handling infected blood and developed oliguria and altered mental status several days later (54); a laboratorian became ill 14 days after accidentally stabbing his finger while preparing a blood smear (5); a prosector who pricked his finger during an autopsy became ill 15 days later (33); and a laboratorian working with cultured parasites became ill 17 days after breaking a capillary hematocrit tube and receiving a minor puncture wound (38).

The possibility of malaria should be considered for persons with unexplained febrile illnesses who may have been exposed in laboratory settings to malaria parasites. Giemsa-stained thick and thin blood smears should be examined at frequent intervals over several days. An IFA assay for antibodies to *Plasmodium* species is performed at the CDC, but examining blood films usually is more helpful than serologic analysis early in infection. The CDC can be contacted for assistance in confirming the diagnosis. When prescribing treatment for confirmed cases, the identity of the infecting species and its drug susceptibilities should be considered.

Pneumocystis carinii

Pneumocystis carinii, which is an important cause of pneumonia in immunocompromised persons (e.g., AIDS patients), is generally classified as a protozoan, but some investigators prefer to consider it a fungus. *Pneumocystis* is not discussed in detail here because no laboratory-acquired infections have been reported, and infections in persons with normal immune systems apparently are asymptomatic or mild. The airborne route appears to be the main mode of transmission (35, 36). Immunocompromised persons should avoid working in laboratories where this organism is handled.

Sarcocystis species

Various species of *Sarcocystis* are infective for humans. Humans are the definitive host (the host for the sexual stage) for *S. hominis* and *S. suihominis,* for which the intermediate hosts (the host for the asexual stage) are cattle and swine, respectively. Persons working with raw beef or pork should guard against accidental ingestion via contaminated fingers. Persons infected with these species generally are asymptomatic but may have various gastrointestinal symptoms. The diagnosis is established by finding oocysts in stool specimens. Unknown species of sarcocysts (the asexual stage of the parasite) have been found in biopsy specimens of cardiac or skeletal muscle from asymptomatic persons; the definitive host(s) is unknown. One case of laboratory-acquired infection with *Sarcocystis* has been reported, but no details were provided (55). No specific therapy is available for treating infection with any of the *Sarcocystis* species.

Toxoplasma gondii

Toxoplasma gondii, the etiologic agent of toxoplasmosis, is transmitted in nature to persons who ingest tissue cysts in infected undercooked meat or sporulated oocysts from infected cat feces. The possibility of swallowing inhaled oocysts has been suggested (73). Congenital transmission also occurs.

The clinical spectrum of toxoplasmosis ranges from the absence of symptoms, to a syndrome of fever and lymphadenopathy, to diffuse organ system involvement. Because persons who are immunocompromised or infected congenitally may have severe disease, immunocompromised persons and women of child-bearing age without demonstrable antibodies to *T. gondii* should avoid working with live *Toxoplasma.* Seropositive persons, including pregnant women, can continue to do such work.

Laboratory workers can become infected through handling human or animal tissue, feline fecal specimens, or cultured organisms. All *Toxoplasma* isolates should be considered pathogenic for humans even if they are avirulent for mice (16). Procedures for separating oocysts from cat feces and for infecting mice have been outlined; fecal flotations should be performed before oocysts sporulate (16). Instruments and glassware contaminated with oocysts should be boiled or autoclaved; oocysts are not readily killed by exposure to chemicals or the environment (16).

Many cases of laboratory-acquired toxoplasmosis have been reported (20, 22, 28, 39, 48, 52, 59, 62, 68). Infections have resulted from ingestion of organisms, skin punctures, and mucous membrane contact. Before oocysts were recognized to be extraordinarily hardy, laboratorians became infected when handling feline feces. Seven cases of toxoplasmosis circumstantially attributed to ingesting oocysts (M-7741 strain) have been reported (48). The persons were essentially asymptomatic, and the diagnoses were confirmed serologically. A pathologist supervising an autopsy on someone with cerebral toxoplasmosis became acutely ill 2 months later with parasitologically confirmed toxoplasmosis, but exposures during the autopsy were not detailed in the report (52).

In 1959, Rawal (59) described his own laboratory-acquired infection and reviewed 17 reported cases. The probable mode of transmission was unknown for 10 of the 18, including for himself; he suspected that organisms contaminated his skin while he performed the Sabin-Feldman dye test. Four of the 18 persons sustained needlestick injuries, and three splashed infective material onto their faces or into their eyes. The incubation period ranged from 3 to 9 days for those with needle pricks and was 3 days for two of those with splashes. One person may have become infected by the bite of an infected rabbit; *T. gondii* has been isolated from rabbit and mouse saliva (37). The most commonly reported symptoms and signs were fever, headache, malaise, lymphadenopathy, and rash, but two persons were asymptomatic. Three persons developed signs of encephalitis, two of whom also developed myocarditis. One who developed both conditions died (68); effective treatment regimens for toxoplasmosis had not yet been established.

A technician who scratched a finger with a needle contaminated with *T. gondii* (RH strain) developed fever, chills, and headache 5 days later (22). Three days later, she noted a lesion at the inoculation site. When she was hospitalized the next day, an upper body rash and regional and cervical adenopathy were noted; dye test antibodies were detected, and *T. gondii* was isolated from blood.

Another researcher developed parasitologically confirmed toxoplasmosis after scratching himself with a contaminated (RH strain) needle (39). He developed myalgias 6 days later and then malaise, headache, generalized lymphadenopathy, a petechial rash, and fever. On day 9 (the day of hospitalization), mice were inoculated with his blood, and *T. gondii* was later recovered from the mice. Dye test antibodies were first detectable 2 days after admission.

A person using a faulty syringe became ill 9 days after splashing one side of her face with peritoneal exudate from a mouse infected with *T. gondii* (RH strain) (20). She developed headache, malaise, fever, unilateral conjunctivitis, pharyngitis, and cervical lymphadenopathy. Dye test antibodies became detectable 2 weeks after the accident, and hemagglutinating antibodies were detectable 2 weeks later.

A laboratorian who squirted a mixture of saline and tachyzoites (BK strain) from a faulty syringe into his left eye developed edema of that eye and of the left side of his face 9 days later (28). Lymphadenopathy was noted on day 11, and the patient became seropositive on day 15. Another person described in the same report accidentally injected parasites (BK strain) into her thumb. Three days later, her hand was painful, and regional lymphadenopathy was noted. The patient became seropositive on day 14. The dye test was not performed for either patient.

In a report of four cases of probable laboratory-acquired toxoplasmosis, two persons recalled accidents (a needle puncture and a needle scratch) and two did not; one of the latter was asymptomatic, but routine yearly screening indicated that the laboratorian had developed antibody to *T. gondii* (62).

The diagnosis of toxoplasmosis can be confirmed serologically, by mouse inoculation, or by tissue cell culture. The most widely used screening tests for detecting *Toxoplasma*-specific immunoglobulin G (IgG) antibodies are the IFA assay and the enzyme immunoassay. A single test demonstrating elevated levels of IgG may reflect a previous infection and therefore is of little value for diagnosing acute infection. If acute infection is suspected and the IgG screening test is positive, an IgM-capture enzyme immunoassay test may be helpful. Although a high titer of *Toxoplasma*-specific IgM suggests that the acute infection occurred within the past several months, IgM may persist for 18 months or more (78).

Persons with evidence of organ involvement or persistent severe symptoms should be treated for weeks to months with pyrimethamine and either sulfadiazine or trisulfapyrimidines in conjunction with folinic acid. However, prophylactic treatment may be indicated even without evidence of infection because the risk of morbidity from toxoplasmosis may be greater than that associated with therapy.

Trypanosoma cruzi

Trypanosoma cruzi, the etiologic agent of Chagas' disease (American trypanosomiasis), is transmitted in the Americas from reservoir hosts to humans by reduviid bugs. Natural transmission occurs when bug feces containing infective metacyclic trypomastigotes contaminate a wound (e.g., the bug's bite wound) or mucous membranes of the eyes, nose, or mouth. After invading host cells, parasites replicate as amastigotes and differentiate into trypomastigotes, which are released when infected host cells rupture. Circulating trypomastigotes may invade other host cells or be withdrawn by an insect vector. Infected persons may be asymptomatic for years before developing cardiac or gastrointestinal manifestations of chronic Chagas' disease.

Laboratorians can become infected by being exposed to the excreta of infected reduviid bugs, by handling culture or blood specimens from infected persons or animals, and possibly by inhaling aerosolized organisms (80). *Trypanosoma cruzi* can infect persons through needlestick injuries or pre-existing skin abrasions or by crossing intact mucous membranes; mice have been experimentally infected by applying parasites to the conjunctiva or the oral mucosa (40). Safety precautions for work with live organisms have been outlined (8, 24, 34).

In 1987, Brener (9) reported being aware of more than 50 cases of laboratory-acquired Chagas' disease, including one in a person who was not treated and died from myocarditis. In an earlier review of 45 cases, Brener noted that infected blood was the source of infection for 15 of the 20 cases with a known source; the most common accident was needle puncture (8). For example, one person was infected when a syringe containing infective blood (Y strain; 800,000 trypomastigotes in 0.4 ml of blood) fell on his foot. Fever, malaise, and regional lymphadenopathy were noted 12 days later. A chagoma (a lesion characterized by erythema, warmth, and swelling) was noted at the inoculation site 4 days later, and trypomastigotes were found in blood smears. Another person in Brener's report had oral mucosal contact with the organism while pipetting.

The course of laboratory-acquired Chagas' disease presumably varies with the mode of transmission and the virulence of the infecting strain. The course in a technician who stuck his thumb with a needle contaminated with blood from an infected (CL strain) mouse has been described in detail (31). Interesting features were that the patient was well for 24 days postinoculation, developed a chagoma proximal to the inoculation site, had high fever (42°C) and relative bradycardia, and never had demonstrable parasites in smears of concentrated blood or in biopsy specimens from a tender axillary lymph node. Other symptoms and findings included headache, malaise, a generalized rash, and leukopenia. Mouse inoculation and serologic techniques (enzyme-linked immunosorbent assay and IFA) simultaneously yielded positive results nearly 5 weeks after the accident; the mouse had been inoculated with the patient's blood 1 week earlier, at which time antibody had been undetectable.

The courses of three persons with laboratory-acquired Chagas' disease have been described and reported (57). Two had symptoms suggestive of meningoencephalitis, and one developed signs of myocarditis; other symptoms and signs included fever, adenopathy, generalized rash, splenomegaly, and facial edema. Of the two who recalled needlestick injuries, one had demonstrable parasitemia on day 8 and the other on day 14.

Other cases of laboratory-acquired Chagas' disease have been reported. A person with conjunctival contact with feces from an infected reduviid bug developed xenodiagnosis-confirmed Chagas' disease (29). A scientist with a needlestick injury developed a chagoma and other unspecified signs of acute Chagas' disease 16 to 18 days after becoming infected with a strain of *T. cruzi* from Brazil (25). A microbiologist developed a diffuse rash but not a chagoma approximately 1.5 weeks after spilling a solution of trypomastigotes (Tulahuen strain) onto slightly abraded skin (4). Two days later, he developed a systolic murmur, a pericardial friction rub, an enlarged heart, and T-wave changes. His course may have been affected by concomitant bacteremia and steroid therapy. Two and a half weeks after the accident, trypomastigotes were seen in smears of his blood and in blood from a mouse inoculated with the patient's blood 6 days earlier.

A person who operated on the peritoneal cavity of an infected (Tulahuen strain) mouse without wearing gloves developed erythema along a cuticle in 2 days and fever and myalgias in 4 days (11). When hospitalized 12 days after the incident, he had splenomegaly and generalized lymphadenopathy. Findings on an electrocardiogram were consistent with myocarditis, and trypomastigotes were seen in blood smears.

A technician working with the Tulahuen strain developed a chagoma on her thumb but did not recall having had a laboratory accident (75). She subsequently developed weakness, headache, fever, night sweats, regional lymphadenopathy, transient pedal edema, intermittent tachycardia, and nonspecific T-wave changes on an electrocardiogram. Blood cultures obtained 7 days after the skin findings were noted were positive for *T. cruzi*, and seroconversion was later observed.

We are aware of three unreported cases of laboratory-acquired Chagas' disease in the United States and of 12 laboratorians who had laboratory accidents (2 in the 1970s, 7 in the 1980s, and 3 in the 1990s) but did not develop Chagas' disease. In 1975, a researcher who had been bleeding mice infected with *T. cruzi* recalled being bitten on his finger by an uninfected [sic] mouse. He developed fever and a lesion on his finger the following day and an enlarged axillary lymph node the next day. A biopsy of the node was obtained 13 days after the incident, and amastigotes were found in the specimen. In 1981, a graduate student injecting mice with trypomastigotes (Brazil strain) was grazed with a contaminated needle after a mouse kicked the syringe. A local skin lesion was noted 10 days later. He was hospitalized 8 days later, at which time he also had fever and a headache. Examinations of blood, buffy coat, and an impression smear of a skin biopsy specimen were negative for *T. cruzi*, but xenodiagnosis, which was performed weeks later, was positive. In 1990, a technician working with *T. cruzi* had symptoms suggestive of a viral illness. Although she did not recall having had a laboratory accident, trypomastigotes were found when wet mounts of the buffy coat of a blood specimen were examined, and an enlarged epitrochlear node was subsequently noted. The diagnosis of Chagas' disease in this case was supported serologically and by the polymerase chain reaction.

We recommend monitoring laboratorians who have had low-risk accidents for clinical and laboratory evidence of infection (see protocol below) rather than treating them presumptively. Close monitoring for signs of acute Chagas' disease is important because treatment during later stages of the disease generally is not effective. Although prompt institution of short-course (i.e., 8 to 10 days) therapy after high-risk accidents has been recommended (8, 9), the efficacy and optimal duration of prophylactic regimens have not been established in controlled clinical trials. The possibility that short-course therapy could suppress parasitemia and mask indicators of infection should be considered if such therapy is given. Insufficient data currently are available on which to base definitive recommendations for prophylactic therapy with the relatively toxic drugs nifurtimox (available through the CDC Drug Service) and benznidazole (not available in the United States). The treatment course of nifurtimox recommended by the CDC for adults with confirmed acute Chagas' disease lasts 120 days and is almost always accompanied by some drug toxicity; persons confirmed to be infected may be more willing than persons being treated prophylactically to continue therapy despite toxicity.

The postexposure monitoring protocol we suggest is as follows. (i) The laboratorian should be advised to report immediately any swelling or redness at or near the site of the exposure and any rash. His or her temperature should be monitored every day for 4 weeks, and febrile illnesses during the next 6 months should be reported. (ii) Serum specimens should be examined for antibody to *T. cruzi*. Both IFA and complement-fixation assays are available through the CDC. Serum should be obtained immediately postexposure for comparison with subsequent specimens and then every week for the first 8 weeks or until seroconversion is noted, every month for the next 4 months, and whenever symptoms or signs suggestive of Chagas' disease are noted. (iii) The patient's blood should be monitored for parasitemia twice a week for 4 or more weeks and whenever symptoms or signs suggestive of Chagas' disease are noted. Fresh blood should be centrifuged (e.g., in a capillary hematocrit tube), and wet mounts of the buffy coat should be examined as soon as possible for motile trypomastigotes. Also, thin smears of the buffy coat should be fixed and stained with Giemsa to provide a permanent record of the findings. Other means of parasitologically confirming the diagnosis include demonstrating the organism histologically, cultivating it in vitro from blood or biopsy specimens, and inoculating animals. Although xenodiagnosis using *T. cruzi*-free reduviid bugs is a sensitive technique for detecting circulating parasites in acute infections, the capability to do xenodiagnosis is not readily available in the United States, and infections usually can be diagnosed by other means. Use of the polymerase chain reaction for diagnostic purposes is investigational.

Trypanosoma rhodesiense and *gambiense*

Trypanosoma (brucei) rhodesiense and *T. (brucei) gambiense,* the etiologic agents of East and West African trypanosomiasis, respectively, are transmitted in portions of sub-Saharan Africa by tsetse flies. Unlike American trypanosomes, African trypanosomes multiply in the bloodstream of their mammalian hosts. East African trypanosomiasis generally follows a more acute course than the West African disease and is characterized by early invasion of the central nervous system (CNS).

Laboratory-acquired African trypanosomiasis is generally caused by contact (e.g., through a break in the skin) with infective blood or tissue from animal or human specimens. A technician became infected with *T. (brucei) gambiense* by scratching his arm with an infected needle (17). When evaluated 1 week later, he had a large chancre (the inflammatory primary lesion) at the inoculation site, fever, headache, anorexia, and fatigue. Many trypanosomes were found in blood smears.

A technician inoculating mice stuck her hand with a needle and thereby became infected with a strain of *T. (brucei) gambiense* (FEO ITMAP-1893) maintained in laboratory animals for decades (60). She became febrile 8 days later; developed erythema, warmth, and swelling of her hand 2 days thereafter; and had splenomegaly and an axillary lymph node noted on examination the following day. Laboratory abnormalities included leukopenia and thrombocytopenia. Trypanosomes were isolated from the chancre and from a blood specimen concentrated on diethylaminoethyl cellulose. Seroconversion was noted by IFA testing 18 days after the accident.

A student separating trypanosomes from the blood of infected rats by column chromatography became infected with *T. (brucei) rhodesiense* from a stock thought to be *T. (brucei) brucei* and therefore not infectious for humans (27, 64); the separation procedure resulted in a concentrated suspension of organisms (about 10^8/ml). Although he did not recall having had an accident, he had many abrasions on his hands through which organisms may have been inoculated. He developed an area of erythema and swelling on a finger, a generalized rash, arthralgias, fever, cervical adenopathy, splenomegaly, vomiting, diarrhea, headache, tinnitus, fatigue, and disorientation. One trypanosome was found in 400 microscopic fields of a Giemsa-stained thick blood smear. His serum IgM level increased markedly after trypanosomiasis had already been diagnosed.

African trypanosomiasis is diagnosed by identifying trypanosomes in a chancre, lymph node aspirates, peripheral blood, cerebrospinal fluid (CSF), or bone marrow aspirates. The ease of finding trypanosomes in various tissues and fluids depends on the infecting species (*gambiense* versus *rhodesiense*) and the stage of infection (hemolymphatic versus CNS). Blood concentration and staining techniques similar to those used for diagnosing Chagas' disease are appropriate. Identifying the morular cell of Mott (a modified plasma cell with large eosinophilic inclusions) in the CSF is pathognomonic of CNS invasion. Animal inoculations (for *T. rhodesiense*) and in vitro cultivation may be used to isolate the parasite. Serologic tests are more useful in epidemiologic surveys than for diagnosing the disease in individual patients. Use of the polymerase chain reaction to diagnose infections is investigational.

The hemolymphatic stage of infection is treated with suramin (available through the CDC Drug Service) or pentamidine isethionate. Infections that have progressed to involve the CNS are treated with the arsenical melarsoprol (available through the CDC Drug Service). Difluoromethylornithine is effective for treating both the hemolymphatic and CNS stages of *T. (brucei) gambiense* infection.

Intestinal Protozoa

The main intestinal protozoa of concern to laboratory workers are *Entamoeba histolytica, Giardia lamblia (intestinalis)*, and *Cryptosporidium parvum,* and to a lesser extent, *Isospora belli.* The cysts or oocysts that are excreted in feces are immediately infectious via the fecal-oral route. Because protozoa multiply within the host, ingesting a small inoculum can result in illness (63). Infections with these protozoa can be asymptomatic or associated with various gastrointestinal symptoms, which may be remitting and relapsing. Lower gastrointestinal symptoms generally predominate. Persons infected with *E. histolytica* may have alternating periods of diarrhea and constipation; invasive amebiasis may be associated with fever and bloody stools. Immunocompromised persons infected with *Cryptosporidium* may have chronic, life-threatening diarrhea. *Dientamoeba fragilis* and *Balantidium coli,* which cause asymptomatic or mildly symptomatic infections, pose relatively little risk to laboratorians.

Routine precautions for work with fecally contaminated material, including careful handwashing after handling specimens, should be observed by laboratory personnel. Commercially available iodine-containing disinfectants are effective against *E. histolytica* and *Giardia* when used as directed, as are high concentrations of chlorine (1 cup of bleach per gallon of water; 1:16 vol/vol). Environmental contamination with *Cryptosporidium* oocysts is problematic, especially for persons working with infected calves shedding large numbers of organisms. Although the oocysts are destroyed by freezing and also by moist heat (warming to 55°C over 15 to 20 min) (1), they are highly resistant to chemical disinfection (7, 10, 53), as are oocysts of other coccidian parasites. Solutions that destroy oocyst viability include 5% ammonia (10), 10% formol saline (10), and 3% hydrogen peroxide (7), none of which are innocuous. A 10- to 15-min contact time with commercial bleach (5% chlorine) may be useful for decontaminating surfaces (70). Contaminated skin should be thoroughly washed; no disinfectant effective against *Cryptosporidium* is safe for skin use. Contaminated clothing and equipment should be autoclaved.

Laboratory-acquired infections with intestinal protozoa are probably relatively common but rarely reported in part because of the relative ease of diagnosis and treatment. Pike (55) reported 23 cases of amebiasis and 2 of giardiasis but did not provide any details. The coccidian parasite *Isospora belli* infected a laboratory technician who examined many stool specimens from an infected patient; the technician became ill approximately 1 week after the first specimens were examined (47). Two researchers feeding a rabbit a capsule containing *Isospora* oocysts were sprayed on their faces with infectious material when the rabbit regurgitated the material and vigorously shook its head; the researchers became ill 11 and 12 days later (26).

Although cryptosporidiosis is a well-recognized occupational hazard of persons exposed to naturally infected calves (2, 43, 61), it has also been reported among persons exposed to experimentally infected animals (6, 32, 58). Five veterinary students with direct (four) or indirect (one) contact with experimentally infected calves became ill 6 to 7 days later and had diarrhea for a median of 5 days (range, 1 to 13); one was hospitalized (58). Also, oocysts were found in a stool specimen from an infected student's spouse. In another instance, a researcher developed gastrointestinal symptoms 5 days after a rabbit, which was infected with oocysts through a gastric tube, coughed droplets of inoculum onto his face when he was removing the tube (6). Oocysts were found in the researcher's stool the day after he became ill. A veterinary scientist wearing extensive protective clothing developed flulike symptoms 7 days after checking the position of a gastric tube in an infected calf by smelling for gastric odor; she did not have any other recognized exposures to *Cryptosporidium* (32). She developed gastrointestinal symptoms 10 days after this exposure, and a stool specimen on day 16 had oocysts. Although airborne transmission of this small (4 to 6 μm) organism is plausible, aerosolization of oocysts from the rumen of a calf is speculative.

Infections with intestinal protozoa are diagnosed by examining stool specimens. Because organisms may be shed intermittently and in low numbers, examining multiple stools collected on different days is recommended. Stools should be preserved in 10% Formalin and in polyvinyl alcohol or alternative fixatives. The specimens should be examined by a concentration technique and with a permanent stain such as trichrome. Immunodiagnostic tests for detecting *Giardia* antigen in stool are available. Detecting *Cryptosporidium* requires special stains (e.g., acid fast) or immunodiagnostic tests for antigen in stool. *Cryptosporidium* and *Isospora,* both of which are acid fast, are distinguishable by size and shape.

Asymptomatic persons passing only *E. histolytica* cysts can be treated with one of the so-called lumenal agents: iodoquinol, paromomycin, or diloxanide furoate; the latter is available through the CDC Drug Service. Symptomatic persons should be treated first with metronidazole or tinidazole (not available in the United States) and then with one of the lumenal agents. Persons with giardiasis are treated with quinacrine or metronidazole. *Isospora* responds to treatment with trimethoprim-sulfamethoxazole. Although drugs for treating cryptosporidiosis are being investigated, none have yet proved to be both safe and efficacious.

Other Protozoa

Microsporidia

Some species of microsporidia have recently been recognized to be pathogens in immunocompromised persons, especially AIDS patients, but thus far have only occasionally been found to cause disease in persons with normal immune systems.

Trichomonas vaginalis

Trichomonas vaginalis, a common cause of vaginitis, generally is spread venereally but occasionally via contaminated fomites. This organism does not pose much risk to laboratorians; ingested organisms would be destroyed in the stomach.

INFECTIONS WITH HELMINTHS

Relatively few laboratory-acquired helminthic infections have been reported. Pike's series reported infections with *Ascaris* (eight), hookworm (two), *Strongyloides* (two), *Enterobius* (one), *Fasciola* (one), and *Schistosoma* (one) but did not provide details about any of the cases (55). Cases of cutaneous larva migrans (creeping eruption) caused by *Strongyloides* species have been reported (45, 65). The scarcity of such reports reflects the fact that helminthic infections generally are less likely than protozoan infections to be acquired in the laboratory. Whereas the cysts of intestinal protozoa are directly infectious to humans, freshly passed helminth eggs and larvae generally are not (see exceptions below). The eggs of most intestinal nematodes (e.g., *Ascaris* species and *Trichuris trichiura*) require an extrinsic maturation period of days to weeks to become infective; flukes (trematodes) and most tapeworms (cestodes) require further larval development in a nonhuman host. Even if a laboratorian became infected by ingesting infective eggs or through skin penetration by infective larvae, significant clinical illness would be unlikely. Most helminths do not multiply in human hosts; therefore, infected laboratory personnel typically would have low worm burdens and few, if any, symptoms.

The eggs of *Enterobius vermicularis* (pinworm) and *Hymenolepis nana* (dwarf tapeworm), neither of which requires an intermediate host, are unusual in that they are infectious immediately or shortly after passage. Therefore, technicians in diagnostic laboratories could become infected by ingesting these organisms if routine laboratory precautions such as handwashing were neglected. Similarly, laboratory personnel exposed to mature filariform larvae of *Strongyloides stercoralis,* which can penetrate intact skin, could become infected. Although the larvae shed in stool typically are noninfective rhabditiform larvae, a few infective filariform larvae may be present. Hyperinfected persons may shed large numbers of such larvae in respiratory secretions as well as in stool.

Laboratory personnel could also become infected with the eggs of *Taenia solium* (pork tapeworm). Humans can serve as the intermediate as well as the definitive host of this parasite. Ingesting eggs from a tapeworm carrier may result in larvae being deposited in brain and other tissues. This condition, which is known as cysticercosis, can have grave consequences, including seizures. Specimens containing *Taenia solium* eggs should be autoclaved or incinerated after they are examined.

Human infection with the tapeworm *Echinococcus granulosus* requires ingestion of contaminated feces from dogs or other carnivores susceptible to infection with *Echinococcus* species. Therefore, infection could be acquired by persons working in veterinary diagnostic or research laboratories.

Trichinella spiralis, the etiologic agent of trichinosis, is the only tissue nematode presenting a substantial risk to laboratory personnel. Fresh tissue preparations and even specimens digested with pepsin hydrochloride may contain *Trichinella* cysts, which are infective if ingested. Because most infected laboratorians would have ingested few organisms, testing for serologic conversion would be preferable to muscle biopsy for confirming the diagnosis. Filarial infections, which also are caused by tissue nematodes, could be acquired by laboratory personnel working with infected arthropods.

The infectivity of flukes depends on development in an intermediate host. Therefore, flukes do not pose a risk to personnel in diagnostic laboratories, but persons in research laboratories who handle competent intermediate hosts should exercise caution. In aquaria for snail intermediate hosts, the infective stage of *Fasciola* species (metacercariae) encysts on aquatic grasses or plants, and schistosome cercariae swim freely. Gloves should be worn while working with such aquaria. Also, persons with potential exposure to cercariae should wear a gown to prevent water droplets from contacting unprotected skin. We are aware of more than 10 unreported cases of schistosomiasis in persons who acquired the infection while working with snail hosts.

CONCLUSION

Cases of laboratory-acquired parasitic infections of varying degrees of severity have been reported. Some infections were asymptomatic and recognized only through serologic screening (48, 62). However, two deaths have been reported: one in a person with myocarditis caused by Chagas' disease (9) and the other in a person with myocarditis and encephalitis caused by toxoplasmosis (68). Other serious consequences have included mucosal leishmaniasis and possibly cerebral malaria (54). The severity of illness may be affected by the species, strain, and virulence of the infecting organism; the inoculum size; the nature of the accident; the actions taken thereafter (e.g., local treatment of the wound); and the laboratorian's immune status. Immunocompromised persons should avoid working with live organisms because these protozoan diseases can be life-threatening in such persons. Women of childbearing age should exercise caution because congenital transmission of parasitic diseases such as toxoplasmosis and American trypanosomiasis can occur.

Although these parasitic diseases generally are treatable, complicating factors may include the drug sensitivity of the parasite, the stage of the disease (e.g., mucosal versus cutaneous leishmaniasis and chronic versus acute Chagas' disease), and side effects of therapy. Despite therapy, organisms such as cysts of *T. gondii* may persist for years and reactivate if the person becomes immunosuppressed. The need for prophylactic chemotherapy after laboratory accidents involving exposure to human parasites should take into account the likelihood of infection based on the nature of the accident and inoculum size, the potential severity of the illness that could result, and the efficacy and toxicity of therapy.

Clearly, preventing laboratory accidents is preferable to managing their consequences. Thorough training of all personnel should precede assignment to research or diagnostic facilities where there is potential for exposure to pathogenic parasites. Laboratorians handling parasites should follow parasite-specific and general laboratory precautions, such as routinely wearing gloves and washing hands, using mechanical pipettes, and adequately sedating animals that will be bled. Special care should be exercised when handling needles and other sharp objects. Aerosolization of organisms and contamination of unrecognized breaks in the skin may account for some infections not attributable to recognized accidents. Persons working with organisms for which serologic tests are available should have a pre-exposure serum specimen with which semiannual or yearly specimens and specimens after accidents can be compared. Laboratories should have established protocols for handling specimens that may contain viable organisms, handling spills of infectious organisms, and responding to accidents.

We thank the following persons for contributing to this chapter: Michael J. Arrowood, William E. Collins, J. P. Dubey, Mark L. Eberhard, Paul J. Edelson, Warren D. Johnson, Felipe Kierszenbaum, Louis V. Kirchhoff, Phyllis L. Moir, Theodore E. Nash, Franklin A. Neva, Francis J. Steurer, Susan L. Stokes, Herbert B. Tanowitz, Govinda S. Visvesvara, Marianna Wilson, and Murray Wittner.

Portions of this chapter were previously published in the *American Journal of Tropical Medicine and Hygiene* (Herwaldt, B. L., and D. D. Juranek. 1993. Laboratory-acquired malaria, leishmaniasis, trypanosomiasis, and toxoplasmosis. *Am. J. Trop. Med. Hyg.* **48:**313–323). Permission to republish them was obtained from the publisher.

References

1. **Anderson, B. C.** 1985. Moist heat inactivation of *Cryptosporidium* sp. *Am. J. Public Health* **75:**1433–1434.
2. **Anderson, B. C., T. Donndelinger, R. M. Wilkins, and J. Smith.** 1982. Cryptosporidiosis in a veterinary student. *J. Am. Vet. Med. Assoc.* **180:**408–409.
3. **Anonymous.** 1990. Drugs for parasitic infections. *Med. Lett. Drugs Ther.* **32:**23–32.
4. **Aronson, P. R.** 1962. Septicemia from concomitant infection with *Trypanosoma cruzi* and *Neisseria perflava.* First case of laboratory-acquired Chagas' disease in the United States. *Ann. Intern. Med.* **57:**994–1000.
5. **Bending, M. R., and P. D. L. Maurice.** 1980. Malaria: a laboratory risk. *Postgrad. Med. J.* **56:**344–345.
6. **Blagburn, B. L., and W. L. Current.** 1983. Accidental infection of a researcher with human *Cryptosporidium. J. Infect. Dis.* **148:**772–773.
7. **Blewett, D. A.** 1988. Disinfection and oocysts, p. 107–115. *In* K. W. Angus and D. A. Blewett (ed.), Proc. 1st Int. Workshop on Cryptosporidiosis. Moredun Research Institute, Edinburgh.
8. **Brener, Z.** 1984. Laboratory-acquired Chagas' disease: an endemic disease among parasitologists?, p. 3–9. *In* C. M. Morel (ed.), *Genes and antigens of parasites: a laboratory manual,* 2nd ed. Fundação Oswaldo Cruz, Rio de Janeiro.
9. **Brener, Z.** 1987. Laboratory-acquired Chagas disease: comment. *Trans. R. Soc. Trop. Med. Hyg.* **81:**527. (Letter.)
10. **Campbell, I., S. Tzipori, G. Hutchison, and K. W. Angus.** 1982. Effect of disinfectants on survival of *Cryptosporidium* oocysts. *Vet. Rec.* **111:**414–415.
11. **Centers for Disease Control.** 1980. Chagas' disease—Michigan. *Morbid. Mortal. Weekly Rep.* **29:**147–148.
12. **Centers for Disease Control.** 1984. *Malaria Surveillance Annual Summary 1982.* Centers for Disease Control, Atlanta.
13. **Chung, H.-L.** 1931. An early case of kala-azar, possibly an oral infection in the laboratory. *Natl. Med. J. China* **17:**617–621.
14. **Coatney, G. R., W. E. Collins, M. Warren, and P. G. Contacos.** 1971. *The primate malarias.* U.S. Government Printing Office, Washington, D.C.
15. **Cross, J. H., M.-Y. Hsu-Kuo, and J. C. Lien.** 1973. Accidental human infection with *Plasmodium cynomolgi bastianellii. Southeast Asian J. Trop. Med. Public Health* **4:**481–483.
16. **Dubey, J. P., and C. P. Beattie.** 1988. *Toxoplasmosis of animals and man.* CRC Press, Inc., Boca Raton, Fla.
17. **Emeribe, A. O.** 1988. Gambiense trypanosomiasis acquired from needle scratch. *Lancet* **i:**470–471. (Letter.)
18. **Evans, T. G., and R. D. Pearson.** 1988. Clinical and immunological responses following accidental inoculation of *Leishmania donovani. Trans. R. Soc. Trop. Med. Hyg.* **82:**854–856.
19. **Eyles, D. E., G. R. Coatney, and M. E. Getz.** 1960. Vivax-type malaria parasite of macaques transmissible to man. *Science* **131:**1812–1813.
20. **Field, P. R., G. G. Moyle, and P. M. Parnell.** 1972. The accidental infection of a laboratory worker with *Toxoplasma gondii. Med. J. Aust.* **2:**196–198.

21. **Freedman, D. O., J. D. MacLean, and J. B. Viloria.** 1987. A case of laboratory acquired *Leishmania donovani* infection; evidence for primary lymphatic dissemination. *Trans. R. Soc. Trop. Med. Hyg.* **81:**118–119.
22. **Frenkel, J. K., R. W. Weber, and M. N. Lunde.** 1960. Acute toxoplasmosis: effective treatment with pyrimethamine, sulfadiazine, leucovorin calcium, and yeast. *JAMA* **173:**1471–1476.
23. **Garnham, P. C. C.** 1967. Malaria in mammals excluding man. *Adv. Parasitol.* **5:**139–204.
24. **Gutteridge, W. E., B. Cover, and A. J. D. Cooke.** 1974. Safety precautions for work with *Trypanosoma cruzi*. *Trans. R. Soc. Trop. Med. Hyg.* **68:**161.
25. **Hanson, W. L., R. F. Devlin, and E. L. Roberson.** 1974. Immunoglobulin levels in a laboratory-acquired case of human Chagas' disease. *J. Parasitol.* **60:**532–533.
26. **Henderson, H. E., G. W. Gillespie, P. Kaplan, and M. Steber.** 1963. The human *Isospora*. *Am. J. Hyg.* **78:**302–309.
27. **Herbert, W. J., D. Parratt, N. Van Meirvenne, and B. Lennox.** 1980. An accidental laboratory infection with trypanosomes of a defined stock. II. Studies on the serological response of the patient and the identity of the infecting organism. *J. Infect.* **2:**113–124.
28. **Hermentin, K., A. Hassl, O. Picher, and H. Aspöck.** 1989. Comparison of different serotests for specific *Toxoplasma* IgM-antibodies (ISAGA, SPIHA, IFAT) and detection of circulating antigen in two cases of laboratory acquired *Toxoplasma* infection. *Zentralbl. Bakteriol. Mikrobiol. Hyg. [A]* **270:**534–541.
29. **Herr, A., and L. Brumpt.** 1939. Un cas aigu de maladie de Chagas contractée accidentellement au contact de triatomes Mexicains: observation et courbe fébrile. *Bull. Soc. Pathol. Exot.* **32:**565–571.
30. **Herwaldt, B. L., and J. D. Berman.** 1992. Recommendations for treating leishmaniasis with sodium stibogluconate (Pentostam) and review of pertinent clinical studies. *Am. J. Trop. Med. Hyg.* **46:**296–306.
31. **Hofflin, J. M., R. H. Sadler, F. G. Araujo, W. E. Page, and J. S. Remington.** 1987. Laboratory-acquired Chagas disease. *Trans. R. Soc. Trop. Med. Hyg.* **81:**437–440.
32. **Hojlyng, N., W. Holten-Andersen, and S. Jepsen.** 1987. Cryptosporidiosis: a case of airborne transmission. *Lancet* **ii:**271–272. (Letter.)
33. **Holm, K.** 1924. [Subtertian malaria acquired from a corpse.] Ueber einen Fall von Infektion mit Malaria tropica an der Leiche. *Klin. Wochenschr.* **3:**1633–1634.
34. **Hudson, L., F. Grover, W. E. Gutteridge, R. A. Klein, W. Peters, R. A. Neal, M. A. Miles, J. E. Williams, M. T. Scott, R. Nourish, and B. P. Ager.** 1983. Suggested guidelines for work with live *Trypanosoma cruzi*. *Trans. R. Soc. Trop. Med. Hyg.* **77:**416–419.
35. **Hughes, W. T.** 1982. Natural mode of acquisition for de novo infection with *Pneumocystis carinii*. *J. Infect. Dis.* **145:**842–848.
36. **Hughes, W. T., D. L. Bartley, and B. M. Smith.** 1983. A natural source of infection due to *Pneumocystis carinii*. *J. Infect. Dis.* **147:**595.
37. **Jacobs, L.** 1957. The interrelation of toxoplasmosis in swine, cattle, dogs, and man. *Public Health Rep.* **72:**872–882.
38. **Jensen, J. B., T. C. Capps, and J. M. Carlin.** 1981. Clinical drug-resistant falciparum malaria acquired from cultured parasites. *Am. J. Trop. Med. Hyg.* **30:**523–525.
39. **Kayhoe, D. E., L. Jacobs, H. K. Beye, and N. B. McCullough.** 1957. Acquired toxoplasmosis: observations on two parasitologically proved cases treated with pyrimethamine and triple sulfonamides. *N. Engl. J. Med.* **257:**1247–1254.
40. **Kirchhoff, L. V., and D. F. Hoft.** 1990. Immunization and challenge of mice with insect-derived metacyclic trypomastigotes of *Trypanosoma cruzi*. *Parasite Immunol.* **12:**65–74.
41. **Lettau, L. A.** 1991. Nosocomial transmission and infection control aspects of parasitic and ectoparasitic diseases. Part I. Introduction/enteric parasites. *Infect. Control Hosp. Epidemiol.* **12:**59–65.
42. **Lettau, L. A.** 1991. Nosocomial transmission and infection control aspects of parasitic and ectoparasitic diseases. Part II. Blood and tissue parasites. *Infect. Control Hosp. Epidemiol.* **12:**111–121.
43. **Levine, J. F., M. G. Levy, R. L. Walker, and S. Crittenden.** 1988. Cryptosporidiosis in veterinary students. *J. Am. Vet. Med. Assoc.* **193:**1413–1414.
44. **Ma, P., G. S. Visvesvara, A. J. Martinez, F. H. Theodore, P.-M. Daggett, and T. K. Sawyer.** 1990. *Naegleria* and *Acanthamoeba* infections: review. *Rev. Infect. Dis.* **12:**490–513.
45. **Maligin, S. A.** 1958. A case of cutaneous form of strongyloidiasis caused by larvae of *S. ransomi*, *S. westeri* and *S. papillosus*. *Med. Parazitol. (Mosk.)* **27:**446–447.
46. **Mandell, G. L., R. G. Douglas, and J. E. Bennett (ed.).** 1990. *Principles and practice of infectious diseases*, 3rd ed. Churchill Livingstone, New York.
47. **McCracken, A. W.** 1972. Natural and laboratory-acquired infection by *Isospora belli*. *South. Med. J.* **65:**800,818.
48. **Miller, N. L., J. K. Frenkel, and J. P. Dubey.** 1972. Oral infections with *Toxoplasma* cysts and oocysts in felines, other mammals, and in birds. *J. Parasitol.* **58:**928–937.
49. **Mintz, E. D., J. F. Anderson, R. G. Cable, and J. L. Hadler.** 1991. Transfusion-transmitted babesiosis: a case report from a new endemic area. *Transfusion* **31:**365–368.
50. **Most, H.** 1973. *Plasmodium cynomolgi* malaria: accidental human infection. *Am. J. Trop. Med. Hyg.* **22:**157–158.
51. **Navin, T. R., F. E. Arana, A. M. de Mérida, B. A. Arana, A. L. Castillo, and D. N. Silvers.** 1990. Cutaneous leishmaniasis in Guatemala: comparison of diagnostic methods. *Am. J. Trop. Med. Hyg.* **42:**36–42.
52. **Neu, H. C.** 1967. Toxoplasmosis transmitted at autopsy. *JAMA* **202:**284–285.
53. **Pavlásek, I.** 1984. Effect of disinfectants in infectiousness of oocysts of *Cryptosporidium* sp. *Cesk. Epidemiol. Microbiol. Immunol.* **33:**97–101.
54. **Petithory, J., and G. Lebeau.** 1977. [A probable laboratory contamination with *Plasmodium falciparum*.] Contamination probable de laboratoire par *Plasmodium falciparum*. *Bull. Soc. Pathol. Exot. Filiales.* **70:**371–375.
55. **Pike, R. M.** 1976. Laboratory-associated infections: summary and analysis of 3921 cases. *Health Lab. Sci.* **13:**105–114.
56. **Pike, R. M.** 1978. Past and present hazards of working with infectious agents. *Arch. Pathol. Lab. Med.* **102:**333–336.
57. **Pizzi, T., G. Niedmann, and A. Jarpa.** 1963. Comunicación de tres casos de enfermedad de Chagas aguda

producidos por infecciones accidentales de laboratorio. *Bol. Chil. Parasitol.* **18:**32–36.

58. **Pohjola, S., H. Oksanen, L. Jokipii, and A. M. M. Jokipii.** 1986. Outbreak of cryptosporidiosis among veterinary students. *Scand. J. Infect. Dis.* **18:**173–178.
59. **Rawal, B. D.** 1959. Laboratory infection with *Toxoplasma. J. Clin. Pathol.* **12:**59–61.
60. **Receveur, M. C., and P. Vincendeau.** 1993. Laboratory-acquired gambian trypanosomiasis. *N. Engl. J. Med.* **329:**209–210.
61. **Reif, J. S., L. Wimmer, J. A. Smith, D. A. Dargatz, and J. M. Cheney.** 1989. Human cryptosporidiosis associated with an epizootic in calves. *Am. J. Public Health* **79:**1528–1530.
62. **Remington, J. S., and L. O. Gentry.** 1970. Acquired toxoplasmosis: infection versus disease. *Ann. N. Y. Acad. Sci.* **174:**1006–1017.
63. **Rendtorff, R. C.** 1954. The experimental transmission of human intestinal protozoan parasites. *Am. J. Hyg.* **59:**209–220.
64. **Robertson, D. H. H., S. Pickens, J. H. Lawson, and B. Lennox.** 1980. An accidental laboratory infection with African trypanosomes of a defined stock. I. The clinical course of the infection. *J. Infect.* **2:**105–112.
65. **Roeckel, I. E., and E. T. Lyons.** 1977. Cutaneous larva migrans, an occupational disease. *Ann. Clin. Lab. Sci.* **7:**405–410.
66. **Sadick, M. D., R. M. Locksley, and H. V. Raff.** 1984. Development of cellular immunity in cutaneous leishmaniasis due to *Leishmania tropica. J. Infect. Dis.* **150:**135–138.
67. **Sampaio, R. N., L. M. P. de Lima, A. Vexenat, C. C. Cuba, A. C. Barreto, and P. D. Marsden.** 1983. A laboratory infection with *Leishmania braziliensis braziliensis. Trans. R. Soc. Trop. Med. Hyg.* **77:**274. (Letter.)
68. **Sexton, R. C., D. E. Eyles, and R. E. Dillman.** 1953. Adult toxoplasmosis. *Am. J. Med.* **14:**366–377.
69. **Smith, R. P., A. T. Evans, M. Popovsky, L. Mills, and A. Spielman.** 1986. Transfusion-acquired babesiosis and failure of antibiotic treatment. *JAMA* **256:**2726–2727.
70. **Soave, R., and W. D. Johnson.** 1988. *Cryptosporidium* and *Isospora belli* infections. *J. Infect. Dis.* **157:**225–229.
71. **Strickland, G. T. (ed.).** 1991. *Hunter's tropical medicine,* 7th ed. The W. B. Saunders Co., Philadelphia.
72. **Terry, L. L., J. L. Lewis, and S. M. Sessoms.** 1950. Laboratory infection with *Leishmania donovani:* a case report. *Am. J. Trop. Med.* **30:**643–649.
73. **Teutsch, S. M., D. D. Juranek, A. Sulzer, J. P. Dubey, and R. K. Sikes.** 1979. Epidemic toxoplasmosis associated with infected cats. *N. Engl. J. Med.* **300:**695–699.
74. **Visvesvara, G. S., and J. K. Stehr-Green.** 1990. Epidemiology of free-living ameba infections. *J. Protozool.* **37:**25S–33S.
75. **Western, K. A., M. G. Schultz, W. E. Farrar, and I. G. Kagan.** 1969. Laboratory acquired Chagas' disease treated with Bay [sic] 2502. *Bol. Chil. Parasitol.* **24:**94.
76. **Williams, J. L., B. T. Innis, T. R. Burkot, D. E. Hayes, and I. Schneider.** 1983. Falciparum malaria: accidental transmission to man by mosquitoes after infection with culture-derived gametocytes. *Am. J. Trop. Med. Hyg.* **32:**657–659.
77. **Wilson, M., and P. Schantz.** 1991. Nonmorphologic diagnosis of parasitic infections, p. 717–726. *In* A. Balows, W. J. Hausler, Jr., K. L. Herrmann, H. D. Isenberg, and H. J. Shadomy (ed.), *Manual of clinical microbiology,* 5th ed. American Society for Microbiology, Washington, D.C.
78. **Wilson, M., D. A. Ware, and D. D. Juranek.** 1990. Serologic aspects of toxoplasmosis. *J. Am. Vet. Med. Assoc.* **196:**277–281.
79. **Wittner, M., K. S. Rowin, H. B. Tanowitz, J. F. Hobbs, S. Saltzman, B. Wenz, R. Hirsch, E. Chisholm, and G. R. Healy.** 1981. Successful chemotherapy of transfusion babesiosis. *Ann. Intern. Med.* **96:**601–604.
80. **Zeledón, R.** 1974. Epidemiology, modes of transmission and reservoir hosts of Chagas' disease, p. 51–85. *In* Ciba Foundation Symposium 20 (new series), *Trypanosomiasis and leishmaniasis with special reference to Chagas' disease.* Associated Scientific Publishers, Amsterdam.

Transmission of Bacterial and Rickettsial Zoonoses in the Laboratory

ARNOLD F. KAUFMANN AND JOHN M. BOYCE

7

INTRODUCTION

Over the years, diseases that affect both humans and animals (zoonoses) have been among the most commonly reported occupational illness of laboratory workers (19, 24, 25, 38, 39). Agents responsible for a variety of bacterial and rickettsial zoonoses accounted for 35 to 40% of all laboratory-acquired infections reported by Pike in 1976 (24). Fully 50% of the fatal bacterial infections listed by Pike were caused by agents normally transmitted from animals to humans. Published information on human infective dose is included in the Agent Summary Statements in Appendix I.

Although some agents such as *Brucella* spp., *Francisella tularensis, Coxiella burnetii,* and *Chlamydia psittaci,* which formerly caused many laboratory-associated infections, are not handled as frequently as they once were, infections caused by such zoonotic agents continue to affect personnel working in research and reference laboratories. In this chapter, we discuss the more common bacterial and rickettsial zoonoses that have been transmitted to laboratory personnel, the modes of transmission for these diseases, and how this transmission can be controlled.

ANTHRAX

Human anthrax has three main clinical forms: cutaneous, inhalation, and gastrointestinal. Cutaneous anthrax is associated with a characteristic skin lesion developing at the site where *Bacillus anthracis* is introduced beneath the skin (e.g., by needle puncture or through a cut). After a 2- to 7-day incubation period, the lesion begins as a papule, which gradually enlarges and develops a central vesicle. The vesicle ruptures shortly after formation, revealing an underlying ulcer. A depressed scab or eschar rapidly forms over the ulcer's surface. The surrounding skin is commonly edematous, and secondary vesicles may develop in the edematous zone. Death from overwhelming septicemia or other complications occurs in about 5 to 20% of untreated patients but is uncommon if effective antibiotic therapy is administered.

Inhalation anthrax results from breathing aerosols of *B. anthracis* spores generated, for example, during specimen centrifugation. The spores are transported from the lungs to the mediastinal lymph nodes, where germination occurs. The disease is characterized by a primary hemorrhagic mediastinitis with secondary septicemia. The clinical disease begins as a mild febrile illness resembling the common cold. After 2 to 4 days, the second stage of acute toxicity begins with sudden onset of dyspnea, cyanosis, and profuse sweating. Death usually ensues 24 hours after symptoms appear, even with therapy.

Gastrointestinal anthrax, a rare form of the disease characterized by severe abdominal distress, fever, and septicemia, results almost exclusively from eating raw or undercooked meats from infected animals. Gastrointestinal anthrax has not occurred in laboratory personnel.

B. anthracis normally resides in soil. Animal anthrax results from grazing on infective pastures or eating contaminated feeds, and human anthrax usually results from exposure to anthrax-infected animals or their by-products.

Laboratory-acquired anthrax has been a problem primarily at facilities conducting anthrax research. Most cases have been cutaneous, with inhalation anthrax being rarely reported. In one series, cutaneous infection occurred at the sites of known preceding trauma in only 5 of 25 cases (8). The largest single episode of laboratory-associated infections resulted from a burst safety filter at a Russian military laboratory in 1979 (17, 27). More than 100 cases of inhalation anthrax occurred in persons working near but not in the laboratory compound, with at least 64 deaths. Since the introduction of the human anthrax vaccine in the late 1950s, no laboratory-acquired anthrax cases have been reported in the United States. Occasional cases have occurred in persons performing necropsies in nonlaboratory settings on animals that died of anthrax.

Persons who are frequently exposed to anthrax-infected animals or pure cultures of *B. anthracis* should be immunized with anthrax vaccine. Biosafety level (BSL) 3 precautions should be followed for procedures involving large volumes, high concentrations, or potential aerosols of *B. anthracis;* otherwise, BSL 2 precautions are adequate.

Antibiotic treatment with penicillin or, alternatively, tetracyclines, erythromycin, ciprofloxacin, or other broad-spectrum antibiotics is usually effective in curing cutaneous anthrax. Treatment is less successful with inhalation and gastrointestinal anthrax. In accidents posing significant risk of inhalation anthrax for unimmunized persons, postexposure vaccination combined with antimicrobial prophylaxis is recommended.

BORRELIA INFECTIONS

Lyme Disease

Lyme disease is a tick-borne spirochetal disease caused by *Borrelia burgdorferi* (4, 36). Lyme disease often occurs in stages: localized erythema migrans (stage 1), disseminated infection (stage 2), and persistent infection (stage 3). From 3 to 32 days after a bite by an infected tick, 60 to 80% of infected patients will develop an expanding annular skin lesion called erythema migrans, which may be accompanied by fever, regional lymphadenopathy, or mild constitutional symptoms.

Disseminated infection (stage 2) begins days to weeks after inoculation and is often associated with episodes of severe headache, malaise, fatigue, and involvement of multiple organ systems. Infection may involve the skin (secondary annular lesions not associated with the bite site), musculoskeletal tissues (myalgia, arthralgia, or acute oligoarticular arthritis), central or peripheral nervous system (meningitis, Bell's palsy, encephalitis, or radiculoneuritis), and heart (atrioventricular block or myopericarditis), as well as the eyes, liver, kidney, or respiratory system. About 60% of patients with Lyme disease will develop brief attacks of arthritis (often affecting the knee) during stage 2.

Persistent infection (stage 3) may be manifested by prolonged arthritis attacks or chronic arthritis (lasting more than 1 year), progressive encephalomyelitis, keratitis, and acrodermatitis chronica atrophicans.

The nymphal stages of *Ixodes dammini* and, to a lesser extent, *Ixodes scapularis* and *Ixodes pacificus* are the most commonly implicated vectors for human *B. burgdorferi* infections in the United States. *I. dammini* is the primary vector in the northeastern and midwestern states, *I. scapularis* is the primary vector in the southeastern states, and *I. pacificus* is the primary vector in the western coastal states. Humans are incidental hosts, with small rodents and larger mammals, particularly the white-tailed deer, being primary hosts at various developmental stages of the ticks.

For early Lyme disease, doxycycline or tetracycline is the recommended treatment. Alternative antibiotics include amoxicillin and penicillin G. In later stages, ceftriaxone or high doses of penicillin G are recommended.

No laboratory-acquired infections have been reported, but animal-to-animal transmission in the absence of insect vectors during experimental studies points to the potential for such occurrences. Personnel conducting field studies have been infected from tick bites.

BSL 2 precautions appear adequate for work with *B. burgdorferi* in the laboratory. Field personnel should take precautions against tick bites. Periodic checking of the body for ticks should be performed. Tick repellents sprayed on socks and pants are recommended. The pants should be tucked into the socks. The risk of infection by the bite of any given tick is low, even if the tick is infected, so antimicrobial prophylaxis is not routinely indicated after a recognized tick bite (33).

Relapsing Fever

Epidemic (louse-borne) and endemic (tick-borne) relapsing fevers are both caused by spirochetes that belong to the genus *Borrelia.* After an incubation period of 5 to 15 days, affected persons experience sudden onset of fever, chills, myalgia, arthralgia, photophobia, and headache. Physical examination often reveals hepatosplenomegaly, injection of the sclerae, petechiae, epistaxis, and occasionally, jaundice, lymphadenopathy, or nuchal rigidity. The initial febrile illness lasts for 3 to 7 days and ends suddenly. The patients remain afebrile for 7 to 10 days, and then the fever and other symptoms return abruptly and persist for several days. Patients with endemic relapsing fever usually have more relapses than those with the epidemic form.

At least 45 cases of laboratory-acquired relapsing fever have been reported, making it the seventh most common bacterial cause of laboratory-associated infections (24).

Laboratory-acquired relapsing fever is most likely to occur when infectious materials from humans or animals (including hemolymph from lice and ticks) are accidentally injected or come in direct contact with broken skin or oral, nasal, or conjunctival mucous membranes of laboratory personnel (9). At least one case of probable relapsing fever occurred when a syringe containing spirochetes suddenly became separated from the attached needle and a worker sprayed infectious materials into his eye. Bites of *Borrelia*-positive ectoparasites have also probably accounted for some cases of relapsing fever among persons employed in research units.

BSL 2 precautions are adequate for almost all work with *Borrelia* spp. Specific preventive measures include the use of syringes with locking hubs and the wearing of water-impervious gloves when handling *Borrelia* cultures or potentially infected blood. Research units that house infected ectoparasites should follow procedures that minimize the chances of employees being bitten by tick vectors.

Tetracyclines are the drugs of choice for both endemic and epidemic relapsing fever. Recent studies have shown that a single 500-mg dose of tetracycline or 100 mg of doxycycline will cure most patients with epidemic relapsing fever. Patients should be watched carefully for signs and symptoms of a Jarisch-Herxheimer reaction, which often occurs within several hours of the institution of antibiotic therapy. Patients who are allergic to tetracycline may be treated with erythromycin or penicillin G.

BRUCELLOSIS

Human brucellosis is characterized by fever that may be irregular, intermittent, or continuous, chills, profuse sweating particularly at night, headache, weakness, anorexia, weight loss, arthralgia, generalized aches, and depression. After an incubation period ranging from 5 days to 1 month or more, the illness begins abruptly or insidiously and lasts for days to months or, occasionally, several years. A variety of complications may occur, including septic arthritis, spondylitis, orchitis, endocarditis, meningitis, and rarely, death.

Hypersensitivity to *Brucella* antigens is also a biohazard in *Brucella* research laboratories. The clinical presentation varies from an intensely burning or itching rash on exposed skin areas to an illness resembling acute brucellosis.

Brucellosis has been the most commonly reported bacterial infection acquired in laboratories. Most cases have occurred in research laboratories and have involved exposure to *Brucella* organisms being grown in large quantities. All four *Brucella* spp. with known human pathogenicity (*B. abortus, B. canis, B. melitensis,* and *B. suis*) have caused illness in laboratory personnel (20, 34).

Human infections can result from exposure to *Brucella* organisms by contact, inhalation, or ingestion. Direct contact with cultures of *Brucella* organisms or with infectious materials such as uterine discharges or blood of infected animals is an important mode of transmission. The intact skin is an effective barrier against invasion by *Brucella* organisms, but even minor scratches and abrasions may act as portals of entry. *Brucella* organisms can also invade the body through the conjunctiva if infectious material is rubbed or sprayed into the eyes.

Aerosol transmission of *Brucella* organisms is a well-documented hazard in laboratories. The largest single laboratory-associated epidemic (45 cases, 1 death) and many smaller outbreaks have been caused by this mode of transmission (16, 21, 35). Ingestion is probably the least frequent mode of transmission among laboratory workers but has resulted from mouth pipetting.

Brucella infections frequently result in long-lasting immunity. A safe and effective human brucellosis vaccine, however, is not available. Protection of laboratory personnel is dependent on adherence to BSL 3 precautions. Because *Brucella* infection or antibiotics administered to patients with brucellosis may cause fetal injury, pregnant women should be counseled about potential hazards and the need to adhere to safety procedures before assignment to work with *Brucella* organisms.

A combination of doxycycline and rifampin is now recommended for treatment of brucellosis. A combination of streptomycin and a tetracycline is also effective. Alternative drugs are ciprofloxacin, chloramphenicol, or a combination of gentamicin and trimethoprim-sulfamethoxazole. Relapses occur in about 5 to 10% of patients, particularly if treatment is delayed or administered for an inadequate period of time. Prophylactic antibiotic therapy is commonly administered to personnel exposed in high-risk accidents, but the optimal treatment regimen for this purpose has not been determined.

GLANDERS AND MELIOIDOSIS

Glanders and melioidosis have clinical and pathologic similarities but differ significantly in their epidemiology. Both diseases are characterized by localized disseminated suppurative lesions involving a variety of organ systems. In glanders, abscess formation tends to be more extensive than in melioidosis, and involvement of the upper respiratory tract is very common. Lung abscesses occur more frequently in melioidosis than in glanders. The clinical course of both diseases ranges from acute to extremely chronic, and some illnesses have been reported to last more than 10 years. Untreated glanders and melioidosis cases commonly result in death. Specific antimicrobial therapy results in a significant reduction of mortality.

The incubation period of both diseases after traumatic inoculation ranges from 1 to 5 days. In glanders, the incubation period after inhalation of infectious aerosols is usually 10 to 14 days. The initial infection in melioidosis and, less commonly, glanders may result in latent disease that becomes clinically apparent months to years later.

Glanders is caused by *Burkholderia mallei* and has its reservoir in horses and other solipeds. Glanders has been eradicated from most areas of the world, and its distribution is currently restricted to Asia. Melioidosis is caused by *Burkholderia pseudomallei*, a soil saprophyte. Although *B. pseudomallei* is widely distributed in tropical and subtropical areas of Asia, Africa, Australia, and the Americas, melioidosis occurs most commonly in Southeast Asia.

Many laboratory-acquired glanders infections have occurred in research and veterinary diagnostic laboratories (11, 26). Cases have resulted from contact exposure of broken skin and nasal mucous membranes, accidental inoculations, and inhalation of aerosols. Sources of infection included cultures as well as infected animal tissues In contrast to glanders, few cases of laboratory-acquired melioidosis have been reported (10). Infections have resulted from accidental inoculation of cultures and inhalation of infectious aerosols.

Vaccines are not available for either glanders or melioidosis. BSL 3 precautions should be followed for all work with *B. mallei* and *B. pseudomallei.*

Optimal antimicrobial therapy regimens have not been determined for human glanders because of its rarity. Sulfadiazine, streptomycin, and tetracycline have been used with apparent success. In melioidosis, antimicrobial therapy should be based on susceptibility studies. Ceftazidime has been recommended as drug of first choice, with trimethoprim-sulfamethoxazole, cefotaxime, imipenem-cilastatin, and amoxicillin clavulanate as alternatives. In Thailand, 80% of strains are resistant to trimethoprim-sulfamethoxazole, and many are resistant to tetracycline, novobiocin, and ceftriaxone.

LEPTOSPIROSIS

Leptospires are motile, finely coiled spirochetes that are cultivable in artificial media. More than 250 serovars (serotypes) of pathogenic leptospires have been recovered from feral and domestic animals. Seven to fourteen days (rarely up to 26 days) after exposure, affected persons experience sudden onset of fever, headache, nausea, conjunctival suffusion, and myalgia. Pharyngitis, evanescent rashes, and cough with pulmonary infiltrates occur in some patients. Jaundice occurs in only 5 to 10% of patients. After 4 to 8 days, the fever usually disappears. One to three days after becoming afebrile, many patients have recurrence of fever that may be accompanied by meningitis or, rarely, uveitis. Several laboratory workers who became infected with *Leptospira interrogans* serovar *ballum* after exposure to Swiss albino mice developed unilateral or bilateral orchitis during the second phase of illness. Subclinical leptospirosis has also been reported in workers who frequently handle Swiss albino mice (37).

About 70 cases of leptospirosis among laboratory personnel have been reported, making it the sixth most common bacterial cause of laboratory-acquired infections (24). The most common sources responsible for such infections are cultures of leptospires and natural or experimentally infected animals. Mice, rats, rabbits, and guinea pigs are the animals most frequently implicated.

Cases have been acquired by needlestick or by accidental spraying of the conjunctiva or other mucous membranes with a leptospiral culture. The latter-type accident usually occurs when a needle suddenly becomes separated from the syringe being used to inoculate animals with leptospires. Mouth pipetting has also resulted in cases of laboratory-acquired leptospirosis. A few laboratory personnel appear to have become infected when bitten by an infected animal. Serovars implicated in cases of laboratory-acquired leptospirosis include *ballum, georgia, icterohaemorrhagiae,* and members of Javanica serogroup.

No leptospiral vaccine is approved for human use in the United States. BSL 2 precautions are recommended for laboratory studies of pathogenic leptospires. Preventive measures should include appropriate training of personnel involved in handling experimental animals and maintaining their awareness that laboratory animals may be asymptomatically infected with leptospires. Use of water-impervious gloves and careful handwashing after contact with known or potentially infected animals should minimize the risks of acquiring leptospirosis from experimental animals.

Both doxycycline and penicillin have been shown in controlled trials to be effective therapy for leptospirosis (18, 42). Doxycycline has also been shown to be effective for chemoprophylaxis against leptospirosis (40). In at least one reported case, prompt administration of prophylactic oral penicillin failed to prevent symptomatic leptospirosis.

PASTEURELLOSIS

Pasteurella spp. are small, nonmotile gram-negative coccobacilli that cause disease in a wide variety of domestic and laboratory animals. The genus includes *P. multocida, P. pneumotropica, P. haemolytica, P. ureae, P. aerogenes,* and *P. dagmatis. Pasteurella* infections in laboratory workers are almost always caused by *P. multocida* and are associated with animal bites. Local swelling, tenderness, and erythema develop in less than 3 days (often within 6 to 8 h) after a bite. The rapidity with which symptoms and signs develop and the severity of the pain, which seems out of proportion to the size of the injury, are characteristic of *P. multocida* bite wound infections. Regional lymphadenopathy and fever occur in some patients. Local suppurative complications include tenosynovitis, arthritis, and osteomyelitis. A few asymptomatic infections (oropharyngeal colonization) have been reported.

Very few human cases of laboratory-acquired pasteurellosis have been reported. Animals (mainly dogs and cats) used in research work represent the main source from which laboratory personnel become infected. About 10 to 65% of dogs and 50 to 90% of healthy cats carry *P. multocida* in their oropharynx. Personnel who sustain a dog or cat bite (or cat scratch) have about 5 to 15% chance of developing a *P. multocida* infection. One case caused by a rat bite and another by a rabbit bite have also been reported.

Avoiding animal bites is the only effective way to minimize the risk of acquiring *P. multocida* infections. There is no vaccine available for animal caretakers or other personnel who must handle laboratory animals. BSL 2 precautions are generally adequate for work with *Pasteurella* spp.

Penicillin is the drug of choice for *P. multocida* infections. Tetracycline, ampicillin, cefotaxime, ceftriaxone, cefazolin, or cephalothin are reasonable alternative drugs. Although there are no adequate studies of the efficacy of penicillin in preventing animal (particularly cat) bite-associated *P. multocida* infections, postexposure prophylaxis with penicillin is reasonable (41).

PLAGUE AND OTHER YERSINIOSES

Yersinia pestis is the etiologic agent of plague. The most common clinical syndromes caused by *Y. pestis* are bubonic, primary septicemic, and pneumonic plague. The incubation period is usually 2 to 6 days in bubonic and septicemic plague and 2 to 5 days in pneumonic plague. When direct contact with infected animal tissue or a flea bite results in percutaneous inoculation of *Y. pestis,* a primary lesion may develop at the site of entry. Although such lesions are often clinically insignificant, bullous or ulcerative lesions may develop. The infection spreads within a few days via the lymphatics to the regional lymph nodes, which become swollen and painful (a bubo). Shortly before or at the time the bubo becomes apparent, fever and nonspecific signs and symptoms of toxemia develop. Bacteremia is common and may result in various secondary complications such as pneumonia and meningitis. Death occurs in 50% or more of untreated patients but is substantially reduced by early appropriate antibiotic therapy.

Primary septicemic plague is similar to bubonic plague except the bubo formation does not occur or is clinically inapparent. The prognosis is worse for primary septicemic than for bubonic plague patients, but cure is possible with early appropriate therapy.

Persons with pneumonic plague experience fever, chills, and malaise that may be accompanied by vague pain in the chest. Eighteen to twenty-four hours after onset, patients develop dyspnea and a cough that later becomes productive of blood-streaked sputum. Roentgenographic evidence of pneumonia is often present on day 1 or 2 of the illness. Unvaccinated, untreated persons often die on day 2 or 3 of the illness.

Surprisingly few cases of laboratory-acquired plague have been reported. Before 1936, only 4 bubonic and 11 pneumonic plague cases were reported as being laboratory associated (45, 46). Since 1936, five cases of laboratory-acquired pneumonic plague, including three in the United States, have been documented (6). A variety of personnel have been affected, including physicians, bacteriologists, students, a chemist, an animal caretaker, and other unspecified laboratory workers.

The sources from which personnel are most likely to acquire plague include cultures of virulent *Y. pestis*, infected animals, and tissues from plague patients. In the laboratory, the plague bacillus can be transmitted to personnel by infectious aerosols created during centrifugation of cultures, inoculation of solid media, and postmortem examination of infected animals. One case has resulted from mouth pipetting. Research projects and field studies dealing with *Y. pestis* or animals with naturally acquired or experimentally induced plague should be conducted in laboratories that meet the containment requirements specified for BSL 3.

Plague vaccines have been used since the late 19th century, but their effectiveness has never been measured precisely. Field experience indicates that plague vaccine reduces the incidence and severity of disease resulting from the bite of infected fleas. Whether comparable protection is afforded against an accidental injection of a larger inoculum than that resulting from a flea bite (>10,000 CFU) is unknown. The degree of protection against primary pneumonic infection is also not known. Because of this uncertainty, persons who have been exposed to known *Y. pestis* aerosols or accidental inoculations should be given a 7- to 10-day course of antibiotic therapy regardless of immunization history. Plague vaccination should be a routine requirement for laboratory personnel who are working with *Y. pestis* organisms resistant to antibiotics that are normally used in plague therapy, laboratory personnel frequently working with *Y. pestis* or *Y. pestis*-infected animals, and persons engaged in aerosol experiments with *Y. pestis*.

Routine bacteriologic precautions, including the use of a biological safety cabinet to isolate procedures that may produce aerosols, are sufficient to prevent accidental infection with plague in clinical laboratory workers at hospitals. No cases acquired during the course of activities normally conducted in hospital-based clinical laboratories have been reported. Immunization might further reduce the already minimal risk but is not clearly indicated.

The primary vaccination series consists of three doses of vaccine, with an interval of 4 weeks between the first two doses and 6 months between the first and third doses. Whenever possible, the first two doses of the primary series should be administered before any anticipated exposures.

Serum antibody to fraction 1 capsular antigen, as measured by the passive hemagglutination (PHA) test, is correlated with resistance to *Y. pestis* infection in experimental animals. Although direct evidence does not exist, a comparable correlation between PHA titer and immunity probably occurs in humans. After the primary immunization series, about 7% of individuals may not produce PHA antibody. Other persons occasionally fail to achieve a PHA titer of 128, the level correlated with immunity in experimental animals. PHA titers should be determined for persons who have unusually high risk of infection or who have a history of serious reactions to the vaccine.

Laboratory workers who develop signs or symptoms compatible with pneumonic or bubonic plague should be hospitalized promptly and placed in strict isolation. As soon as appropriate cultures have been obtained, streptomycin or tetracycline therapy should be instituted without waiting for bacteriologic confirmation of the diagnosis. Chemoprophylaxis is recommended for persons who have had close contact with a patient with suspected or confirmed primary pneumonic plague or plague pneumonia. If a researcher develops pneumonic plague after a recognized accident, other similarly exposed individuals should be given chemoprophylaxis, usually tetracycline, for 7 days. Persons allergic to tetracycline may be given an appropriate oral sulfonamide. However, sulfonamides are not recommended for patients with symptomatic pneumonic plague.

Although *Yersinia pseudotuberculosis* and, to a lesser extent, *Yersinia enterocolitica* are capable of causing disease in feral and domestic animals and in humans, these agents have apparently caused very few laboratory-acquired infections.

PSITTACOSIS

Psittacosis is an acute systemic disease caused by *Chlamydia psittaci*. After an incubation period of 4 to 15 days, the illness usually begins abruptly with fever, chills, headache, backache, and photophobia. Although respiratory signs and symptoms tend to be limited to a nonproductive cough, X rays commonly reveal a patchy lobular pneumonia resembling the interstitial pneumonia caused by various viral agents. A variety of other clinical manifestations such as splenomegaly, hepatitis, myocarditis, thrombophlebitis, and meningoencephalitis may occur. The case-fatality rate is currently about 1%.

Psittacosis has its reservoir in many species of wild and domestic birds. Human infections are acquired primarily from pet birds and domestic poultry. Infected birds shed *C. psittaci* in their respiratory secretions and feces, with the latter having particular epidemiologic significance. Infection is primarily spread by aerosolized bird feces (31). In the laboratory, infectious aerosols may be produced while dissecting infected animals or processing diagnostic specimens. Direct contact with or traumatic inoculation of infectious material may also result in disease. Person-to-person transmission has occurred but is rare.

Prevention of laboratory-acquired psittacosis is based on control of potentially infectious aerosols. BSL 3 precautions are usually appropriate, but BSL 2 precautions may suffice when working with small specimens or materials that contain low concentrations of *C. psittaci.* No vaccine is available. Recovery from a psittacosis infection does not confer immunity, and multiple infections have occurred in persons with ongoing exposure. Tetracycline or doxycycline are drugs of choice for treatment of psittacosis, with erythromycin and chloramphenicol being alternatives.

RAT-BITE FEVER

Rat-bite fever (RBF) is actually two different diseases with clinical and epidemiologic similarities. Streptobacillary RBF is caused by *Streptobacillus moniliformis,* and spirillary RBF (soduku) is caused by *Spirillum minus.*

The clinical spectra of streptobacillary and spirillary RBF overlap, preventing reliable clinical differentiation. Streptobacillary RBF has a short incubation period (3 to 10 days, rarely longer). If a rat bite has occurred, the wound usually heals uneventfully, although swelling and tenderness may be present initially. The illness begins abruptly with fever, chills, vomiting, headache, and myalgia. Shortly after onset, a maculopapular rash, most pronounced on the extremities, commonly develops. Arthritis, often involving multiple joints, occurs in more than 70% of cases. Endocarditis and focal abscesses may be late complications of untreated cases.

Spirillary RBF has a longer incubation period (1 to 3 weeks, usually more than 10 days). The bite wound often heals initially, but an inflammatory recrudescence occurs at the site coincidental with onset of systemic illness. The systemic illness is characterized by fever, headache, myalgia, nausea, vomiting, regional lymphadenopathy, and rash. The rash may be maculopapular or consist of red to purple plaques. Arthritis is unusual in spirillary RBF.

Untreated, both forms of RBF tend to run a chronic course, frequently taking months to resolve. About 7 to 10% of untreated cases result in death. Both diseases respond promptly to penicillin therapy, with tetracycline and streptomycin being effective alternative drugs.

Most cases of streptobacillary and spirillary RBF, including those acquired in the laboratory, result from rat bites (3, 7). Cases have infrequently resulted from the bite of other animals such as mice, weasels, cats, and dogs. Occasional cases have occurred in persons with no history of animal bites or scratches. Experimental treatment of syphilis with spirillary RBF induced by inoculation of naturally infected rodent blood has demonstrated that accidental inoculation of rodent blood could also potentially lead to disease. Streptobacillary fever has resulted from ingestion of contaminated milk. Aerosol transmission has not been documented, but available evidence suggests that this mode of transmission is a potential hazard.

Prevention is based on eliminating the infection from laboratory rodent colonies and on training personnel in appropriate animal handling techniques. A vaccine is not available. Prophylactic antimicrobial therapy for personnel who are bitten by rats from known infected colonies has been recommended, but the efficacy of prophylactic therapy has not been evaluated.

RICKETTSIAL INFECTIONS

Rocky Mountain Spotted Fever

Rickettsia rickettsii is the etiologic agent of Rocky Mountain spotted fever (RMSF). The incubation period for laboratory-acquired RMSF varies from 1 to 8 days. Affected personnel usually have sudden onset of fever, chills, photophobia, myalgia, and moderate to severe headaches. Although patients often develop a maculopapular rash that later becomes petechial, some individuals with laboratory-acquired RMSF never develop a rash (13). Rhinorrhea, sneezing, and cough may occur in personnel who acquire RMSF via the respiratory route. About 75 cases of laboratory-acquired RMSF have been reported (24). Eleven of the cases, including two fatal cases, occurred in the 1970s. Most of the affected personnel worked directly with *R. rickettsii* or infected ticks, whereas others merely entered laboratories where the organism was being handled. Cultures of *R. rickettsii* grown in yolk sacs or chicken embryo fibroblast cultures are the source implicated in most cases. Modes of transmission that have been responsible for laboratory-acquired cases include spread via droplets or the airborne route, direct inoculation during accidents involving needles and syringes, and bites inflicted by infected ticks. Direct person-to-person spread via droplet nuclei produced by the cough of patients with respiratory involvement has been suggested as a possible means of transmission, but this theory has never been proved.

Guidelines for containment of BSL 3 agents should be observed. Inoculation of cell cultures or yolk sacs of embryonated eggs, harvesting and centrifugation of infectious suspensions, and animal inoculation studies should be carried out in biosafety cabinets. Mouth pipetting should be strictly forbidden.

For many years, exposed personnel were immunized with a killed-cell yolk sac vaccine; however, direct-challenge studies have shown that this vaccine is relatively ineffective. An improved formalin-inactivated vaccine prepared in cell cultures is currently being evaluated; if this vaccine is shown to be efficacious, laboratory personnel at high risk of acquiring RMSF should be immunized.

Anecdotal reports suggest that prophylactic administration of tetracycline may protect persons involved in laboratory accidents with *R. rickettsii*. Exposed persons who develop signs and symptoms compatible with RMSF should be treated with tetracycline or chloramphenicol for 5 to 7 days.

Q Fever

Coxiella burnetii is the rickettsia that causes Q fever. The incubation period in laboratory-associated diseases varies from 13 to 28 days (usually 14 to 18) in cases acquired via the respiratory route but may be as short as 1 day in persons who have been accidentally inoculated with the organism. Affected individuals often have sudden onset of fever, chills, headache, myalgia, cough, and pleuritic chest pain. About 50% of patients have roentgenographic evidence of pneumonitis. Granulomatous hepatitis occurs in some patients. Relatively mild cases are often mistaken for influenza because affected persons do not develop a rash. Asymptomatic infections have been documented during serosurveys of exposed laboratory personnel. Weil-Felix agglutinin titers, which become elevated in other rickettsial diseases, remain negative in Q fever.

Coxiella burnetii is one of the most common causes of laboratory-acquired infections, with more than 400 cases having been reported (1, 24). Individuals handling the organism, as well as support personnel (e.g., maintenance workers, clerical personnel) employed in the same building, are often affected in institutional outbreaks (12, 32). Sources that have been implicated in cases of laboratory-associated Q fever include yolk sac suspensions of *C. burnetii*, naturally infected sheep and goats and possibly experimentally infected guinea pigs (or their excreta), contaminated laundry and clothing, and infected human tissues handled during postmortem examination.

Q fever is usually spread to laboratory workers via infectious aerosols produced when yolk sac suspensions are emulsified, centrifuged, pipetted, or injected into animals. If yolk sac antigens are prepared in rooms under positive air pressure, persons in adjacent laboratories or in remote areas of the same building may acquire the disease via airborne transmission. In the past decade, many research personnel and animal caretakers exposed to sheep (and their excreta) have acquired Q fever, presumably by inhaling the organism. Occasionally, recognized laboratory accidents, such as breaking glass vials containing *C. burnetii*, or sudden leaks in aerosol chambers or animal exposure bags have caused Q fever in exposed personnel. At least one laboratory worker has become infected by accidentally inoculating himself with viable organisms.

Adherence to recommended safety precautions is of paramount importance in preventing laboratory-associated Q fever. Inoculation and harvesting of yolk sac antigens should be carried out in laboratories that meet BSL 3 containment standards. Personnel working with suspensions of viable organisms must be familiar with established procedures for handling hazardous agents. Carcasses of infected animals should be placed in impervious containers until disposal, and animal caretakers should use measures that prevent aerosolization of dried animal excreta. Autoclaving contaminated media, glassware, and laundry before removal from the laboratory is desirable.

No Q fever vaccine is approved for use in humans in the United States. Prophylactic administration of tetracycline to persons involved in a known laboratory accident is of questionable value. Exposed employees who develop signs and symptoms suggestive of Q fever should be treated with tetracycline, doxycycline, or chloramphenicol. Ciprofloxacin is recommended for treatment of chronic infections.

Epidemic and Endemic Typhus

Rickettsia prowazekii and *Rickettsia typhi* are the etiologic agents of epidemic and endemic typhus, respectively. The incubation period in cases of laboratory-acquired typhus varies from 4 to 14 days. Affected personnel often have a sudden onset of fever, chills, headache, and myalgia. A macular rash may appear on the trunk on day 4 to 7 of the illness in unimmunized workers but is frequently absent in immunized individuals. Occasionally, patients will complain of upper quadrant tenderness and may have elevated serum aspartate aminotransferase values.

Approximately 110 cases of laboratory-acquired typhus have been reported (24). Infections occur primarily in persons working with the organisms and, to a lesser extent, among employees with other responsibilities in laboratories where the agents are being handled. Suspensions of rickettsia prepared in yolk sacs or passaged in guinea pigs are usually the source of laboratory-acquired typhus. Dried feces from infected lice have also been suggested as a possible source. Exposure to infectious droplets and airborne transmission account for most cases (44). Intranasal inoculation of mice and emulsifying suspensions of rickettsia in blenders are procedures that are likely to create infectious aerosols. A few laboratory personnel have acquired typhus by accidentally splashing or spraying suspensions of virulent organisms into their eyes.

To prevent laboratory-acquired typhus, research projects and vaccine production should be carried out in facilities that meet BSL 3 containment standards. For many years, persons at risk were immunized with typhus vaccines prepared from formaldehyde-inactivated *R. prowazekii*. The efficacy of this vaccine, however, had not been established in appropriate controlled trials. American and Canadian manufacturers no longer produce these vaccines. Laboratory personnel exposed to *R. prowazekii* or *R. typhi* should be treated for 4 to 5 days with tetracycline or chloramphenicol if they develop signs and symptoms suggestive of typhus. Recent studies have shown that a single dose of doxycycline may be effective therapy for epidemic typhus.

Scrub Typhus

Rickettsia tsutsugamushi is the etiologic agent of scrub typhus. After an incubation period of 1 to 3 weeks, affected persons experience sudden onset of fever, chills, headache, and myalgia and sometimes have a nonproductive cough. An eschar may develop at the site where the rickettsiae have entered the skin but is absent in patients who acquired the disease via the respiratory route. A macular or maculopapular rash appears 5 to 8 days after onset of fever in some, but not all, patients.

Thirty-five cases of laboratory-acquired scrub typhus have been reported (24). Most cases have occurred in personnel who worked directly with the organism. Suspensions of virulent *R. tsutsugamushi* are almost always the source of laboratory-acquired infections, although tissues obtained during postmortem examination of affected patients or experimentally infected animals have also served as sources. The disease has been transmitted to research workers via exposure to droplet nuclei with subsequent inhalation of the organism, direct contact (contamination of conjunctival mucosa or abraded skin), accidental inoculation during accidents involving needles or glassware, rat bites, and possibly mite bites (23).

BSL 3 precautions should be followed for work with *R. tsutsugamushi*. A satisfactory scrub typhus vaccine is not available.

Although prophylactic administration of chloramphenicol to exposed personnel will prevent the disease, chemoprophylaxis is not indicated in most situations. Personnel who develop an illness compatible with scrub typhus should be treated with tetracycline or chloramphenicol for 7 to 14 days. Relapses are fairly common if either drug is given for only 3 to 4 days but can be avoided by giving a second short course of therapy 4 to 5 days after stopping the initial treatment. A single 200-mg dose of doxycycline also appears to be effective.

TULAREMIA

Francisella tularensis may cause several clinical syndromes, including ulceroglandular, glandular, typhoidal, oculoglandar, primary pulmonary, and pharyngeal tularemia. In the ulceroglandular form, a papular lesion develops at the site where organisms enter the skin. The incubation period varies from 1 to 5 days. The papule frequently ulcerates and is accompanied by painful regional lymphadenopathy. Immunized laboratory personnel may or may not develop concomitant fever, myalgia, headache, and diarrhea. In primary pulmonary and typhoidal tularemia, fever, sore throat, substernal discomfort, and cough develop after an incubation period of 1 to 10 days. Occasionally, affected individuals also complain of cervical lymphadenopathy. Although 30 to 45% of laboratory workers with typhoidal tularemia have evidence of bronchopneumonia by chest roentgenogram, physical examination of the chest is often unimpressive. Oculoglandular tularemia is characterized by painful, purulent conjunctivitis with preauricular or cervical lymphadenopathy. Pharyngeal tularemia produces an acute exudative pharyngotonsillitis with cervical lymphadenopathy.

Multiple episodes of laboratory-acquired tularemia (reinfections) have been documented in persons who received killed-cell vaccines and in unimmunized individuals. In some cases of reinfection, the severity and duration of symptoms may approach that seen in initial infections.

In the past 50 years, tularemia has been the third most common bacterial cause of laboratory-acquired infections, with at least 225 cases having been associated with facilities involved in tularemia research as well as a few cases among personnel in clinical microbiology laboratories (24). Bacteriologists, physicians, and laboratory technologists who work with the organism are at greatest risk, but cases have also occurred in animal caretakers, maintenance personnel, and employees responsible for washing glassware used in research laboratories.

Cultures of *F. tularensis* are the source most commonly implicated in cases of laboratory-acquired tularemia (22). As few as 10 to 50 organisms can result in disease after inhalation as an aerosol or inoculation beneath the skin (28, 29). On a few occasions, naturally or experimentally infected animals have been the source. Laboratory workers may acquire tularemia in several ways. Infection may occur when organisms come into direct contact with cuts, cracks, or abrasions on the skin. Several cases have occurred when persons cut themselves with contaminated autopsy instruments, glassware, or ampoules containing *F. tularensis*. Accidental needle punctures sustained during animal inoculation studies have caused ulceroglandular tularemia, and at least one immunized worker acquired oculoglandar disease when a needle came off a syringe and virulent organisms were sprayed into his eye. Accidents associated with mouth pipetting may cause pharyngeal tularemia.

Many laboratory workers have acquired typhoidal tularemia via infectious droplets and by the airborne route. The respiratory tract is usually the portal of entry when these modes of transmission are implicated. Personnel who perform high-risk procedures such as centrifugation and lyophilization of cultures are usually affected, but several persons who merely entered laboratories have acquired the disease via airborne transmission.

An investigational live attenuated tularemia vaccine is currently available. This vaccine does not prevent overt clinical illness but does reduce the severity of illness (5). Prevention of infection is based on adherence to BSL 3 standards.

The role of chemoprophylaxis for immunized persons exposed to a laboratory accident involving infectious material is unknown. Studies conducted before the availability of the live attenuated vaccine demonstrated that a 5-day course of streptomycin usually prevented laboratory-acquired tularemia. Tetracycline has also been demonstrated to be an effective chemoprophylactic drug (30). Because typhoidal tularemia is uncommon among immunized personnel, postexposure chemoprophylaxis is probably not indicated.

Persons who develop laboratory-acquired tularemia should be treated with streptomycin given intramuscularly for 7 to 10 days. Gentamicin, tetracycline, and chloramphenicol are acceptable alternatives, but clinical relapses occur more frequently when the latter two drugs are used.

NATURALLY INFECTED LABORATORY ANIMALS

Naturally infected laboratory animals are another potential biohazard for laboratory personnel. However, the reported spectrum of bacterial infections acquired from the commonly used laboratory animals is limited.

Rodents and Rabbits

Rodents and rabbits comprise more than 99% of all animals used in laboratories. Although many thousands of people are directly exposed each year, few bacterial infections have been reported as being acquired from these animals in the laboratory.

RBF is the most commonly reported bacterial disease associated with laboratory animals. Leptospirosis has been reported slightly less frequently. Most leptospirosis infections have resulted from mouse or rat bites. The infecting Leptospira serovar has been *ballum* in cases associated with mice and *icterohaemorrhagiae* in cases associated with rats.

Other bacterial infections are so rare as to hardly merit discussion. For example, two cases of possible *Bordetella bronchicanis (bronchiseptica)* infection have been reported in animal caretakers with, respectively, a chronic rhinorrhea and an influenzalike illness, and a single case of *Pasteurella multocida (lepiseptica)* wound infection caused by a rabbit bite has been reported (2, 43).

Nonhuman Primates

Tuberculosis and bacterial gastroenteritis are the primary bacterial diseases associated with nonhuman primates (14). Tuberculosis has been reported as a spontaneous disease in most primate species commonly used in the laboratory. The incidence in recently imported animals varies with their geographic origin, being most common (1 to 2% incidence) in primates from Asia, less frequent in African species, and rare in New World species. Exposure to tuberculous primates in holding facilities after importation, however, may result in a high infection rate regardless of the group's original geographic origin.

Simian tuberculosis may be caused by either *Mycobacterium tuberculosis* or *Mycobacterium bovis*. Infections by other mycobacteria such as *M. kansasii* have been reported, but secondary transmission of these infections to laboratory personnel has not been documented.

Simian tuberculosis is primarily spread by aerosols of infectious droplet nuclei. Shedding of *M. tuberculosis* or *M. bovis* via the respiratory tract usually precedes the development of tuberculin hypersensitivity, a factor that complicates control programs. Transmission via bites, ingestion, and fomites such as tattoo needles and thermometers has been reported, but these mechanisms of spread are of lesser importance. Primates are extremely susceptible to tuberculosis, and the disease spreads rapidly once introduced into a colony.

Persons working with primates are at high risk of tuberculous infection. In the United States, the annual tuberculin conversion rate in persons occupationally exposed to primates is about 70 per 10,000 compared with less than 3 per 10,000 in the general population (15). Most infections appear to result from inhaling infectious aerosols while working in areas where tuberculous animals are housed. At least one case of cutaneous tuberculosis caused by a scalpel puncture during necropsy of a monkey has been reported.

Simian tuberculosis control is based on routine quarantine and tuberculin testing of all new additions to the colony, as well as periodic tuberculin testing of all primates in the maintenance colony (15). All tuberculin-positive animals should be immediately killed or placed in strict isolation if further evaluation is indicated. Treatment is not recommended except in unusual circumstances. Because tuberculosis can be transmitted from nonhuman primates to their handlers as well as the reverse, all persons who work around these animals should be included in a tuberculosis screening program. High-efficency face masks or other suitable respiratory protective devices should be routinely worn while working in primate-holding areas.

Bacterial gastroenteritis, primarily caused by *Shigella, Salmonella,* and *Campylobacter jejuni,* is the leading cause of morbidity and mortality among laboratory primates. Anecdotal information suggests that persons working with primates are also at high risk of acquiring diarrheal disease; at least 71 *Shigella,* 2 *Salmonella,* and 1 *C. jejuni* infections have been reported as being acquired from these animals. The fecal-oral route of transmission is of primary importance. Control is based on adherence to scrupulous personal hygiene practices while working with primates. Water-impervious gloves should be worn while handling primates and their excretions, and no eating or smoking should be allowed in primate-holding areas.

References

1. **Bernard, K. W., G. L. Parham, W. G. Winkler, and C. G. Helmick.** 1982. Q fever control measures: recommendations for research facilities using sheep. *Infect. Control* **3:**461–465.
2. **Boisvert, P. L., and M. D. Fousek.** 1941. Human infection with *Pasteurella lepiseptica* following a rabbit bite. *JAMA* **116:**1902–1903.
3. **Brown, T. M., and J. C. Nunemaker.** 1942. Rat-bite fever: a review of the American cases with reevaluation of etiology; report of cases. *Bull. Johns Hopkins Hosp.* **70:**201–328.
4. **Buchstein, S. R., and P. Gardner.** 1991. Lyme disease, p. 103–116. *In* A. N. Weinber and D. J. Weber (ed.), *Animal-associated human infections. Infectious disease clinics North America,* vol. 5. The W. B. Saunders Co., Philadelphia.
5. **Burke, D. S.** 1977. Immunization against tularemia: analysis of the effectiveness of live *Francisella tularensis* vaccine in prevention of laboratory-acquired tularemia. *J. Infect. Dis.* **135:**55–60.
6. **Burmeister, R. W., W. D. Tigertt, and E. L. Overholt.** 1962. Laboratory-acquired pneumonic plague: report of case and review of previous cases. *Ann. Intern. Med.* **56:**789–800.
7. **Cole, J. S., R. W. Stroll, and R. J. Bulger.** 1969. Rat-bite fever: report of three cases. *Ann. Intern. Med.* **71:**979–981.
8. **Ellingson, H. V., P. J. Kandull, H. L. Bookwalter, and C. Howe.** 1946. Cutaneous anthrax: report of twenty-five cases. *JAMA* **131:**1105–1108.
9. **Felsenfeld, O.** 1971. *Borrelia: strains, vectors, human and animal borreliosis.* Warren H. Green, Inc., St. Louis.
10. **Green, R. N., and P. G. Tuffnell.** 1968. Laboratory-acquired melioidosis. *Am. J. Med.* **44:**599–605.
11. **Howe, C., and W. R. Miller.** 1947. Human glanders: report of six cases. *Ann. Intern. Med.* **26:**93–115.
12. **Johnson, J. E., III, and P. J. Kandull.** 1966. Laboratory-acquired Q fever: a report of fifty cases. *Am. J. Med.* **41:**391–403.
13. **Johnson, J. E., III, and P. J. Kandull.** 1967. Rocky Mountain spotted fever acquired in a laboratory. *N. Engl. J. Med.* **277:**842–847.

14. **Kaufmann, A. F.** 1972. Nonhuman primate zoonoses surveillance in the United States, p. 58–67. *In* E. I. Goldsmith and J. Moor-Jankowski (ed.), *Medical primatology 1972*, part III. S. Karger, Basel.
15. **Kaufmann, A. F., and D. C. Anderson.** 1978. Tuberculosis control in nonhuman primate colonies, p. 227–234. *In* R. J. Montali (ed.), *Mycobacterial infections of zoo animals.* Smithsonian Institution Press, Washington, D.C.
16. **Kaufmann, A. F., M. D. Fox, J. M. Boyce, D. C. Anderson, M. E. Potter, W. J. Martone, and C. M. Patton.** 1980. Airborne spread of brucellosis. *Ann. N. Y. Acad. Sci.* **335:**105–114.
17. **Leitenberg, M.** 1992. Anthrax in Sverdlovsk: new pieces to the puzzle. *Arms Control Today* **22**(3):10–13.
18. **McClain, B. L., W. R. Ballou, S. M. Harrison, and D. L. Steinweg.** 1984. Doxycycline therapy for leptospirosis. *Ann. Intern. Med.* **100:**696–698.
19. **Miller, C. D., J. R. Songer, and J. F. Sullivan.** 1987. A twenty-five year review of laboratory-acquired human infections at the National Animal Disease Center. *Am. Ind. Hyg. Assoc. J.* **48:**271–275.
20. **Morisset, R., and W. W. Spink.** 1969. Epidemic canine brucellosis due to a new species, *Brucella canis. Lancet* **ii:**1000–1002.
21. **Olle-Goig, J. D., and J. Canela-Soler.** 1987. An outbreak of *Brucella melitensis* infection by airborne transmission among laboratory workers. *Am. J. Public Health* **77:**335–338.
22. **Overholt, E. L., W. L. Tigertt, P. J. Kandull, M. Salzman, and M. Stephens.** 1961. An analysis of forty-two cases of laboratory-acquired tularemia. *Am. J. Med.* **30:**785–806.
23. **Philip, C. B.** 1948. Tsutsugamushi disease (scrub typhus) in World War II. *J. Parasitol.* **34:**169–191.
24. **Pike, R. M.** 1976. Laboratory-associated infections: summary and analysis of 3921 cases. *Health Lab. Sci.* **13:**105–114.
25. **Pike, R. M., S. E. Sulkin, and M. L. Schulze.** 1965. Continuing importance of laboratory-acquired infections. *Am. J. Public Health* **55:**190–199.
26. **Redfearn, M. S., and N. J. Pallerone.** 1975. Glanders and melioidosis, p. 110–128. *In* W. T. Hubbert, W. F. McCulloch, and P. R. Schnurrenberger (ed.), *Diseases transmitted from animals to man*, 8th ed. Charles C Thomas, Publisher, Springfield, Ill.
27. **Rich, V.** 1992. Russia: anthrax in the Urals. *Lancet* **339:**419–420.
28. **Saslaw, S., H. T. Eigelsbach, H. R. Wilson, J. A. Prior, and S. R. Carhart.** 1961. Tularemia vaccine study. I. Intracutaneous study. *A.M.A. Arch. Intern. Med.* **107:**689–701.
29. **Saslaw, S., H. T. Eigelsbach, H. R. Wilson, J. A. Prior, and S. R. Carhart.** 1961. Tularemia vaccine study. II. Respiratory challenge. *A.M.A. Arch. Intern. Med.* **107:**702–714.
30. **Sawyer, W. D., H. G. Dangerfield, A. L. Hogge, and D. Crozier.** 1966. Antibiotic prophylaxis and therapy of airborne tularemia. *Bacteriol. Rev.* **30:**542–548.
31. **Schacter, J., and C. R. Dawson.** 1977. *Human chlamydial infections*, p. 9–43. PSG Publishing Co., Littletown, Mass.
32. **Schacter, J., M. Sung, and K. F. Meyer.** 1971. Potential danger of Q fever in a university hospital environment. *J. Infect. Dis.* **123:**301–304.
33. **Shapiro, E. D., M. A. Gerber, N. B. Holabird, A. T. Berg, H. M. Feder, G. L. Bell, P. N. Rys, and D. H. Persing.** 1992. A controlled trial of antimicrobial prophylaxis for Lyme disease after deer-tick bites. *N. Engl. J. Med.* **327:**1769–1773.
34. **Spink, W. W.** 1956. *The nature of brucellosis*, p. 106–108. University of Minnesota Press, Minneapolis.
35. **Staszkiewicz, J., C. M. Lewis, J. Covillle, M. Zervos, and J. Band.** 1991. Outbreak of *Brucella melitensis* among microbiology laboratory workers in a community hospital. *J. Clin. Microbiol.* **29:**287–290.
36. **Steere, A. C.** 1989. Lyme disease. *N. Engl. J. Med.* **321:**586–596.
37. **Stoenner, H. G., and D. Maclean.** 1958. Leptospirosis (ballum) contracted from Swiss albino mice. *Arch. Intern. Med.* **101:**606–610.
38. **Sulkin, S. E., and R. M. Pike.** 1951. Survey of laboratory-acquired infections. *Am. J. Public Health* **41:**769–781.
39. **Sullivan, J. F., J. R. Songer, and I. R. Estrem.** 1978. Laboratory-acquired infections at the National Animal Disease Center 1960–1976. *Health Lab. Sci.* **15:**58–64.
40. **Takafugi, E. T., J. W. Kirkpatrick, R. N. Miller, J. J. Karwacki, P. W. Kelley, M. R. Gray, K. M. McNeil, H. L. Timboe, R. E. Kane, and J. L. Sanchez.** 1984. An efficacy trial of doxycycline chemoprophylaxis against leptospirosis. *N. Engl. J. Med.* **310:**497–500.
41. **Tindall, J. P., and C. M. Harrison.** 1972. *Pasteurella multocida* infections following animal injuries, especially cat bites. *Arch. Dermatol.* **105:**412–416.
42. **Watt, G., L. P. Padre, M. L. Tuazon, C. Calubaquib, E. Santiago, C. P. Ranoa, and L. W. Laughlin.** 1988. Placebo-controlled trial of intravenous penicillin for severe and late leptospirosis. *Lancet* **i:**433–435.
43. **Winsser, J.** 1960. A study of *Bordetella bronchiseptica. Proc. Anim. Care Panel* **10:**87–104.
44. **Wright, L. J., L. F. Barker, I. D. Mickenberg, and S. M. Wolff.** 1968. Laboratory-acquired typhus fevers. *Ann. Intern. Med.* **69:**731–738.
45. **Wu, L. T.** 1926. *A treatise on pneumonic plague*, p. 100–105. League of Nations, Geneva.
46. **Wu, L. T., J. W. Chun, R. Pollitzer, and C. Y. Wu.** 1936. *Plague: a manual for medical and public health workers*, p. 516–521. The Mercury Press, Shanghai.

Transmission and Control of Viral Zoonoses in the Laboratory

WILLIAM G. WINKLER AND DONALD C. BLENDEN

8

INTRODUCTION

Accidentally acquired infections are an inherent risk in any laboratory that intentionally or unknowingly handles infectious microorganisms. Laboratory-acquired viral zoonoses are undoubtedly underreported; still, there is ample evidence to show that such infections occur with disturbing frequency, and some of these have the potential for causing very serious or fatal illness.

This chapter discusses the risks associated with zoonotic viruses in the laboratory and ways in which these risks may be minimized. Many of the viruses discussed here, as well as other zoonotic viruses, cause infection and disease in nonlaboratory settings beyond the purview of this text.

In addition to the risk levels attributable to the organism, accidents in the laboratory increase the risk of laboratory infections. Accident-prone persons are at innately higher risk of infection than others; employers should be cognizant of this fact in assigning employees to work areas (39).

The Centers for Disease Control and Prevention (CDC) and the National Institutes of Health manual *Biosafety in Microbiological and Biomedical Laboratories* (44), reprinted as Appendix I in this book, provides safety recommendations for any microbiological laboratory. It should be a required reference in laboratories handling infectious microorganisms. Also included in the CDC/NIH manual are summaries of the infectious agents associated with those viral zoonoses most often found in the laboratory, human disease surveillance information, and shipping guidelines. The recommended biosafety levels (BSL) for the type of work and the microorganisms handled should be understood by all laboratory personnel working with infectious agents. This is an important requirement in preparing laboratory personnel to work in the appropriate manner to minimize risk to themselves and others.

The sources of laboratory-acquired infections can be conveniently divided into two types: those in which infection results from exposure to known agents being manipulated in the laboratory and against which specific protective measures should be in place, and those in which unknown or unsuspected agents are introduced into the laboratory and against which specific protective measures very likely may not have been taken. The former type of infection is likely to occur in research laboratories working with zoonoses if appropriate barriers are not used. The second type of infection is most likely to occur in diagnostic laboratories where material of unknown background is regularly introduced into the laboratory (29, 34). In the first instance, in which the spectrum of zoonotic viruses in use in the laboratory is known, it is appropriate to use specific protective steps against those agents including personnel training as well as biological and physical barriers to infection. One potential problem often seen in this type of laboratory is the complacency that may develop as personnel become accustomed to working with known zoonotic agents, even highly pathogenic organisms (e.g., "familiarity breeds contempt"). In the second instance, in which the zoonotic viruses are unknown or unsuspected, it is necessary to create a more generic protective shield and to remind personnel routinely of the continuing risks.

This chapter reviews some common zoonotic viruses and the risks associated with their use in the laboratory setting. For organizational purposes, without regard to frequency or severity of infection, these zoonoses are reviewed by virus families listed in alphabetical order.

ARENAVIRIDAE

The small family *Arenaviridae* includes 10 viruses related by similarities in structural and physicochemical properties (14, 23, 28). Those of laboratory safety importance include lymphocytic choriomeningitis (LCM) virus, Lassa fever virus, Korean hemorrhagic fever (KHF) virus, Machupo virus, and Junin virus. These latter two viruses are the etiologic agents for Bolivian hemorrhagic fever and Argentinian hemorrhagic fever, respectively.

A new strain of KHF virus, previously reported mainly from Asia, has been associated with a fatal respiratory disease in previously healthy young adults in the Four Corners region of Colorado. LCM virus occurs throughout much of the world, Lassa fever virus occurs in Africa, Machupo virus and Junin virus occur in Latin America, and the remaining Tacaribe complex viruses occur in Latin America or occasionally Florida. The serologic relationship between the arenaviruses, to some extent, reflects their geographic proximity. The Old World viruses are identifiable by immunofluorescence as a group as are the New World viruses, and all are recognizably related to the ubiquitous LCM virus.

Rodents serve as the important animal reservoirs for arenaviruses, often with only a single species of rodent serving as the natural host for a given virus type. Although rodents are considered the natural hosts of most arenaviruses, it should be understood that in the laboratory setting other species may become infected. For example, although the house mouse (*Mus musculus*) is the natural host of LCM virus, guinea pigs, dogs, swine, primates, and chick embryos have all been shown to be susceptible to LCM infection in the laboratory environment (35, 46).

Some arenaviruses, most notably LCM virus, may produce nonclinical disease with virus shedding. LCM infection is well recognized for its propensity for inducing these persistent, inapparent infections in rodent hosts and for producing virus carrier-shedder states that may last for the life of the infected rodent. This carrier state is thought to be established mainly when infection occurs in newborn or young mice (20–22, 26).

LCM virus transmission to other rodents usually results from exposure to infectious urine, feces, contaminated aerosols, or fomites as dust, food, water, or bedding associated with the infected animals. Bite wounds and vertical transmission may also occur. Modes of transmission to humans are similar, with the exception of vertical transmission, which has yet to be documented. Naturally infected rodents (*Mus musculus*) also may invade laboratory rodent colonies, especially feed storage areas, and constitute a potential exposure risk to both resident animals and laboratory personnel.

Other arenaviruses may be transmitted by routes similar to those seen with LCM. The subclinical infections common with LCM do not appear to occur with many arenaviruses. The exceptions include the Lassa fever virus, which may produce a persistent, inapparent infection in young Mastomys rats (32), the Machupo virus infection in its natural host, the Cricetine rodent *Calomys callosus,* which does not produce acute disease nor, in some circumstances, does Junin virus infection in its natural hosts, Cricetine rodents. Recent studies also indicate that the new strain of KHF virus appears asymptomatic in the natural host *Peromyscus peromyscus* (5–7).

Lassa fever virus is among the more pathogenic of the arenaviruses and is particularly infectious for laboratory rodents and personnel (30, 31). Its natural host, the multimammate rat *Mastomys natalensis,* lives in sub-Saharan Africa, where it is found in rural areas adjacent to and, at times, within human dwellings (32, 47). Transmission of the virus from rodents to humans may occur through handling of the infected rodents or contaminated fomites or by inhalation of infected aerosols. The severity of illness appears to depend on the mode of transmission (32). Like LCM infection, early exposure of susceptible rodents will often produce an infected carrier animal that may shed virus for its entire lifetime without developing any overt illness (32). Medical personnel have become infected after contact with human cases of Lassa fever.

The virus of KHF is not well studied, and its importance as a source of laboratory infections is not yet known, but a few cases of laboratory-acquired infections have been reported (37, 43). Experimental studies have found that the virus is strongly species-specific to its natural host, the rat (*Apodemus agraria*), and to tissue cultures developed from this host (43). Thus, the importance of KHF virus in the laboratory would seem to be very limited. However, several hemorrhagic uremic diseases are similar to KHF, and these may, in fact, be caused by the same or closely related viruses with broader host and geographic ranges, and their laboratory importance is as yet unknown. The recent outbreak of KHF in the United States has been related to a different natural host, *Peromyscus* spp., and transmission has been associated with contact with aerosols of mouse urine. The CDC has published guidelines to protect those who must go into the field to trap the natural host (5). When the agent is cultured in a reference laboratory, it is handled with BSL 4 practices (5).

Among the arenaviruses, LCM virus is the one most commonly found in rodent laboratory colonies as an adventitious agent. It may well be the most common pathogen of all rodent viruses found as a contaminant in laboratory colonies. Virus has been introduced to colonies by infected feral rodents that had access to the animal colony by introduction in contaminated tissue cultures and contaminated tumor tissues and by introduction of infected laboratory rodents into the animal colony. Arthropods (roaches) have been reported as a mechanical vector for the spread of LCM. Several outbreaks occurred in the United States in the 1970s after intermixing of hamsters in the pet trade with those in laboratory animal colonies (16). Many cases of human illness resulted, in pet owners, laboratorians, and even office personnel who had casual contact with laboratory animals (17). Introduction of LCM virus into a rodent colony may lead to widespread infection in the colony without overt clinical sign and symptoms. Subsequent stress of the animals, as in experimental procedures or immunosuppressed animals, may trigger clinical illness (9).

At least two cases of laboratory-origin Lassa fever have been recorded; one of these was a fatal case. One fatal case of Machupo virus infection of laboratory origin has been reported. At least 21 human cases of Junin virus infection have been transmitted in the laboratory setting; one of these was fatal. With all arenaviruses, as with LCM virus, all tissues, fluids, excreta, and secretions from infected laboratory animals should be presumed infected and disposed of in an appropriate manner.

Prevention and Control

The arenaviruses, especially LCM virus, Lassa fever virus, and the new strain of KHF virus, should be considered highly infectious and easily transmitted in the confines of the laboratory. Steps should be taken to minimize the creation of aerosols that might be contaminated with urine, feces, or other contaminated materials. Bedding and excreta should be disposed of in a safe and sanitary manner (i.e., autoclave cages before removal of bedding). In colonies where infection is known to exist or considered a likely occurrence, it may be desirable to maintain colony animals in filter-top cages until such time as the infection can be eliminated. Laboratory personnel, similarly, should be protected from exposure to aerosols and contaminated fomites. Immunization is not available for prevention of LCM, Lassa fever, or KHF. Plasma from recovered Lassa fever cases has been used in treatment of clinical human cases, but the value of such treatment is unclear. Various antiviral agents have also been used with similarly uncertain results.

LCM virus requires BSL 3 practices and facilities for risk management. All other arenaviruses require BSL 4, the highest containment category (44). Details of the criteria defined by these levels are described in Appendix I and in Chapter 13.

BUNYAVIRIDAE

More than 200 viruses are classified in the *Bunyaviridae* family, and these are subdivided into 18 subgroups (1, 3, 41). Some 80 of the members of the group belong to the genus *Bunyavirus*. Other genera in the *Bunyaviridae* family include *Phlebotomus* (sandfly fever group), *Nairovirus* (Nairobi sheep disease or Ganjam disease), and *Ukuvirus* (Uukuniemi and related viruses) (Table 1). Biologically, all are arthropod-borne viruses (arboviruses), and most are transmitted by mosquitoes or midges. Most are pathogenic for animals; a few are pathogenic for humans (1, 35). Those of importance in the laboratory environment include Rift Valley fever (RVF) and Crimean-Congo hemorrhagic fever (CCHF). The former causes illness in humans and animals; the latter produces clinical illness in humans but not in animals.

Viruses of the *Bunyaviridae* family infect various warm- and cold-blooded vertebrates and arthropods; humans are an accidental host. In animals, infection may occur without overt illness or may result in severe illness and death (1, 3, 41). The same spectrum of clinical disease is also seen in humans. Major outbreaks of RVF have occurred in recent

TABLE 1 Bunyaviruses of major zoonotic importance

Genus/subtype	Virus	Vector	Human disease
Bunyavirus	Bunyamwera	Mosquito	Fever
	Tensaw	Mosquito	Encephalitis
	California	Mosquito	Encephalitis
	LaCrosse	Mosquito	Encephalitis
	Keystone	Mosquito	Encephalitis
Phlebotomus	Sicilian SF	Phlebotomine	Fever, rash
	Candira	Phlebotomine	Fever, rash
	RVF	Mosquito	Fever, encephalitis
Nairovirus	CCHF	Tick	Hemorrhagic fever
	Nairobi sheep disease (Ganjam)	Tick	Fever

years in hot sub-Saharan Africa (its traditional distribution) as well as in Upper and West Africa, areas not previously noted as endemic (13, 42). In both areas, the infection in sheep and cattle resulted in heavy livestock mortality and was accompanied by human disease and deaths.

RVF infects livestock (sheep, cattle, goats, buffalo, camels) and some wild hoofed animals (antelope) as well as wild rats (*Arvicanthus* spp.) (11, 12). CCHF affects only humans. Nairobi sheep disease affects only livestock.

Transmission of virus of the *Bunyaviridae* family in nature is most often via mosquitoes (RVF) and in some cases by ticks (CCHF) or biting flies (phlebotomus fever) (1, 15, 37, 41). Several encephalitogenic bunyaviruses are mosquito-transmitted.

In the laboratory, infection through contact with infected blood or other tissues or through aerosols containing infectious material are common routes of transmission. Laboratory cases of RVF after inhalation of contaminated floor dust have been described (25, 40). In addition to RVF virus laboratory infections, Germiston and Apeu virus laboratory infections have also been described. It is appropriate to assume that all bunyaviruses are capable of causing human infections in the laboratory settings.

Prevention and Control

Vaccine is available for persons who work with RVF virus in the laboratory. No vaccines are generally available against the other *Bunyaviridae* viruses. General preventive measures should include special attention to personal hygiene, appropriate outer clothing, and careful handling of animals and their excreta. Procedures should minimize the risk of exposures to potentially infectious aerosols. Control of arthropod vectors is essential. Although seldom a problem in the contained laboratory setting, arthropod transmission may be a problem in laboratories that hold livestock in outdoor pens.

The American Committee on Arboviruses has registered more than 400 viruses as arboviruses and placed them all in BSL 2 through 4 as reported in that committee's publication *Laboratory Safety for Arboviruses and Certain Other Viruses of Vertebrates* (37). Ninety-four are assigned in BSL 3 or 4; 7, which are commonly used in the laboratory, are assigned to BSL 2.

HERPESVIRIDAE

The *Herpesviridae* family forms a large group of morphologically similar viruses, and representatives of the group probably can be isolated from all vertebrate species. Based on genome characteristics and properties such as host range, reproductive cycle, and antigenic relationships, three subfamilies have been established: *Alphaherpesvirinae*—herpes simplex viruses 1 and 2, varicella zoster virus, bovine mammilitis virus, pseudorabies virus of swine, equine abortion virus, herpesvirus simiae (B virus), and others as probable members of the group; *Betaherpesvirinae*—cytomegalovirus of humans and animals; and *Gammaherpesvirinae*—Epstein-Barr virus of humans and several other probable and possible members including herpesvirus saimiri, herpesvirus ateles, and Marek's disease virus of chickens (28).

Herpesviruses are relatively labile, and close contact is usually required between infected individuals and susceptibles to effect transmission. Virus is commonly shed in large quantities in vesicular lesions but has also been reported in ororespiratory and urogenital secretions, conjunctival sacs, milk, and urine. Venereal transmission is well documented. Transmission from mother to offspring may occur in the birth canal at parturition (15, 36).

A feature of many viruses, but particularly important for herpesviruses, is the ability of the virus infection to remain latent and unrecognized in its host for years and to exacerbate into clinical disease as a result of stress, immune deficiency, or other factors altering the host's immune system. Virus may be shed by asymptomatic individuals. Another characteristic of importance in herpesvirus infections is the severe disease produced in aberrant hosts by viruses that cause only mild disease in their usual hosts. In fact, the herpesvirus of greatest concern in the laboratory is herpesvirus simiae (B virus) because of the severe and often fatal outcome of human infection (15, 24, 45). Herpesvirus simiae, normally found in Old World monkeys, especially macaques, produces mild to asymptomatic disease in macaques. The clinical disease in macaques is analogous to herpes simplex virus infection in humans; it produces inapparent or mild illness and may cause oral ulcers. In the animal facility and laboratory setting, the appearance of oral ulcers in macaques is often taken as a sign of herpesvirus simiae (B virus) infection, although infection with shedding may occur with or without the presence of oral lesions.

Nonhuman primates have been responsible for severe and fatal disease in laboratory workers infected with herpesvirus simiae (2). At least 25 cases of herpesvirus simiae (B virus) infection have been reported in persons working with nonhuman primates or their tissues. Twenty of these cases were fatal; three of the survivors suffered severe neurologic impairment. The modes of transmission for these cases were primarily monkey bites, scratch wounds, and penetration of intact skin by contaminated needles or broken glassware.

Other herpesviruses including pseudorabies and herpes simplex and varicella zoster viruses are recognized human pathogens and may be of some concern in the animal care laboratory because of their potential to cause disease in nonhuman primates. Humans infected with herpes simplex virus have been a source of severe disease for laboratory primates, including a fatal outbreak in gibbon apes (1, 36, 45). Transmission of other herpesviruses between humans and nonhuman primates in the laboratory setting is less well defined and seemingly of lesser importance.

Prevention and Control

There is no vaccine or pharmacological product available for prevention of herpesvirus infection. Human plasma containing herpesvirus antibodies has been recommended for treatment of persons with clinical herpesvirus simiae (B virus) disease, but the value of this therapy remains unproved. Physical separation of laboratory workers from infectious material is the most effective preventive measure available. BSL 2 practices, containment equipment, and facilities are recommended for any activities involving the handling or manipulation of nonhuman primate tissues and fluids (44). Nonhuman primates should be handled only by properly trained personnel wearing appropriate protective clothing. These precautions are especially important for prevention of transmission of herpesvirus simiae (B virus) to laboratory workers.

ORTHOMYXOVIRIDAE

The *Orthomyxoviridae* family includes both the human and animal influenza viruses. This antigenically complex group of viruses is divided into three types, A, B, and C, according to nucleoprotein antigenic structure (27, 28). These are further subdivided into many subtypes based on surface antigens (i.e., neuraminidase and hemagglutinin) (27, 38). The main mammalian hosts for type A include, in addition to humans, domestic and wild birds, swine, and horses. Ferrets are also easily infected in the laboratory. Type B influenza is reported only in humans. Type C influenza, also found in humans, has not been adequately studied to establish its species limitations or other characteristics.

Influenza viruses characteristically produce respiratory and systemic manifestations in both humans and animals that may range from mild to severe. Recovery is usually rapid but may be complicated by bacterial superinfection.

Transmission of type A influenza between animals and humans has been reported in the laboratory, but such events are uncommon either because transmission is rare or because the resulting infection is seldom severe and may be difficult to differentiate from disease acquired naturally outside of the laboratory. The ferret, which because of its susceptibility to human influenza viruses is the laboratory animal of choice for much in vivo influenza work, is also a significant potential source of laboratory infections. Ferrets infected experimentally or via natural exposure to infected humans or animals are a recognized source of human infection (10).

Transmission of influenza viruses in the laboratory mimics that which occurs in nature. Most often, sneezing and coughing by infected individuals releases large numbers of virus particles into the atmosphere, which as aerosols may then be inhaled by susceptible individuals. In the laboratory, aerosols also may be generated by sonication, homogenization, or centrifugation of infected material. Transmission also may occur by contact with infected fomites, but as these viruses are rapidly inactivated by heat, sunlight, and desiccation, such transmission is less important.

Although infection may confer immunity in immunologically competent individuals, the frequent drift (small changes) or shift (major changes) in antigenic structure that is characteristic of the myxoviruses tends to circumvent this acquired protection, and reinfection with related viruses is not uncommon.

Prevention and Control

Vaccine is not available for animals; vaccine is available for humans, and its effectiveness in preventing infection may vary with the strains of influenza that are currently prevalent, the antigenic composition of the vaccine used, and the immunologic competence of the vaccinated individual. Vaccination is recommended for anyone working in the laboratory with influenza virus and for any person older than 60 years of age. BSL 2 practices, containment equipment, and facilities are recommended when working with orthomyxoviruses in the laboratory (44).

PARAMYXOVIRIDAE

The *Paramyxoviridae* family contains three genera: *Paramyxovirus* (parainfluenza, mumps, Newcastle disease viruses [NDV]); *Morbillivirus* (measles, canine distemper, rinderpest viruses); and *Pneumovirus* (respiratory syncytial virus). All family members share a common general morphology, but antigenic relationships exist only within the genera (28).

The only paramyxovirus of significance as a laboratory risk for humans is the NDV. Most members of the family are essentially human pathogens, usually of limited pathogenicity, and are not normally infectious for laboratory animals. Some parainfluenza viruses can infect both humans and animals. There is a close antigenic relationship between some parainfluenza viruses of humans and parainfluenza viruses that infect sheep, horses, water buffalo, and some deer species (18, 19). Antibodies against respiratory syncytial viruses have been identified in nonhuman primates, cattle, dogs, horses, and swine.

NDV, normally a pathogen of birds, is readily transmitted to humans in work environments where infected birds are housed. These include poultry processing plants and laboratories or animal colonies where birds, especially gallinaceous birds, are used. The primary clinical manifestation of NDV in humans is conjunctivitis, often unilateral (1, 8). The foreign strains of NDV (viscerotropic velogenic strains), so devastating to domestic poultry flocks, appear to be no more pathogenic for humans than the domestic (lentogenic) strains.

Prevention and Control

Prevention of human infection with NDV is best accomplished by excluding the virus from the laboratory environment. Prevention of human infection in facilities contaminated or working with the virus requires protection against infectious aerosols and protection of mucous membranes against exposure to the virus.

PICORNAVIRIDAE

The *Picornaviridae* family is divided into four genera: *Enterovirus* (poliovirus, coxsackievirus, swine vesicular disease, echovirus, and animal enteroviruses); *Cardiovirus* (encephalomyocarditis virus); *Rhinovirus* (common cold, animal rhinoviruses); and *Aphthovirus* (foot-and-mouth disease virus). Collectively, the *Picornaviridae* viruses have a broad host range, infecting most vertebrate species; however, individually the viruses are often species-specific.

The enteroviruses include more than 70 viruses, primarily as inhabitants of the gastrointestinal tract, many causing no clinical manifestations. Although most enteroviruses are species-specific, a few can cross species barriers and may become important concerns in the laboratory setting. Great apes are susceptible to human poliovirus, with a resultant clinical illness similar to that of humans. Similarly, the swine vesicular disease virus can infect humans and cause aseptic meningitis.

The rhinoviruses cause mild upper respiratory disease such as the common cold in humans and other vertebrates. They may also cause bronchitis and bronchopneumonia.

The cardioviruses usually are associated with inapparent infection but can be pathogenic for some species including humans. Rodents and some large mammals (including cattle, swine, and elephants) are susceptible to infection with encephalomyocarditis virus and, as the name suggests, may develop clinical encephalitis or myocarditis.

The aphthoviruses cause foot-and-mouth disease in cloven hoofed animals and may also infect humans, producing a relatively benign febrile illness with similar symptomatology including vesicular oral and skin lesions.

Transmission of the picornaviruses is facilitated by the relatively high resistance of this group of viruses to inactivation both by natural environmental factors and by specific chemical and physical agents. Most picornavirus infections result in shedding of large amounts of virus in body secretions and excretions from lesions. Also, picornaviruses often cause subclinical infections that result in development of inapparent carriers that may shed virus for the remainder of the life of the host.

Transmission of enteroviruses is mostly by fecal contamination of hands and fomites, with food, water, and insects serving as mechanical vectors. However, those enteroviruses that replicate first in the pharynx and respiratory tract (polio, echo, coxsackie) may also transmit through aerosols, droplets, or contact with saliva. Cardioviruses are shed in the feces and urine of infected animals, presumably the source of infection for other susceptible hosts. Rhinoviruses are shed in respiratory secretions; infection occurs when these contaminated aerosols are inhaled. Aphthoviruses, which are shed in large quantities by infected animals, may be transmitted via fomites, by direct contact of mucous membranes with infectious material, and by various mechanical means including arthropod bites.

Among the enteroviruses, only polio appears important in the animal care and laboratory setting, primarily because persons shedding the virus may infect nonhuman primates. The reverse, transmission from nonhuman primates to humans, has not been described. Theoretically, nonhuman primates infected naturally or artificially with poliovirus could serve as a source of disease for susceptible humans. Enteroviruses, shed in urine and feces, probably enter the susceptible host through oral or other mucous membranes. Cardioviruses have the potential to be transmitted between humans and various laboratory animals, especially rodents and swine. This could cause serious illness if cross-transmission from animals to humans occurs with encephalomyocarditis virus. Rhinoviruses are not known to be important as laboratory hazards. Aphthoviruses can be transmitted from animals to humans through aerosols and multiple mechanical means, but the human disease is generally mild and transient.

Prevention and Control

The risk of human infection with *Picornaviridae* viruses in the laboratory setting is relatively low; it is recommended that laboratories using any of these viruses use BSL 2 precautions (44).

POXVIRIDAE

The poxviruses of vertebrates are divided into eight genera; seven infect mammals and one infects avian species. Also, a subfamily of poxviruses infects insects. The genera are classified on the basis of morphology, antigenic relationships, and host range. The zoonotic poxviruses are contained in three genera: *Orthopoxvirus*, *Parapoxvirus*, and *Yatapoxvirus* (Table 2).

Smallpox disease has been eradicated in nature. Possession and use of the smallpox (variola) virus is now restricted to the Maximum Containment Laboratory of the World Health Organization Collaborating Center for Smallpox and Other Pox Virus Infections located at the CDC in Atlanta, Georgia,

TABLE 2 Major poxviruses having potential as zoonoses

Genus	Virus	Host range
Orthopox	Monkeypox	Humans, monkeys, flying squirrels
	Cowpox	Humans, felines, rodents
	Buffalopox	Humans, livestock, buffalo
	Vaccinia	Many vertebrates including humans
Parapox	Pseudocowpox	Humans, cattle
	Orf	Humans, livestock
	Milker's nodule	Humans, livestock
Yatapox	Tanapox	Humans, monkeys
	Yabapox	Humans, monkeys
Unclassified	Molluscum contagiosum	Humans, chimpanzees(?)

and to the Research Institute for Virus Preparations in Moscow.

Poxviruses tend to be species-specific, although several of them are able to infect multiple species, including humans. The vaccinia virus is capable of infecting humans, nonhuman primates, and livestock. At least three other poxviruses infect nonhuman primates and can cause disease in humans. Monkeypox virus has caused smallpox-like illness in nonhuman primates and humans. Monkeypox virus is a rare sporadic zoonosis found in the rain forest of Africa. Transmission of monkeypox between nonhuman primates, flying squirrels, and humans was suggested. Benign epidermal monkeypox (or OrTeCa) appears to be identical to the virus that causes tanapox in African children (2). Finally, molluscum contagiosum, which causes a nonfatal papillomatous skin lesion in humans, has been reported in chimpanzees but never confirmed and appears likely to be strictly a human disease (45).

Variolalike whitepox virus referred to in the literature was at one time presumed to be a distinct virus but appears to have been the result of inadvertent contamination of monkeypox cultures with smallpox virus. The parapox viruses are the most prevalent of the zoonotic poxviruses. Diseases caused by parapox viruses include orf and milker's nodule, both commonly transmitted from livestock to humans in nature.

The severity and clinical manifestations of poxvirus infections may range from inapparent and mild to severe and fatal. The clinical course may be affected by route and severity of exposure, immune state of the individual, and the type of infecting agent. Poxvirus illness in humans and animals is characterized by the development of pock lesions on the skin and at times also on the mucous membranes of the mouth and throat. Typically, the lesions progress through stages of macules, papules, vesicles, and pustules. These eventually dry up, forming scabs, sometimes with mild to severe scarring after healing. Diseases caused by parapox viruses are usually localized and mild in humans and animals, with the possible exception of orf, which may be complicated by secondary infection in goats and sheep. Rare cases of generalized but uneventful cases of orf in humans have been reported (1, 15, 33).

Poxviruses are shed in large quantities in vesicular fluid, pus, skin scrapings, and crusts from lesions. The virus is extremely stable in the environment, especially in the desiccated state, and may be resistant to common chemical and physical inactivants. Transmission of poxvirus between natural hosts is usually the result of contact with skin lesions or secretions, but transmission may also be affected by mechanical means through arthropod vectors and fomites. Transmission of poxviruses from animal to human is usually restricted to direct contact. Virus enters the susceptible host through skin abrasions, lesions, and bite wounds. The respiratory route of infection may be important in dusty environments.

In the laboratory, only infection with vaccinia, Yaba pox, orf, milker's nodule, and tanapox have been reported since the eradication of variola (smallpox). Laboratory transmission has resulted from needlesticks and contamination of scratches and skin lacerations. In laboratories handling variola (only two laboratories in the world are so authorized) or closely related poxviruses, the risk of transmission of monkeypox between humans and nonhuman primates should be of concern.

Prevention and Control

The elimination of variola (smallpox) from the natural environment has resulted in the widespread discontinuation of human vaccination against smallpox. This may, however, have increased the susceptibility of unvaccinated persons to zoonotic poxvirus infections sometimes found in nonhuman primates and other animals. For these reasons, continued vaccination with vaccinia vaccine should be considered for persons working directly with zoonotic orthopox viruses. Work with currently available poxviruses should be conducted under BSL 2 precautions; work with variola is severely restricted and does not require description here.

There are no antiviral drugs or vaccines available for humans at risk to parapox, tanapox, or molluscum contagiosum. Distribution of smallpox vaccine is controlled by the CDC in Atlanta and is restricted to personnel directly handling orthopox viruses. Vaccine should be administered by a qualified physician experienced in this method of vaccination. The recommendations of the CDC's Advisory Committee on Immunization Practices for smallpox should be followed (44).

REOVIRIDAE

The *Reoviridae* family includes a ubiquitous group of viruses that may infect mammals, arthropods, and plants. The animal viruses included in *Reoviridae* belong to the genera *Orbivirus*, *Reovirus*, and *Rotavirus*.

All orbiviruses are arboviruses and may be transmitted by various vectors including mosquitoes, ticks, and biting flies. Orbiviruses infect various mammals, but only Colorado tick fever virus is known to infect humans and be a potentially important source of cross-infection in the laboratory environment. In nature, the Colorado tick fever virus has been found in various rodent species and cervids. Transmission to humans and other mammalian hosts is via tick bite. Infected rodents may experience long periods of viremia without clinical disease. Orbivirus infection typically results in an acute febrile, denguelike illness, which may occasionally produce severe encephalitis and rarely death.

The genus *Reovirus* includes several serotypes that infect humans, other mammals, and avian species. Reovirus infection typically results in a mild and usually transient upper respiratory illness.

Rotaviruses are ubiquitous throughout the world and infect most mammalian species with some serotypes able to infect multiple mammalian and avian host species. Rotavirus infections characteristically produce gastroenteritis, which in children and young animals may be severe or fatal. Death may occur as a result of severe untreated dehydration.

Transmission of orbiviruses is commonly via an arthropod; transmission of reoviruses is commonly by the fecal-oral route, although some reoviruses are transmitted via the reparatory route. Rotaviruses are transmitted primarily by the fecal-oral route.

Laboratory transmission of Colorado tick fever is a recognized hazard. Rarely, laboratory infections have occurred without tick bite transmission. Both reoviruses and rotaviruses may possibly be transmitted in the laboratory setting, but such transmission is difficult to prove.

Prevention and Control

In the case of orbiviruses, prevention of exposure to biting arthropods (ticks, mosquitoes, midges, gnats, and phlebotomine flies) should be stressed. To prevent infection by the reoviruses, steps should be taken to minimize risk of aerosol contamination. Prevention of fecal contamination of food and water is the primary method for eliminating risk of rotavirus spread. Good handwashing practices play an important role in the prevention of laboratory infection by agents normally transmitted by the fecal-oral route.

No vaccines are available for immunization against the *Reoviridae* viruses. Laboratories using viruses of the *Reoviridae* family should follow BSL 2 recommendations.

RHABDOVIRIDAE

The *Rhabdoviridae* family includes several diverse RNA viruses in part grouped together on the basis of morphologic similarity. All are "bullet-shaped," although the plant viruses in this family are often more bacilliform than bullet-shaped. The natural hosts of rhabdoviruses include vertebrates, arthropods, and plants. Rhabdoviruses are divided into two genera that have vertebrate hosts: 6 *Lyssavirus* species (rabies and related viruses) and 4 *Vesiculovirus* species (vesicular stomatitis [VS] and related viruses); a group of 36 unclassified animal viruses, which include mammalian and arthropod viruses; and approximately 25 various plant and other invertebrate viruses (Table 3).

The rhabdoviruses of greatest importance in the laboratory include rabies and related viruses and VS and related viruses. The rabies and rabies-related viruses are found in mammalian species, especially carnivores (rabies) and bats (Lagos bat, Duvenhage) but may infect any warm-blooded species. The VS viruses are found chiefly in horses, cattle, and swine.

The diseases associated with the rhabdoviruses vary considerably in presentation and severity. Rabies virus is an encephalitogenic virus with an almost always fatal outcome. The rabies-related viruses may produce similar illness (e.g., Duvenhage virus) or may produce a less serious neurologic illness (Mokola) or no illness at all (Obodhiang and Kotonkan). The diseases in nonhuman vertebrates appear similar to those seen in humans, although some of the lesser studied viruses may have unknown disease potential. The vesiculoviruses affect primarily epithelial tissue. In cattle, VS infection begins with a nonspecific febrile prodrome, followed

TABLE 3 Partial list of selected rhabdoviruses

Genus	Vertebrate and invertebrate viruses	
Lyssavirus (6)	Rabies[a]	Lagos bat[a]
	Duvenhage[a]	Mokola[a]
	Kotonkan	Obodhiang
Vesiculovirus (4)	VS[a]	Piry
	Chandipura	Isfahan
None established (36)	Flanders	Mt. Elgon bat
	Hart Park	Bovine ephemeral fever
	Egtved	Sigma
	Kern Canyon	Many others

[a]Proven human disease.

by development of vesicles on gums, tongue, and lips that rupture in about 24 h, leaving shallow reddish areas of ulceration. Lesions may also appear on teats and coronary bands of the feet. In swine, lesions are most often found on the snout and feet. In humans, VS infection is characterized by an acute influenzalike syndrome, with fever, malaise, myalgia, headache, and perhaps, nausea and vomiting. The illness in humans is short, and complete recovery in 3 to 6 days is usual.

In the laboratory setting, only VS and rabies have proved to be of significant concern. VS may be found in the saliva and vesicular fluids of infected livestock. The route of transmission to humans or other animals is probably multiple, by inhalation and direct inoculation into mucous membranes or abraded skin. Rabies is normally transmitted to humans or other animals by the bite of an infected animal, by a puncture wound with a contaminated needle, glassware, or sharp instrument, by contamination of mucous membranes or abraded skin, or on rare occasions, by inhalation of infectious droplets or aerosols. In recent years, the risk of exposure to VS virus in the laboratory has increased as this virus has become widely used in recombinant virus studies.

Two cases of laboratory-acquired rabies have been reported, and both occurred under very unusual circumstances. One presumably resulted from inhalation of huge quantities of aerosolized fixed vaccine virus during preparation of animal rabies vaccine. The other presumably resulted from inhalation of very high titered modified viruses that were being evaluated for vaccine potential. The events surrounding these two cases are unlikely to occur in the normal laboratory environment, and inhalation infection should be of concern in the laboratory only when large quantities of high titered viruses are manipulated in such a way as to result in the possible inhalation of high doses of infectious material.

Prevention and Control

In the case of rabies, pre-exposure immunization is strongly recommended for any person knowingly working with rabies virus and also for persons working with animals of unknown disease history but considered likely to be infected with the rabies virus. Public health authorities can provide the most current recommendations on immunizations. Also, if laboratory procedures include the potential for creation of infectious aerosols, appropriate physical barriers should be used. Conventional immunization against rabies (pre- or postexposure) is not considered adequate to protect against airborne exposure. In the event of exposure to rabies by bite wounds, puncture, or skin injuries, the affected area should be flushed copiously and scrubbed with soap and water immediately. Personnel should contact health officials for current postexposure treatment recommendations. This is true whether the victim has had pre-exposure immunization or not. BSL 2 is recommended for any laboratory using rabies virus. However, if aerosols or droplets of viruses are likely to be created by laboratory procedures or if large quantities of viruses are being manipulated in the laboratory, BSL 3 is recommended. Risk of exposure to the rabies-related viruses is not fully understood; thus, the safety procedures followed for rabies virus should be used.

The VS group of viruses constitute limited risk to personnel. Viruses are present in saliva and vesicular fluids of infected animals. Transmission in nature is usually by direct contact. In the laboratory, transmission may be by direct contact, inhalation of infectious aerosols or dust, inoculation by sharp objects, or exposure of mucous membranes or abraded skin. No vaccine is available for humans against the vesiculoviruses. Vaccine for use in animals has had limited use and is not well studied. Laboratories engaged in work with the vesiculoviruses should use BSL 2 procedures (44).

TOGAVIRIDAE

More than 80 viruses belong to the family *Togaviridae.* They are divided into four genera: the *Alphavirus* (group A arboviruses), *Flavivirus* (group B arboviruses), *Rubivirus* (rubella virus), and *Pestivirus* (hog cholera and other animal viruses). Within each genus, there are serologic and morphologic relationships and similarities in replication and composition. Between genera, there are none, and the grouping of all viruses included within *Togaviridae* is somewhat artificial. Only about 20 viruses, contained within the two genera *Alphavirus* and *Flavivirus,* are important as potential or proven zoonoses.

Most infections of humans and animals with togaviruses are asymptomatic, but a few can cause severe and often fatal disease in both humans and animals. The clinical syndromes may be variable and complex. All the viruses in Table 4 can cause fever of variable severity, but some also cause rash, arthralgia, hemorrhage, and at times, encephalitis, hepatitis, nephritis, and hypotensive shock. The most important of the togavirus zoonoses are the viruses of eastern equine, western equine, Venezuelan equine, and St. Louis encephalitis and dengue fever. Eastern equine encephalitis (EEE) usually causes low morbidity but high mortality, whereas

TABLE 4 Some togaviruses of zoonotic importance

Genus	Virus	Vector	Reservoir hosts
Alphavirus	Venezuelan equine encephalomyelitis	Mosquito	Rodents, equines
	WEE	Mosquito	Birds
	EEE	Mosquito	Birds
	Chikungunya	Mosquito	Humans, monkeys
	Sindbis	Mosquito	Mammals, birds
	Semliki Forest	Mosquito	Rodents, chimpanzees
Flavivirus	Kyasanur Forest	Tick	Monkeys, rodents
	Dengue 1–4	Mosquito	Humans, monkeys, livestock
	Louping Ill	Tick	Humans, sheep
	West Nile	Mosquito	Birds

western equine encephalitis (WEE) produces high morbidity and low mortality. Venezuelan equine encephalitis seldom causes mortality, and morbidity is somewhat less than that seen with WEE. St. Louis encephalitis is somewhat less severe than EEE, but in older human populations it may cause significant mortality. Dengue fever may cause a broad spectrum of disease from mild to fatal, the latter especially true when hemorrhagic fever or shock syndrome is part of the clinical picture.

In the laboratory, togavirus infections are well recognized. Those of greatest risk include EEE, WEE, Venezuelan equine encephalitis, and St. Louis encephalitis. Semliki Forest virus is a less likely risk to laboratorians, with only one case reported. However, all the togaviruses should be considered potential human pathogens in the laboratory. There are multiple potential routes of infection with togaviruses. Needlestick or similar penetration of intact skin with an infected object is a proven mode of transmission. Exposure of mucous membranes to viruses (including airborne exposure) is an accepted risk. Bites by infected arthropods, the usual means of spread in nature, can also occur in the laboratory.

Prevention and Control

The togaviruses are a well-recognized, if sometimes poorly understood, risk to laboratory workers. Immunization is available against some of the togaviruses, and some cross-protection against other members of the group may be afforded in some cases. Care should be taken to avoid exposure by needlestick (or similar exposure) and airborne exposure. Laboratories using any of the zoonotic togaviruses should operate under BSL 2 or 3, depending on several factors (e.g., agent in use, amount of virus being handled, procedures being used). The appropriate levels are described in Appendix I.

UNCLASSIFIED POTENTIAL AND EMERGING ZOONOSES

The ongoing development of improved diagnostic and research methodologies has and will continue to reveal new zoonotic viruses and to differentiate among old groupings to produce more and better-defined classifications. Greater interaction between humans and animals in ways not previously used will tend to select new zoonotic viruses. Increasingly rapid movement of persons and animals around the world is increasing exposure risks to exotic viruses new to civilized humans. Genetic manipulation has the potential to produce myriad new agents, both good and bad. Changing characteristics in the human population may also have great effect on emerging zoonoses (e.g., increase in immunodeficiency disease).

Examples of relatively new and sometimes quite pathogenic viruses include the as-yet-unclassified Marburg and Ebola viruses and the more recently emerged and poorly understood Reston virus (a filovirus of nonhuman primates whose physical similarities to the Ebola virus caused great consternation in recent years, although no human illness has yet been associated with infection). Marburg and Ebola viruses are part of the African hemorrhagic fever disease complex, and as the name implies, the clinical disease is characterized by severe, often fatal hemorrhagic fever. The fatality rate in primary disease has ranged from 29% to more than 90%. Marburg infection has been linked to association with African green monkeys (*Cercopithecus aethiops*); no such animal association has yet been proved for Ebola virus infection. Close contact with infected persons is clearly a high risk factor, although the route of transmission has yet to be firmly established. Immunization is not available for protection of humans against infection with these agents. Laboratories working with either Marburg or Ebola viruses should follow BSL 4 recommendations (44).

The so-called slow viruses, including the unclassified agents of scrapie and transmissible mink encephalopathy and, more recently, sporadic bovine encephalitis, are not known to be the cause of laboratory-acquired infections. However, these agents are resistant to most disinfectants. BSL 2 practices are recommended, with special care in the disposal of tissues and the cleaning of equipment.

Among other emerging zoonoses, the coronaviruses are widespread in nature and infect humans and a variety of domestic and wild mammals and birds. At present, the host range appears to be restricted to the species of origin, but experimentally, some other species have been infected with some of these viruses. Transmission of coronaviruses may occur via the fecal-oral route as well as by aerosols and fomites.

References

1. **Acha, P. N., and B. Szyfres.** 1980. *Zoonoses and communicable diseases common to man and animals.* Scientific Pub. No. 354. Pan American Health Organization, Washington, D.C.
2. **Adams, S. R. (ed.).** 1987. Biohazards associated with natural and experimental diseases in nonhuman primates. *J. Med. Primatol.* **16**(2):51–138.
3. **Bishop, D. H. L., C. H. Calisher, J. Casals, M. P. Chumakov, S. Y. Gaidanovich, C. Hannoun, D. K. Lvov, I. D. Marshal, N. Oker-blom, R. F. Petterson, J. S. Porterfield, P. K. Russell, R. E. Shope, and E. G. Westaway.** 1980. Bunyaviridae. *Intervirology* **14:**125–143.
4. **Bremen, J. G.** 1980. Human monkeypox 1970–1979. *Bull. W.H.O.* **58:**165–182.
5. **Centers for Disease Control.** 1993. Hantavirus infection—southwestern United States: interim recommendations for risk reduction. *Morbid. Mortal. Weekly Rep.* **42** (Suppl. RR-11).
6. **Centers for Disease Control.** 1993. Update: hantavirus infection—southwestern United States. *Morbid. Mortal. Weekly Rep.* **42:**441–443.
7. **Centers for Disease Control.** 1993. Update: hantavirus pulmonary syndrome—United States, 1993. *Morbid. Mortal. Weekly Rep.* **42:**816–820.
8. **Chanock, R. M.** 1979. Parainfluenza viruses, p. 611–623. *In* E. H. Lennette and N. J. Schmidt (ed.), *Diagnostic procedures for viral, rickettsial, and chlamydial infections,* 5th ed. American Public Health Association, Washington, D.C.
9. **Dykewitz, C. A., V. M. Dato, S. P. Fisher-Hoch, M. V. Horwath, G. I. Perez-Oronoz, S. M. Ostroff, H. Gary, Jr., L. B. Schonberger, and J. B. McCormick.** 1992. Lymphocytic choriomeningitis outbreak associated with nude mice in a research institute. *JAMA* **267:**1349–1353.
10. **Easterday, B. C.** 1975. Animal influenza, p. 449–481. *In* E. D. Kilbourne (ed.), *The influenza viruses and influenza.* Academic Press, Inc., New York.
11. **Easterday, B. C.** 1965. Rift Valley fever. *Adv. Vet. Sci.* **10:**65–127.
12. **Easterday, B. C., and L. C. Murphy.** 1963. Studies on Rift Valley fever in laboratory animals. *Cornell Vet.* **53**(3):423–433.
13. **Eisa, M., H. M. A. Obeid, and A. S. A. El Sawi.** 1977. Rift Valley fever in Sudan. *Bull. Anim. Health Prod.* **25**(4):343–347.
14. **Frame, J. D., J. M. Baldwin, D. J. Gocke, and J. M. Troupe.** 1970. Lassa fever, a new virus disease of man from West Africa. *Am. J. Trop. Med. Hyg.* **19**(4):670–676.
15. **Gillespie, J. H., and J. F. Timoney.** 1981. *Hagan and Brunner's infectious diseases of domestic animals,* 7th ed. Cornell Univ. Press, Ithaca, N.Y.
16. **Gregg, M. B.** 1975. Recent outbreaks of lymphocytic choriomeningitis in the U.S.A. *Bull. W.H.O.* **52:**549–554.
17. **Hinman, A. R., D. Fraser, R. G. Douglas, G. S. Bowen, A. L. Kraus, W. G. Winkler, and W. W. Rhodes.** 1975. Outbreak of lymphocytic choriomeningitis virus infections in medical center personnel. *Am. J. Epidemiol.* **101**(2):103–110.
18. **Hore, D. E., and R. G. Stevenson.** 1967. Influenza in sheep. *Vet. Rec.* **80:**26–27.
19. **Hore, D. E., and R. G. Stevenson.** 1969. Influenza in sheep. *Res. Vet. Sci.* **10:**342–350.
20. **Hotchin, J.** 1962. Lymphocytic choriomeningitis. *Cold Springs Harbor Symp. Q. Biol.* **27:**479–483.
21. **Hotchin, J., L. Bensom, and J. Seamer.** 1962. Factors affecting the induction of persistent tolerant infection of newborn mice with lymphocytic choriomeningitis. *Virology* **18:**71–78.
22. **Hotchin, J., and H. Weigand.** 1961. Studies of LCM in mice, I. Relationship between age at inoculation and outcome of infection. *J. Immunol.* **86**(4):392–400.
23. Howard, C. R., and D. I. H. Simpson. 1980. The biology of the arenaviruses. *J. Gen. Virol.* **51:**1–14.
24. **Hull, R. N.** 1973. The simian herpesviruses, p. 390–426. *In* A. S. Kaplan (ed.), *The herpesviruses.* Academic Press, Inc., New York.
25. **Kitchen, S. F.** 1934. Laboratory infections with the virus of Rift Valley fever. *Am. J. Trop. Med. Hyg.* **14:**547–564.
26. **Larsen, J.** 1969. On the induction of immunological tolerance to a self-reproducing antigen. *Immunology* **16:**15–23.
27. **Laver, W. G., and G. Air.** 1980. *Structure and variation in influenza virus.* Elsevier, Amsterdam.
28. **Matthews, R. E. F.** 1982. *Classification and nomenclature of viruses (report of ICTV).* S. Karger, New York.
29. **Mayr, A.** 1980. New emerging viral zoonoses. *Vet. Rec.* **106:**503–506.
30. **Monath, T. P.** 1973. Lassa fever. *Trop. Doc.* **10:**155–171.
31. **Monath, T. P., and J. Casals.** 1975. Diagnosis of Lassa fever and management of patients. *Bull. W.H.O.* **52:**707–711.
32. **Monath, T. P., V. F. Newhouse, G. E. Kemp, H. W. Setzer, and A. Cacciapuoto.** 1974. Lassa fever isola-

tion from Mastomys natalensis rodents during an epidemic in Sierra Leone. *Science* **19:**263–265.

33. **Nakano, J. H.** 1979. Poxviruses, p. 257–308. *In* E. H. Lennette and N. J. Schmidt (ed.), *Diagnostic procedures for viral, rickettsial and chlamydial infections,* 5th ed. American Public Health Association, Washington, D.C.
34. **Pike, R. M., and J. H. Richardson.** 1979. Prevention of laboratory infections, p. 49–63. *In* E. H. Lennette and N. J. Schmidt (ed.), *Diagnostic procedures for viral, rickettsial and chlamydial infections,* 5th ed. American Public Health Association, Washington, D.C.
35. **Porterfield, J. S., and A. J. Della-Porta.** 1981. Bunyaviridae: infection and diagnosis, p. 479–508. *In Comp. diagnosis of viral diseases,* vol. 4. Academic Press, Inc., New York.
36. **Rawls, W. E.** 1979. Herpes simplex virus types 1 and 2 and herpesvirus simiae, p. 309–373. *In* E. H. Lennette and N. J. Schmidt (ed.), *Diagnostic procedures for viral, rickettsial and chlamydial infections,* 5th ed. American Public Health Association, Washington, D.C.
37. **Scherer, W. F.** 1980. Laboratory safety for arboviruses and certain other viruses of vertebrates (The Subcommittee on Arbovirus Laboratory Safety of the American Committee on Arthropod-Borne Viruses). *Am. J. Trop. Med. Hyg.* **29:**1359–1381.
38. **Schild, G. C., and W. R. Dowdle.** 1975. Influenza diagnosis, p. 316–368. *In* E. D. Kilbourne (ed.), *The influenza viruses and influenza.* Academic Press, Inc., New York.
39. **Schnurrenberger, P. R., J. K. Grigor, J. F. Walker, and R. J. Martin.** 1978. The zoonosis-prone veterinarian. *J. Am. Vet. Med. Assoc.* **173:**373–376.
40. **Schwentker, F. F., and T. M. Rivers.** 1934. Rift Valley fever in man: report of a fatal laboratory infection complicated by thrombophlebitis. *J. Exp. Med.* **59:**305–313.
41. **Shope, R. E., and G. Sather.** 1979. Arboviruses, p. 767–814. *In* E. H. Lennette and N. J. Schmidt (ed.), *Diagnostic procedures for viral, rickettsial, and chlamydial infections,* 5th ed. American Public Health Association, Washington, D.C.
42. **Swanepol, R., B. Manning, and J. A. Wyatt.** 1979. Fatal Rift Valley fever in man in Rhodesia, Central African. *J. Med.* **25**(1)**:**1–8.
43. **Traub, R., and C. L. Wisseman, Jr.** 1978. Korean hemorrhagic fever. *J. Infect. Dis.* **138**(2)**:**267–272.
44. **U.S. Department of Health and Human Services.** 1993. *Biosafety in microbiological and biomedical laboratories.* HHS Publication No. (CDC) 93-8395. U.S. Government Printing Office, Washington, D.C.
45. **Whitney, R. A., Jr.** 1976. Important primate diseases (biohazards and zoonoses), p. 23–50. *In* M. L. Simmons (ed.), *Biohazards and zoonotic problems of primate procurement, quarantine, and research.* DHEW Publication No. 76-890. U.S. Dept. of Health, Education, and Welfare, Washington, D.C.
46. **Wilsnack, R. E.** 1966. Lymphocytic choriomeningitis. *Natl. Cancer Inst. Monogr.* **20:**77–92.
47. **Wulff, H., B. McIntosh, D. B. Hamner, and K. Johnson.** 1977. Isolation of an arenavirus. *Bull. W.H.O.* **55:**441–445.

Biological Safety in the Biotechnology Industry

GERARD J. McGARRITY AND CAROL LAX HOERNER

9

INTRODUCTION

Since the first commercial introduction of a pharmaceutical product manufactured by recombinant DNA technology applications, manufacturing processes and regulatory requirements have rapidly evolved. The biotechnology industry continues to build on the safety record of the pharmaceutical industry. The existing commercial biotechnology industry, composed of production of monoclonal antibodies as well as production of recombinant proteins in *Escherichia coli, Bacillus subtilis, Saccharomyces cervisiae,* and mammalian cell cultures, is little more than a decade old. Although speculative concerns regarding safety have been plentiful, the number of controlled safety studies, especially involving laboratory and production workers, is relatively small. However, some data are available.

In recent years, human gene therapy has received attention from federal regulatory and nonregulatory agencies. Research on animal models and existing human clinical studies have yielded insight into the safety of this newest genetic engineering technology.

Several federal regulatory agencies and international associations have issued regulations and guidelines for the biotechnology industry. Government directives, both direct and indirect, have been aimed at all aspects of the biotechnology process, from laboratory research to large-scale operations, commercial production, waste treatment, and disposal. Some government agencies require complete occupational health and environmental reviews before products may be introduced in commerce. For example, more than 30 environmental health and safety statutes and regulatory programs were directly related to some aspect of Actimmune and Activase commercialization at Genentech, Inc., as reported in environmental assessments including establishment-licensing applications (34).

The issuance of these regulatory directives and programs as well as industry-initiated safety guidelines helps ensure that safety is built into the facilities and procedures of the operations. In addition to general requirements such as pressure differential zones, air filtration and containment systems, and waste treatment systems, each manufacturing process must be completely standardized, typically by using good manufacturing practices, standard operating procedures, preventive maintenance schedules, and validated quality control procedures.

The purpose of this chapter is to review the field of laboratory safety as it pertains to the field of biotechnology, reviewing, in particular, commercial applications and human gene therapy technology and identifying the appropriate existing standards applicable to biotechnology laboratory and production systems. This review is predominantly confined to microbiological safety. Other aspects of safety are cited when integral to a microbiological safety assessment.

SOURCE OF EXPOSURE

Theoretically, workplace exposure to potentially hazardous microbiological hosts can occur beginning at the research laboratory level and continuing through development, scale-up, and full-scale manufacturing, through all processing steps in which viable cells may be present. Typically, more significant exposures will occur during intermediate cell processing or purification steps. Exposure to cellular components and the recombinant DNA product may occur during cultivation of the host at small and large scale. Also, personnel exposure to the product may occur during cell processing and purification steps, formulation, and filling. Theoretical potential exposure increases as processes are increased in scale and products such as human proteins or human gene therapy vectors become concentrated during manufacturing operations. Exposure can occur as a result of an inadvertent accidental spill or leak, possibly from faulty design, inadequate procedure, or lack of adherence to safety practices. The principal routes of exposure to concentrated biological agents and their products occur via aerosols through inhalation or by absorption through the skin. The received dose may theoretically exceed the therapeutic or established safe dosage.

In manufacturing operations, wholly unrelated to current biotechnology, the pharmaceutical industry has experienced adverse occupational incidents. For example, Harrington et al. (32) reported gynecomastia among workers involved in the manufacture of oral contraceptives. In other studies, the frequency of sister chromatid exchange was increased among workers who were occupationally exposed to elevated concentrations of ethylene oxide (28, 74). Worker exposure was decreased to safe levels through engineering changes and use of personnel protective equipment. These studies and others illustrate that inappropriate occupational exposures can occur in what was believed at the time to be proper, well-designed, and well-controlled facilities.

How can acute or chronic effects be detected in workers, especially in a new and rapidly expanding field such as biotechnology? Many of the classic signs of exposure to microorganisms may not be fully sufficient. In a timely occupational medicine review of recombinant DNA technology, Cohen and Hoerner (15) stressed the use of medical surveillance and the value of the interview and targeted medical history to discover potential exposures and their sources. This methodology is proper for workers who handle potentially biohazardous agents and their products and other hazardous material such as carcinogens, mutagens, and radioactive isotopes. Data analysis is important. Cohen and Hoerner recommended a grouped trend analysis for workers having similar exposures. Such analysis in a noncontrolled occupational setting requires great care and should include an appropriate control group, considering such factors as pay grades, smoking status, age, and sex (15).

This chapter includes a review of the available safety literature in the field of proteins produced through large-scale commercial recombinant DNA technology, hybridoma cell culture cultivation, and human gene therapy. For hybridoma cell culture systems, worker microbiological safety issues for the most part are limited to adventitious microorganisms that can cause gross contamination of the growth vessel (bioreactor) or to components of the cell culture medium, including animal serum, or medium arginine that can serve as an irritant. The following is an analysis of the regulatory and nonregulatory government oversight of rDNA technology.

RECOMBINANT DNA TECHNOLOGY

A variety of federal regulations and statutes govern production of biotechnology products and processes. The Food and Drug Administration (FDA) has the lead role in product regulation, and certain requirements are identified as codified regulations in the *Code of Federal Regulations* 21; parts 1 to 1299, April 1, 1992. Also, since 1976, use of recombinant DNA even in laboratory scale is de facto regulated by the document *NIH Guidelines for Research Involving Recombinant DNA Molecules* (59). These guidelines apply to all organizations that receive any support for recombinant DNA research from the National Institutes of Health (NIH). Also, most organizations have chosen to follow these guidelines, because adherence may assist in eventual product-licensing approval. The guidelines, among other things, define physical and biological containment levels for recombinant organisms. The NIH guidelines define recombinant DNA molecules as either "(i) molecules which are constructed outside living cells by joining natural or synthetic DNA molecules that can replicate in a living cell, or (ii) DNA molecules that result from the replication of those described in (i) above" (59).

Industrial applications of recombinant DNA production will typically involve the propagation of the host organism or cell culture in large scale. The NIH guidelines define *large scale* as in excess of 10 liters.

TABLE 1 GLSP criteria

1. Well-characterized, nonpathogenic host
2. Host with an extended history of safe use
3. Host without adventitious agents
4. Host that contains well-characterized inserts
5. Inserts that are free from harmful sequences
6. Vectors that are limited in size
7. Vectors that are poorly mobilizable
8. Host vector inserts unable to transfer DNA to wild-type hosts

The complete details of physical containment for research or production of recombinants using volumes greater than 10 liters are described in Appendix K of the NIH guidelines (58).

Three biosafety levels were initially established in 1981 for large-scale containment: biosafety level 1 large-scale (BSL-1LS), BSL-2LS, and BSL-3LS. To date, most biotechnology manufacturing operations have used BSL-1LS. The NIH Recombinant DNA Advisory Committee (RAC) has recommended that, as appropriate, the large-scale levels should be used for organisms that are handled in those respective levels (BSL 1, BSL 2, and BSL 3) at laboratory scale. More recently, the NIH RAC, having identified the inherent biological containment of certain safe existing commercial hosts, has moved toward harmonizing the guideline's large-scale containment practices and the Organization for Economic Cooperation and Development good industrial large-scale practices. The NIH guidelines now identify a fourth large-scale level, good large-scale practices (GLSP), that meets the criteria presented in Table 1.

The FDA and other agencies have endorsed the use of containment levels lower than BSL-1LS such as good (industrial) large-scale practices (8, 14). Some differences between SBL-1LS and GLSP are listed in Table 2.

In addition to the host organisms themselves, products of the host including those encoded by the transgene must be considered in any risk assessment. The recombinant product can be potentially hazardous, especially in concentrated form. The NIH RAC has placed genes encoding for vertebrate toxins in a special category for review. Certain toxins of demonstrated potency cannot be produced by recombinant DNA technology, even in laboratory scale, until review by an ad hoc committee to advise

TABLE 2 Comparison of select GLSP features to BSL-1LS[a]

GLSP	BSL-1LS
Cultures of viable organisms containing recombinant DNA molecules should be handled in facilities intended to safeguard health during work with microorganisms that do not require containment.	Cultures of viable organisms containing recombinant DNA molecules shall be handled in a closed system or other primary containment equipment.
Cultures containing viable recombinant organisms shall be handled in accordance with applicable environmental regulations.	Culture fluid . . . shall not be removed from a closed system or other primary containment equipment unless the viable organisms have been inactivated. . . .
Addition of materials to a system, sample collection, transfer of culture fluids within/between systems, and processing of culture fluids shall be conducted in a manner that maintains employee exposure to viable organisms containing recombinant DNA molecules at a level that does not adversely affect the health and safety of the employees.	Sample collection from the closed system and the transfer of culture fluids shall be done in a manner which minimizes the release of aerosols or contamination of exposed surfaces.
Not relevant	Exhaust gases removed from a closed system or other primary containment equipment shall be treated by filters which have the efficiencies equivalent to HEPA filters or by other equivalent procedures. . . .
Not relevant	A closed system or other primary containment equipment that has contained viable organisms containing recombinant DNA organisms shall not be opened for maintenance or other purposes unless it has been sterilized by a validated sterilization procedure.
The facility's emergency response shall provide provisions for handling spills.	Emergency plans . . . shall include methods and procedures for handling large losses of culture on an emergency basis.

[a]From references 34 and 59, with permission.

TABLE 3 Airborne endotoxin levels[a]

Operation	No. of samples	Range (ng/m^3)	Mean (ng/m^3)
Fermentation	13	0.06–2.14	0.33
Continuous centrifugation	23	0.08–12.8	1.39
Mixing-homogenizing	20	0.07–4.54	0.86
Experimental batch harvest	34	0.47–1812	162.55

[a]From reference 62, with permission.

on safety. These toxins include diphtheria and other toxins that have a 50% lethal dose of less than 100 ng/kg body weight (60). More detail regarding specific toxins can be obtained from the Office of Recombinant DNA Activities of the NIH. Gill (29) reviewed the toxicities of various microbial toxins.

Care must be taken in the overall planning of new processes and facilities for commercial production. For protection of public health and environment, an environmental checklist must be prepared (65). Components of the process are evaluated for the checklist including the host, cellular products, transgenic products, chemicals used in production, cleaning compounds, and bioproducts and waste. For a biopharmaceutical, the checklist would be used to ensure that the following items are appropriately planned and accomplished:

Containment facilities
Air emissions control and treatment
Liquid effluent treatment
Solid waste management
Hazardous waste management
Pharmaceutical waste management
Spill contingencies
Written standard operating procedures
Personnel training

In such an analysis, toxic substances, including the commercial product, must be identified. For environmental assessment purposes, the FDA defines a toxic substance as

> a substance that is harmful to some biological mechanism or system. A substance is considered to be a toxic substance if it is harmful to appropriate test organisms at expected environmental concentrations even though it may be without effect to humans. . . . If the maximum concentration at any point in the environment, i.e., at any point of entry or any point where higher concentrations are expected as a result of bioaccumulation . . . exceeds the concentration of the substance that causes any adverse effect in a test organism species (minimum effect level) or exceeds 1/100 of the concentration that causes 50% mortality in a test organism, whichever concentration is less (cited in reference 34).

Individual host vector constructs and any hazardous materials used in a particular process should be examined on a case-by-case basis from the worker safety viewpoint, especially in commercial and large-scale operations. Palchak et al. (62) reported on detection of airborne endotoxin in large-scale fermentation facilities using gram-negative hosts. Air sampling was performed during fermentation, continuous centrifugation, and mixing-homogenizing. Airborne endotoxin levels ranged from 0.07 to 12.8 ng/m^3 for normal processing steps such as fermentation, continuous centrifugation, and mixing-homogenizing (Table 3). The levels for experimental batch harvest of an inactivated culture ranged from 0.47 to 1812 ng/m^3 in 34 different samples ($\bar{a}$ = 162.85) (62).

In a literature review, Palchak et al. noted that the parameters most often observed among workers who were exposed to significant airborne levels of endotoxin were fever development and a decrease in forced expiration volume at 1 second. They noted that a clinically significant decrease in forced expiration volume at 1 s was defined operationally as a change of 5% or greater. Several groups have recommended an exposure limit of 30 ng/m^3 as an 8-h time-weighted average (43, 64). This level represents a threshold to trigger medical surveillance, periodic industrial hygiene monitoring, and possible engineering, procedural, or administrative controls. Palchak et al. (62) recommend the use of suitable respiratory protection when handling cultures grown in large scale in ways that potentiate significant levels of airborne endotoxin. Endotoxin exposure effects are transitory, lasting less than 24 h, and have not contributed to reportable worker illness (15).

LARGE-SCALE CULTIVATIONS

The most commonly used microbial hosts are *E. coli*, *B. subtilis*, and *S. cervisiae*. Most commercial production has been performed with *E. coli* K-12 strains, which are very well characterized, completely attenuated, and nonpathogenic. The FDA has approved many pharmaceuticals derived from this host strain. These products include Protropin, Humulin, Actimmune, Proleukin, Neupogen, Roferon, Intron A, and Epogen.

Escherichia coli Hosts

The inability of most *E. coli* host strains used in commercial production to colonize the human intestinal tract has been documented (40, 72). Levine et al. (39), in a human risk assessment study, failed to achieve significant colonization after feeding volunteers *E. coli* strain HS-4 containing a poorly mobilizable plasmid pBR 325 in a carefully controlled 21-day study. In this study, mobilizable plasmid pJBK5 successfully transmitted tetracycline resistance, pointing out the difference between various levels of mobilization of plasmids. The HS-4 strain was selected for this study because it was known to be capable of transmitting to human gut (39). The study was therefore a measurement of the plasmids under investigation. The authors noted the difficulty in establishing survival of *E. coli* by ingestion; $NaHCO_3$ was needed in this study. They suggested that small inocula would be unlikely to survive without food or buffer; other work has shown the effect of anaerobic growth on plasmid function (12, 51). An overlooked item from this study was the authors' question of whether laboratory workers taking antibiotics should work in recombinant DNA laboratories (39). Gorbach (30) also performed a similar risk assessment of strain K-12. Bogosian and Kane (11) showed that K-12 strains did not survive in the environment for significant periods. The plasmids used in commercial production are predominantly derived from pBR322, a poorly mobilizable, nonconjugative plasmid.

The NIH RAC has declared that *E. coli* K-12 host vector systems that do not contain conjugation proficient plasmids or generalized transducing phages, and when nonconjugative plasmids are used as vectors, pose no quantifiable risk to humans or animals and are exempt at small or laboratory scale from the NIH guidelines (59). It is likely that most systems used in commercial production of recombinant proteins meet these criteria. Because no adverse health effects have been found attributable to recombinant *E. coli* K-12 strain hosts to date (5, 15) and no incidence of pathogenicity or infectivity has been reported (23), GLSP is the appropriate containment for large-scale activities in which the particular *E. coli* K-12 host system qualifies under the criteria previously noted in Table 1 (34).

Bacillus subtilis Hosts

The genus *Bacillus* has been widely used in the fermentation and biotechnology industries. Several reviews have been recently published (19, 67). Priest (67) published an extensive review of the biotechnology of the species *B. subtilis*, which is the most widely used. However, other species have also been used, mainly *B. amyloliquefaciens* and *B. licheniformis*. *B. subtilis* has no pathogenic potential in immunocompetent humans, although it has been implicated in infections in immunosuppressed patients. *B. subtilis* is ubiquitous in the environment. In one environmental survey, Finch et al. (22) reported that *Bacillus* spp. and *Micrococcus* spp. were the most frequently isolated organisms. A survey of the literature cited by DeBoer and Diderichsen (19) noted that most cases of infections caused by *Bacillus* spp. occurred in immunosuppressed patients or was related to drug abuse.

B. subtilis strain 168 has been a widely used host to clone DNA. Like *E. coli*, nonsporulating strains of *B. subtilis* are exempt from the NIH guidelines for small scale. Commercial *B. subtilis* systems are likely to qualify for manufacture under GLSP. *B. subtilis* has been used commercially to produce recombinant proteins in the United States, Japan, and Denmark. It is generally regarded as a safe organism. CPC International filed a Generally Regarded as Safe petition to the FDA for *B. subtilis* in 1986 (18). The submitters showed that an α-amylase from *B. stereothermophilus* produced by *B. subtilis* was safe for use in food. *B. subtilis* is consumed in large quantities in the Japanese food natto (19).

Saccharomyces cerevisiae

Saccharomyces cerevisiae and other yeast organisms offer an attraction for the production of recombinant DNA-produced proteins. Yeasts offer the advantage of eukaryotic post-translational modifications. These are reviewed in the publication of Bisson (10). These modifications are not necessarily identical to certain modifications (e.g., glycosylation) achieved in mammalian cells (5, 68). Yeasts have a longer history in production of nonrecombinant products than either *E. coli* or *B. subtilis*, most notably in the brewing industry. Vandame (79) reviewed the production of some products by *Saccharomyces*. Because of the long history of successful industrial use of *Saccharomyces* spp., this organism will not be reviewed in detail in this presentation.

Animal Cell Cultures

A variety of cell cultures have been propagated in large scale. The main uses have been for production

TABLE 4 Putative risk factors for cell-culture-derived products[a]

Putative risk factor	Sources
Intact cells	Cells
Adventitious agents (bacteria fungi, mycoplasma, viruses)	Cells, raw materials
Endogenous retroviruses	Cells
Residual cellular nucleic acid	Cells
Residual cellular proteins	Cells
Other foreign proteins	Raw materials, antibodies used in purification
Microbial contaminants	In-process bioburden
Endotoxin	
Proteins	
Process chemicals	Raw materials
Antibiotics	
Ligands	
Solvents	
Cleaning compounds	
Inducers	
Nutrients	

[a]From reference 49, with permission.

of monoclonal antibodies, recombinant proteins, and viral vaccines. Many reviews and a critical text have been published (1, 6, 37, 45, 49, 57, 61). Some factors that affect product consistency have been reported (46, 49, 70, 73). Lubiniecki et al. (49) listed some risk factors for production variation. These are listed in Table 4.

In most instances, the main risks will originate in microbial agents present in the culture. These may be the agents being deliberately cultured in the system or may be present as adventitious agents.

If adventitious agents are present in the culture, they may grow to high concentrations if undetected. Mycoplasmas are the most frequently detected adventitious agents in cell cultures. Mycoplasmas have been reported in approximately 12 to 15% of continuous cell lines (52, 54, 55). The main sources of the cell culture mycoplasmas are bovine and other animal sera, laboratory personnel, and most importantly, other mycoplasma-infected cultures. The problem of mycoplasma infection of cell cultures is exacerbated by (i) the high concentrations of mycoplasmas in infected cell cultures, on the order of 10^7 to 10^8 organisms per milliliter of supernatant medium; (ii) the ease of droplet formation during routine cell culture manipulations such as trypsinization and passage; and (iii) the resistance of mycoplasmas to dehydration. These facts indicate that mycoplasmas from an infected cell culture can contaminate the immediate environment for prolonged periods if effective disinfection procedures are not undertaken. Mycoplasmas could be readily detected by environmental sampling during work with an infected culture and for as long as 10 days after work (52). Some results of this environmental sampling are presented in Table 5.

It is the goal of the typical large-scale production of cell cultures to use serum-free media. Of the mycoplasmas most frequently encountered in cell cultures, *Mycoplasma orale, M. arginini,* and *M. fermentans* require cholesterol (54). Another mycoplasma, *Acholeplasma laidlawii,* does not require cholesterol. Although the need of *Mycoplasma* species for cholesterol would seem to limit their ability to grow in cell cultures propagated in serum-free medium, we have isolated several *Mycoplasma* spe-

TABLE 5 Environmental sampling for mycoplasmas[a,b]

Sample site	No. positive/ no. tested	$\bar{a}$ CFU[c] (range)
Lip of culture flask	20/20	TN[d]
Used pipette	10/10	TN
Outside of flask	3/7	250 (4–TN)
Cap of flask	1/1	6
Hands	2/15	TN
Droplet on work surface	2/2	TN
Work surface	8/12	7(1-22)
Propipet	1/7	150
Used disinfectant solution	0/10	0
Outside of disinfectant pan	4/4	157 (7–TN)
Outside of used pipette	1/1	110
Glassware partially submerged in disinfectant	1/11	75
Glassware partially submerged in disinfectant	2/2	185 (70–TN)
Trypan blue + infected cells	2/2	TN
Used hemacytometer	2/2	TN
Tube containing culture	4/4	65 (10–200)
Settling plates	8/32	85 (1–TN)
Air samples[e]	0/11	0

[a]From reference 52, with permission.

[b]Samples were taken before, during, and after trypsinization and passage of 3T-6 mouse fibroblast cell cultures infected with *A. laidlawii*, 4.3×10^7 CFU/ml ($\bar{a}$ of 10 tests). Samples obtained with Rodac and settling plates over a 15-week period.

[c]Average CFU per plate of positive samples.

[d]Greater than 200 CFU per Rodac plate and 300 CFU per settling plate.

[e]Obtained with Andersen and all glass impinger (AGI-4) samplers.

cies that grew under these conditions. We have shown that these organisms obtained cholesterol from the plasma membranes of the host cells (53).

Fortunately, no laboratory infections have been traced to cell culture mycoplasmas. It has been generally believed that the mycoplasmas that typically infect cell cultures are not pathogenic for humans. However, in recent years, results of some studies have postulated that *M. fermentans* may be pathogenic for humans (42). The organism has been detected from AIDS patients by indirect methods. Although these results are controversial, they do suggest a potential risk. The same general methods used to prevent and detect mycoplasmas in laboratory-scale operations can, with modification, be applied to large-scale cultivation of cells.

Results of such studies can yield valuable information about the methods of transmission of mycoplasmas. Also, they can serve as models for other adventitious agents in cell cultures. If mycoplasmas can be disseminated into the immediate environment, what about other agents that can be present in cell cultures? This topic is covered extensively in reference 16.

Herpesviruses

Some of the exogenous agents detected in cell cultures in recent years may have more relevance to general laboratory safety rather than large-scale operations specifically. For example, human herpesviruses 6 and 7 (HHV-6 and HHV-7) have been isolated from peripheral blood lymphocytes of healthy donors (27). Results suggested that HHVs resided latently in the lymphocytes and were induced from latency by T-cell activation. It appeared that HHV-7 could act as a helper for activation of HHV-6.

HHV-6 was first isolated in 1986 from the blood of AIDS patients (69). HHV-7 was first isolated from the blood of a 25-year-old healthy individual (26). HHV-6 causes roseola infantum and can cause severe sequelae in bone marrow and transplantation patients who are undergoing immunosuppressive therapy (27). HHV-7 has not been shown to cause human disease.

Laboratory workers should be aware of agents such as HHV-6 and HHV-7, especially when handling peripheral blood and T-cell lines. Universal precautions established by the Centers for Disease Control and Prevention for handling blood specimens must be used (13).

Filoviruses

Filoviruses were first recognized after African green monkeys were discovered to carry the etiological agents of human disease. The first and largest outbreak occurred in Marburg, Germany, and the city gave its name to the newly discovered agent and disease (71). New isolates were made, and the agents were classified as a new family, the *Filoviridae,* in 1988. The pathogenesis of the human disease is poorly understood; a review has been published (63).

These single-stranded RNA viruses grow well in a wide variety of mammalian cell cultures, causing minimal cytopathogenicity. Vero cells have been widely used for primary isolation of filoviruses. Peters et al. (63) reviewed the implications for cell culture production. According to this review, significant replication occurs within 2 days, and most cells were positive for viral antigen by immunofluorescence by day 4 with minimal cytopathic effects. Titers of different strains ranged from approximately 10^3 to 10^6 PFU/ml.

Filovirus contamination of cell cultures will not be readily apparent due to the minimal cytopathic effects. Specific tests such as immunofluorescence or electron microscopy would have to be performed. Being lipid-enveloped RNA viruses, they can be readily inactivated by heat (56°C for 1 to 2 h), solvents, and common disinfectants. Inactivation occurred in 1-min exposures to each of the following: 5% Lysol, 2% Staphene, 1% Roccal, 1% Chloramine T, 1% Nolvasan, or 10% Formalin. Solutions of 2% Omega or 2% Micro Quat did not inactivate the Ebola strain of filovirus even after a 5-min exposure. Peters et al. (63) concluded that there does not seem to be a health hazard for research users of nonhuman primates or for vaccine substrate production with filoviruses.

Bovine Spongiform Encephalopathy

In 1985, cattle in the United Kingdom were shown to have bovine spongiform encephalopathy (BSE), even though only sheep in the United Kingdom were known to have scrapie. In 1987, Wells et al. (80) postulated a scrapie-like illness in cattle. The only reliable assay to prove the presence of the agents of scrapie and BSE is by the parenteral inoculation of susceptible strains of mice. The early epidemiology of this outbreak has been reported by Tyrrell (78). It was recognized early in the outbreak that the only consistent risk factor was the feeding of meat and bone meal to the affected cattle, usually calves. The outbreak has been hopefully controlled by a ban on ruminant feeding. However, because the inoculation period of spongiform encephalopathy is long, data are still inconclusive.

Tyrrell reported a total of 21,636 cases of BSE as of January 1, 1991, in 9,942 affected herds in the United Kingdom. He also noted there had been zero cases in cattle not exposed to meat and bone meal since July 1988 (78).

Of immediate concern is the potential of contamination of cell cultures by BSE via the use of fetal or newborn bovine sera as a medium supplement. Unfortunately, there is no laboratory test for BSE or for the presence of infection. Government agencies in the United Kingdom have dealt with this by placing strict requirements on the use of bovine materials that originated in infected herds.

The potential risk to cell culture laboratory workers and ultimately to recipients of products derived from cell culture systems is unclear. According to Tyrrell (78), a species barrier seems to exist for these agents. Further, there does not seem to be any proof of transmission of these agents across species barriers by inoculation, eating, or other exposures.

Parvoviruses

Parvoviruses are efficient in establishing latent infections in cell cultures (7). Hallauer et al. (31) were the first to detect these agents in cell cultures during extraction of yellow fever virus hemagglutin. They noted a second hemagglutin to be present, which was distinct for yellow fever virus. Over the years, surveys of cell cultures for parvovirus infection have been performed (21). In one of these surveys, Hallauer et al. (31) detected the agents in 38 of 43 cultures assayed. The source of these agents in cell culture is unclear, although porcine pancreas, used as a source of trypsin used in cell culture, is one possibility.

One of the members of the parvovirus group, the dependiviruses, require helper virus for detection. The most prominent members of this group are the adeno-associated viruses. Hoggan (35) showed that the incidence of adeno-associated virus latent infection ranged from 1 to 20% of cell cultures assayed by introduction of helper adenoviruses.

The risk of parvoviral latent infection to production workers is unknown but thought to be minimal. The serologic prevalence of one human parvovirus, B19, is 60% in the population reported by Anderson (2). Threat of infection may be increased in immunosuppressed patients.

Human Gene Therapy

In recent years, much progress has been made in the field of human gene therapy. At this writing, more than 300 patients have received foreign genes for marker or therapeutic trials in the United States, France, Italy, Germany, England, and the Peoples Republic of China. More trials are being initiated. In response to these developments, regulatory agencies have established a framework to ensure patient safety with these products. In the United States, most clinical protocols are reviewed by the FDA and the NIH. Each of these agencies have published separate Points to Consider documents to serve as a guide to investigators submitting protocols (25, 59). As the field of human gene therapy has continued to evolve, so has the regulatory framework. This is expected to continue.

Several reviews on human gene therapy have recently been published (3, 77). Presently, safety discussions have focused on the vector delivery systems of the genes. To date, most of the trials have used murine retroviral vectors that were developed from the Moloney murine leukemia virus (17). More recently, the NIH RAC has approved several human trials that use vectors generated from human adenoviruses. Two trials have used liposomes. The journal *Human Gene Therapy* publishes protocols approved by the NIH RAC, and readers can refer to the journal for details of specific protocols.

The main safety concerns with retroviral vectors are the characteristic of the vectors to integrate randomly into the host genome as well as the possibility of generating replication-competent viruses. The possible consequences of such random integration include mutation of some essential gene that would result in death of the cell. Alternately, the vector could insert into a protooncogene or into a tumor suppressor gene. However, these events alone would not necessarily be responsible for oncogenic conversion. Multiple events are needed (3).

Much of the early work in the field of safety assessment of murine retroviral vectors has been to demonstrate the absence of replication-competent retroviruses (RCR). Viral vectors, retroviral and adenoviral, have been carefully designed to be replication incompetent. Several studies have been performed on the safety of retroviral vectors. These have included inoculation into monkeys. Some results of retroviral vector studies are summarized below.

An early assessment by both the NIH RAC and the FDA resulted in the requirement that retroviral vector supernatants for use in clinical trials be assayed for the presence of RCR.

It is important to quantitate the relative effectiveness of the RCR assay systems used. The probability of detecting RCR will be equivalent to other types of sterility assays and will depend, in part, on the number of vessels, the level of contamination (both the number of contaminated vessels and the concentration of the contaminant), and the sample size. In addition to these general considerations, the safety margin afforded by RCR assay systems also needs to be assessed, based on information obtained in animal models.

Most retroviral vectors have been developed from murine retroviruses that are part of the *Oncoviridae* (which cause malignancies, especially in rodents), one of the three groups that comprise the family *Retroviridae* (36). Retroviral vectors have been designed to be replication incompetent.

RCRs most likely result from homologous or nonhomologous recombinational events involving the sequences of the retroviral vector and the sequences providing the structural components of a retroviral vector particle. The components needed in *cis* for a retrovirus are the 5′LTR, psi or encapsulation sequence, and 3′LTR. Components needed in *trans* are the genes coding for the structural genes *gag/pol* and *env*.

In the earliest generation retroviral vectors, these components were not separate. The intact retrovirus genome (*gag*, *pol*, and *env*) were present on the same RNA strand as the gene for transfer. For gene transfer purposes, retroviral systems were developed in which there was physical separation of vector sequences from retrovirus structural sequences. The early generation of vectors and packaging systems had significant areas of homology between the structural and vector sequences. To generate an RCR, only a single recombinational event was needed to occur between packaging and vector sequences.

To assess the potential risk of exposure to RCR, studies were performed on normal and immunosuppressed monkeys exposed to amphotropic RCR (17). These safety studies were designed as a risk assessment for patients receiving gene therapy. The risk to laboratory workers should be significantly less. However, these animal studies are still appropriate. Five monkeys were exposed to RCR derived from two sources, murine amphotropic retrovirus 4070A and the RCR present in SAX vector supernatant.

To reduce host defense mechanisms and therefore maximize the likelihood of toxicity, some of the animals were immunosuppressed using high doses of cyclosporine A. One immunosuppressed and three normal monkeys were injected intravenously with a mean RCR exposure of 7.2×10^7 focus-forming units (FFU) and mean vector exposure of 11.5×10^7 CFU. The volume of retroviral supernatant injected into the animals represented 12 to 22% of total blood volume of the animals. No animal was viremic for more than 15 minutes after the completion of the infusion, at least in part due to the ability of the complement system to inactivate murine retroviruses (17). Although no RCR could be detected by biological-based assays, lymph nodes were found by RCR to contain 4070A *env* sequences. Other than transient acute lymphadenopathy, no animal has shown any acute or chronic effects during a follow-up period exceeding a mean of 6.5 years.

These five monkeys have been followed for more than 6 years after exposure to RCR. Results have shown that exposure of these primates to large quantities of RCRs did not result in acute or chronic pathology. The immunosuppressed animals did show evidence of RCR gene transfer into lymphocytes, but no pathology or persistent retroviremia could be detected. RCRs intravenously administered to primates were readily cleared from the circulation.

Other monkeys have been exposed to retroviral vectors or RCR in a variety of protocols, including in utero marrow transplantation. These fetal studies involved animals with immature immune functioning. Fetal animals can accept allogeneic fetal hematopoietic cells without graft-versus-host or host rejection of graft. In these experiments, no in utero toxicities have been observed, and the animals have been followed after birth and have had no signs of retroviremia for the 4 years since their birth. The total monkey years of follow-up since exposure is approximately 130. No retroviremia or other toxicities have been detected (56).

A2 (N763A2) represents an N2-based amphotropic retroviral vector producer cell developed for use in primate bone marrow gene transfer experiments. The A2 vector lacks safety factors of later vectors that reduce recombination that could produce RCR. As a result of the A2 sequence homology with packaging cell sequences, two recombinational events can result in the generation of RCR. Production of high titers of the A2 vector was accompanied by generation of RCR with a titer in the range of 10^4 to 10^6 FFU/ml. The gene transfer protocol used in one primate study exposed CD34-enriched autologous primate cells in the presence of polybrene to the A2 vector and accompanying 10^4 to 10^6 FFU/ml of RCR over an 80- to 86-h period. During the exposure period, various growth factors or producer cells were also present. After lethal irradiation of the monkeys, the autologous precursor cells were infused into the monkeys.

A2 producer cells were used to achieve gene transfer into monkey bone marrow. It was known that the vector was RCR contaminated, but the level of RCR was not higher than previous monkey marrow gene transplant experiments. The A2 vector was not anticipated to be clinically useful because of the RCR present. Three animals of the 10 exposed to A2 vector developed lethal cell lymphomas about 200 days after transplantation (20). These three animals; but not the others, developed a chronic retroviremia and failed to generate detectable antibodies to RCR. These lymphomas appeared to be causally related to RCR.

In the study that resulted in lymphoma, monkeys were severely immunosuppressed. Exposure of bone marrow cells to 2×10^7 FFU of RCR for 80 to 86 h (together with other protocol modifications) resulted in the development of a T-cell lymphoma in 3 of 10 animals (20). From these data, one can estimate some risks: Under severe immunosuppressive conditions, a monkey exposed to 2×10^7 FFU of RCR has a 30% probability of developing a T-cell lymphoma, whereas under less severe conditions, primates remain disease-free after exposure of 10^8 to 10^9 FFU of RCR.

What do the lymphoma monkeys tell us about the risks in human gene transfer clinical protocols and to laboratory workers? Most important, these animals begin to show us what it takes to get a malignancy from RCR. Fortunately, the conclusion appears to be that clinical protocols are relatively safe. There are several safety nets, all of which would need to be violated to lead to the conditions that resulted in lymphoma in the monkeys. Immunosuppression may be a key factor in establishment of these lymphomas. An extensive review of RCR has been published (4).

Certification and Validation of Cell Culture Systems

The above sections review the potential hazards of agents that are used for production of commercial products or that have been detected as adventitious agents in cell culture systems. Additional agents may also be present, but detection methods may not be adequate for their detection. Therefore, in addition to known and validated assays to detect these agents, the commercial producer laboratory must also be able to demonstrate that the entire manufacturing system has been validated to achieve minimal exposure levels for the worker and the community. Typically, cultured cells will be significantly more sensitive to environmental stresses and disinfectants. Abreo et al. (1) showed less than 0.1% of a mammalian cell culture survived a 2-min exposure to 60°C.

Some of these methods will be similar for microbial and cell culture systems. Some of these have been reviewed, stressing the purity and consistency of the product as the ultimate parameter (14, 37). Also, some indirect measurements may also be helpful, such as the medium feed rate, oxygen uptake, amount of carbon dioxide evolved, cell density, and as noted, the concentration of the product. Low product yield is of concern. Low product yields may require concentration efforts beyond the capacity of the validated procedures. Also, the product run may not be cost-effective.

Liu et al. (41) reviewed some of the quality control assays needed for recombinant DNA products. In addition to the quality control assays needed to detect adventitious agents, assays for the final product should include peptide mapping and at least partial sequence analysis of the protein produced. Also, physicochemical analysis should be performed. Further assays should include polyacrylamide gel electrophoresis, isoelectric focusing, and high-pressure liquid chromatography. Circular dichroism and optical rotatory dispersion should also be performed.

The assays for adventitious agents should include the standard tests for sterility to exclude the potential for bacterial, yeast, fungi, mycoplasma, and viral contamination. The FDA, in its Points to Consider documents, also recommends that assays sensitive to detect DNA contamination at the 10-pg level be instituted (24). Finally, testing on the final product should include tests for pyrogen and antigenicity. Where appropriate, these assays should include animal testing.

All assays that are to be performed in both quality control and production must be of known efficiency and must be validated (49). Manufacturers must ensure that the stock culture used for each production lot is of the same consistency. Master and working cell banks can help ensure such consistency. Some areas for characterization have been reviewed. This standardization should help minimize risk to workers as well. The use of DNA fingerprinting to further characterize mammalian cell lines has been reported (75). This may be effective to distinguish one species of origin from another or one line from another. However, its effectiveness to characterize stability of a line through production remains to be validated. Arathoon and Birch (6) and Lubiniecki et al. (48, 49) have listed some factors that influence standardization of cell cultures through a production run.

Ironically, one aspect of mammalian cell banks used to produce recombinant proteins is validation of procedures to reduce the probability of retroviral contamination (33, 44, 66). CHO cells used in production of recombinant proteins have been shown to harbor particles resembling type A and type C retroviral particles (47, 50, 76, 81). However, CHO cells are resistant to infection by several retroviruses studied (66). However, use of cell culture systems to generate replication-incompetent retroviruses has been focused on methods to improve vector production (38).

The integration of effective biological assays with carefully designed facilities and well-thought-out validated procedures can reduce the possibility of exposures to production workers and the community to minimal levels (8). To produce a harmful effect, the organisms used in the manufacturing process must have the capability to infect humans or other animals, they must escape the manufacturing container or containment vessel, they must spread and multiply in the environment, and finally, they must establish themselves on or in humans. Of the organisms used in biotechnology manufacturing processes, *Bacillus* spp. and *Saccharomyces* spp. have the greatest potential to survive in the environment for prolonged periods. *E. coli* and mammalian cells are susceptible to dehydration. Andrup et al. (5) reviewed some safety factors. These authors concluded that no diseases or accidents have been attributed to recombinant DNA. Cohen and Hoerner (15) agreed with this conclusion.

In younger start-up biotechnology companies, multiuse facilities have been used in the manufacture of biologicals (73). Multiuse facilities are those that use areas for production of two or more products. A report on multiuse facilities by a committee of the Pharmaceutical Manufacturer's Association stressed the use of engineering controls, operating procedures, and temporal segregation to minimize the potential for cross-contamination (9). The lesson from multiuse facilities as well as smaller laboratory operations is that safety, whether focused on product, the worker, or the community, is still reliant on a thorough knowledge of the agent or culture being handled, well-trained personnel, adequate facilities and engineering controls, and a strict adherence to efficient, well-defined aseptic practices.

The authors thank Marietta Toal for editorial assistance.

References

1. **Abreo, C., C. Hoerner, A. Lubiniecki, W. Wales, and M. Wiebe.** 1989. Recombinant DNA containment considerations for large scale mammalian cell culture expression system, p. 93–96. *In* R. E. Spier, J. B. Griffith, J. Stepheme, and P. J. Crooy (ed.), *Advances in animal cell biology and technology for bioprocesses.* Butterworths, London.
2. **Anderson, L. J.** 1990. Human parvoviruses. *J. Infect. Dis.* **161:**603–608.
3. **Anderson, W. F.** 1993. Human gene therapy. *Science* **256:**808–812.
4. **Anderson, W. F., G. J. McGarrity, and R.C. Moen.** 1993. Report to the NIH Recombinant DNA Advisory Committee on murine replication-competent retrovirus (RCR) assays. *Hum. Gene Ther.* **4:**311–321.
5. **Andrup, L., B. H. Nielson, and S. Kolvraa.** 1990. Biosafety considerations in industries with production methods based on the use of recombinant deoxyribonucleic acid. *Scand. J. Environ. Health* **16:**85–95.
6. **Arathoon, W. R., and J. R. Birch.** 1986. Large scale cell culture in biotechnology. *Science* **232:**1390–1395.
7. **Archetti, I., E. Bereckzy, and D. Bocciarelli.** 1966. A small virus associated with the simian adenovirus. SVII. *Virology 29:*671–675.
8. **Avallone, H. L., M. G. Beatrice, and T. T. Sze.** 1991. Food and Drug Administration inspection and licensing of manufacturing facilities, p. 315–340. *In* Y. H. Chiu and J. L. Gueriguian (ed.), *Drug biotechnology regulation, scientific basis and practices.* Marcel Dekker, Inc., New York.
9. **Bader, F. G., A. Blum, B. Garfinkle, D. MacFarlane, T. Massa, and T. Copmann.** 1992. Multiuse manufacturing facilities for biologicals. *Biopharm* **5:**32–40.
10. **Bisson, W. F.** 1986. Strategies for improvement of industrial strains of *Saccharomyces cerevisiae. World Biotechnol. Rep.* **2:**103–116.
11. **Bogosian, G., and J. F. Kane.** 1991. Fate of recombinant *Escherichia coli* K-12 strains in the environment. *Adv. Appl. Microbiol.* **36:**87–131.
12. **Burman, L. G.** 1977. Expression of R-plasmid functions during anaerobic growth of an *Escherichia coli* K-12 host. *J. Bacteriol.* **131:**69–75.
13. **Centers for Disease Control/National Institutes of Health.** 1988. *Biosafety in microbiological and biomedical laboratories.* Centers for Disease Control, Atlanta.
14. **Chiu, Y.** 1988. Validation of the fermentation process for the production of recombinant DNA drugs. *Pharm. Technol.* **12:**132–138.
15. **Cohen, R., and C. L. Hoerner.** 1994. Occupational health perspective: recombinant DNA technology. *Biopharm* **7:**28–40.
16. **Continuous Cell Lines—An International Workshop on Current Issues, Bethesda, Md.** 1991. *Developments in biological standardization,* vol. 76. Karger, Basel.
17. **Cornetta, K., C. Moen, K. Culver, R. A. Morgan, J. R. McLachlin, S. Sturm, J. Selegue, W. Lindon, R. M. Blaese, and W. F. Anderson.** 1990. Amphotropic murine leukemia retrovirus is not an acute pathogen for primates. *Hum. Gene Ther.* **1:**15–30.
18. **CPC International.** 1986. GRAS Petition 4G0293 proposing that alpha-amylase from a strain of *Bacillus subtilis* (ATCC 39705) containing the gene for alpha-amylase from *B. stereothermophilus* inserted by recombinant DNA techniques be affirmed as GRAS as a direct food ingredient. Notice of filing. *Fed. Regist.* **51:**10571.
19. **DeBoer, A. S., and B. Diderichsen.** 1991. On the safety of *Bacillus subtilis* and *B. amyloliquefaciens:* a review. *Appl. Microbiol. Biotechnol.* **36:**1–4.
20. **Donahue, R. E., S. W. Kessler, D. Bodine, K. McDonagh, C. Dunbar, S. Goodwin, B. Agricola, E. Byrne, M. Raffeld, R. Moen, J. Bacher, K. M. Zsebo, and A. W. Nienhuis.** 1992. Helper virus induced T cell lymphoma in nonhuman primates after retroviral mediated gene transfer. *J. Exp. Med.* **176:**1125–1135.

21. **Fikrig, M. K., and P. Tattersall.** 1992. Latent parvoviral infection of continuous cell lines. *Dev. Biol. Stand.* **76:**285–293.
22. **Finch, J. E., J. Prince, and M. Hawksworth.** 1978. A bacteriological survey of the domestic environment. *J. Appl. Bacteriol.* **45:**364–375.
23. **Food and Drug Administration.** 1987. *Fermentation produced chymosin,* p. 1.0055–1.0056. GRAS Affirmation Petition (GRASP 86 0337), Pfizer Central Research. Food and Drug Administration, Washington, D.C.
24. **Food and Drug Administration.** 1987. *Points to consider in the characterization of cell lines to produce biological products.* Food and Drug Administration, Washington, D.C.
25. **Food and Drug Administration.** 1991. Points to consider in human somatic cell therapy and gene therapy. *Hum. Gene Ther.* **2:**251–256.
26. **Frenkel, N., E. C. Schirmer, L. S. Wyatt, G. Katasafanas, E. Roffman, R. M. Danovich, and C. H. June.** 1990. Isolation of a new herpesvirus from human CD4 and T cells. *Proc. Natl. Acad. Sci. USA* **87:**748–752.
27. **Frenkel, N., and L. S. Wyatt.** 1992. HHV-6 and HHV-7 as exogenous agents in human lymphocytes. *Dev. Biol. Stand.* **76:**259–265.
28. **Galloway, S. M., P. K. Berry, W. W. Nichols, K. A. Soper, P. D. Stolley, and P. Archer.** 1986. Chromosome aberrations in individuals occupationally exposed to ethylene oxide, and in a large population. *Mutat. Res.* **170**(1–2):55–74.
29. **Gill, D. M.** 1983. Bacterial toxins: a table of lethal amounts. *Microbiol. Rev.* **46:**86–94.
30. **Gorbach, S. L.** 1978. Risk assessments protocols for recombinant DNA experimentation. *J. Infect. Dis.* **137:**704–708.
31. **Hallauer, C., G. Kronauer, and G. Siegl.** 1971. Parvoviruses as contaminants of permanent human cell lines. Virus isolation from 1960–1971. *Arch. Gesamte Virusforsch.* **35:**80–90.
32. **Harrington, J. M., G. F. Stein, R. O. Rivera, and A. V. de Morales.** 1978. The occupational hazards of formulating oral contraceptives: a survey of plant employees. *Arch. Environ. Health* **33**(1):12–14.
33. **Hartley, J. W.** 1992. Detection of murine leukemia viruses—*in vitro* infectivity tests. *Dev. Biol. Stand.* **76:**179–186.
34. **Hoerner, C. L.** 1992. Environmental considerations in commercial biopharmaceuticals, p. 94–109. Proc. Biopharm. Conf., San Francisco.
35. **Hoggan, M. D.** 1970. Adeno-associated viruses. *Prog. Med. Virol.* **12:**211–239.
36. **Katz, R. A., and A. M. Skalka.** 1990. Generation of diversity in retroviruses. *Annu. Rev. Genet.* **24:**409–445.
37. **Kaugman, R.** 1990. Use of recombinant DNA technology for engineering mammalian cells to produce proteins, p. 15–69. *In* A. Lubiniecki (ed.), *Large-scale mammalian cell culture technology.* Marcel Dekker, Inc., New York.
38. **Kotani, H., P. B. Newton III, S. Zhang, Y. L. Chang, E. Otto, L. Weaver, R. M. Blaese, W. F. Anderson, and G. J. McGarrity.** 1994. Improved methods of retroviral vector transduction and production for gene therapy. *Hum. Gene Ther.* **5:**19–28.
39. **Levine, M. M., J. B. Kaper, H. Lockman, R. E. Black, M. L. Clements, and S. Falkow.** 1978. Recombinant DNA risk assessment studies in man: efficacy of poorly mobilizable plasmids in biological containment. *J. Infect. Dis.* **148**(4):699–670.
40. **Levy, S. B., B. Marshall, and D. Rowse-Eagle.** 1980. Survival of *Escherichia coli* host-vectors systems in the mammalian intestine. *Science* **209:**391–394.
41. **Liu, D. T., F. T. Gates III, and N. D. Goldman.** 1985. Quality control of biologicals produced by r-DNA technology. *Dev. Biol. Standard.* **59:**161–166.
42. **Lo, S.-C., J. W.-K. Shih, B. B. Newton III, D. M. Wong, M. M. Hayes, J. R. Benisn, D. J. Wear, and R. Y.-H. Wang.** 1989. Virus-like infectious agent (VLIA) is a novel pathogenic mycoplasma: *Mycoplasma incognitus. Am. J. Trop. Med. Hyg.* **41:**586–600.
43. **Losada, E., M. Hinojosa, S. Quirce, M. Sanchez-Cano, and I. Moneo.** 1992. Occupational asthma caused by alpha-amylase inhalation: clinical and immunologic findings and bronchial response patterns. *J. Allergy Clin. Immunol.* **89:**118–125.
44. **Losikoff, A. M., J. A. Poiley, R. Raineri, R. E. Nelson, and T. Hillesund.** 1992. Industrial experience with the detection of retroviruses. *Dev. Biol. Stand.* **76:**187–200.
45. **Lubiniecki, A.** 1990. *Large-scale mammalian cell culture technology.* Marcel Dekker, Inc., New York.
46. **Lubiniecki, A., K. Anumula, J. Callaway, J. L'Italien, M. Oka, B. Okita, G. Wasserman, D. Zabriskie, R. Arrathoon, S. Builder, R. Garnick, M. Wiebe, and J. Browne.** 1992. Effects of fermentation on product consistency. *Dev. Biol. Stand.* **76:**105–115.
47. **Lubiniecki, A., M. Dinowitz, E. Nelson, M. Wiebe, L. May, J. Ogez, and S. Builder.** 1988. Endogenous retroviruses of continuous cell substrates. *Dev. Biol. Stand.* **70:**187–191.
48. **Lubiniecki, A., and L. May.** 1985. Cell bank characterization for recombinant DNA mammalian cell lines. *Dev. Biol. Stand.* **60:**141–146.
49. **Lubiniecki, A., M. E. Wiebe, and S. E. Builder.** 1990. Process validation for cell culture-derived pharmaceutical proteins, p. 515–541. *In* A. Lubiniecki (ed.), *Large-scale mammalian cell culture technology.* Marcel Dekker, Inc., New York.
50. **Marcus-Sekura, C. J.** 1992. Validation of removal of human retroviruses. *Dev. Biol. Stand.* **76:**215–223.
51. **Marshall, B., S. Schuluederberg, S. Tachibana, and S. B. Levy.** 1981. Survival and transfer in the human gut of poorly mobilizable (pBR322) and transferable plasmids from the same carrier *E. coli. Gene* **14:**145–154.
52. **McGarrity, G. J.** 1976. Spread and control of mycoplasmal infection of cell cultures. *In Vitro* **12:**643–648.
53. **McGarrity, G. J.** Unpublished data.
54. **McGarrity, G. J., and H. Kotani.** 1985. Cell culture mycoplasmas, p. 353–390. *In* S. Razin and M. F. Barile (ed.), *The mycoplasmas,* vol. IV. Academic Press, Inc., New York.
55. **McGarrity, G. J., J. Sarama, and V. Vanaman.** 1985. Cell culture techniques. *ASM News* **51**(4):170–183.
56. **Moen, R. C.** 1993. Personal communication.
57. **Moyer, M. P.** 1989. Human cell models for genetic engineering. *Med. Prog. Technol.* **15:**83–100.
58. **National Institutes of Health.** 1987. Recombinant DNA research, actions under guidelines, appendix K physical containment for large-scale uses of organisms containing recombinant DNA molecules. *Fed. Regist.* **56**(138):333174–333183.

59. **National Institutes of Health.** 1994. Guidelines for research involving recombinant DNA molecules. *Fed. Regist.*, July 5, 1994, Separate Part IV.
60. **National Institutes of Health—Recombinant DNA Advisory Committee.** 1984. Toxins classified under Appendix F of the guidelines. *Recombinant DNA Bull.* 7(1):44.
61. **Oka, M. S., and R. G. Rupp.** 1990. Large-scale animal cell culture: a biological perspective, p. 71–92. *In* A. Lubiniecki (ed.), *Large-scale mammalian cell culture technology.* Marcel Dekker, Inc., New York.
62. **Palchak, R. B., R. Cohen, and C. L. Hoerner.** 1988. Airborne endotoxin associated with industrial-scale production of protein products in gram-negative bacteria. *Am. Ind. Hyg. Assoc. J.* **49:**420–421.
63. **Peters, C. J., P. B. Jahrling, T. G. Ksiazek, E. D. Johnson, and H. W. Lupton.** 1992. Filovirus contamination of cell cultures. *Dev. Biol. Stand.* **76:**267–274.
64. **Petsonk, E. L., S. A. Olencock, R. M. Csatellan, D. E. Banks, J. C. Mull, J. L. Hankinson, H. H. Perkins, and J. B. Cocke.** 1986. Human ventilatory response to washed and unwashed cottons from different growing areas. *Br. J. Ind. Med.* **43:**182–187.
65. **Pharmaceutical Research and Manufacturers Association.** 1991. Draft interim guidance to the pharmaceutical industry for environmental assessment compliance with FDA. Pharmaceutical Manufacturers Association, Washington, D.C.
66. **Poiley, J. A., R. E. Nelson, T. Hillesund, and R. Raineri.** 1991. Susceptibility of CHO-K-1 cells to infection by eight adventitious viruses and four retroviruses. *In Vitro* **4:**1–12.
67. **Priest, F. G.** 1990. Products and applications, p. 293–320. *In* C. R. Harwood (ed.), *Bacillus.* Plenum Press, New York.
68. **Rinderknecht, E., B. O'Connor, and H. Rodriquez.** 1984. Natural human interferon-gamma: complete amino acid sequence and determination of sites of glycosylation. *J. Biol. Chem.* **259:**6790–6797.
69. **Salahuddin, S. Z., D. V. Ablashi, P. D. Markham, S. F. Josephs, S. Sturzenberger, M. Kaplan, G. Halligan, P. Biberfeld, F. Wong-Staal, B. Kramarsky, and R. C. Gallo.** 1986. Isolation of a new virus, HBLV, in patients with lymphoproliferation disorders. *Science* **234:**596–601.
70. **Seamon, K.** 1992. Genetic and biochemical factors affecting product consistency: introduction to the issues. *Dev. Biol. Stand.* **76:**63–67.
71. **Siegert, R.** 1972. Marburg virus. *Virol. Monogr.* **11:**97–153.
72. **Smith, H. W.** 1969. Transfer of antibiotic resistance from animal and human strains of *E. coli* in the alimentary tract of man. *Lancet* **i:**1174–1176.
73. **Srigley, W.** 1990. Design and construction of manufacturing facilities for mammalian cell-derived pharmaceuticals, p. 567–596. *In* A. Lubiniecki (ed.), *Large-scale mammalian cell culture technology.* Marcel Dekker, Inc., New York.
74. **Stolley, P. D., K. A. Soper, S. M. Galloway, W. W. Nichols, S. A. Norman, and S. R. Wolman.** 1984. Sister chromatid exchange in association with occupational exposure to ethylene oxide. *Mutat. Res.* **129**(1)**:**89–102.
75. **Thacker, J. M., B. T. Webb, and P. B. Debenham.** 1988. Fingerprinting cell lines: use of hypervariable DNA probes to characterize mammalian cell cultures. *Somat. Cell Mol. Genet.* **14:**519–525.
76. **Tihon, C., and M. Green.** 1973. Cyclic AMP-amplified replication of RNA tumor virus-like particles in Chinese hamster ovary cells. *Nature* **744:**227–231.
77. **Tolstoshev, P., and W. F. Anderson.** 1993. Gene transfer techniques for use in human gene therapy, p. 35–50. *In* K. W. Adolph (ed.), *Genome research in molecular medicine and virology.* Academic Press, Inc., New York.
78. **Tyrrell, D. A. J.** 1992. An overview of bovine spongiform encephalopathy (BSE) in Britain. *Dev. Biol. Stand.* **76:**275–284.
79. **Vandamme, E. J.** 1992. Microbial production of vitamins and biofactors: an overview. *J. Nutr. Sci.* Special No.:244–247.
80. **Wells, G., A. Scott, C. Johnson, R. S. Gunning, R. Hancock, M. Jeffrey, J. Dawson, and R. Bradley.** 1987. A novel progressive spongiform encephalopathy in cattle. *Vet. Rec.* **121:**419–420.
81. **Wurm, F. M., M. G. Pallavicini, and R. Arathoon.** 1992. Integration and stability of CHO amplicons containing plasmid sequences. *Dev. Biol. Stand.* **76:**69–82.

Toxic and Carcinogenic Chemicals in Biomedical Laboratories

WILLIAM A. TOSCANO, JR., AND SUSAN E. MCNULTY

10

INTRODUCTION

One prominent laboratory chemical supply company lists approximately 27,000 items in their catalog, 4,000 of which were added in the past year (2). Coupled with biologically active agents and toxins and the many agents being created by organic chemists and molecular biologists, laboratory personnel could easily encounter more than 50,000 different xenobiotic chemicals in their career. A xenobiotic is a foreign chemical that usually has no nutritional significance; however, for employees in the biomedical laboratory, many xenobiotics have potential toxicant action. The amount of toxicant to which one is exposed, the duration of that exposure, and other factors such as the route of exposure (i.e., ingestion, inhalation, or dermal absorption) affect the severity of the toxic response. To minimize the potential for a toxic response, exposure must be minimized. This can be achieved by engineering controls, facility design, and "good laboratory practice." These protective controls, discussed in more detail in Chapters 11, 12, and 13 of this book, include wearing protective clothing such as coats, aprons, proper gloves, and respirators and working in a properly ventilated area. Laboratory workers should use universal precautions for all potentially hazardous materials by wearing gloves and other protective clothing when handling all chemicals.

RELEVANT TOXICOLOGIC LITERATURE

Several computerized data bases have been established by the National Library of Medicine and others to serve as resources on the potential toxicity of xenobiotic chemicals (Table 1). Detailed studies of the toxicity of a limited number of chemicals are given in a series of reports prepared for the Agency for Toxic Substances and Disease Registry and the Environmental Protection Agency. Each toxicology profile examines, summarizes, and interprets the latest available data from toxicologic and epidemiologic studies. Ideally, the studies will be updated as the need arises or every 3 years, whichever comes first. The World Health Organization, under the auspices of the International Agency for Research on Cancer, publishes a series of monographs called *IARC Monographs on the Evaluation of Carcinogenic Risk to Humans* that detail the current state of knowledge and expert opinion on suspected carcinogens. Another very useful and readily available source of information on potentially hazardous xenobiotics is the material safety data sheet that is supplied by the manufacturer of laboratory chemicals. The Occupational Safety and Health Administration laboratory standard now requires employers to keep a file of material safety data sheets for the chemicals being used in the laboratory and stipulates that laboratory workers be trained about the hazards before under-

TABLE 1 Toxicology data bases

Data base	Comment
MEDLARS	Medical Literature Analysis and Retrieval System
TOXLINE TOXLIT TOXLIT-65	Data bases at the National Library of Medicine for Toxicology; for further information, contact: National Library of Medicine (NLM), MEDLARS Management Section, 8600 Rockville Pike, Bethesda, Md. 20814, (800) 638-8480
TOXNET HSDB CCRIS RTECS CDF IRIS GENTOX ASTDR	Toxicology data network; contains data bases relevant to hazardous material handling, such as Hazardous Substance Data Bank, Chemical Carcinogen Research Information System, Registry of Toxic Effects of Chemical Substances, Chemical Directory File, plus specialized systems developed by the Environmental Protection Agency, such as Integrated Risk Information System, Genetic Toxicology, and Agency for Toxic Substances and Disease Register Toxicology Profiles. Further information can be obtained from NLM, Specialized Information Services Division, (301) 496-6531 or 496-1131
ETIC	Environmental Teratology Information Center, established at the Oak Ridge National Laboratory; contains information on substances that could have potential teratogenic and developmental biology effects

taking experiments or assays that use a particular chemical.

Some exposure guidelines are published by the American Conference of Government Industrial Hygienists as annual guides listing threshold limit values (TLV) of many chemicals encountered in the laboratory setting. The TLV indicates the airborne concentrations of a substance to which a worker may be exposed daily without an adverse effect. These values give some indication of the limits to which one may be exposed in the work week. The TLV is based on the best available information obtained from industrial experience and animal and human exposure data but is only a guide for using a particular compound.

TOXIC RESPONSES

Toxic responses can be caused by an acute high dose exposure or from chronic low dose exposures to certain chemicals. In humans, these responses can range from simple rashes to complex diseases such as cancer. A toxic response may be specific or general, or it may be systemic or local. Acute overexposure to potentially toxic xenobiotics can often cause alterations in common vital signs, such as body temperature, pulse rate, and respiratory rate. Body temperature may be increased or decreased, depending on the nature of the chemical to which one is exposed. Three alterations in pulse rate are possible: bradycardia (decreased pulse rate), tachycardia (increased pulse rate), or arrhythmia (irregular pulse rate). Alteration in respiratory rate or blood pressure is often the early sign of exposure in humans.

Exposure to toxic levels of some chemicals can result in unnatural body odors when the xenobiotic or its metabolite is secreted through the skin or is exhaled. Such symptoms of poisoning include aromatic odors from various hydrocarbons (e.g., aroma of violets from ingestion of turpentine, bitter almond from HCN, rotten eggs from H_2S, garlic breath from selenium compounds, pears from chloral hydrate, acetone breath from ketones, or a shoe polish smell from nitrobenzene exposure). The eyes reveal exposure to potential toxicants in several ways: excessive or prolonged pupil contraction (miosis), excessive pupil dilation (mydriasis), inflammation of the conjunctiva (conjunctivitis), and the side-to-side involuntary movement of the eyeballs (nystagmus). Changes in the moisture content of the mouth may also signal toxicity, ranging from excessive moisture and salivation to dryness of the mouth. Pain and vomiting or paralytic ileus (slowed peristalsis of the small intestine) has been observed after exposure to some toxicants. Toxicity to the nervous system can result in convulsions, paralysis, hallucinations, ataxia, agitation, hyperactivity, disorientation, delirium, and coma. In less severe reactions, nervous system effects can include dizziness or headache that disappears after the person is removed from the source of the intoxicant (16, 33). Immediate medical attention is required if any of these symptoms are observed in laboratory personnel.

The skin, the first line of defense against many xenobiotics, often shows evidence of exposure as altered coloration and moisture. Skin may become flushed when exposed to high concentrations of some toxicants; the classic blue color of cyanosis may be a symptom of oxygen deprivation. Intoxication by arsenicals, iron, or carbon tetrachloride will impart a jaundiced (yellow) color to skin; antihistamines will cause excessive dryness of skin; and some metals and organophosphates will increase skin moisture. Chronic irritant dermatitis is the most common skin complaint (29). Many of the actions of potential toxicants on skin may be avoided by using proper gloves or other apparel. Heavy rubber or neoprene gloves are recommended for inorganic and organic acids, whereas nitrile gloves are preferred for handling aromatic, aliphatic, and chlorinated solvents (29). Polyethylene gloves are not recommended for use in handling most solvents used in the biomedical laboratory.

Xenobiotic metabolism plays an important role in modulating the toxicity of many solvents and potentially carcinogenic compounds by the xenobiotic metabolizing enzymes usually located in the endoplasmic reticulum of cells. The metabolism of xenobiotics often occurs in two stages: phase I and phase II metabolism, respectively. Phase I metabolism can involve hydration, oxidation, reduction, or hydrolysis and can possibly result in a more toxic chemical than the parent compound. In that case, the xenobiotic to which one is exposed is a protoxicant, and the metabolite is often the actual toxicant. Phase II metabolism involves the conjugation of a parent compound containing a reactive functional group or a metabolite from phase I metabolism to a sugar, sulfur compound, or amino acid. Metabolites from phase II pathways are often more hydrophilic than the parent xenobiotic and are readily excreted by the kidney or in the bile.

ACUTE TOXICANTS

Many compounds in common use in biomedical laboratories are acutely neurotoxic (5) or metabolic toxins. Such compounds are also used as pesticides or as poison gases. Some inhibit ATPase activity, whereas others (uncouplers) act to dissociate electron transport from ATP formation. Examples of compounds in this class include dicoumarol, seconal, and rotenone.

Many common laboratory reagents found in biomedical laboratories are potential neurotoxins (12, 43). Examples include solvents, serine protease inhibitors, acrylamide, metals, gases, and some agents from marine organisms and plants that are commonly used as probes of various ion channels. Acrylamide is routinely used to prepare polymerized gels used in protein separations. Polyacrylamide is not neurotoxic, but acrylamide monomer is. Peripheral neuropathy has been observed in humans as a result of cumulative exposures to acrylamide. Toxic manifestations of acrylamide exposure are characterized by weakness in the limbs, tremors, and ataxia (5, 13, 20). Toxicity resulting from exposure to acrylamide is reversible after removal from the source (20).

Methyl mercuric hydroxide, diethylpyrocarbonate, and guanidine thiocyanate are extremely toxic compounds used in RNA isolation. Methyl mercuric hydroxide is an extremely toxic denaturant that reacts with guanosine and uridine via the imino bonds of RNA; work should be carried out in a fume hood with concentrations exceeding 10 mM (26). Formaldehyde solutions are also used as denaturants and should be used in a fume hood. Exposure to formaldehyde causes irritation of skin, eyes, and mucous membranes (34). Gel electrophoresis experiments using formaldehyde should be set up and run in a fume hood (26). Guanidine thiocyanate, a denaturant used to inhibit RNases, is an irritant that is harmful if inhaled. Diethylpyrocarbonate, also used to inhibit RNases, is used to treat water and other solutions and to rinse equipment and labware used in isolating RNA; it reacts with amines, carboxymethylates purine residues, and when used with solutions containing ammonium ion, is converted to ethylcarbamate, a suspected carcinogen (11). Skin protection should be used when working with diethylpyrocarbonate or guanidine thiocyanate (11).

Dichlorodimethyl silane, used to silanize glass and plastic ware, is toxic by several routes: poisonous by inhalation, mildly toxic by ingestion, corrosive by dermal exposure, and a severe eye irritant (34). It is also volatile and highly flammable, reacts violently with water, and should only be handled in a fume hood (26, 34).

SOLVENTS

Solvents of various kinds are used in the biomedical laboratory, the most common of which are aliphatic and aromatic hydrocarbons; oxygenated compounds such as alcohols, aldehydes, ketones, and ethers; and chlorinated compounds such as chloroform and methylene chloride. Because many organic solvents have a high vapor pressure, they are highly volatile; therefore, the most likely pathway of exposure is by inhalation, but dermal absorption also occurs. For these reasons, solvents must only be used in a properly ventilated area, and personnel must wear appropriate protective clothing. The site of toxicity resulting from exposure to common laboratory solvents is summarized in Table 2.

Aliphatic hydrocarbons (class C_6 to C_8, hexane to octane) are often used as solvents. These com-

TABLE 2 Common laboratory solvents[a]

Substance	TWA[c] 8 h (ppm)	Affected organ[b]							
		Skin	Eye	Liver	Kidney	CNS	CVS	Blood	Respiratory
Acetaldehyde	200	√			√				√
Acetic acid	10		√						√
Acetone	1,000	√	√						√
Acetonitrile	40			√	√	√	√		√
Allyl alcohol	1	√	√						√
Ammonia	50		√						√
Aniline	5			√	√	√		√	
Benzene	10 (50[d])	√	√			√		√	√
Bromoform	0.5	√		√	√	√			√
2-Butanone	200					√			√
n-Butyl alcohol	100	√	√						√
sec-Butanol	150	√	√			√			
tert-Butanol	100	√	√						
Carbon disulfide	20 (30[d])	√	√		√	√	√		
Carbon tetrachloride	10 (25[d])	√	√	√	√	√			√
Chloroacetaldehyde	1[d]	√	√						√
Chlorobenzene	75	√	√	√		√			√
Chloroform[e]	50	√	√	√	√				
Chloromethane (methylene chloride)	500 (1,000[d])	√	√			√	√		
Crotonaldehyde	2	√	√						√
Cyclohexane	300	√	√						√
Cyclohexanol	500	√	√						√
Cyclohexanone	50	√	√			√			√
Cyclohexene	300	√	√						√
o-Dichlorobenzene	50	√	√	√	√				√
p-Dichlorobenzene	75	√	√	√	√				√
Diisobutylketone	50	√	√						√
Dimethylformamide	10	√		√	√		√		
Dinitrotoluene	1.5 mg/m^3			√			√	√	
Dioxane	100	√	√	√	√				
Ethanolamine	3	√	√						√
Ethyl acetate	400	√	√						√
Ethylamine	10	√	√						√
Ethylbenzene	100	√	√			√			√
Ethyl butyl ketone	50	√	√						√
Ethylene dibromide	10 (30[d])	√	√	√	√				√
Ethylene dichloride	50 (100[d])	√	√	√	√	√			
Ethyl ether	400	√	√			√			√
Formaldehyde	3 (5[d])	√	√						√
Formic acid	5	√	√	√	√				√
Heptane	500	√							√
Hexane	500	√	√						√
2-Hexanone	100	√				√			√

TABLE 2 *Continued*

Substance	TWA[c] 8 h (ppm)	Affected organ[b]							
		Skin	Eye	Liver	Kidney	CNS	CVS	Blood	Respiratory
Isoamyl acetate	100	√	√						√
Isoamyl alcohol	100	√	√						√
Isobutyl alcohol	100	√	√						√
Isopropyl acetate	150	√	√						√
Isopropanol	400	√	√						√
Isopropyl ether	500	√							√
Ketene	0.5	√	√						√
Methyl acetate	200	√	√						√
Methyl alcohol[f]	200	√	√			√			
Methyl cellosolve	25	√	√		√	√			
Methyl cyclohexane	500	√							√
Methyl cyclohexanol	100	√	√	√[e]	√[e]	√[e]			√
o-Methylcyclohexanone	100	√		√[e]	√[e]				√[e]
Methylene chloride	500 (1,000[d])	√	√			√	√		
Nitrobenzene	1	√		√	√		√		
Octane	500	√	√						√
Phenol	5	√		√	√				
Phenyl ether	1	√	√						√
Propyl alcohol[f]	200	√	√						√
Pyridine[f]	5	√		√	√	√			
Stoddard solvent	500	√	√			√			√
Tetrahydrofuran	200	√	√			√			√
Toluene	200 (300[d])	√		√	√	√			
o-Toluidine	5	√	√	√	√		√	√	
Turpentine	100	√	√		√				√
o-, *m*-, *p*-Xylene[f]	100		√	√	√	√		√	

[a]Adapted from reference 27.

[b]CNS, central nervous system; CVS, cardiovascular system.

[c]TWA, Time-weighted average for maximal allowable workplace air level.

[d]Ceiling value.

[e]From animal studies.

[f]Gastrointestinal tract also affected.

pounds were previously thought to be relatively safe due to their lack of chemical reactivity. Contact dermatitis is one of the first signs of exposure to aliphatic hydrocarbon solvents. However, humans who inhale high concentrations of these compounds can become dizzy and lose coordination because of central nervous system damage. In studies of sandal makers in Japan, neurotoxicity was found to occur among those employees who were chronically exposed to hexane at levels ranging from 500 to 2,000 ppm, apparently the result of demyelination of nerve sheaths (4). Apparently, the neurotoxic actions of *n*-alkane solvents such as hexane are a result of oxidative metabolism. 2-Hexanone and 2,5-hexanedione produced in the metabolism of hexane are highly potent neurotoxins; for heptane and octane, 2,5-heptanedione and 3,6-octane dione exert a similar neurotoxicity as 2,5-hexanedione. These toxic responses are probably a result of the γ-diketone structure, because structural analogs 2,3- and 2,4-hexanedione and 2,6-heptanedione are not known to be neurotoxic (4).

Benzene is unique among solvents in attacking the hemapoetic systems, causing aplastic anemia and leukemia, rather than the nervous system (9, 37). Much of the evidence linking cancer to benzene exposure is, however, correlative or from retrospective epidemiology studies. Also, there are no good animal models demonstrating leukemia induced by benzene, and there is no direct evidence of the ability of benzene to be a mutagen from in vitro tests (1). Catechols and hydroquinones, produced as metabolites of benzene, are thought to play a major role in the suspected genetic damage observed with benzene (19). Interestingly, toluene protects against benzene-metabolite-induced sister chromatid exchange, an assay of genetic damage (14, 19). The TLV for benzene is 1 ppm (3), but the National Institute for Occupational Safety and Health has recommended an ambient air exposure limit of 0.1 ppm (1). Acute exposure to benzene at levels between 1 and 10 ppm can result in drowsiness, dizziness, and headaches (1).

Exposure of up to 200 ppm of toluene in ambient air does not result in significant symptoms, but at 500 ppm headache and nausea are observed, and with massive exposures toluene acts as a narcotic. Toluene and other alkylbenzenes are not as toxic as benzene because the side chain is oxidized and metabolized into products that are readily excreted (benzoic acid and hippuric acid) (25).

Alcohols are metabolized predominantly by alcohol dehydrogenase to a corresponding aldehyde (44). Catalase and the microsomal ethanol oxidative system (cytochrome P450 IIE1) can also metabolize alcohols to aldehydes, but they apparently play only a minor role (10). Aldehydes formed from alcohol metabolism are converted to a corresponding organic acid by aldehyde dehydrogenases. Aldehydes are usually more toxic than the parent alcohol because they can form Schiff bases with various proteins, peptides, and amine neurotransmitters to alter important biochemical processes in the body. Relatively low levels of alcohols apparently act as neurotoxic agents by interfering with important receptors and second messengers in the hippocampal region of the brain (17, 24). Ethylene glycol is metabolized to oxalic acid, the agent responsible for the toxicity of ethylene glycol (46).

Chlorinated solvents such as chloroform and methylene chloride are widely used in biomedical laboratories. Chloroform is an anesthetic that is metabolized to phosgene, which may be responsible for observed kidney and liver toxicity (23). Prolonged respiratory exposure to chloroform can result in paralysis, cardiac respiratory failure, and death. Work with this volatile chemical should be carried out in a fume hood, and bottles should be capped when not in use. Chloroform is an experimental teratogen and carcinogen and, therefore, a suspected human carcinogen (34). Methylene chloride is a central nervous system depressant. Overexposure to methylene chloride can result in dizziness, numbness, and nausea. Methylene chloride is readily metabolized to carbon monoxide, resulting in the formation of carboxyhemoglobin (23).

Dimethyl sulfoxide (DMSO) is a solvent commonly used in biomedical laboratories. This compound is a skin and eye irritant (34). High concentrations of DMSO (50 to 80%) can cause skin irritation. DMSO has unique biochemical and pharmacologic properties that allow compounds dissolved in DMSO to enter the skin freely and to distribute rapidly throughout the body; therefore, laboratory workers should use extra precautions when working with DMSO (32).

One solvent of serious concern in the molecular biology laboratory is phenol, which is used, along with chloroform, in the extraction and purification of nucleic acids. Because of its low vapor pressure, phenol is not a major inhalation hazard. Rather, most phenol toxicity occurs as a result of exposure to skin. Even small skin exposures can cause painful burns, but the local anesthetic action of phenol makes it possible to overlook an exposure. The victim should be removed from the source of the spill, and appropriate first aid should be given (21). Exposure to large doses of phenol can be fatal within 30 min (34). Prolonged flushing with copious amounts of water and timely medical assistance should be sought in the event of any contact with phenol. Early signs of phenol toxicity include headache, dizziness, and nausea. Low doses can cause extremely painful burns; damage to kidneys, spleen, liver, and pancreas; and edema of the lungs. Chronic doses cause liver and kidney damage, which may be fatal. Phenol is also carcinogenic under experimental conditions (34). Contact with skin or eyes should be avoided. It is advisable to use nitrile gloves when working with phenol or chloroform.

Phenol, used for extraction and purification of nucleic acids, is often treated with 8-hydroxyquinolone to prevent the formation of oxidation products that can subsequently break nucleic acid chains. 8-Hydroxyquinolone is a suspected carcinogen that should only be used with adequate ventilation. Contact with eyes and skin, as well as inhalation of dust and vapors, which may be toxic, should be avoided.

CHEMICAL CARCINOGENESIS

Carcinogenesis is a multistage process of at least three stages: initiation, promotion, and progression

(30). Xenobiotics are involved in initiation and promotion and possibly in progression. Initiation is a mutagenic event, thought to be irreversible; thus, initiators are genotoxic. The action of a chemical initiator is to mutate an oncogene or tumor suppressor gene (38, 45). Compounds that are *complete carcinogens* can mediate both initiation and promotion. Many carcinogenic compounds (i.e., benzo[α]pyrene, aflatoxin, and *N*-acetylaminofluorene) are not complete carcinogens but must first be activated (usually by phase I metabolism) to the ultimate carcinogen. After activation to a reactive metabolite (electrophile) such as an epoxide, these compounds may then interact with DNA to cause a mutation. However, promoters or epigenitic carcinogens do not interact with DNA and are not genotoxic (45). Tumor-promoting agents act by altering the expression of growth factors or important signal transducing systems and thereby alter the proliferation of cells (6–8). Not all animal carcinogens have been shown to cause tumors in humans. Lists of human carcinogens have been compiled by the National Toxicology Program (28) and the International Agency for Research on Cancer. Substances are classified as (i) associated with cancer in humans; (ii) probable carcinogens; and (iii) no evidence of human carcinogenicity. Although these compounds or their metabolites are the ultimate carcinogen, it is wise to treat compounds that are suspected human carcinogens as though they were human carcinogens (i.e., by using the National Institutes of Health guidelines for chemical carcinogens) (42).

ONCOGENIC VIRUSES

In laboratories where oncogenic viruses and recombinant DNA molecules are used, an environmental control and safety program should be developed to protect laboratory workers and to prevent release of biohazardous organisms into the environment. Physical and biological containment may be required. Physical containment, as exemplified by the Centers for Disease Control and Prevention/National Institutes of Health guidelines in Appendix I, includes controlling laboratory access, use of laboratory clothing and shoes or shoe covers, decontamination of all waste and clothing, use of laminar flow hoods, carts or pans for transportation of breakable containers of cultures, and methods for the prevention of aerosol release during centrifugation or other activities. Biological containment is designed to impair or eliminate the agent by altering the ability of either a genetically engineered organism or a recombinant DNA molecule or its host to survive or replicate outside the laboratory (31).

The two main groups of oncogenic viruses are (i) DNA tumor viruses, which include the herpesviruses, the adenoviruses, and the papovaviruses; and (ii) RNA tumor viruses, which are retroviruses and include the human T-cell leukemia virus and the human immunodeficiency virus. The criteria for evaluation of risk from these viruses depend on several factors, including the infectivity of the virus, the oncogenic potential of the virus in humans, and the experimental procedure used. Two publications from the U.S. Public Health Service delineate the guidelines for safe handling of oncogenes (40, 41). To be classified as high risk, viruses must be shown to cause cancer in humans, but thus far, no oncovirus has been so classified. Moderate-risk viruses are those, such as herpesvirus, that have been linked to cancer in humans. Some viruses may, however, fall into the moderate-risk category because of a combination of characteristics (40).

As mentioned in Chapters 11, 12, and 13, the risk level determines the stringency of the containment level required for use of these viruses. Accordingly, government regulatory or advisory agencies may recommend the use of a basic facility or a containment facility.

Molecular biology is a rapidly growing area of research that has wide-ranging implications at all levels of biomedical science; all biomedical laboratories may be doing some molecular biology in the near future. Because molecular biology is closely associated with the study of nucleic acids, it is not surprising that many of the agents used in molecular biology experiments are potentially mutagenic.

Several broad classes of toxic or carcinogenic agents are used in molecular biology: dyes and intercalating agents, alkylating agents, mutagens and carcinogens, ionizing and UV irradiation, solvents used for the extraction and purification of nucleic acids, antibiotics, phage, and oncoviruses. Appropriate protective equipment must be worn to protect employees working with these agents and materials.

Dyes and intercalating agents are used to visualize proteins and nucleic acids. Ethidium bromide, 4′,6′-diamidino-2-phenylindole, and Hoeschst 33258 dye are commonly used intercalating agents (15, 26). These compounds insert between base pairs of nucleic acids and fluoresce when irradiated. Although they are useful in protein and nucleic acid work, the intercalating agents can disrupt replication and transcription, thereby leading to toxic responses in the exposed individual. Ethidium bromide is also used to separate closed circular DNA from nicked circular or linear DNA in a cesium chloride gradient. Besides its ability to intercalate, ethidium bromide interacts with the phosphate backbone of nucleic acids. It interferes with both DNA and RNA polymerases and nucleic acid synthesis, causes frame shift mutations, and is a powerful mutagen (22); thus, the handling or mixing of stock solutions of intercalating agents must be carried out in a fume hood. The use of double gloves is recommended when working with solutions or gels containing intercalating agents. Ethidium bromide can penetrate the skin even through gloves; spills should be cleaned promptly and hands thoroughly washed after gloves are removed.

N^1,*N*-dimethylnitrosoguanidine, N-methyl-*N*-nitrosourea, and alkyl-sulfonates produce O^6 methyl or alkyl guanidines, as well as other base modifications. Although they can be dealkylated by mammalian repair systems, such alkyl transferases have not been shown to be inducible in mammals (35), and thus genetic damage may accumulate. Methylation can also occur with the use of dimethyl sulfate. In addition to direct mutagenesis via base changes, the methylation state of DNA is involved in gene regulation, and alteration of this state may operate in tumor promotion (36, 39).

DNA sequencing methods that use chemicals which cause specific chemical modifications of DNA bases, dimethyl sulfate, piperidine, and hydrazine should be used with extreme care (26). Historically, other chemicals that have been used as mutagens in molecular biology include sodium bisulfite, nitrous acid, formic acid, hydrazine, and hydroxylamine. These older mutagenesis methods have now been replaced by newer methods using the polymerase chain reaction (18, 26).

Further information on chemical safety can be found in Chapter 16.

References

1. **Agency for Toxic Substances and Disease Registry.** 1989. *Toxicological profile for benzene.* Department of Health and Human Services, Centers for Disease Control, Atlanta.
2. **Aldrich Chemical Co.** 1991. *Chemical catalog.* Aldrich Chemical Co., Milwaukee, Wis.
3. **American Conference of Government Industrial Hygienists.** 1991. *Threshold limit values for chemical substances and physical agents and biological exposure indices for 1991–1992.* American Conference of Government Industrial Hygienists, Cincinnati, Ohio.
4. **Andrews, L. S., and R. Snyder.** 1991. Toxic effects of solvents and vapors, p. 681–722. *In* M. O. Amdur, J. Doull, and C. D. Klaassen (ed.), *Casarett and Doull's toxicology: the basic science of poisons,* 4th ed. McGraw-Hill Book Co., New York.
5. **Auld, R. B., and S. F. Bedwell.** 1967. Peripheral neuropathy with sympathetic overactivity from industrial contact with acrylamide. *Can. Med. Assoc. J.* **96:**652–654.
6. **Choi, E. J., D. G. Toscano, J. A. Ryan, N. Riedel, and W. A. Toscano, Jr.** 1991. Dioxin induces transforming growth factor-α in human keratinocytes. *J. Biol. Chem.* **266:**9591–9597.
7. **Choi, E. J., and W. A. Toscano, Jr.** 1988. Modulation of adenylate cyclase in human keratinocytes by protein kinase C. *J. Biol. Chem.* **262:**17167–17172.
8. **Coffey, R. J., Jr., N. J. Sipes, C. C. Bascom, R. Graves-Deal, C. Y. Pennington, B. E. Weissman, H. L. Moses, and M. R. Pittelkow.** 1988. Growth modulation of mouse keratinocytes by transforming growth factors. *Cancer Res.* **48:**17167–17172.
9. **Cole, P., and F. Merletti.** 1980. Chemical agents and occupational cancer. *J. Environ. Pathol. Toxicol.* **3:**399–417.
10. **Damgaard, S. E.** 1982. The D(VK) isotope effect of the cytochrome P450-mediated oxidation of ethanol and its biological application. *Eur. J. Biochem.* **125:**593–603.
11. **Ehrenberg, L., I. Fedorcsak, and F. Solymosy.** 1976. Diethyl pyrocarbonate in nucleic acid research, p. 189–362. *In* W. Cohen (ed.), *Progress in nucleic acid research and molecular biology.* Academic Press, Inc., New York.
12. **Estrin, W. J., and G. J. Parry.** 1990. Neurotoxicology, p. 267–274. *In* J. LaDou (ed.), *Occupational medicine.* Appleton-Lange, Norwalk, Conn.
13. **Garland, T. O., and M. W. H. Patterson.** 1967. Six cases of acrylamide poisoning. *Br. Med. J.* **4:**134–138.
14. **Goldstein, B. D.** 1977. Hematotoxicity in man. *J. Toxicol. Environ. Health* Suppl. **2:**69–105.
15. **Harlow, E., and D. Lane.** 1988. *Antibodies: a laboratory manual.* Cold Spring Harbor Laboratory Press, Cold Spring Harbor, N.Y.
16. **Hodgson, E.** 1987. Diagnosis and treatment of toxicity, p. 319–333. *In* E. Hodgson and P. E. Levi (ed.), *Modern toxicology.* Elsevier Science Publishing, Inc. New York.
17. **Hoffman, P. L., C. S. Rabe, F. Moses, and B. Tabakoff.** 1989. *N*-methyl-D-aspartate receptors and ethanol: inhibition of calcium flux and cyclic GMP production. *J. Neurochem.* **52:**1937–1940.
18. **Innis, M. A., D. H. Gelfand, J. J. Sninsky, and T. J. White.** 1990. *PCR protocols.* Academic Press, Inc., New York.
19. **Irons, R. D., W. F. Greenlee, D. Wierda, and J. S. Bus.** 1982. Relationship between benzene metabolism and toxicity: a proposed mechanism for the formation of reactive intermediates from polyphenol metabolites, p. 229–243. *In* R. Snyder, D. V. Park, J. J. Kocsis, D. J. Jollow, G. G. Gibson, and C. M. Witmer (ed.), *Biological reactive intermediates—II: chemical mechanisms and biological effects.* Plenum Press, New York.
20. **Kaplan, M. L., and S. D. Murphy.** 1975. Effect of acrylamide on rotarod performance and sciatic nerve β-glucuronidase activity of rats. *Toxicol. Appl. Pharmacol.* **22:**259–268.
21. **Lefevre, M. J., and S. A. Conibear.** 1989. *First aid manual for chemical accidents,* 2nd ed. Van Nostrand-Reinhold, New York.
22. **LePecq, J. S., and C. Paoletti.** 1967. A fluorescent complex between ethidium bromide and nucleic acids. *J. Mol. Biol.* **27:**87–106.
23. **Levi, P.** 1987. Toxicity of chemicals, p. 185-232. In E. Hodgson and P. E. Levi (ed.), *Modern toxicology.* Elsevier Science Publishing, Inc., New York.

24. **Lovinger, D. M., G. White, and F. Weight.** 1989. Ethanol inhibits NMDA-activated current in the hippocampal neurons. *Science* (Washington, DC) **243:**1721–1724.
25. **Low, L. K., J. R. Meeks, and C. R. Mackerer.** 1989. Health effects of the alkylbenzenes: II. Xylenes. *Toxicol. Ind. Health* **5:**86–105.
26. **Maniatis, T., E. F. Fritsch, and J. Sambrook.** 1989. *Molecular cloning: a laboratory manual.* Cold Spring Harbor Laboratory Press, Cold Spring Harbor, N.Y.
27. **National Institute of Occupational Safety and Health.** 1990. *Pocket guide to chemical hazards.* National Institute of Occupational Safety and Health, Washington, D.C.
28. **National Toxicology Program.** 1989. *Fifth annual report on carcinogenesis summary 1989.* National Institutes of Environmental Health Sciences, Research Triangle Park, N.C.
29. **Nethercott, J. R.** 1990. *Occupational skin disorders,* p. 209–220. Appleton-Lange, Norwalk, Conn.
30. **Pitot, H. C.** 1988. *Cancer—an overview,* p. 1–18. American Chemical Society, Washington, D.C.
31. **Rayburn, S. R.** 1990. *The foundations of laboratory safety: a guide for the biomedical laboratory,* p. 123–139. Springer-Verlag, New York.
32. **Rubin, L. F.** 1983. Toxicologic update of dimethyl sulfoxide, p. 6–10. *In* J. C. de la Torre (ed.), *Biological actions and medical applications of dimethyl sulfoxide.* New York Academy of Science, New York.
33. **Rumack, B. H., and F. H. Lovejoy, Jr.** 1991. Clinical toxicology, p. 924–946. *In* M. O. Amdur, J. Doull, and C. D. Klaassen (ed.), *Casarett and Doull's toxicology: the basic science of poisons,* 4th ed. McGraw-Hill, New York.
34. **Sax, N. I., and R. J. Lewis, Sr.** 1987. *Hazardous chemicals desk reference.* Van Nostrand-Reinhold Co., New York.
35. **Singer, M., and P. Berg.** 1991. *Genes and genomes,* p. 107–113. University Science Books, Mill Valley, Calif.
36. **Smith, S. S.** 1991. DNA methylation in eukaroytic chromosome stability. *Mol. Carcinogenesis* **4:**91–92.
37. **Snyder, R.** 1984. The benzene problem in historical perspective. *Fundam. Appl. Toxicol.* **4:**692–699.
38. **Stanbridge, E. J., and W. K. Cavenee.** 1989. Heritable cancer and tumor suppressor genes: a tentative connection, p. 281–306. *In* R. A. Weinberg (ed.), *Oncogenes and the molecular origins of cancer.* Cold Spring Harbor Laboratory Press, Cold Spring Harbor, N.Y.
39. **Strobl, J. S.** 1990. A role for DNA methylation in vertebrate gene expression. *Mol. Endocrinol.* **4:**181–183.
40. **U.S. Public Health Service.** 1974. *Biological safety manual for research involving oncogenic viruses.* National Cancer Institute, National Institutes of Health, Bethesda, Md.
41. **U.S. Public Health Service.** 1974. *Safety standards for research involving oncogenic viruses.* National Cancer Institute, National Institutes of Health, Bethesda, Md.
42. **U.S. Public Health Service.** 1981. *NIH guidelines for the laboratory use of chemical carcinogens.* 81-2385 D. P. N. National Institutes of Health, Bethesda, Md.
43. **Valciukas, J. A.** 1991. *Foundations of environmental and occupational neurotoxicology.* Van Nostrand-Reinhold, New York.
44. **von Wartberg, J. P., J. L. Bethuen, and B. L. Vallee.** 1964. Human alcohol dehydrogenase: kinetics and physicochemical properties. *Biochemistry (USA)* **3:**1775–1782.
45. **Weinberg, R. A.** 1989. Oncogenes and multistep carcinogenesis, p. 307–326. *In* R. A. Weinberg (ed.), *Oncogenes and the molecular origins of cancer.* Cold Spring Harbor Laboratory Press, Cold Spring Harbor, N.Y.
46. **Weiner, H. L., and K. E. Richardson.** 1988. The metabolism and toxicity of ethylene glycol. *Res. Commun. Subst. Abuse* **9:**77–87.

Hazard Control in the Laboratory

Primary Barriers and Personal Protective Equipment in Biomedical Laboratories

RALPH W. KUEHNE, MARK A. CHATIGNY, BRUCE W. STAINBROOK, ROBERT S. RUNKLE, AND DAVID G. STUART

11

PRIMARY BARRIERS

As the term implies, *primary containment* is the first line of defense when working with infectious organisms. The concept of a primary barrier can perhaps best be depicted as a protective "envelope" that results in an encapsulation of the infectious agent, animals, or personnel. This "envelope" can be a rigid or semirigid enclosed container, or a field of rapidly moving directional air, or a combination of both of these. An infectious agent in a stoppered bottle is, for example, already in a primary container. Obviously, the safety problem arises when the integrity of the container is damaged inadvertently or when one intentionally wishes to open or penetrate the container to transfer an aliquot to a new system, host, or vector. During these circumstances, the resulting release of agent must be contained to not result in an exposure or infection of the employee working directly with the agent, of other personnel within the laboratory room, of others in the facility, and of those outside the building. It is more effective and, therefore, "safer" to contain the aerosol as close to the site of release as possible to avoid having to install and rely on secondary barriers, and it is far less costly. An example of the most elementary form of primary containment would be enclosing a tissue-grinding operation in a sealed bag. The next level would be the use of a method to safely evacuate the contents or the potentially contaminated air within the bag by means of a mechanical device or engineering control. More sophisticated forms of primary barriers are a stepwise progression of one or both of these elemental forms. The most common primary barriers are in the general categories of glove boxes, biological safety cabinets, and animal caging equipment. Primary barriers include physical barriers of steel, plastic, glass, or other similar materials; air barriers or air "curtains"; and exhaust and supply barriers such as HEPA filters, charcoal filters, or air incinerators. The general principles applicable to containment within a primary barrier system include the following:

1. Minimize the volume to be contained.
2. Provide safe (i.e., noncontaminating) transfer of material into and out of the container without destroying the barrier.
3. Provide means for decontamination of the enclosure and effluents.

If primary containment fails or is inadequate, clothing and other items of personal protective equipment (PPE) often become an important line of defense against a physical, biological, chemical, or radiological exposure. Such items may be required to prevent introduction of hazardous materials or infectious microorganisms through mucous membranes, broken skin, the circulatory system, or the respiratory or digestive tracts and are often used in combination with biological safety cabinets and other primary containment devices. The current view is that total containment should be a "system" encompassing clothing, mechanical devices, laboratory design, and work practices. The objective of this chapter is to provide the laboratory worker with information and advice regarding selection, use, and efficacy of various physical containment methods, equipment, and apparel used to isolate the worker from the biological agent in use. Further information on work practices can be found in Chapter 13 and on laboratory design in Chapter 12.

Process Containment Techniques

Centrifuge Safety

One readily recognizable hazard that had been addressed for many years by a simple containment procedure is the microbiological centrifuge, for which construction of safety cups has provided one method of containment. These containers range from individual sealed tubes to larger screw-capped buckets and sealed rotors. Because the primary containers used in a centrifuge may be subject to extremely high stresses, careful attention must be paid to the quality of the seal. If an aerosol or fluid containing an infectious agent escapes from a rotor or cup during high-speed operation, the potential for extensive contamination and multiple exposures or infections would be great and the consequences could be severe. Some of the early tube closures that depended on expansion of O-rings were not satisfactory. Today, most manufacturers produce effective closures that prevent leakage of small-batch materials under low-, medium-, or even high-speed centrifugation. Examples of several such devices are shown in Fig. 1. However, screw-capped buckets are not available for all models of centrifuges, and many of the commercially available plastic tubes and bottles leak. Therefore, when appropriate safety buckets are not obtainable, it is recommended that the chamber be evacuated after centrifuging infectious materials before the centrifuge is opened. This can be accomplished by means of a vacuum pump and Tygon tubing hose inserted into one of the available capped ports located on the side of some models or by drilling an access hole into the chamber in the side of models not so equipped. A disinfectant trap and/or in-line HEPA filter should be used to protect the pump from contamination. Large bulk or zonal rotors and continuous-flow centrifuges are particularly difficult to seal, and extreme care should be taken in their use. The simple primary barriers described can be effective, but one must also consider the possibility of a major accident (e.g., rotor rupture). Work with large volumes of infectious agents may merit putting the entire centrifuge in a ventilated enclosure (17).

FIGURE 1 Safety trunnion cups.

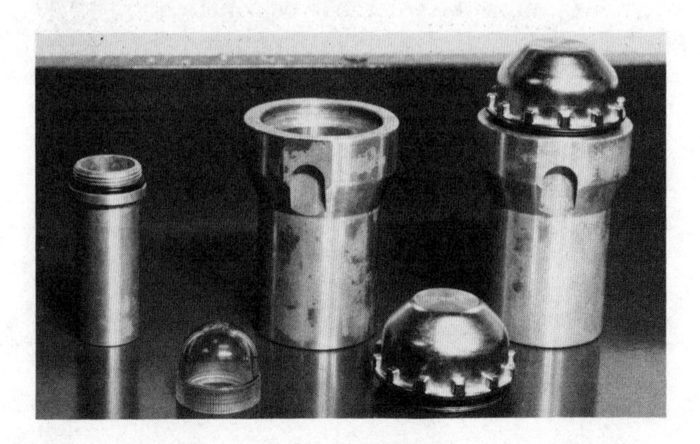

Blenders

Blenders are also well-known producers of aerosols. Without special sealing design, they can, like centrifuges, rapidly contaminate spaces and spread high levels of surface contamination. Workers at the U.S. Army Biological Laboratories designed an autoclavable safety blender cup, improvements of which are now marketed commercially (Waring Products Division, New Hartford, Conn.) (Fig. 2).

FIGURE 2 Safety blender container.

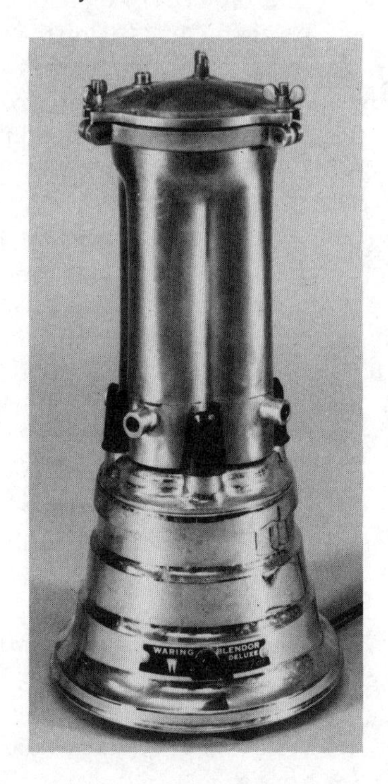

FIGURE 3 Laminar-flow animal cage enclosure. (Courtesy Lab Products, Inc., Maywood, N.J.)

Animals

Simple but effective primary containment can be achieved for animal care and use. Spun-molded polyester or polycarbonate filter-top animal cages and ventilated racks are examples of caging systems applicable in rodent housing. Laminar-flow, HEPA-filtered, negative-pressure rack enclosures (Fig. 3) can also be used in a positive-pressure mode as a laboratory animal clean-air quarantine station. Larger animals can be housed in ventilated cages or in cages within negative-pressure cubicles or rooms with filtered nonrecirculating room exhaust air.

Although the discussion below deals primarily with three recognized "classes" of cabinets, many of the specialized cabinet enclosures now available, particularly in plastic fabrications, can play a useful role in laboratory and health care facilities. These include simple plastic bags with built-in glove access, rigid "glove boxes" (Fig. 4), flexible-film isolators, and "static" enclosures often used on a bench top. Stretcher patient isolators (Fig. 5), aircraft transport patient isolators, and even patient bed isolators are modifica-

FIGURE 4 Glove box.

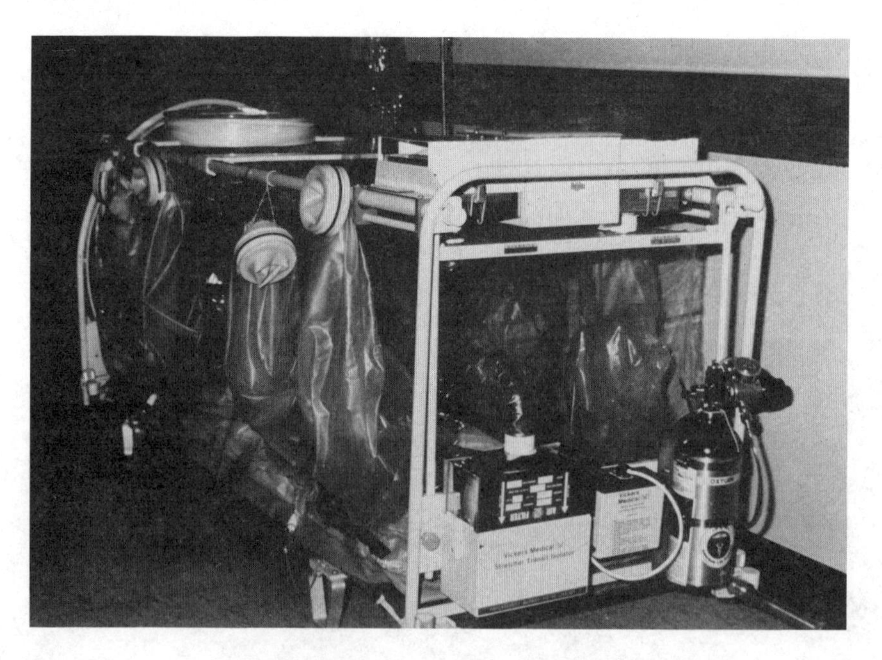

FIGURE 5 Stretcher patient isolator.

tions of these devices, which allow the provision of medical care under containment conditions.

Biological Safety Cabinets

Regardless of the effectiveness of individual containment devices, it is usually necessary to have some kind of cabinet or enclosure in which to load and unload centrifuge rotors, sealed containers, and blender bowls. The safety hood or biosafety cabinet is most widely accepted for the purpose and available in a variety of sizes and designs. It is perhaps the single most useful safety device in the microbiology laboratory and is second in importance only to safe work practices in maintaining control of the environment.

Primary barriers, in one form or another, have been used in microbiology since Louis Pasteur demonstrated that sealed or plugged vessels did not become contaminated with airborne microbes. As microbiology advanced, so did containment methods. Although attention to containment lagged somewhat behind laboratory worker infection experience, the latter was a strong driving force (see also Chapter 2). The biological safety cabinet was built in 1909 by W. K. Mulford Pharmaceutical Co. (Glenolden, Pa.) as a ventilated hood to prevent infection with *Mycobacterium tuberculosis* during the preparation of tuberculin (61). The earliest publication describing a biological safety cabinet was in 1943 by Van den Ende (58), who designed a cabinet in response to the high incidence of infection in scrub typhus laboratory workers. Later, when work with the rickettsiae of Q fever became the source of many infections, an open-front cabinet was used (51) to contain the work with *Coxiella burnetii*. The exhaust air from these early cabinets was either passed through a disinfectant or incinerated by means of an electric furnace or gas burner. The first cabinet to use exhaust filters (spun-glass fiber) was fabricated in 1948 (60). The evolution of the biological safety cabinet was reviewed by Wedum (60), Chatigny (16), and more recently, Kruse et al. (31). Only the infection experiences were unique to the microbiology laboratory, because parallel efforts were in progress to protect workers from chemical and radiological hazards in laboratory work in those disciplines. Some of the early versions of the biological safety cabinets were variants of the familiar chemical fume hood. After some years of use, workers observed that special requirements for the biological safety cabinet differed from those for the chemical fume hood. They included a need for a fixed view screen that would protect against splatter of large droplets, a requirement to close the cabinet for decontamination, and the ability to provide a flow of air through the cabinet with effective decontamination of the effluent airstream. The need for decontamination required minimizing the quantity of airflow through the cabinet, and a front access opening of 10 in. (25 cm) or less height was accepted. By 1960, there was broad agreement that a hinged, sloped window cabinet with air drawn in the front and exhausted out the top thorough high-efficiency filters was a necessary adjunct to every laboratory working with pathogens. Several reviews have described the types of cabinets and rationales for their use (16, 20, 31, 60). An early development in clean-air biological safety cabinets was reported in 1969 (1).

At present, three general classes of biological safety cabinets are defined. These are class I, the open-front air inflow cabinet, usually with a fixed height opening and sloped view window; class II, open-front, vertical airflow cabinets, of which there are several subtypes; and class III, cabinets hermetically sealed with access through gas-tight air locks and work access through fixed, heavy-duty, arm-length rubber gloves. A summary description of the essential characteristics of each of these classes of cabinets can be found in Table 3 of Appendix I of this book. A descriptive review has been published (53).

Biological safety cabinets are heavily dependent on the availability of the HEPA filter, first developed by the U.S. Army Chemical Corps, Atomic Energy Commission, and others during World War II. No less important is the instrumentation for measurement of the very high efficiency of these filters, light-scattering photometers used to demonstrate retention of 99.97% or greater of near-monodisperse airborne particles of di-octyl-*n*-phthalate approximately 0.3 μm in diameter. Particles in this size range are stated to be the most difficult to retain. The filters collect the larger particles by simple interception and smaller particles, such as bacteriophage and animal viruses, by virtue of both impaction and diffusion collection.

In evaluating the parameters affecting containment characteristics of safety cabinets, Barkley (11) used a design "containment factor" of 10^5, as does the British Standard 5726 (13). It is convenient to use the reciprocal (e.g., 10^{-5}) as a "leakage factor." This is an expression of the fraction of aerosol in the cabinet that is expected to leak out. The factor is a model number on which to base a cabinet design analysis, as done by Barkley, but it is also a practical estimate of expected cabinet performance. In fact, a range of leakage factors should be expected when there is activity in the cabinet (21, 55). The position of the operator, how far into the cabinet the operations are performed, and the rapidity of movements all affect this leakage. One should expect a range of 10^{-4} to 10^{-7} for open-front class I cabinet leakage factors. The highest of these (10^{-4}) shows the effect of poor practice (e.g., work within 4 in. [10 cm] of the front and moving arms out of the cabinet). The lowest leakage factor (10^{-7}) requires good opening design and little activity at the workspace. Class II cabinets would have approximately the same range, whereas class III cabinets should have leakage factors of 10^{-8} or less.

The three classes of cabinets are not directly related to the biosafety level (BSL) required for agent use. Class I and class II cabinets can be used for work at BSLs 1 to 3. Class III cabinets are usually reserved for work at BSL 4, although a class II cabinet may be used for that purpose if the worker is provided protection such as the use of a ventilated suit.

Class I Biosafety Cabinets

The earliest cabinets, as stated above, were class I cabinets as shown in Fig. 6 and diagrammed in Fig. 7. These cabinets depend on a flow of air into the front work opening, across the work surface, and out through a decontamination device, usually a high-efficiency filter, and an exhaust blower. The cabinet can provide good protection of the operator from the work and allows the use of electronic incinerators, small gas (i.e., Touch-A-Matic) burners, small centrifuges, and other equipment without seriously degrading the containment effectiveness. The cabinets may be constructed of stainless steel or fire-resistant reinforced plastic, with glass or clear optical-grade plastic for view windows. They are usually available in lengths from 3 ft (0.9 m) to about 6 ft (1.8 m). Materials and equipment may be moved in and out through the front opening, through a hinged view window, or via air-lock doors added to

FIGURE 6 Class I biological safety cabinet with airfoil formed front entry. (Courtesy Baker Co., Sanford, Maine.)

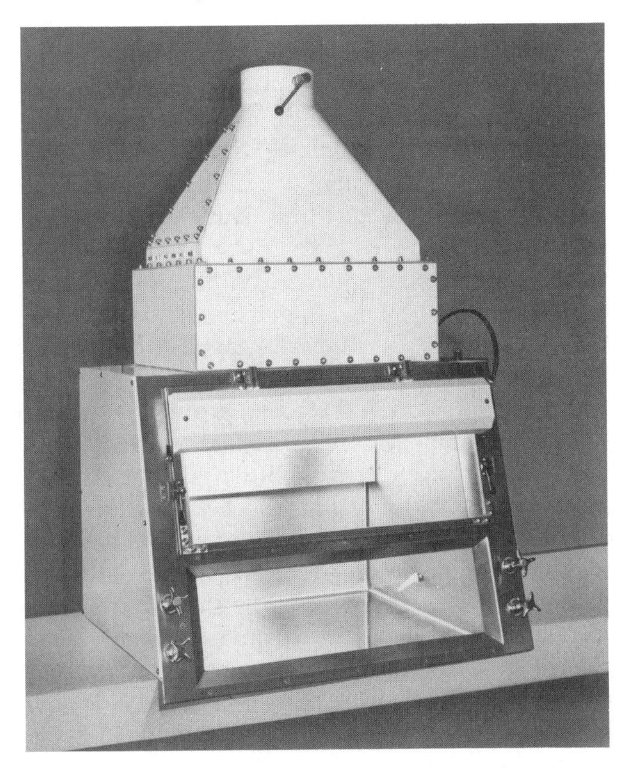

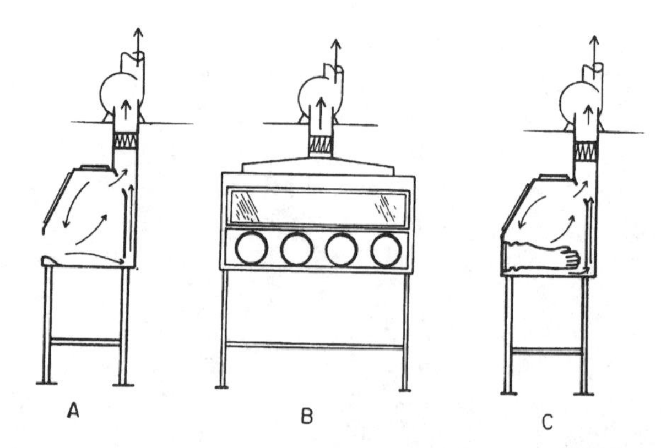

FIGURE 7 Diagram of class I open-front biological safety cabinet. (A) Conventional use; (B) with armhole plate attached; (C) with rubber gloves attached to armhole ports.

the cabinet end. There have been extensive tests and discussion of the required input air velocity needed for safe operation of these cabinets. Inlet air velocity is a grossly inadequate measure of containment effectiveness; nonetheless, it is quoted in the commercial literature. The current consensus is that inlet air velocities from 75 to 125 linear feet per minute (lfpm; 0.38 to 0.63 m/s), depending on inlet design, provide optimal operator protection. Such a cabinet can facilitate day-to-day operations because work access is excellent. However, this ease of access tends to encourage operators to make rapid motions at the front of the unit and to move their arms in and out of the cabinet; these kinds of motions, plus those of room air currents and drafts at 60 lfpm (0.25 m/s) or higher can degrade the containment provided by the inflowing air. Many room ventilating systems provide from 75 to 150 lfpm (0.38 to 0.76 m/s) air velocity in the workspaces. The use of a slow diffuser can sometimes alleviate the problem.

There is general agreement that the cabinet should have an interior rear baffle to provide a smooth airflow across the work surface while permitting some air to be removed from the upper section. The front opening design is also important, and the user should ensure that this aspect of design has been resolved satisfactorily. The addition of an air foil or suction slots at the opening can be effective. The class I cabinet is an example of the use of directional airflow, compared with the popular concept of "negative pressure" for aerosol capture and containment, because the degree of negative pressure immediately inside such a cabinet is probably less than 0.01 in. (0.25 cm) of water negative to the room atmosphere. In fact, the thermal pressure head caused by a large-sized laboratory burner can overcome this differential pressure and cause backflow out the front opening if a portion of the ventilating air is not removed near the top of the cabinet. Obviously, such large burners are not recommended for use in the cabinet.

Performance of this cabinet can be substantially improved by the addition of a closure panel bearing circular ports for arm holes, as shown in Fig. 7B. The panel need not be airtight. Because the blower is not changed, the resultant air velocity through the armhole ports may be increased to 150 to 250 lfpm (0.74 to 1.25 m/s). Because the area of the opening (and periphery) is substantially reduced, the turbulence created by this high-velocity airflow is not critical, and substantially less leakage usually occurs. Additional containment can be afforded with this cabinet by use of arm-length rubber gloves added to this closure panel (Fig. 7C). In this case, the airflow is now restricted only to the leakage points of the cabinet, and the cabinet may become substantially negative in pressure (ca. 2 in. [5 cm] of water static pressure) with very little exchange rate of air unless inlet air pressure relief is provided, usually through HEPA filters.

Class I cabinets offer no protection of the work from the operator or the environment. In a laboratory that does not supply clean air or in a cell culture operation in which contamination from the worker may affect the work product, the class I cabinet may be contraindicated. However, to their advantage, class I cabinets are simple and economical, easily installed, can be used with radioisotopes and some toxic chemicals, and can be adapted in various forms to meet the unique needs of special processes. For example, it is not at all uncommon to see modifications of class I cabinets added to the tops of centrifuges (10), used as enclosures around fermenter devices, or used in work involving both biological and chemical agents of low to moderate risk.

Most class I cabinet installations incorporate a high-efficiency filter in the cabinet or in a sealed housing immediately before the exhaust fan, which provides discharge (usually to the outdoors). An air incinerator may supplement or be substituted for the air filter. However, the use of an incinerator in the exhaust of a class I cabinet is not usually warranted. Activated carbon or other chemical treatment filters may be used in conjunction with the HEPA filters if the work involves significant quantities of carcinogenic or toxic materials.

The installation requirements for this type of cabinet are similar to those of the class II cabinets described below and are discussed with the decontamination and general use procedures.

Clean-Air Benches

Laminar-flow clean-air benches were developed from the observation that a stream of air at approxi-

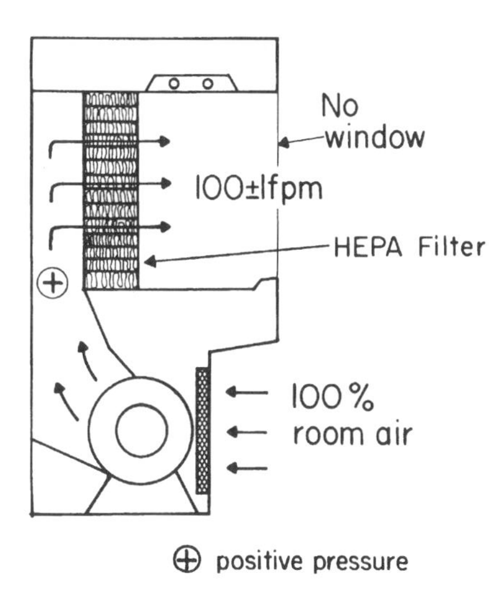

FIGURE 8 Diagram of horizontal laminar-flow clean bench.

mately 100 lfpm (0.5 m/s) forced through a HEPA filter provides a particle-free environment for several feet downstream of the filter if there are no obstructions. This has been termed a *laminar-flow, essentially nonmixing, air stream* and is used in "clean rooms" and in "clean benches" to protect the work product. Laminar airflows are widely used in the preparation of precision microelectronics and machine parts needed for the U.S. space program. They do not all provide worker protection.

The horizonal-flow clean-air bench (Fig. 8) has been used in cell culture laboratories for many years to provide near-sterile work areas. However, it does not provide operator protection and, in fact, can expose the worker to aerosols of allergenic or infectious materials. This type of cabinet can be useful in microelectronics fabrication, in the hospital pharmacy laboratories for final preparation of parenteral solutions, for final packaging of reprocessed devices, or for other applications in which the product is unlikely to have any ill effect on the cabinet user. However, the horizontal-flow clean-air bench is considered unsuitable for microbiological laboratory work. Because vertical-flow clean-air benches blow their air out into the room just as horizontal-flow clean-air benches do, they must be differentiated from and not confused with biological safety cabinets. In this section, laminar flow will be interpreted to indicate an airflow of clean, filtered air over the work surface with minimal mixing with the airstream coming into the cabinet via the work opening.

Class II Biosafety Cabinets

The main difference between the class I and class II cabinets of consequence to the user is that the class II vertical laminar-flow biological safety cabinet can afford protection for the operator and the work being performed.

Two general subtypes of class II cabinets are available. They are designated IIA and IIB in the U.S. National Sanitation Foundation (NSF) standard number 49 (38) and in the British standard number 5726 (13). In general, the class IIA cabinet maintains a minimum of 75 lfpm (0.4 m/s) inflow velocity thorough the work opening and recirculates a major part of the air traversing the work surface (ca. 70%), whereas the class IIB cabinet maintains an inlet flow velocity of 100 lfpm (0.5 m/s) and exhausts most (70%) or all (100%) of the air traversing the work surface to the outdoors.

Class IIA biosafety cabinets. Figure 9 shows a typical class IIA cabinet. The airflow characteristics of these cabinets are generally as shown in Fig. 10. The air drawn into and over the blower and then, under pressure, up to the recirculating or exhaust filters and through the exit filter is contaminated both from the work and from the room. Therefore, this air plenum must be airtight and leakproof. NSF 49 calls for a halide gas (e.g., R-12 refrigerant) test of this plenum, allowing less than 10^{-7} ml/s of R-12, when mixed with air by pressurizing the cabinet to 2 in. (5 cm) of water static pressure. Because a sub-

FIGURE 9 Class IIA biological safety cabinet. Exhaust arranged to discharge to laboratory. (Courtesy Baker Co., Sanford, Maine.)

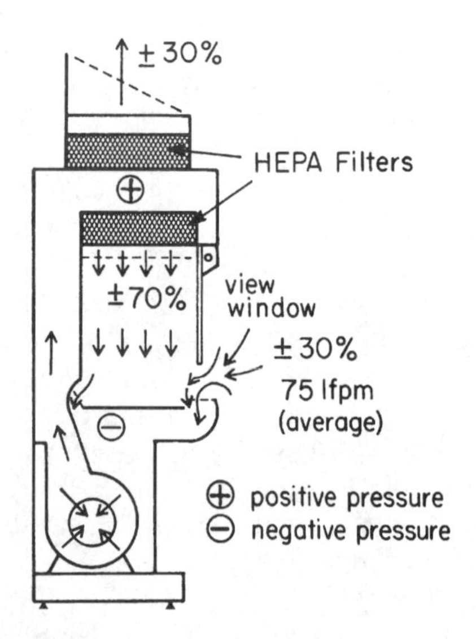

FIGURE 10 Simplified diagram of class IIA biological safety cabinet airflow scheme.

stantial fraction of the air in the cabinet (up to 70%) is recirculated through the supply filter, the type A cabinet is generally not considered suitable for use with high-activity radioactive materials or with toxic or carcinogenic chemicals. The essential elements of class IIA cabinets are HEPA-filtered laminar-flow recirculated air, traveling downward over the work surface, air inlet into the front with immediate conveyance away from the work surface, and discharge of excess air from the cabinet via a HEPA filter to the room or outdoors (Fig. 10 and 11). As the simplified diagram shows, the blower in the cabinet forces the air both through the recirculating air filter and the exhaust air filter, and a careful balance must be achieved to obtain the expected performance.

The concept of sealing the type A positive-pressure plenum Freon tight is good, but testing it is a difficult task. An alternative to guaranteeing that the positive-pressure plenum is leaktight is to surround the plenum with a negative-pressure area, as shown in Fig. 11. The performance characteristics of this type of cabinet are equal to those of the conventional class IIA cabinet. At the confluence of the downflowing stream of air over the workplace and the air entering the work opening, there is a joining of two airstreams that may be of substantially different velocities. This offers some difficult aerodynamic problems, and this cabinet can suffer from the same problems of turbulence around the work opening as the simple class I cabinet has (18). The containment characteristics are approximately the same for the two cabinets, but if the cabinet is vented directly to

FIGURE 11 Simplified diagram of class IIA biological safety cabinet with positive-pressure plenum enclosed within negative-pressure space.

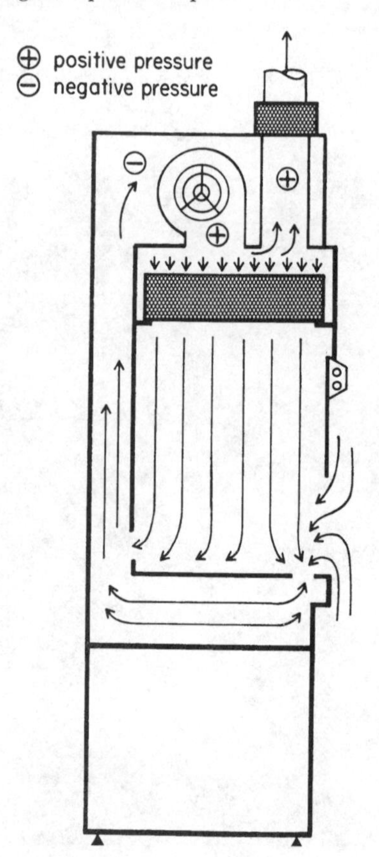

FIGURE 12 Another version of a class IIA biological safety cabinet configuration. (Adapted from Bionomics, Inc., Sanford, Maine.)

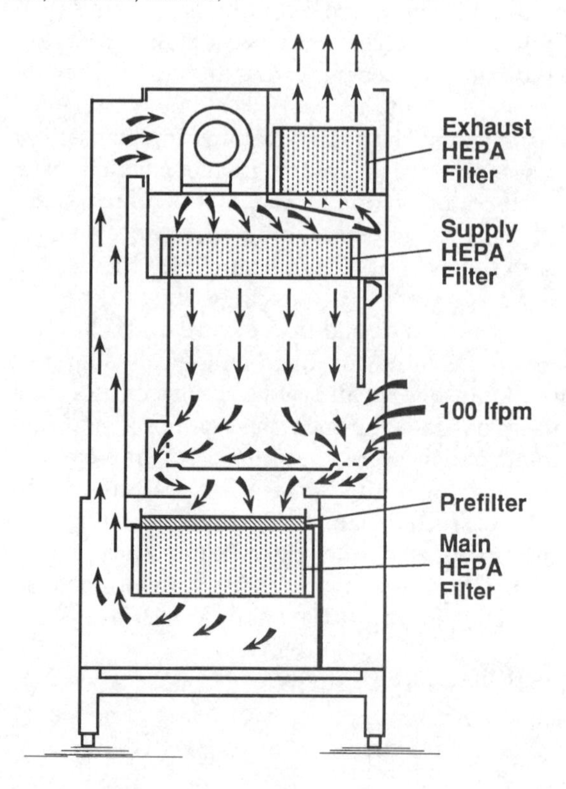

the room, an overall leakage factor of 10^{-5} (exhaust filter penetration) may be expected. The cabinet is particularly sensitive to interruption of the vertical airflow by gas burners. Use of disposable sterile transfer loops or electronically heated incinerators is recommended (6).

An innovative class IIA cabinet produced by one manufacturer is a variation of the traditional configuration (Fig. 12). Air enters the cabinet at a velocity of 100 lfpm (0.5 m/s) and is HEPA-filtered immediately downstream of the room-workspace air boundary. The position of this filter prevents contamination of the blower motor with hazardous particulates and eliminates contaminated positive-pressure areas in the cabinet. Both exhaust and supply air are HEPA-filtered a second time, increasing protection for personnel, product, and environment.

Class IIB biosafety cabinets. Class IIB biosafety cabinets maintain a minimum average inflow velocity of 100 lfpm (0.5 m/s) through the work area access opening and should be hard-ducted to a dedicated external exhaust that discharges outside the building at a height and location that permit no recirculation (31). There are three subtypes of these cabinets, B1, B2, and B3. The class IIB1 cabinet (Fig. 13) has filtered downflow air that is composed largely of uncontaminated recirculated inflow air, exhausts the air going to the rear of the work area directly outdoors after HEPA filtration, has all biologically contaminated ducts and plena under negative pressure, and has an inflow air velocity averaging 100 lfpm (0.5 m/s). The cabinet can be useful for microbiological work and for work with low-level radioisotopes and limited amounts of toxic chemicals. However, the degree of air mixing and recirculation in the cabinet requires that use of such materials be restricted to levels not considered toxic to the work product. Further, this class of cabinet will not usually meet the air inflow standards for work with carcinogens in chemical fume hoods. The type IIB2 "total exhaust" cabinet (Fig. 14) is similar in design, but all air entering the cabinet makes only one pass through the cabinet before being discharged through a HEPA filter to the outdoors. The work opening inlet air velocity averages 100 lfpm (0.5 m/s) or higher. This air is prevented from contaminating the work by a protective flow of HEPA-filtered room air entering the top of the cabinet. The diagram shows enclosure of the positive-pressure exhaust duct within the negative-pressure volume established by the exhaust fan suction. Class IIB2 cabinets are designed to be used for work with limited quantities of toxic chemicals or radionuclides

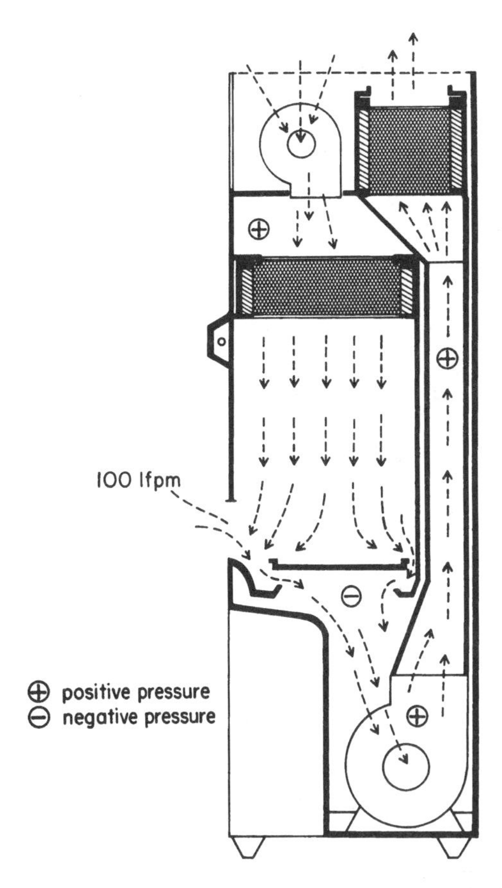

FIGURE 14 Simplified diagram of class IIB2 biological safety cabinet. (Adapted from Germfree Laboratories, Miami, Fla.)

FIGURE 13 Simplified diagram of class IIB1 biological safety cabinet showing airflow directions. Note requirement for external exhaust blower.

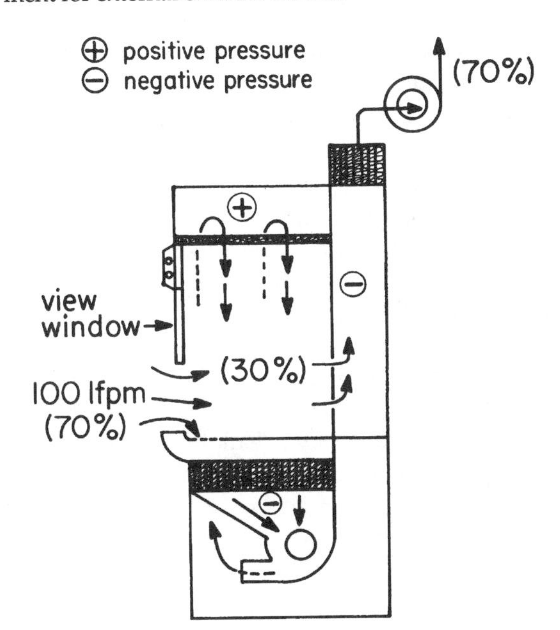

required in microbiological studies. At least one manufacturer claims the total exhaust cabinet will meet Environmental Protection Agency and Occupational Safety and Health Administration (OSHA) standards (125 to 150 lfpm [0.6 to 0.75 m/s] inlet velocity) for work with chemical carcinogens in chemical fume hoods. Cabinets of this design meet NSF 49 standards for biocontainment and product protection. If air velocities (downward and inward) are maintained similar to those in the IIB1 configuration, the containment performance should be equal. The class IIB3 cabinet definition by NSF is essentially that of a type IIA cabinet. In use, it must be considered equivalent to the front portion of the IIB1, where the downflow air is coming to the front perforated grill.

A major drawback to the class IIB cabinet is the requirement for passthrough of relatively large quantities of room air and subsequent discharge to the atmosphere. As is done with fume hoods, some unconditioned air may be supplied directly from outdoors by separate ducting. This can be expensive and difficult to accomplish. Another drawback common to most class IIB cabinets is that they must have at least two fans (supply and exhaust) operating in balance (e.g., with the exhaust always exceeding the input to provide the necessary work opening inflow and negative pressure within the cabinet). Considering that the exhaust and supply filters are subject to differing rates of dirt loading, airflow at the inlet can vary with usage. This added complication in installation and setup should be examined by the prospective users of these cabinets.

It is essential that there be little or no leakage through the recirculating or exhaust air filters. Many of the recirculating filters are large (as big as 24 × 72 × 8 in. deep [61 × 183 × 20 cm]) and are extremely fragile. For many years, the incidence of failure caused by damage to these filters after shipping was quite high. This problem has been reduced by improving the handling, as well as the design, of the filter mounting, but the possibility of such leakage after movement of the cabinet persists. Figure 15 shows details of a typical HEPA filter construction. The filter medium is installed in the frame in a pleated manner, with corrugated metal or fiber separators. It is fragile, being very thin, and is subject to puncture or cracking by handling or shock in shipping. The paper is usually attached to the edges of the frame by a cement that hardens. The paper becomes highly stressed at the point of attachment, such that dropping the filter or handling it with any degree of roughness causes cracks at this juncture. In many cases, these small cracks are not readily visible and cannot be located by simple inspection. The

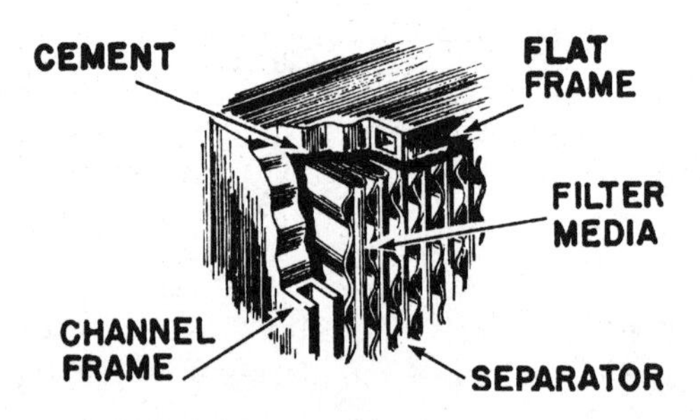

FIGURE 15 Detail of typical HEPA filter construction. Two types of edge seals are shown.

procedure for testing is to generate a fine-particle (0.1 to 3.0 μm) di-octyl-*n*-phthalate or equivalent aerosol upstream of the filter. Penetration of particulates is then measured immediately downstream by using a light-scatter-type aerosol penetrometer to scan the face of the filter in a pattern ensuring coverage of all edges and pleats. The edge gasket, on which the filter is seated within the frame in the cabinet, has also been a source of many leaks and can be tested in a similar manner. Leakage greater than 0.01% at any one point, measured by the di-octyl-*n*-phthalate/penetrometer method, is unacceptable. If the leaks in the filter or gasket are small, they may be repaired readily by the use of room-temperature vulcanizing silicone sealant (Dow-Corning). The edge groove seal shown in Fig. 15 (Flanders Filters, Inc., Washington, N.C.) is used with the groove filled with a silicone grease meeting the on-edge mating flange. It can be very effective, particularly in large installations, but should be used with caution in locations where the grease can become contaminated.

Both the circulating air velocities and the inflow exhaust air rates are important in the class II cabinets. For the most part, the recirculating air velocity is affected by the pressure drop across the HEPA supply filter at the top or bottom of the cabinet and by dampers for deflectors. The velocity of the vertical downflow and the incoming airstreams may be adjusted by setting the correct speed of the exhaust and recirculating blowers or resetting dampers. The measurements are made with an air velocity meter (anemometer) with suitable sensitivity. The front inlet air velocities cannot usually be measured directly in the class II-type cabinets; this is customarily done by measuring the quantity of air exhausted out of the cabinet and then, by calculation, determining the average inflow air velocity through the front opening. The velocity profile of downflow air throughout the cabinet should be within ±5% of the manufacturer's stated operation, but it is sometimes a function of the cabinet and cannot be adjusted readily in the field. Exceptions would occur if a tester or operator were to permit an object to block part of a filter or blower; paper towels have been retrieved from blower inlets due to careless use.

One other criterion, not often tested but perhaps of equal importance, is the degree of turbulence around the work opening. The user can move a smoke stick (Mine Safety Appliance Co., Pittsburgh, Pa.) slowly around the periphery of the front opening to determine and ensure that the air is flowing smoothly into the cabinet. The smoke stick may be held and traversed a few inches above the work surface so that, again, the desired direction of air is observed. Other tests (e.g., vibration, noise, illumination intensity), along with the tests described above and the details of the certification test used by the manufacturer, are fully described under the section on biosafety cabinet use procedures.

Selection of Class I and II Biosafety Cabinets

When selecting one of the class I or class II biosafety cabinets, there are several factors to be considered. These include the operations to be conducted in the cabinet: the classification of the etiologic agents to be used; the protection required for the work product; the possibility of use of radioisotopes or toxic or carcinogenic material in the course of the work; the funds available; and the need to use cabinets that meet accepted standards for housing the work to be performed. Although cost is very often a strong factor in cabinet selection, the purchaser should be aware of the "marginal cost" of including improvements or additions when ordering the cabinet. These options may provide increased flexibility in the use of the cabinet and ensure a longer life and wider applicability in the laboratory. For example, the 4-ft (1.22-m)-long cabinet appears to be a fairly standard length for many laboratories. It is suitable for a single operator; however, a recommended procedure in the use of a cabinet is to minimize the movement of the operator's arm in and out of the cabinet. This can only be done by loading as much of the work material as possible (e.g., dilution blanks, petri plates, test tubes) into the cabinet before work commences and then using the materials in a well-designed work pattern (6). These materials frequently require extra space to not interrupt the airflow over the work area. For these reasons, a 6-ft (1.8-m) cabinet is recommended if space and funds permit. The cost for the increased size is not usually proportional to the increase in usable work space. The 4- or 6-ft measurement is the length of the work surface and not the length outside of the cabinet.

Cabinet designs can be varied to meet special needs. One of the rather specialized kinds of use that can be met by cabinets is that of small animal necropsy. Cabinets are available with two-sided openings, facing each other. These are often very helpful when one operator can prepare the animal and the other can retrieve the desired samples.

In general, the criteria shown in Table 3 of Appendix I of this book and in the National Institutes of Health (NIH) slide-cassette package (8) provide a guide for a rational selection of cabinets. It is unfortunate that the class I cabinet has not been receiving as much attention from manufacturers and distributors as the class IIA and IIB cabinets, because it is intrinsically a low-cost simple device and affords adequate operator protection for workers using moderate-risk (BSL 2 and 3) microbial agents and for animal work in which work product protection is not required. Furthermore, it can be useful for toxic or carcinogenic chemicals and radioisotopes with the proper decontamination of exhaust air. The disadvantage that it permits contamination of the work by room air may not always be of major concern. Newsome (39, 40) has shown that a well-engineered cabinet gives some degree of product protection, and the "Code of Practice" of the Howie Working Party in England (26) strongly recommends the class I cabinet for routine bacteriological work. Nonetheless, the class II cabinets have achieved rapid acceptance in the laboratory community in the United States. With the consideration of space having been made as described above, the choice between a type A and a type B cabinet can usually be made on the basis of the need for use of radioisotopes, toxic chemicals, or carcinogens in the course of the work. The type B cabinet is the desired cabinet to cope with these hazards. It should be noted that the original type IIB (NSF B1) cabinet is useful only for very low-level short-lived radioisotopes and limited quantities of toxic or carcinogenic material. These limitations can be relaxed substantially by use of the class IIB2 cabinet, which provides a single pass of 100% room or outside air through the workspace.

The IIA and the IIB cabinets offer nearly equal protection for the work. One factor that must not be overlooked in selection of safety cabinets is the requirement for exhaust air. It is apparent that if a 6-ft (1.8-m) class I or class IIB2 cabinet is installed in a room, approximately 350 to 1,200 ft^3/min (0.17 to 0.57 m^3/s) of air will need to be exhausted from the room. Smaller quantities are needed for some class II cabinets, but the volume of air to be exhausted can be substantial. This disadvantage is not found with the class IIA, which exhausts filtered air directly back into the room.

If all class II cabinets considered for purchase by an institution are required to meet the standards of NSF 49 (38), the purchaser can be assured of performance and reliability. However, the NSF standard is a minimum set of requirements and should not be used as a purchase specification. During the acquisition process, the proposed use of the cabinet should be described clearly to the suppliers. It is not unreasonable to ask for test data demonstrating performance claims and clear statements that the equipment proposed can provide the needed product and operator protection. The purchaser should also require the supplier of the cabinet to ensure that it will be put into place and demonstrated to be operating according to the stated performance specifications before acceptance. Although most manufacturers do not provide full installation services, they can make arrangements with local servicing companies to provide installation assistance and in-place testing. This is important unless the purchaser has full mechanical installation and test capabilities at hand and is willing to accept responsibility for the performance of the cabinet "as received" at the loading dock. Damage in shipping is not infrequent, components may be missing, and controls or dampers may have become misadjusted during shipment. On-site inspection and initial "filed" certification are essential. Similarly, provision for periodic retest should be discussed with the supplier. The Centers for Disease Control and Prevention (CDC)/NIH (1993) guidelines (see Appendix I of this book) recommend annual retesting at a minimum.

In summary, the selection and acquisition procedure for class I and II cabinets should include the steps listed below:

1. Establish the class and type of cabinet to be used, commensurate with the operator and product protection required. The classification of microbial agents used and definition of the tasks to be performed should be described in a "proposed application" statement in the purchase document. The supplier should provide statements and test data, if possible, in substantiation of the suitability of the equipment proposed.
2. Establish the size of the cabinet based on the number of workers using it, the space available, and the complexity of the work process (i.e., amount of supplies and materials to be loaded for a given operation).
3. Evaluate both current and future needs. Early stages of exploratory work frequently require high levels of worker protection due to lack of knowledge of the toxicity or pathogenicity of the work material. In later phases, protection of the work material from contamination may be more important. A cabinet that affords both worker and product protection satisfies both needs.
4. Ensure that the size of the cabinet selected allows it to be brought into the facility, through doorways to the final point of use. Be sure that there is adequate headspace for installation and for the maintenance of the filter boxes, exhaust ducts, alarm systems, and fans. Availability of service connections, such as electric and gas, must also be considered.
5. If work is to be performed with agents that must be handled at BSL 2 and higher, the cabinet should be certified to national standards (NSF 49 for biologic).
6. Specify that the cabinet is to be purchased with the condition that acceptance will be based on satisfactory acceptance tests in place, in the final location. (*Note:* This often entails an increased cost.) Assistance may also be needed in interlocking room ventilation blowers and providing vent failure alarms.
7. The manufacturer should provide complete circuit diagrams, operation instructions, maintenance instructions, spare parts listing, and a listing of expendable replacement parts (e.g., filters) that are available on a continuing basis or from other recognized specialty manufacturers. The user should be aware of the location of this information and make it available to the installer and the certifier.
8. The biological safety cabinet is a basic device for primary containment in BSL 2, 3, and 4 laboratories. The cost of this containment equipment should be considered an essential part of the cost of the laboratory project. Stay within a budget. If one cannot meet the technical requirement of the operation within the budget, then either the operation or the budget will have to be changed.

Class III Cabinets (Specialized Glove Boxes)

For the most part, the class III cabinet system comprises a hermetically sealed cabinet system suitable for extremely hazardous work (e.g., usually in the containment laboratories meeting the CDC/NIH BSL 4 requirements) (55). The cabinets are gastight, and all operations within the cabinet are conducted through arm-length rubber gloves. Entry into the cabinet is usually through a sealed air lock, and exit of material may be through an autoclave, a decontamination-type air lock, or a "dunk tank" filled with liquid disinfectant. Figure 16 shows a class III cabinet system installation. These cabinets are often built as modules and assembled into specialty lines

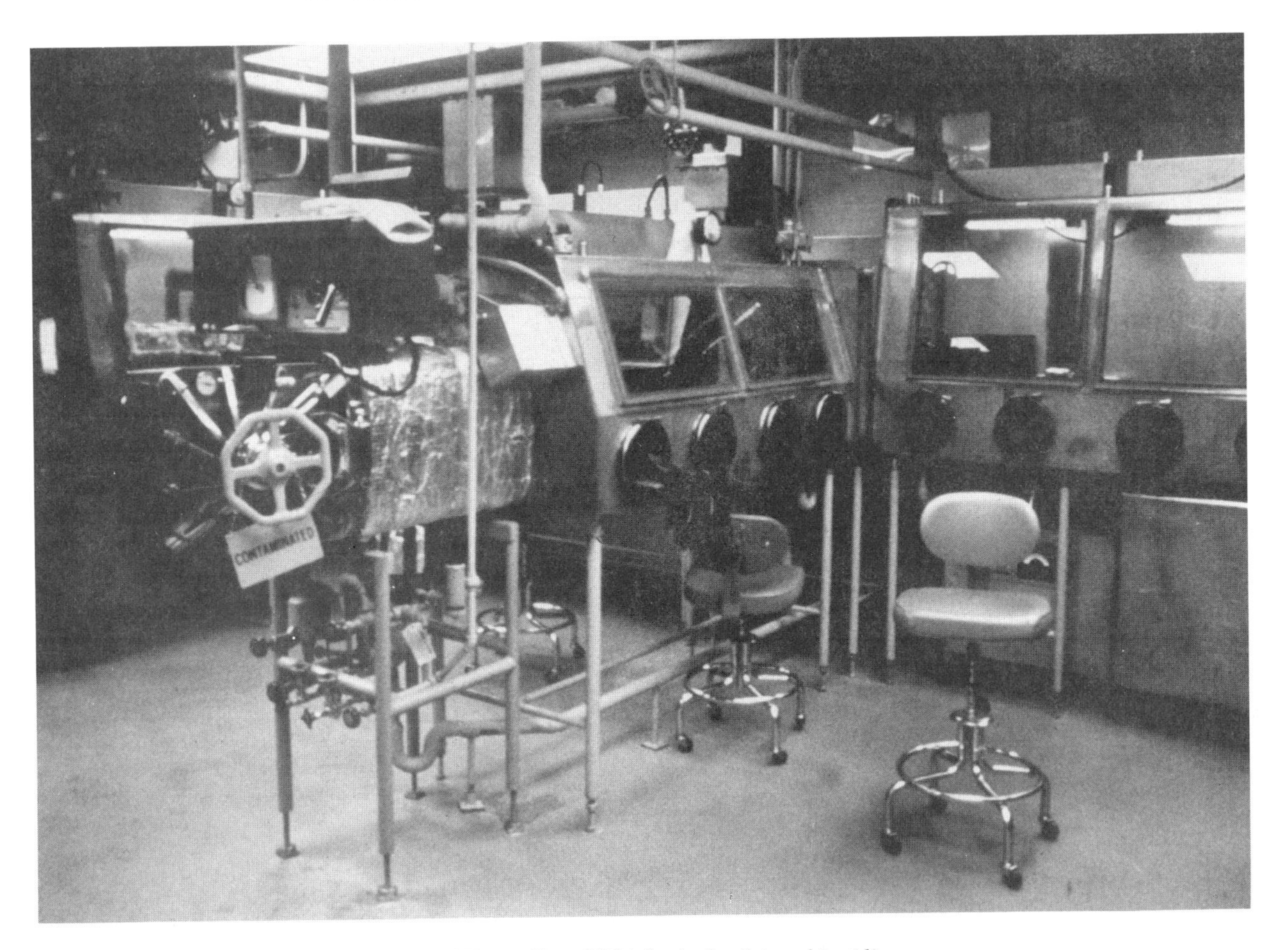

FIGURE 16 Class III biological safety cabinet line.

or systems encompassing a full set of operations in the laboratory (32). Ideally, one should be able to put all the necessary raw materials into the cabinet system, conduct the work, and remove only waste products. For example, some cabinets have been made for such uses as animal inoculation by syringe or aerosol challenge; others may accommodate centrifuges, fixed microscopes, incubators, refrigerators, and other equipment. Most of such cabinet assemblies are made of stainless steel, although some are made of plastic. The latter are often used for controlled-atmosphere protective systems (e.g., anaerobic chambers, germ-free animal isolators). Class III cabinets or "glove boxes" may be provided with strippable or removable liners and additional shielding if the work involves the use of high-activity or long-life radioisotopes.

Disadvantages of class III cabinets include the initial expense of the equipment, as well as the installation and maintenance. The preparation before actual work in the cabinet line is extensive, and the work is made more difficult by the use of the relatively thick arm-length gloves. However, the cabinets are extremely useful when a very high level of protection is required for the operator and the environment. With appropriate training, the operator can become accustomed to the limitations afforded by working through fixed gloves. The gloves provide both the aerosol containment and protection from hand and arm contamination, which can be a main source of contamination release from class I or II cabinets. However, these gloves can be punctured, and thus they constitute the weakest part of the class III cabinet system protection.

Although there are commercially available modular units and a variety of plastic special-purpose chambers, many of the class III installations are fabricated or assembled to order and may involve special engineering. Some design factors to be considered in selection and use of these cabinets are as follows:

1. *Ventilation.* The ventilation rate of these cabinets may be varied to minimize cross-contamination inside but is usually minimal, although at least one manufacturer produces a unit offering near-laminar flow, similar to a class II cabinet within the class III system (Baker Co., Sanford, Maine). Air enters

class III cabinets through at least a single HEPA filter and is exhausted through specially tested ultra-high-efficiency HEPA filters (99.999%) in tandem, an exhaust incinerator, or both. It is usual to maintain approximately 0.5 to 0.75 in. (1.3 to 1.9 cm) of water negative pressure inside the cabinet. More pressure tends to make the gloves distend and lose tactile sensitivity. The blowers should be selected and arranged to provide inlet air velocities of at least 100 lfpm (0.5 m/s) through any glove port in the event of glove rupture or through any air-lock entry in case both doors of an air lock are opened at the same time. The air exhaust blowers from these cabinets are often equipped with dual fans, each of which is adequate to maintain the necessary pressure in the system. The fans may be electrically interlocked to provide increased airflow in the event of glove failure or other inadvertent opening to the system.

2. *Accessories.* A class III cabinet is rarely used by itself but should incorporate the necessary equipment for conduct of the work. If a class III cabinet is used in conjunction with a centrifuge, it should enclose the whole centrifuge if the material to be used warrants that level of containment (17). A cabinet system mounted only on the top of a centrifuge, refrigerator, or other device is subject to leakage of the barrier construction of the centrifuge to which it is attached or possible failure of the centrifuge chamber or connecting piping (vacuum system).

3. *Access.* The cabinet system should have entry and exit locks, a double-door autoclave (preferably with hydraulic or electrically operated doors), and chemical dunk tanks or fumigation chambers. A series of very effective entry-exit air-lock chambers designed to exchange with a standard glove porthole is available (Central Research Laboratories, Inc., Redwing, Minn.).

4. *Containment requirements.* If the cabinets are to be used for work on the CDC/NIH BSL 3 or 4 etiologic agents (55), they should be tested to be gastight under positive pressure (31). The test sets a standard that is difficult to meet, but the equipment must be tested to ensure worker protection. Air exhausts should be directed clear of occupied spaces. Although class III cabinets are usable in both containment and high-containment facilities, their use in a BSL 3 containment facility must be given careful consideration. It is apparent that a BSL 4 agent should be confined entirely within the cabinet system or in a secure container. Thus, in a BSL 3 laboratory, it is desirable to conduct as much preparation as possible outside the cabinet under less highly contained conditions and to conduct all open manipulations with the pathogen in the cabinet. The possibility and consequences of a major failure of the system or transport containers should be considered. This factor alone may make consideration of BSL 4 facilities desirable if the hazard warrants. If not, a class II cabinet should be re-evaluated for suitability. Because the selection and installation of class III cabinets are highly specialized procedures, the material provided here is only an indication of important factors in selection of equipment. The reader is cautioned to define the work requirements very carefully before contracting to install a class III cabinet line. Serious consideration should be given to a "suit protection" area as described below.

Biosafety Cabinet Installation Recommendations

The installation of class III cabinets is highly specialized. Although it is possible to use only a single element of class III modular cabinetry, such equipment is usually installed as a "system," and specialized design requirements often include use of continuous spaces for animal holding and other activities, as described below. The space within which a class III cabinet system is used must be suitable for containment in the event of failure of the cabinet.

Class I and II open-front cabinets are more frequently used in BSL 2 and 3 laboratories under a variety of conditions. They are basic tools for use in the microbiology laboratory. The best location for such cabinets is at the end of a U-configuration, where there will be a minimum of cross-traffic in front of the work surface to interrupt the airflow or to disrupt the operation, and at the same time work bench space will be available at either end for materials (see also Chapter 12). Positioning the workspace against an outside wall permits ready installation of duct work to the outside. An inside wall adjacent to service chases can permit connection to ventilation exhausts or a duct to the roof. It is important to avoid high-velocity air drafts from room air inlets, either in the ceiling or walls or from air conditioners in windows, or even such items as swinging doors, which create transient, high-velocity airflow across the face of the cabinet. Leakage, both into and out of the cabinets, has been shown to be proportional to the velocity of air crossing in front to the cabinet (45). Class I cabinets can be exhausted through a HEPA filter to the laboratory; however, a direct outdoor duct will permit use of the cabinets for chemicals and radioisotopes and is the preferred installation.

Class IIA cabinets do not specifically require ventilation of the exhaust directly to the outside. These cabinets, if they have blowers and exhaust filters incorporated within them, may be exhausted

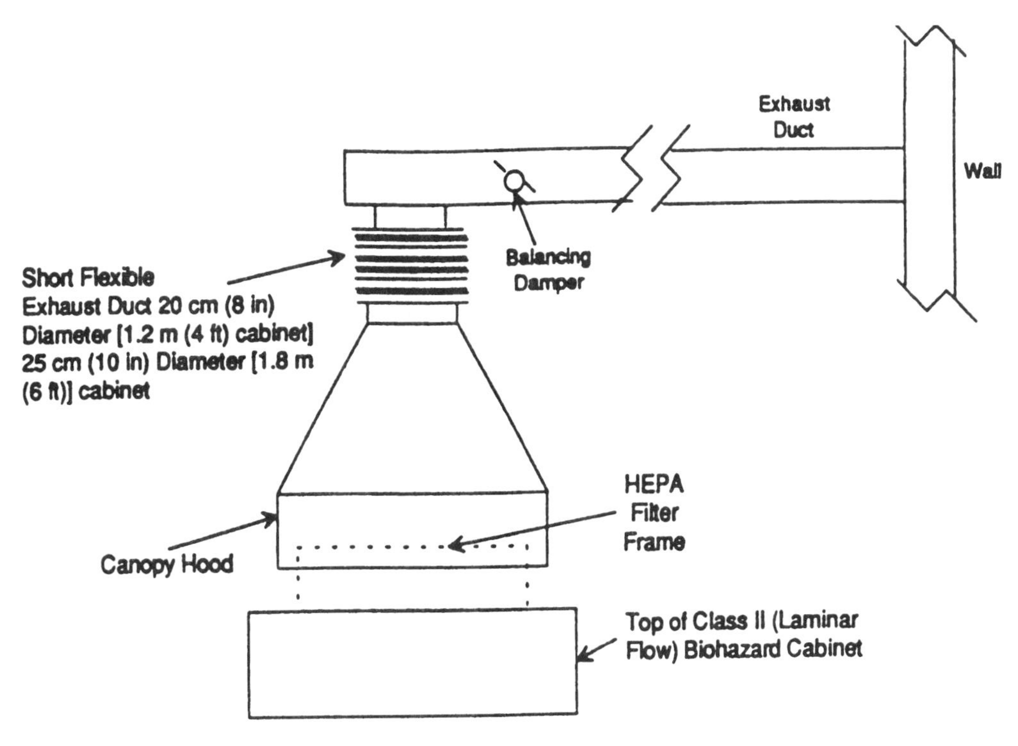

FIGURE 17 Suggested class II (laminar flow), type A biohazard cabinet venting system.

through the filter into the laboratory. However, this is not always the best practice. Exhaust filters occasionally develop leaks; furthermore, discharge and ventilation of the waste formaldehyde used to decontaminate the cabinet is considerably easier when the cabinet has an exhaust duct to the outdoors. The cabinet should have its own exhaust blower and be exhausted directly outside, with an antibackflow damper in the exhaust to ensure that there is no flow back through the filter and cabinet into the "negative pressure" laboratory when the cabinet is shut off. The exhaust should be run to an area clear of and at least 7 ft (ca. 2.1 m) above the roof line, so that workers do not come near the outlet. It should not discharge out into a courtyard or where it may be drawn into other parts of the building. In some cases, it may be possible to exhaust the cabinet into a building exhaust system that does not recirculate to other parts of the building. This is frequently done with a loose connection to the exhaust called a "thimble piece" or variation thereof, as shown in the diagram (Fig. 17) from the current revision of NSF 49. Decontamination will require that the cabinet exhaust be substantially blocked during decontamination gassing; thus a hinge on the thimble or a flexible ducting will be needed for access to the exhaust filter area to be sealed. If necessary, even a "hard" connection can be used if the ductwork system is dedicated to a limited number of cabinets or exhaust systems. Suitable provisions must be made to prevent backflow and to shut down the cabinet in the event that exhaust flow is lost.

With space at a premium in most microbiology laboratories, adequate room for removal and exchange of the cabinet filters from the class I and II cabinets may be overlooked. Provision must made for ready access for periodic maintenance and recertification.

Class IIB cabinets may require connection to a separate exhaust system because many such cabinets do not have an internal exhaust blower. Even if the cabinet has been installed with a dedicated exhaust fan, the use of radioactive or toxic chemicals requires the discharge of the exhaust to be clear of occupied spaces. Most toxic chemical collectors that are available for use within the cabinet (e.g., carbon, molecular sieve) permit penetration as they load up. The chemical filters are difficult to test to determine effective life remaining. Pyrolysis or similar destructive mechanisms may be required for some toxic chemicals. However, the most obvious installation problem for IIB series cabinets is the requirement to provide sufficient inflow of air and lack of cross-drafts. This and the effect on room ventilation balance are similar to the requirements for the class I cabinet.

Biosafety Cabinet Use Procedures

The installation of a biological safety cabinet within a laboratory is usually an indication that careful work practices are needed. The cabinets are not substitutes for good practice and can only complement a careful worker.

Correct operating procedures for open-front cabinets are described in a National Cancer Institute filmstrip (6) slide-cassette set. The operator should wear a closed-front overgarment (e.g., surgical gown with full-length sleeves) and latex or vinyl gloves. For typical cabinet operation, use of bare hands and a button-front laboratory coat is not advised. Many procedures, considered only minor aerosol-producing operations, can contaminate hands, arms, and surfaces. The material for a given operation is to be placed in the cabinet before the work is initiated to minimize in-and-out motions.

In class II cabinets, the operator should work well within the cabinet and not out close to the front. Substantial leakage from the cabinet can occur when work is performed within 4 in. (ca. 10 cm) of the cabinet opening. The condition is almost unique to class II cabinets because of the relatively low inlet air velocities at the top edge of the work opening and the air removal slot or grill immediately along the front of the work surface of some cabinets. When these cabinets are in use, particularly in a small laboratory room, the entry door to the laboratory should be posted to minimize traffic through the room, vigorous swinging of any entry doors (which causes sudden drafts), and distraction of the worker.

When working in class III cabinets, latex or vinyl gloves must always be worn under the long neoprene port gloves. Personnel performing standard techniques using conventional equipment face a major hazard from penetrating wounds from sharp objects. Procedures should be revised to substitute plastic for glass materials and to eliminate pointed instruments within the cabinets (32). The substitution of heavier-gauge port gloves should be considered in areas of the class III cabinet system where good manual dexterity is not required.

Provision must be made for decontamination of waste materials, discard of pipettes and supplies, and disposal of excess material. Also, the cabinet itself will need periodic gaseous decontamination as described below. The cabinet, in any case, should be decontaminated to the degree possible with an effective liquid disinfectant at the end of each work operation, or at least at the end of each day. Most cabinets of stainless steel or durable plastic will withstand the periodic use of 500 to 5,000 ppm available chlorine (1:100 and 1:10 vol:vol dilutions of household bleach in water). However, this material is corrosive, and other disinfectants (e.g., 70% ethanol, quaternary ammonium compounds) may also be satisfactory for use against specific microbes. The selection of appropriate decontaminating agents is addressed in Chapter 14.

UV irradiation may also be used inside the cabinet. Its benefit is primarily for decontamination of physically "cleaned" surfaces after work use. UV is not an effective mode of purification of the air within a cabinet because the time of air contact with the UV tube is usually low. It is not recommended for most installations (15), but it can be useful in disinfecting susceptible bacteria and viruses from surfaces. If the cabinet has a stainless steel interior, a light placed in almost any location within the upper part of the cabinet will bounce 2.537 Å (ca. 254 nm) illumination around the inside and out the opening of the cabinet. Care should be taken to ensure that the materials of the view window will withstand the UV and ozone and that the operator and others in the room are adequately protected against the UV illumination that can cause painful skin and eye burns.

Biosafety Cabinet Decontamination and Recertification

Biosafety cabinets should be certified in place initially by the manufacturer, by the manufacturer's representative, or by a person who has demonstrated competence in biosafety cabinet testing. The NSF has recently begun providing certification tests for cabinet certifiers.

The biosafety cabinet should be recertified annually or at approximately 1,000 h of service, if the filters develop excessive pressure loss, or if the cabinet is moved, even within the same room. Recertification involves testing the specified air velocities and readjusting them to the correct values, leak-testing filters and cabinets, repairing or replacing the filters, or any other malfunctioning elements of the cabinet. Because the personnel performing these tasks may not be fully familiar with the laboratory operation or immunized against the agents used, it is desirable that cabinets be decontaminated before any repairs or certification testing is conducted.

Procedures for decontaminating cabinets with formaldehyde vapor are described in a National Cancer Institute slide-cassette presentation (7). This procedure should only be carried out by trained personnel wearing appropriate PPE. Biosafety cabinet decontamination involves sealing off the cabinet, including both inlets and exhausts, and vaporizing dry paraformaldehyde (0.3 g/ft^3) to provide a concentration of 8,500 ppm. (Recently, vapor-phase hydrogen peroxide has been used as an alternate method.) The overall volume of the cabinet is calculated to allow for takeup in the supply and exhaust filters. The formaldehyde vapor is held in the cabinet for 4 h or overnight. It is very important to ensure that the temperature remains in the 20 to 25°C range and humidity is at least 60% for maximum effectiveness. After sufficient contact time, the formaldehyde gas may be discharged through the exhaust filter to the outdoors if permitted by local regulations or neutralized with ammonium bicarbonate (0.3 g/ft^3) or other appropriate agent. Decontamination effectiveness can be estimated by placing *Bacillus subtilis* spore strips (10^6 to 10^8 per strip) in the cabinet before decontamination. These spore strips are then incubated on Trypticase soy agar to validate spore kill, as a worst case scenario.

Although formaldehyde is the best choice for most cabinet or space decontamination procedures, the effectiveness of any vapor-phase decontaminant against the specific agents used in the cabinet should be ensured. For example, formaldehyde is not effective against many of the so-called slow viruses such as the agents of scrapie or Creutzfeldt-Jakob disease. In such cases, vigorously applied liquid decontaminants may be required. It may be desirable to wet down, remove, and autoclave or incinerate filters in such cases. In fact, considering the relatively poor penetrating capability of formaldehyde vapor, it is prudent to autoclave or incinerate HEPA filters after use in certain infectious disease laboratories.

Recommended Tests for Recertification

Considering the class I and II biosafety cabinet in-place test requirements described previously, the following is the minimum test and certification that should be performed in the laboratory. It is illustrated in a slide-cassette package (5) from the National Audiovisual Center, Washington, D.C.

1. Decontaminate as required.
2. Replace filters as required.
3. Test filters and cabinets for leakage; repair as required.
4. Set inlet and downflow air velocities as specified by the manufacturer.
5. For testing procedures, see NSF 49 or manufacturer's instructions.
6. Test at least annually, at 1,000 h of use, or whenever the unit is moved. Post on the unit with the date tested and the name of the person doing the testing. The list of tests passed satisfactorily and the actual testing results should be available for review by facility engineers and safety personnel, especially biosafety officers.

(*Note:* Testing for "clean bench" operation as performed under FS209E does not require a challenge of the filter with aerosol. Thus, "clean" certification is not an acceptable substitute for safety cabinet certification under NSF 49 because in the latter the filters must be challenged with a test aerosol.)

The Ventilated Suit as a Primary Barrier

Discussions dealing with the philosophy of primary barrier protection are often limited to containment of the work process, either in minimal-volume packages or housings or in work cabinets that allow for the housing of whole operations. A third choice may be needed in cases in which high-hazard materials are used under conditions, such as handling primates, in which the physical manipulations within small containers or within cabinets may be extremely difficult. This is a method wherein the operators are packaged, rather than the work or the work materials. The accepted method for this containment is to provide the operator with a one-piece ventilated suit, usually of a thermoplastic film, with fresh air supplied from a breathing air supply. Such an arrangement requires that there be very little leakage into the suit, which is maintained at 0.5 in. (1.25 cm) of water positive pressure to the room. The suited worker must have clean, cooled, oil-free air of fairly low humidity, supplied at approximately 10 to 12 ft^3/min (ca. 0.3 to 0.4 m^3/min). There must be a backup system, usually cylinders of breathing air, in the event of compressor or pump failure, to allow the operator time to leave the hazardous area. Supplemental air cooling (e.g., by Vortex coolers) may be required for heavy work or work in warm spaces. Because the noise level within such suits tends to be very high due to the rushing of air, it is recommended that ear protection be worn or that the suit be provided with an audiocommunications system. A full alarm system, both audio and visual, should also be installed in the area so that the operator is quickly notified in the event of failure of any of the supporting elements. The protection factor offered by such suits can be high, and they are often used for work with microorganisms requiring BSL 4 containment (55). Supplied-air head masks or hoods or full-face respirators may also be used under these conditions for less hazardous operations (see PPE section). As with class III biological safety cabinets, the gloves are the weakest part of the system, subject to puncturing by sharps, points, and animal bites unless a heavier protective glove can be worn.

The room or area in which the suit is used must, in effect, be equivalent to a large class III cabinet. Accordingly, it must be operated under a negative pressure, and supply and exhaust air systems must be interlocked to ensure inward (or zero) airflow at all times. All air exhausted from these spaces must either pass through two HEPA filters in tandem or be incinerated. Provisions must be made for discard of waste material through an autoclave or other system. An entry air lock, change room, and shower access and egress must be installed. The capability for complete decontamination of the area must also be provided. The disinfectant used in the exit shower from the suited area should be one that is noncorrosive, that emits vapors that are not hazardous to breathe, and that has proved to be effective against the agent(s) in use. In one author's laboratory, 2 to 5% of a quaternary ammonium-based compound has been used for this purpose for many years without problems.

The internal finish of suit areas must permit ready decontamination. Finishes such as epoxy or polyester resins have been quite successful. Such areas are usually constructed with recessed lighting and utilities that may be serviced from outside the space and with provisions for remote control of paraformaldehyde-vaporizing heaters.

Suit rooms may be used for housing animals that have been exposed to pathogens within the class III cabinet system. Animals may be held in filter-topped cages or in open cages placed in ventilated enclosures.

Despite the excellent protection afforded by the ventilated suit, all procedures in which infectious agents are handled should still be conducted in class I or II biological safety cabinets. As for the class III cabinet system, detailed engineering and test procedures are necessary for installation and use of such space. Conversion of existing space of standard construction to such use is difficult and costly. The containment equipment can function reliably only when adequate electrical and mechanical support services are available. For example, redundancy in electrical services, exhaust fans, supply fans, heater systems, exhaust filtration, and breathing air must be provided in addition to communications systems, solid and liquid waste decontamination, and disinfectant shower to ensure adequate protection both to the worker and to the surrounding environment. Maintenance and operation of such equipment should be considered. There is a need for highly trained and well-motivated support staff who can provide the necessary mechanical and instrumentation services.

Application of Primary Barriers to Microbiology Laboratory Research

Three classes of cabinets and an equivalent ventilated suit containment method have been described above. With knowledge of the containment systems available, the user must match the containment equipment to the requirements. In Appendix I of this book, Table 3 summarizes information on the use of biological safety cabinets.

1. *BSL 1.* For materials designated by the CDC/NIH (55) as BSL 1, there are no generalized requirements for containment systems.
2. *BSL 2.* For materials designated by the CDC/NIH as BSL 2 etiologic agents, by the National Cancer Institute as low- to moderate-risk viruses, or by the NIH recombinant DNA guidelines (36) as requiring BSL 2 containment, the primary barriers should be applied to those operations known to produce appreciable aerosols. Class I and II biological safety cabinets can be used to contain aerosol-producing equipment such as blenders, lyophilizers, sonicators, table-top centrifuges, and open-vessel devices used for growth or propagation of microorganisms by aeration or shaking. The biosafety cabinet may not be required with aerosol-producing equipment as contained by other primary barriers; for example, the centrifuge can be operated in the open if a sealed head or effective safety cup is used.
3. *BSL 3.* The class I or II biological safety cabinet is required in the BSL 3 laboratory as defined by the NIH in the recombinant DNA guidelines (36) and by the CDC/NIH for work with BSL 3 etiologic agents (15, 55). Some oncogenic viral agents designated as moderate-risk agents should also be handled within such cabinets. In addition to the operations described above, such aerosol-producing operations as pipetting, serial dilutions, culture transfer operations, plating, flaming, grinding cultures, blending dry materials, and virtually any operation in which high concentrations of microbial agents may be used—or, alternatively, there is a potential for contamination of the product or even a xenotropic contaminant within the work product (e.g., as found in some cell lines)—should be handled under containment such as that provided by a biological safety cabinet. Animals infected with agents in these categories can be housed in partial-containment caging systems such as open cages placed in ventilated enclosures and solid-wall and -bottom cages placed on holding racks equipped with UV lamps and reflectors if the etiologic agent(s) has been shown to be inactivated readily by UV irradiation. A class III cabinet system may be used within a BSL 3 laboratory to enclose an operation that can be a heavy aerosol producer or to house biological agents that are very poorly defined with respect to routes of infection, infectious dose, or persistence or are known to be particularly dangerous.

4. *BSL 4.* The use of extensive class III cabinet systems or suit rooms is substantially as defined for BSL 4 physical containment facilities (55) for work with high-hazard agents and procedures. Laboratory animals held in the facility should be housed either in cages contained within class III cabinets or in partial-containment caging systems such as open cages placed in ventilated enclosures or solid-wall and -bottom cages covered by filter bonnets and located in a specially designed suit room.

PERSONAL PROTECTIVE EQUIPMENT

The laboratory environment encompasses many known and unknown hazards that pose potential risks to personnel. Some of these hazards, which include physical, biological, chemical, and radiation exposure, can be reduced or minimized through the proper use of PPE. Such equipment includes gloves, respirators, eye and ear protection, garments, and other devices worn by the employee to guard against hazards in the laboratory. PPE may be required to reduce the risk of exposure of an employee by contact, inhalation, or ingestion of an infectious agent, toxic substance, or radioactive material involved in the work at hand. OSHA legislation makes it mandatory for an employer to furnish employees with a working environment free from the recognized hazards that could cause death, injury, or illness to the employees. Certain state agencies have been given precedence over OSHA as long as their requirements equal or exceed the OSHA requirements. OSHA requires that the hazard be controlled through engineering means when feasible (i.e., the facility must provide some device that removes the hazard from the employees' work area and thereby eliminates the necessity for PPE). In many cases, however, it is not feasible to eliminate the hazard. Both OSHA and the state agencies recognize that some hazards require a cost-benefit risk analysis, both of the process and of the laboratory procedure, to determine the practicality of eliminating the problem. Protective clothing is a legitimate solution in such instances. Common hazards have been neglected in the laboratory environment, which presents many of the same hazards found in the industrial environment. Employees can encounter flying particles that could cause physical injury and can face exposure to dangerous chemicals such as solvents, vapors, or fumes. Many laboratories pose even more serious problems to the health and welfare of the persons working in the facility (2, 37, 57). In assessing the hazards, personal protective equipment may not only be necessary but also may be the most practical, cost-effective means available to prevent employee injury or illness.

Once the hazard has been identified, appropriate protective equipment must be selected for laboratory use. Two criteria should be included: (i) the degree of protection that a particular piece of equipment affords under varying conditions, and (ii) the ease with which it may be used. Considerable time can be saved by checking with manufacturers, appropriate government agencies, and private testing laboratories for their solutions to similar problems. Information should be supplied with the product or on request from the manufacturer. Questionable data should be verified by contacting OSHA or the appropriate state agency directly.

Once the need for PPE has been determined and the appropriate device(s) identified, the employee must be motivated to use the equipment properly. Employees are now required by law to wear appropriate PPE once hazards have been identified (37, 57). Some factors that help to get the employee to comply include the following:

1. The employee must understand the procedure or process that is hazardous and, therefore, the need for such equipment.
2. The employee must realize that the job can be performed safely using the PPE with a minimum interference with his or her normal work procedures.
3. The employee should have maximum familiarity with the PPE; it should fit properly and be easy to wear under the specific work conditions.
4. Simple instructions for the care of such equipment should be provided, and supervisory personnel should be responsible for the initial demonstration and periodic follow-up of proper use and maintenance of the PPE.

The following sections cover specific types of PPE in some detail. More detailed information may be found in the references to technical source materials.

Laboratory Clothing

Each BSL requires some type of protective laboratory clothing. These include laboratory coats, smocks, gowns, total body suits, coveralls or jump suits, aprons, or two-piece scrub suits, all of which are commercially available. These items come in reusable or disposable models made from a variety of materials including cotton, Dacron, nylon, polyester, olefin, rayon, vinyl, modacrylic, polyvinyl chloride (PVC), or rubber or trade names such as Tyvek (plain, polyethylene-coated, or Saranex-laminated),

Safeguard, Duraguard, and Disposagard. Some materials are designed to protect against specific hazards such as biological, radioactive, chemical, or physical, including heat or cuts. Some materials feature antistatic and flame, caustic, oil, or acid resistance. The selection of the optimum configuration and material depends on the potential hazards, the regulatory requirements, types of operations to be performed, types of decontamination and reprocessing possible and available, the work environment, and personal preferences (34).

Laboratory Coats or Gowns

The laboratory coat can be used to protect street clothing against biological or chemical spills as well as to provide some additional body protection (24). The degree of protection provided by the common, cost-effective laboratory coat is frequently misunderstood. The specific hazards(s) and the degree of protection required must be known before selecting coats for laboratory personnel. The CDC/NIH guidelines for biocontainment practices recommend the use of a laboratory coat, gown, smock, or uniform while working in BSL 2 laboratories. They recommend solid-front or wrap-around gowns, scrub suits, or coveralls, but not front-buttoning laboratory coats, in BSL 3 laboratories. An evaluation must be performed to determine whether the laboratory coat or gown is actually sufficient to protect the wearer from the immediate danger. For example, the material must be impervious enough to protect an employee from a spill or splash when such an event can be expected from the work. When there is a potential for exposure to a flame in the laboratory, the laboratory coat should be made from a fire-resistant material. Because a polyester-cotton blend material is flammable and will melt on the skin after contact with a spark, heat source, or some corrosive materials, a 100% cotton laboratory coat, which is nonreactive to many chemicals as well as flame resistant, may be a better choice (34, 43, 44). The laboratory coat or gown itself should cover the arms as well as most of the middle body. It is a good laboratory practice to keep the laboratory coat buttoned at all times in the laboratory.

Scrub Suits

A complete change of clothing, from street clothes to laboratory clothing, is recommended for those entering BSL 4 areas. A cotton or cotton-polyester blend, two-piece scrub suit, with a long-sleeved shirt, featuring integral ribbed wristlets is recommended. The shirt should close in front for rapid removal in the case of spill or splash. Some BSL 3 facilities also require changing into scrub suits, and some also stipulate that the suits be a unique color and not be worn outside the containment areas.

When wearing a scrub suit in a BSL 3 laboratory and performing operations that may generate infectious aerosols outside a biological safety cabinet, an additional long-sleeved, solid-front, wrap-around gown can be worn to minimize the contamination of the basic laboratory outfit (i.e., the scrub suit). Examples of such procedures are inoculating animals with infectious materials, bleeding viremic animals, and otherwise handling or caring for animals that may be shedding hazardous viruses or bacteria in the urine or feces. When exiting the work area, the gown is removed and left in the room for reuse or steam sterilization before reprocessing. Disposable gowns are more frequently used and can be disposed of with the regular medical waste.

Heavy-duty rubber aprons or other specialized items of apparel can also be worn if there is a significant possibility of contamination from hazardous chemicals or radioisotopes.

Head Coverings

Head coverings are not usually necessary in biohazardous areas except in those containment areas where a complete clothing change is required or where product protection is also required. In situations in which a total body shower is mandatory on exiting, employees are usually given the option of either washing their hair during the shower or wearing a head covering during the time spent in the containment area. Several styles of head coverings are available: a simple cap, various hood styles, or a bouffant style for long hair. Head coverings come in washable or disposable models in a variety of fabrics including cotton-polyester blends, cotton, polyolefin, or Tyvek.

Shoes and Shoe Covers

Industrial safety shoes should be worn in any area where there is a significant risk of dropping heavy objects on the foot. When used in containment areas, these shoes, like all others, should be left within the area or decontaminated before removal. A large variety of types and configurations can be found in safety equipment supply catalogues. For general biological use, comfortable shoes such as tennis shoes or nurses shoes are used extensively. Sandals are not allowed in laboratories using biohazards, due to the potential exposure to infectious agents or materials as well as physical injuries associated with the work. A change from street shoes is strongly advised for those working in BSL 3 areas and should also be considered for BSL 2 areas, especially for work with infected animals in animal rooms. Shoe covers can be used in BSL 2 or BSL 3 areas when a complete change of clothes and a dedicated pair of shoes is not required. Such shoe covers are available in vinyl, polyethylene, Saranex, or Tyvek and are usually considered disposable items. It is important to test shoe covers under the actual conditions of use to ensure that they provide slip resistance. In animal rooms and other areas where the wearer may encounter the splashing of large amounts of water from the hosing of cages, racks, or floors, the wearing of butyl rubber, neoprene, or PVC boots is strongly advised to reduce slipping hazards.

Gloves

Gloves are used in the laboratory for protection against a wide variety of hazards including heat, cold, acids, solvents, caustics, toxins, infectious microorganisms, radioisotopes, cuts, and animal bites. Unfortunately, there is no ideal glove that will protect against all hazards. The selection of proper gloves is essential when hazardous tasks are to be performed.

Gloves are made of such a variety of materials, including rubber, neoprene, neoprene-latex, Viton, polyurethane, fluoroelastomer, nitrile, polyethylene, PVC, polyvinyl alcohol, and for such special uses that the worker should seek advice regarding the best glove for the task to be performed. For example, chemical solvents such as xylene, toluene, benzene, perchloroethylene, dichloroethane, and carbon tetrachloride normally degrade rubber, neoprene, and PVC (56); it is important to avoid these materials in favor of polyvinyl alcohol or Buna-N. A detailed discussion of gloves for handling chemicals is beyond the scope of this chapter; the reader is advised to consult the references (50, 52, 54, 56) for further information.

For protection of hands, wrists, and forearms against steam or for handling hot objects, insulating gloves or mittens made of Zetex aluminoborosilicate fibers or Kevlar aramid fibers have replaced the traditional asbestos gloves. For handling very cold materials, Zetex or insulated latex or neoprene gloves are available, and for liquid nitrogen, loose-fitting gloves are preferred.

The use of protective gloves in laboratories where microbiological hazards may be present often creates the situation in which protection is sacrificed in the name of dexterity. Most surgical gloves (e.g., latex, rubber, vinyl) range from 8 to 10 mils in thickness, offering the good tactile sensitivity and dexterity required for many procedures. Unfortunately, they offer very little, if any, protection against needlesticks, sharps, or animal bites. This has resulted in laboratory exposures and infections caused by the false sense of security associated with the wearing of surgical gloves. At one of our research institutes, 7 of the 10 significant potential exposures to high hazard infectious microorganisms (BSL 4), as well as most laboratory accidents using agents of lower hazard levels, resulted from sharp items (commonly hypodermic needles) and animal bites, usually involving the hands. Some of these accidents involved workers who were wearing ventilated suits with integral 0.018-in. latex gloves or working at class III biological safety cabinets that incorporated arm-length 0.015-in. neoprene gloves. It is important to remember that gloves are the weakest component of the PPE spectrum.

A nonabsorbent surgical glove should always be selected to prevent contamination of the hands when working with infectious materials. Gloves should overwrap the cuff and the lower sleeve of the laboratory clothing. Recent reports have been published on the failure of a certain percentage of surgical gloves to exclude virus particles (9, 28, 47). Even though these reports have been disputed by others (19, 27, 62), they have induced many health care workers to adopt the policy of wearing more than one pair of surgical gloves (9, 35). It is also important to recognize the necessity for handwashing after gloves are removed.

In addition to surgical gloves, the wearer is advised to select the heaviest or the thickest glove possible that does not sacrifice the touch sensitivity or dexterity needed for the work. For example, 0.030-in. neoprene gloves can be substituted for the 0.015-in. gloves on some areas of class III cabinet lines where animals are routinely handled to reduce penetration from bites. Workers in ventilated suits should wear an additional pair of leather gloves when performing certain tasks and handling animals.

Other specialty gloves include gauntlet-type leather gloves for handling monkeys and Kevlar aramid, Kevlar and stainless steel, and stainless steel mesh gloves to be worn during necropsies of infected animals to prevent accidental cuts from contaminated scalpels and surgical saws.

Respiratory Protection

The risk of inhalation of toxic or infectious materials that can occur in the laboratory environment poses a significant potential health hazard to the employee. Engineering controls are designed to eliminate employee exposure to the hazardous agent by providing adequate rates of ventilation or by redesigning the ventilation system to a completely closed system. When engineering controls are not feasible or are inadequate, PPE becomes mandatory. With the possible exception of a maximum containment area designed for the wearing of ventilated suits, respiratory protection equipment should be worn only when required by a regulation or when engineering or administrative controls are not achievable.

There are two general types of respiratory protection equipment: tight-fitting and loose-fitting. Both types can be either atmosphere-supplying (supplied air from pressurized tanks or air compressors) or air-purifying (particle-removing filters). Atmosphere-supplying equipment can be operated in constant- or continuous-flow, demand-flow, or pressure-demand modes. Air purifying respirators are usually breathing-demand but are also available as constant-flow or pressure-demand items that incorporate a battery-powered fan and motor assembly.

Tight-fitting respirators include half or full face, both of which require quantitative fit testing. According to OSHA 29 CFR 1910.134 (e)(59)(916), "Respirators shall not be worn when conditions prevent a good face seal. Such conditions may be a growth of beard, sideburns, or temple pieces on glasses" (41). The American National Standards Institute (4) recommends that a person who has hair (moustache, beard, stubble, sideburns, low hairline, bangs) that passes between the face and the sealing surface of the facepiece of the respirator shall not be permitted to use such a respirator.

When circumstances require the use of respiratory protection in areas such as infected animal rooms where inoculations, excretions, bleeding, or agitation of contaminated bedding may produce infectious aerosols, half-face or full-face respirators with HEPA filters are recommended if a good fit can be attained. Full-face masks (or half-face masks plus eye protection) should be worn in atmospheres that pose an infectious or toxic hazard by the ocular as well as the respiratory route and in other specific instances such as handling monkeys, especially male monkeys, that have been treated with an infectious or toxic agent that is excreted in the urine.

Single-use, paper "dust" masks are not classified as true respirators, are difficult to fit properly, and are used primarily for the prevention of fibrosis and pneumoconiosis in industry (14). They do not offer adequate respiratory protection in infected animal rooms or other areas where infectious aerosols may be present because one cannot be ensured of an adequate fit. They may be worn in "clean" animal rooms to reduce possible irritation from allergens such as fur and dander and to help keep hands away from the mouth and nose. Also, they are acceptable for use during necropsies and other procedures on noninfectious animals to help maintain a sterile surgical field. When used for these purposes, the appropriate size must always be worn to ensure an optimal fit for maximum effectiveness. The CDC has recommended particulate respirators for protection at BSL 3.

Loose-fitting respirators include hoods, helmets, bonnets, blouses, and full suits in which proper fit is not important, and thus the wearing of beards is not prohibited. In areas where the agent cannot be adequately contained, the worker can be isolated by being encapsulated in a ventilated suit (Fig. 18) as previously described in this chapter. Although full suits are usually recommended for use in BSL 4 or maximum containment areas, other configurations may be acceptable depending on the circumstances and operations to be performed. One of the more versatile and newer items of respiratory protection equipment is the airflow hood (Fig. 19). These constant-flow air-purifying hoods operate for 8 h on rechargeable battery packs and can be equipped with a variety of particulate and chemical filters or a combination of both. They are lightweight and comfortable and have the advantage of eliminating fit testing. They can be worn by individuals with beards as well as those who wear eyeglasses. The

FIGURE 18 Ventilated suit. (Courtesy ILC Dover, Inc., Frederica, Del.)

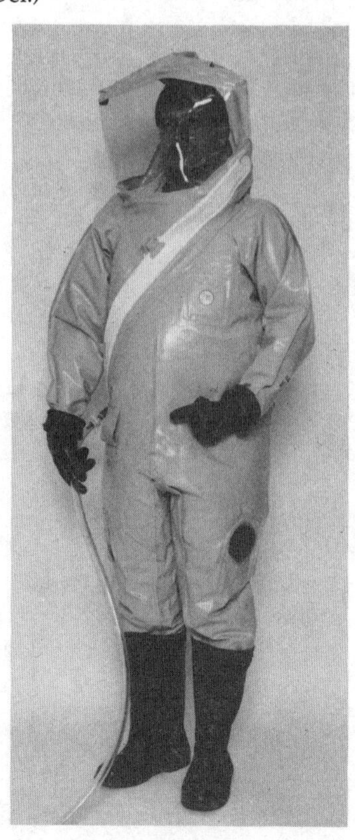

FIGURE 19 Powered air-purifying system. (Courtesy Racal Health and Safety, Inc., Frederick, Md.)

devices have potential for use in high-hazard areas in situations in which splashing of infectious blood or urine is not expected (e.g., no hosing of infected animal cages). They also have application in high-hazard areas for emergency egress or emergency assistance, in high-hazard field situations, during large-scale fermentations using live microorganisms, or in atmospheres containing objectionable concentrations of irritating chemicals such as formaldehyde gas after an area decontamination. The substitution of any configuration other than a full suit or the use of any powered air-purifying respiratory protection equipment must be compatible with both the work requirements and exit requirements. The requirement for a disinfectant shower on exiting a BSL 4 area may negate some of the alternative options. Continuous or pressure-demand flow air line or powered air-purifying helmets, hoods, or suits offer greater protection than demand equipment and are preferred for BSL 4 use. Demand or mechanical respirators are normally unacceptable in high-hazard areas for routine use because of the possibility of leakage during inhalation. Demand units or tight-fitting mechanical filter respirators with HEPA filters might conceivably be acceptable for BSL 4 areas for emergency escape or rescue if, and only if, a good fit is ensured. Positive-pressure or pressure-demand self-contained breathing apparatuses provide a higher degree of protection than demand or closed-circuit equipment, but because of their weight and limited air supply (5 to 60 min) are only practical for escape and rescue from high-hazard areas.

Whenever supplied air equipment is used, the respiratory protection program should include a back-up provision in the event of compressor failure. Either an auxiliary compressor or bottled breathing air is recommended. If either of these is not feasible or is unavailable, a combination pressure-demand breathing apparatus should be worn or provided that permits escape from the dangerous atmosphere in case the primary air is interrupted. Such equipment usually incorporates an approved rated 5- to 10-min escape device.

Federal regulations (41) state that "the wearing of contact lenses in contaminated atmospheres with a respirator should not be allowed." The American National Standards Institute (4) interprets this to mean any respirator with a full facepiece, helmet, hood, or suit. The use of contact lenses in a laminar environment such as a ventilated suit may aggravate the dehydration of the tear layer on which the lens rides, giving rise to subsequent corneal abrasions and other related phenomena (48). Also, foreign bodies or contaminants that penetrate the respirator may get into the eye and cause severe discomfort, compelling the wearer to remove the respirator (42). Finally, full facepieces can pull at the side of the eye and pop out the lens (12).

All employees who are required to use respirators should be enrolled in the institutional Occupational Health Respirator Program before using a respirator or ventilated suit or hood. All ventilated suit wearers should also be enrolled in the facility Hearing Conservation Program because the noise level in these suits sometimes exceeds 85 dB.

Eye or Face Protection

Eye protection in the biological laboratory is important for several reasons. Concentrated acids, alkalis, or other corrosive or irritating chemicals are often used routinely. Concentrated disinfectants, including phenolics and quaternary ammonium compounds, can cause severe damage and blindness if splashed in the eye. Infection can also occur through the conjunctiva if certain pathogenic microorganisms are splattered into the eye. According to Krey et al. (29), the mature retina is vulnerable to virus invasion and susceptible to infection. Virus infection can lead to retinal dysplasias and chronic retinal degeneration. At least one virus, herpesvirus, has been shown to propagate in the brain after intraocular inoculation (30). Full-face respirators or half-face respirators plus splash goggles are recommended when operations with specific microorganisms or toxins may result in the generation of respirable aerosols or droplets that may enter the eye.

Safety glasses are sometimes requested and used by personnel working in biological laboratories because this work often involves the handling of hazardous chemicals. Safety glasses are intended to provide impact protection against projectiles and broken glass but should not be used to protect against chemical splashes in lieu of approved acid or chemical splash goggles or face shields. Although ordinary spectacles offer better splash protection than wearing nothing at all, goggles or shields that are designed for this purpose offer maximum eye protection. Such protective devices are relatively inexpensive and are to be readily accessible in all laboratories where such eye hazards as chemicals are used.

Safety shields, or face shields, also provide good protection against a chemical or biological splash and are recommended for workers handling nonhuman primates because of the potential for exposure to *Herpesvirus simiae* (B virus).

Any discussion of eye protection would not be complete without some comments and cautions on contact lenses. Whereas most authors agree that contact lenses should not be substituted for required eye protection, the position that contact lenses do not provide adequate protection against eye damage after exposure to chemicals, fumes, or foreign bodies is not universally accepted. There are reports (3, 23) that support the idea that contact lenses present an increased hazard to the wearer, particularly in the event of a chemical splash to the eye. In contrast, however, others (25, 46) claim that more protection is provided when hard or soft contact lenses are worn than when absent, including one study with rabbits using hard lenses in one eye and exposed to an acid, a base, and a solvent (25). The authors postulate that the lens may act as a barrier to an irritant, suggesting that an irritant will cause lid spasm to occur, causing the lens to tighten against the cornea, thereby effectively sealing off the area under the lens. This theory is supported by actual cases described by Rengstorff and Black (46). However, Rowe (49) cited the observation that rigid lenses are soluble in or swollen by many organic solvents and that aqueous chemical solutions are readily soluble in the water phase of contact lenses. Ennis and Arons (22) argued that lenses will increase the concentration of chemicals in contact with the eye because the chemicals become trapped underneath the lens and prevent thorough irrigation of the eye, thus accentuating corneal damage.

In a biological laboratory, the situation may be a little more complicated. In the event of an accident of any kind or even the need to reposition a lens, sufficient decontamination of the hands must be performed before the removal or repositioning of the contact lenses. In the hurry to remove the lens after a chemical splash, this hygiene step may be neglected. Also, as stated earlier, contact lenses should never be worn when wearing a respirator equipped with a full facepiece, helmet, hood, or suit.

Contact lenses were never intended to be substitutes for properly designed safety eyewear in hazardous situations. Wearers of contact lenses, as well as wearers of ordinary spectacles, should use the same approved protective devices, such as splash goggles or safety shields, worn by other workers.

For further information on PPE for biohazards, the reader is referred to the NIH *Laboratory Safety Monograph* (36) and other technical references listed below.

References

1. **Akers, R. L., R. J. Walker, F. L. Sabel, and J. J. McDade.** 1969. Development of a laminar flow biological cabinet. *J. Am. Ind. Hyg. Assoc.* **30:**177–185.
2. **Allen, R. W., M. D. Ellis, and A. W. Hart.** 1976. *Industrial hygiene.* Prentice-Hall, Inc., Englewood Cliffs, N.J.
3. **American Medical Association Council on Occupational Health.** 1964. Contact lenses in industry. *JAMA* **188:**397.
4. **American National Standards Institute.** 1982. *Practices for respiratory protection.* Z88.2–1980. American National Standards Institute, Inc., New York.
5. **Anonymous.** 1976. *Certification of class II (laminar flow) biological safety cabinets.* (Slide-cassette.) NAC No. 003134 and No. 009771. National Audiovisual Center, Washington, D.C.
6. **Anonymous.** 1976. *Effective use of the laminar flow biological safety cabinet.* (Slide-cassette.) NAC No. 00971 and No. 003087. National Audiovisual Center, Washington, D.C.
7. **Anonymous.** 1976. *Formaldehyde decontamination of the laminar flow biological safety cabinet.* (Slide-cassette.) NAC No. 005137 and No. 003148. National Audiovisual Center, Washington, D.C.
8. **Anonymous.** 1976. *Selecting a biological safety cabinet.* (Slide-cassette.) NAC No. 00709 and No. 01006. National Audiovisual Center, Washington, D.C.
9. **Arnold, S. G., J. E. Whitman, Jr., C. H. Fox, and M. N. Cottler-Fox.** 1988. Latex gloves are not enough to exclude viruses. *Nature* **335:**19.
10. **Baldwin, C. L. (ed.).** 1973. *Centrifuge biohazards. Proc. Nat. Cancer Inst. Sym. Centrifuge Biohazards.* DHEW Publication No. (NIH) 78–373. National Institutes of Health, Bethesda, Md.
11. **Barkley, W. E.** 1972. Ph.D. thesis. University of Minnesota, Minneapolis.

12. **Birkner, L. R.** 1980. *Respiratory protection: a manual and guideline.* American Industrial Hygiene Association, Akron, Ohio.
13. **British Standards Institution.** 1979. *Specifications for microbiological safety cabinets.* Brit. Std. 5726. British Standards Institution, London.
14. **Bureau of Mines.** 1972. *Respiratory protection devices; tests for permissibility; fees.* 30 CFR 11. U.S. Department of the Interior, Washington, D.C.
15. **Center for Disease Control.** 1976. *Classification of etiologic agents on the basis of hazard.* Center for Disease Control, Atlanta.
16. **Chatigny, M. A.** 1961. Protection against infection in the microbiology laboratory: devices and procedures. *Adv. Appl. Microbiol.* **3:**131–192.
17. **Chatigny, M. A., S. Dunn, K. Ishimaru, J. A. Eagleson, and S. B. Prusiner.** 1979. Evaluation of a class III biological safety cabinet for enclosure of an ultracentrifuge. *Appl. Environ. Microbiol.* **38:**131–135.
18. **Clark, R. P., and B. J. Mullan.** 1978. Airflows in and around downflow safety cabinets. *J. Appl. Bacteriol.* **48:**131–135.
19. **Dagleish, A. G., and M. Malkovsky.** 1988. Surgical gloves as a mechanical barrier against human immunodeficiency virus. *Br. J. Surg.* **75:**171–172.
20. **Darlow, H. M.** 1969. Safety in the microbiological laboratory, p. 169–204. *In* J. R. Norris and D. W. Ribbons (ed.), *Methods in microbiology,* vol. 1. Academic Press, Inc., New York.
21. **Dimmick, R. L., W. F. Fogl, and M. A. Chatigny.** 1973. Potential for accidental microbial transmission in the biological laboratory, p. 246–266. *In* A. Hillman, M. N. Oxman, and R. Pollack (ed.), *Biohazards in biological research.* Cold Spring Harbor Laboratory, Cold Spring Harbor, N.Y.
22. **Ennis, J. L., and I. Arons.** 1979. Caution to contact lens users. *Chem. Eng. News* **57:**4,84. (Letter.)
23. **Fox, S. L.** 1967. Contact lenses in industry. *J. Occup. Med.* **9:**18–21.
24. **Goldstein, L.** 1980. How to protect against radiation. *Natl. Safety News* **October:**46–47.
25. **Guthrie, J. W., and G. F. Seitz.** 1975. An investigation of the chemical contact lens problem. *J. Occup. Med.* **17:**163–166.
26. **Howie, J.** 1978. *A code of practice for the prevention of infection in clinical laboratories (report of the Working Party).* Her Majesty's Stationery Office, London.
27. **Huppert, J., and G. Mathe.** 1988. Latex gloves off in virus porosity dispute. *Nature* **336:**317.
28. **Klein, R. C., E. Party, and E. L. Gershey.** 1990. Virus penetration of examination gloves. *BioTechniques* **9:**197–199.
29. **Krey, H. F., H. Lidwig, and P. Rott.** 1979. Spread of infectious virus along the optic nerve into the retina in Borna disease virus-infected rabbits. *Arch. Virol.* **61:**189–201.
30. **Kristensson, R., B. Ghetti, and H. M. Wisniewski.** 1974. Study on the propagation of herpes simplex virus (type 2) into the brain after intraocular injection. *Brain Res.* **69:**189–201.
31. **Kruse, R. H., W. H. Puckett, and J. H. Richardson.** 1991. Biological safety cabinetry. *Clin. Microbiol. Rev.* **8:**207–241.
32. **Kuehne, R. W.** 1961. Biological containment facility for studying infectious disease. *Appl. Microbiol.* **26:**239–243.
33. **Kuhn, H. S.** 1961. Questions and answers. *JAMA* **178:**1055.
34. **Lynch, P.** 1982. Matching protective clothing to job hazards. *Occup. Health Safety* **1:**30–34.
35. **Matta, H., A. M. Thompson, and J. B. Rainey.** 1988. Does wearing two pairs of gloves protect operating staff from skin contamination? *Br. Med. J.* **297:**597–598.
36. **National Institutes of Health.** 1978. *NIH guidelines for recombinant DNA research supplement: laboratory safety monograph.* National Institutes of Health, Bethesda, Md.
37. **National Safety Council.** 1974. *Accident prevention manual for industrial operations,* 7th ed. National Safety Council, Chicago.
38. **National Sanitation Foundation.** 1976 (revised 1987). *NSF standard no. 49 for class II (laminar flow) biohazard cabinetry.* National Sanitation Foundation, Ann Arbor, Mich.
39. **Newsome, S. W. B.** 1979. Class II (laminar flow) biological safety cabinet. *J. Clin. Pathol.* **32:**505–513.
40. **Newsome, S. W. B.** 1979. Performance of exhaust protective (class I) biological safety cabinets. *J. Clin. Pathol.* **32:**576–583.
41. **Occupational Safety and Health Administration.** 1981. *Respiratory protection.* General Industry Standards 29 CFR 1910.134. U.S. Department of Labor, Washington, D.C.
42. **Occupational Safety and Health Administration.** 1984. *Respiratory Protection.* OSHA 3079. U.S. Government Printing Office, Washington, D.C.
43. **Penland, W. Z., and A. S. Levine.** 1980. Protection from flame and heat. *Natl. Safety News* **October:**48–49.
44. **Peraldi, D. M.** 1971. Role of work clothes in the extension of burns. *Arch. Mal. Prof.* **32:**407–410.
45. **Rake, B. W.** 1978. Influence of cross drafts on the performance of a biological safety cabinet. *Appl. Environ. Microbiol.* **36:**278–283.
46. **Rengstorff, R. H., and C. J. Black.** 1974. Eye protection from contact lenses. *J. Am. Optom. Assoc.* **45:**270–276.
47. **Richardson, J. M., L. K. Redford, H. Morton, R. James, and J. Porterfield.** 1988. Protective garb. *N. Engl. J. Med.* **318:**1333.
48. **Rosenstock, R.** 1988. Contact lenses: are they safe in industry? *Prof. Safety* **January:**18–21.
49. **Rowe, R. D.** 1979. Caution to contact lens users. *Chem. Eng. News* **57:**84. (Letter.)
50. **Sansone, E. B., and Y. B. Tewari.** 1978. The permeability of laboratory gloves to selected solvents. *Am. Ind. Hyg. Assoc. J.* **39:**169–174.
51. **Sheppard, C. C., C. W. May, and N. H. Topping.** 1945. A protective cabinet for infectious disease laboratories. *J. Lab. Clin. Med.* **30:**712–716.
52. **Stainbrook, B. W., and R. S. Runkle.** 1986. Personal protective equipment, p. 164–172. *In* B. M. Miller, D. H. M. Groschel, J. H. Richardson, D. Vesley, J. R. Songer, R. D. Housewright, and W. E. Barkley (ed.), *Laboratory safety; principles and practices.* American Society for Microbiology, Washington, D.C.
53. **Stuart, D. G., J. J. Grenier, R. A. Rumery, and J. M. Eagleson.** 1982. Survey, performance and use of biological safety cabinets. *Am. Ind. Hyg. J.* **43:**265–270.
54. **Thomas, S. M. G.** 1970. The use of protective gloves. *Occup. Health* **22:**281–284.
55. **U.S. Department of Health and Human Services.** 1993. *Biosafety in microbiological and biomedical laboratories.* HHS Publication No. (CDC) 93-8395. U.S. Government Printing Office, Washington, D.C.

56. **U.S. Department of Health, Education and Welfare, Office of Research Safety, National Cancer Institute, and the Special Committee of Safety and Health Experts.** 1978. *Laboratory safety monograph: a supplement to the NIH guidelines for recombinant DNA research.* National Institutes of Health, Bethesda, Md.
57. **U.S. Department of Health, Education and Welfare, Public Health Service, Center for Disease Control, and National Institute for Occupational Safety and Health.** 1973. *The industrial environment—its evaluation and control.* U.S. Government Printing Office, Washington, D.C.
58. **Van den Ende, M.** 1943. An apparatus for the safe inoculation of animals with dangerous pathogens. *J. Hyg.* **43:**189–194.
59. **Vogl, W. F., and M. A. Chatigny.** 1973. A simplified nomograph system for estimation of risk in the microbiological laboratory. In *Reinraumtechnik I: Berichte des internationalen Symposiums Reinraumtechnik gehalten Zurich,* Schweiz 18–30 Oct. 1972.
60. **Wedum, A. G.** 1953. Bacteriological safety. *Am. J. Public Health* **43:**1428–1437.
61. **Wedum, A. G.** 1975. History of microbiological safety: personnel and otherwise, abstr. 18th Annu. Meeting Biol. Safety Conf., Lexington, Ky.
62. **Zbitnew, A., K. Greer, J. Heise-Qualtiere, and J. Conly.** 1989. Vinyl versus latex gloves as barriers to transmission of viruses in the health care setting. *J. Acquired Immune Defic. Syndr.* **2:**201–204.

Design of Biomedical Laboratory Facilities

JONATHAN T. CRANE AND JONATHAN Y. RICHMOND

12

INTRODUCTION

The design of biomedical research laboratories is an exercise in making choices often between competing ideas and needs. This chapter is not intended to be a manual on the design of a laboratory without competent professional assistance. Rather, it is intended to provide information on choices appropriate for the needs of the project. If the architect and engineers make decisions without local input and informed consent, it is unlikely that the completed laboratory will be satisfactory. However, if the potential users become an active, integral part of the design team, the facility will likely meet current needs and future requirements.

This chapter deals with basic biomedical and clinical laboratories with major emphasis on biosafety level (BSL) 2 and on containment laboratories with main emphasis on BSL 3. These comprise the majority of biomedical laboratories.

APPROACH AND PROCESS

Laboratories are specialized facilities in which clinical, research, and development work, even with hazardous materials, can be performed safely. An assessment of the hazards expected to be present in each laboratory is a necessary part of the design process. The assessment of the risk of working with the hazardous material must come from the user of the laboratory. Engineering out the risk of such work is a main component of the thinking that goes into the design of the laboratory.

Common Sense

Most successful laboratory designs are based on simple common-sense solutions to technological challenges. The biggest challenge to the laboratory design team is to keep the design simple, not to overdesign the laboratory and make it too complex for the systems to work (6).

The team must keep from being caught up in a glorification of engineering and architectural technology. It is the nature of architects and engineers to seek new and inventive solutions and to use the latest, most complex technology available. Everyone wants to say that they designed the "state-of-the-art laboratory." Improvements in laboratory design have occurred but have generally evolved from recognition of basic needs. Most laboratorians do not care about the latest design; they just want it to work. If a low-technology solution provides equivalent performance, it should be seriously considered before considering a high-technology one.

A consensus on the level of performance of systems must be reached before designing the facility. Rarely are systems designed for 100% performance except in life-critical situations. For example, standard air-conditioning design criteria provide comfort for 80% of the population. If you expect a system that will make 100% of the people comfortable, your designer must know, and you must be prepared to bear the (possibly extraordinary) additional costs of this level of design. Even taking it to the 100% level does not ensure that everyone will be satisfied all the time. The low-technology solution is a sweater for the person who likes it hot and shirt sleeves for the person who likes it cold. There are some laboratories in which Tyvek suits or air-supplied suits are routinely worn. These laboratories need more cool air for user comfort than a routine laboratory.

Level of complexity must be considered. Complex systems may offer energy savings, better control of conditions, and more responsive systems. However, complex systems demand higher maintenance to retain their reliability, have higher initial design and construction costs, and take longer to construct and commission. A mistake often made in laboratory projects is to design complex systems, then back off on control and monitoring capabilities when costs come in. Complex systems with inaccurate controls rarely work well. In general, as the need for system reliability increases, complexity should decrease.

The Building as an Organism

A building could be looked at as an organism, constantly changing and evolving as the people and projects within it change. It should be structured to allow for adaptation. User needs should be evaluated on an ongoing basis, with building modifications made as required. The design of the facility should anticipate the ease of future modifications. Short-term needs versus long-term goals should be balanced. Unless unusual circumstances dictate, a laboratory should not be designed to meet only specific short-term needs; it should be generic enough to meet long-term needs as well.

Reliability can be achieved in many ways. Systems can be made simple for easy maintenance and operations. Duplicate or redundant systems can be provided to ensure continuance of critical services and to replace components in failure as required for BSL 4. Systems can be overlapped as primary and secondary systems to provide a cumulative approach to reliability. Also, critical systems should be designed to fail in a position that minimizes the threat to life or property.

It is advisable to plan and budget some "forgiveness" into your project, as there is a gap between theoretical design and installed construction. Architecturally, this may mean leaving a little latitude in the space around equipment. Space allotment is especially important for biological safety cabinets (BSC), which have outside dimensions larger than the work surface dimensions by which they are described. It may also mean making rooms a little more generous than necessary and not planning to put 8 ft of base cabinets in an 8-ft-wide room. (With acceptable construction tolerances, the room might end up 7 ft 11¾ in.) Mechanically, this might mean installing an exhaust fan that can be adjusted to move more air than the design calls for. This can allow for changes during the construction process in duct configuration due to space conditions, higher filter loadings, and changes in equipment operating specifications that vary from manufacturer to manufacturer. This will also give you more flexibility in fine-tuning the air balancing of the facility. Plan a facility that will allow the easy resolution of problems that will occur; allow for additional equipment requiring increased supply or exhaust air.

Function, both current and future, is the basis for laboratory design. An operational systems engineering approach should be taken in the design of your facility (22). Determine the movement of people and materials in and out of the facility and from room to room. Detail the steps of how supplies move into the facility, how they are used, and how they need to be handled as waste. What are storage and handling requirements along the way? What are alternative options? Analyze how the facility will be used from date of occupancy on into the future. As Winston Churchill once said, "We shape our buildings and then they shape us." The facility should not dictate the method of operation.

Visualization

It is difficult to visualize the look and feel of the laboratory while reviewing the architectural plans. As the facility planning begins, some of the following techniques can be used to allow the entire design team to visualize the final project and to assess the needs of your laboratory.

Do a walk-through of the facility in use, even if the current facility is totally inadequate, as a source of information for the design team.

Discuss the advantages and disadvantages of the layout, size, casework, lighting, and noise.

Discuss the current operation and the proposed operation.

Identify equipment to be moved to the new laboratory.

Identify the way hazards are currently handled and decide if they should be handled differently in the new facility.

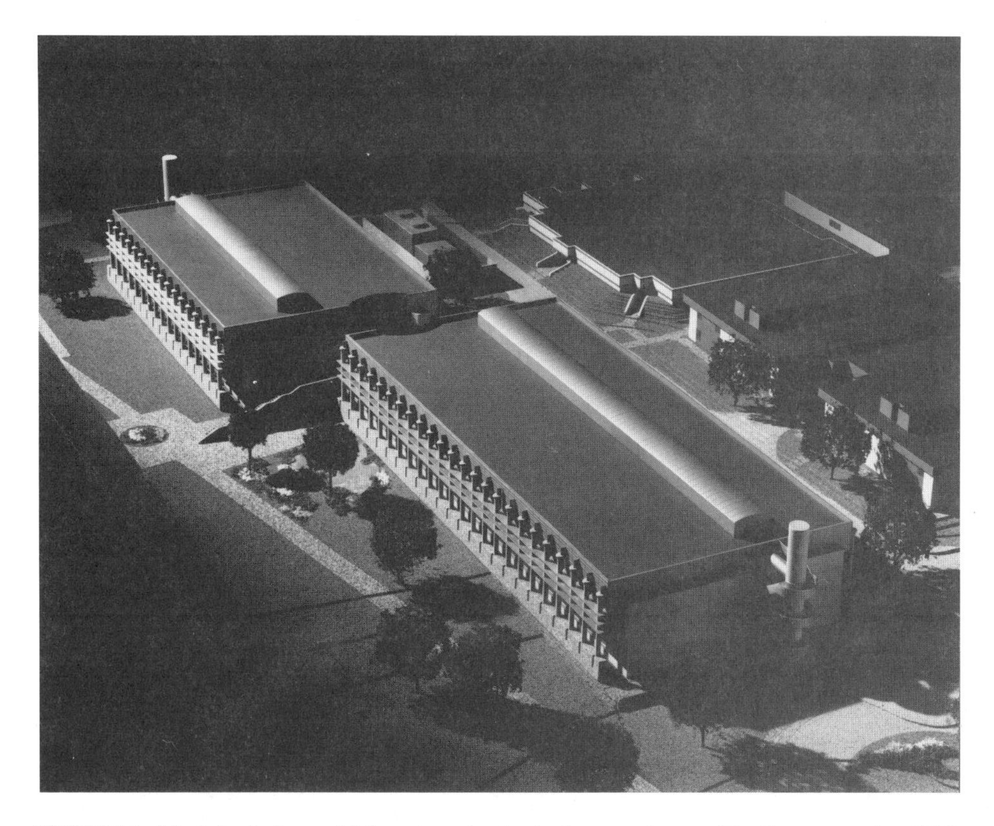

FIGURE 1 Model of planned laboratory for toxicology and parasitic diseases at the CDC. (Photo by Jonathan Hillyer, courtesy of Lord, Aeck & Sargent, Inc., Atlanta, Georgia.)

Key members of the project team should tour comparable facilities to learn quickly about the state of the art and gain a common basis for understanding the vocabulary used. This also can give the design team an understanding of some of the options in laboratory design.

Mockups of proposed or alternative designs can be invaluable in fine-tuning the design of the facility. These can be simple: Use masking tape to outline the size and shape of the room, with cardboard cutouts representing casework and equipment. This allows all parties to see the layout of the room, ensure that equipment will fit, and determine if room size will be comfortable. More complex mockups using actual construction to simulate the final project can be expensive, but it is the best way to make sure the design is correct to the smallest detail. The design team can simulate various operations in the mockup to identify any weakness in the design. This also can be helpful when heating, ventilating, and air-conditioning (HVAC) systems need to be verified. Smoke testing can be performed to ensure proper airflow. This approach can be very cost-effective when several identical rooms such as animal rooms and laboratory or containment modules are planned.

Scaled models can be built to show buildings or detailed components (Fig. 1). Computer modeling (CAD) offers a powerful new tool. Computer programs allow rooms with casework and equipment to be viewed from any vantage point in a very realistic way. This type of modeling allows easy exploration of a variety of options.

Questioning

The entire design process is a process of searching for answers and questioning those answers until the team is comfortable. Questions include: What are the customer's real needs? What has the customer not identified that is necessary? Will the design work? How long will it work? Where has it worked before? Are current methods of operating the best? What is the answer for *this* project?

PREPLANNING: WHAT YOU MUST KNOW BEFORE YOU BEGIN

Program

Space requirements are determined before beginning a laboratory design. The following laboratory requirements are the focus of this chapter.

Basic Research Laboratories

Basic laboratories should be generic and straightforward to allow occupancy by many programs with minimal changes while still usable. The biosafety facility design requirements of basic laboratories at BSL 2 as described in Appendix I (17) are

The laboratory is designed to be easily cleaned.

Bench tops are impervious to water and resistant to acids, alkalis, organic solvents, and moderate heat.

Laboratory furniture is sturdy, and spaces between benches, cabinets, and equipment are accessible for cleaning.

Each laboratory contains a sink for handwashing.

If the laboratory has windows that open, they are fitted with fly screens.

An autoclave for decontaminating infectious laboratory wastes is available in the facility.

BSCs can be exhausted through HEPA filters back into the laboratories.

Although not a specific requirement of the guidelines for biosafety, both the National Institutes of Health (NIH) and the Centers for Disease Control and Prevention (CDC) like to recommend directional, inward airflow at BSL 2 to provide control for fumes and vapors as well as bioaerosols.

Containment Research Laboratories

Although containment laboratories should be designed to handle the specific needs of the program to be housed, they should also allow for as many different programs as possible to occupy them with minimal changes during their functional life. The facility design requirements of containment laboratories at BSL 3 as described in Appendix I and the recombinant DNA guidelines (16) are

The laboratory is separated from areas that are open to unrestricted traffic flow within the building.

Passage through two sets of doors is the basic requirement for entry into the laboratory from access corridors or other contiguous areas.

Physical separation of the containment laboratory from access corridors or other laboratories or activities may also be provided by a double-doored clothes change room, airlock, or other access facility that requires passage through two sets of doors before entering the laboratory.

The interior surfaces of walls, floors, and ceilings are water-resistant so that they can be easily cleaned. Penetrations in these surfaces are sealed or capable of being sealed to facilitate decontaminating the area.

Bench tops are impervious to water and resistant to acids, alkalis, organic solvents, and moderate heat.

Laboratory furniture is sturdy, and spaces between benches, cabinets, and equipment are accessible for cleaning.

Each laboratory contains a sink for handwashing. The sink is foot, elbow, or automatically operated and is located near the laboratory exit door.

Windows in the laboratory are closed and sealed.

An autoclave for decontaminating infectious laboratory wastes is available, preferably within the laboratory.

A ducted exhaust air ventilation system is provided. This system creates directional airflow that draws air into the laboratory through the entry area. The exhaust air is not recirculated to any other areas of the building, is discharged to the outside, and is dispersed away from occupied areas and air intakes. Personnel must verify that the direction of the airflow is proper.

The exhaust air from the laboratory room can be discharged to the outside without being filtered or otherwise treated.

The HEPA-filtered exhaust air from BSCs is discharged directly to the outside or through the building exhaust system. Exhaust air from class I or class II BSCs may be recirculated within the laboratory if the cabinet is tested and certified at least every 12 months.

If the HEPA-filtered exhaust air from class I or class II BSCs is to be discharged to the outside through the building exhaust system, it is connected in such a manner that avoids any interference with the air balance of the cabinets or building exhaust system.

Clinical Laboratories

Clinical laboratories should be designed for handling the collection and processing of biological samples. Risk assessments for clinical activities related to handling human blood or other body fluids generally result in assigning BSL 2 for facility design (20). Certain laboratories that process aerosol-transmissable agents (e.g., a tuberculosis laboratory) usually require BSL 3. Clinical laboratories are assigned one or more of the following functions: hematology, immunology, clinical chemistry, urinalysis, microbiology (bacteriology, mycobacteriology, virology, mycology), anatomic pathology, cytology, and blood banking. The main requirements for clinical laboratories as outlined by *Guidelines for Construction and Equipment of Hospital and Medical Facilities* (1) are as follows:

Laboratory work counters should have space for microscopes, appropriate chemical analyzers, incubators, and centrifuges.

Work areas should include sinks with water and access to vacuum, gas, and electrical services as needed.

Refrigerated blood storage facilities for transfusions should be provided. Blood storage refrigerator should be equipped with temperature monitoring and alarm signals.

Lavatories or counter sinks equipped for handwashing should be provided.

Counter sinks may also be used for disposal of nontoxic fluids.

Storage facilities, including refrigeration, for reagents, standards, supplies, and stained specimen microscope slides should be provided.

Specimen (blood, urine, and feces) collection areas should be provided. Blood collection areas should have work counter, space for patient seating, and handwashing facilities. Urine and feces collection room should be equipped with water closet and lavatory. This facility may be located outside the laboratory suite.

Chemical safety provisions including emergency shower, eye flushing devices, and appropriate storage for flammable liquids should be made.

Facilities and equipment for terminal sterilization (autoclave or electric oven) of contaminated specimens before transport should be provided. (Terminal sterilization is not required for specimens that are incinerated on-site.)

If radioactive materials are used, facilities should be available for long-term storage and disposal of these materials. Special provisions will normally not be required for body waste products from most patients receiving low-level isotope diagnostic material. Requirements of authorities having jurisdiction should be verified.

Lounge, locker, and toilet facilities should be conveniently located for male and female laboratory staff. These may be outside the laboratory area and shared with other departments.

Also, automated processors and some procedures such as tissue processing can release solvents into the work area. Local (capture) ventilation systems must be provided to reduce concentrations of the solvents in the air to acceptable levels because it is not always practical to place this equipment or perform these procedures in a fume hood (19, 21).

Small Clinical Laboratory

Many clinical laboratories are in the physician's office setting without the benefit of appropriate laboratory air supply and exhaust systems. Most physicians' office buildings have heating, ventilating, and air-conditioning systems that recirculate air in the building and have minimum capabilities for supply and exhaust air. Although these laboratories generally work at BSL 2, they almost exclusively handle human tissues and body fluids. All the recommendations for clinical laboratory design should be followed. Particular care should be taken in use and placement of aerosol- or fume-generating equipment and procedures to ensure that aerosols are not introduced into the work area or into air recirculated in the building. Appropriate containment should be provided along with required specimen and biomedical waste storage, including sharps containers.

Budget or Cost Constraints

Each project should be approached with a realistic budget; constraints on design resulting from the budget need to be reconciled early in the process. The more complex and specific the laboratory is, the more it will cost. Specialized systems, such as emergency power generators or central purified water systems, will have a major impact on the budget. Browner (3) suggested reasonable laboratory costs for new construction and renovation (Table 1).

The percentage of building construction cost for the main components in a new laboratory building is shown in Fig. 2. Note the high percentage of cost that typically is allocated for mechanical and electrical systems.

Factors affecting the range of costs include complexity, size, and geographic location. Aesthetics of a laboratory have minimal impact on the overall costs of a laboratory facility. A high-quality design generally represents less than 5% of the construction cost of a laboratory facility. Many psychological benefits from a high-quality design are translated into actual benefits in the operation of the facility.

Schedule or Time Constraints

Laboratories take more time to plan, design, and construct than most other facilities. The design and

TABLE 1 Average costs ($) per gross square foot (GSF) for new research facilities or per net square foot (NSF) for laboratory renovations

	Low GSF cost	High GSF cost	Low NSF cost	High NSF cost
New facility				
Base building	93	206		
Equipped building	111	247		
Building and site development	121	287		
Project cost	151	347		
Renovation				
Base laboratory			65	221
Equipped laboratory			77	251
Project cost			109	338

construction process needs to fit the required schedule, but care should be taken not to sacrifice major needs of a long-term facility to meet short-range schedule requirements. Approximate time frames for design and construction of various types of projects are

Minor renovations: 3 to 6 months
Major renovations: 1 to 2 years
Minor new projects: 1 to 2 years
Major new projects: 3 to 5 years

Figure 3 represents a typical project schedule.

Operational Issues

Operational goals and constraints should be identified for both the long and short term. Operational protocol should be developed for normal laboratory support services, janitorial, maintenance, and waste handling operations. Protocol for emergency situations that may be anticipated should also be developed.

Maintenance

The maintenance capabilities of the organization must be assessed to match the design of the facility

FIGURE 2 Percentage of building construction cost per new laboratory component.

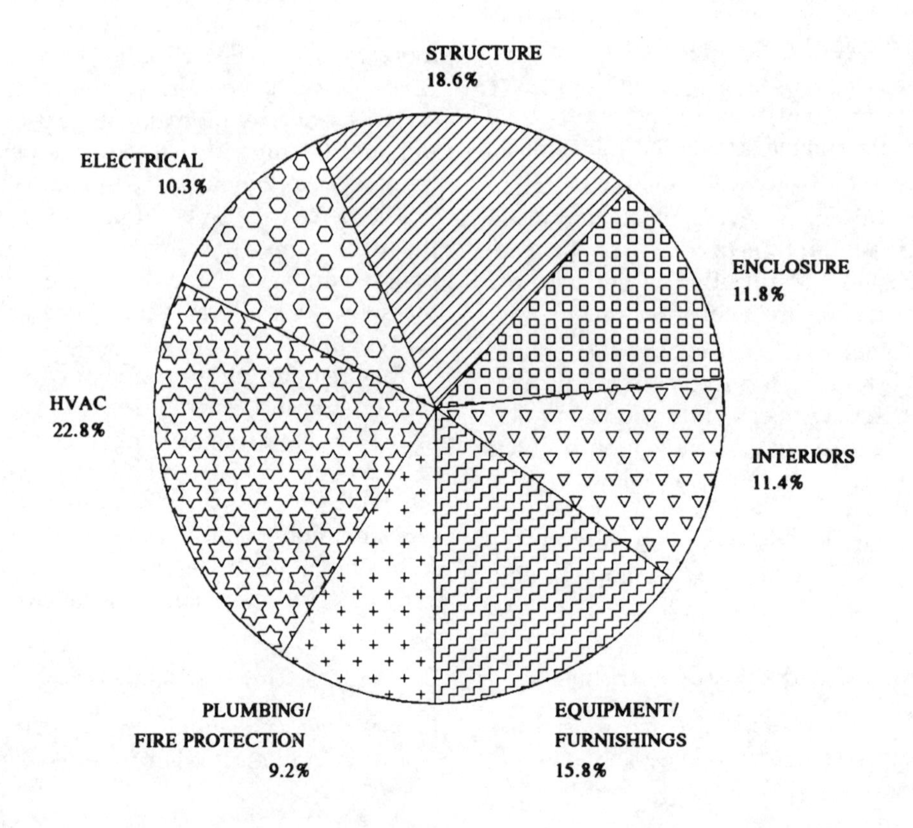

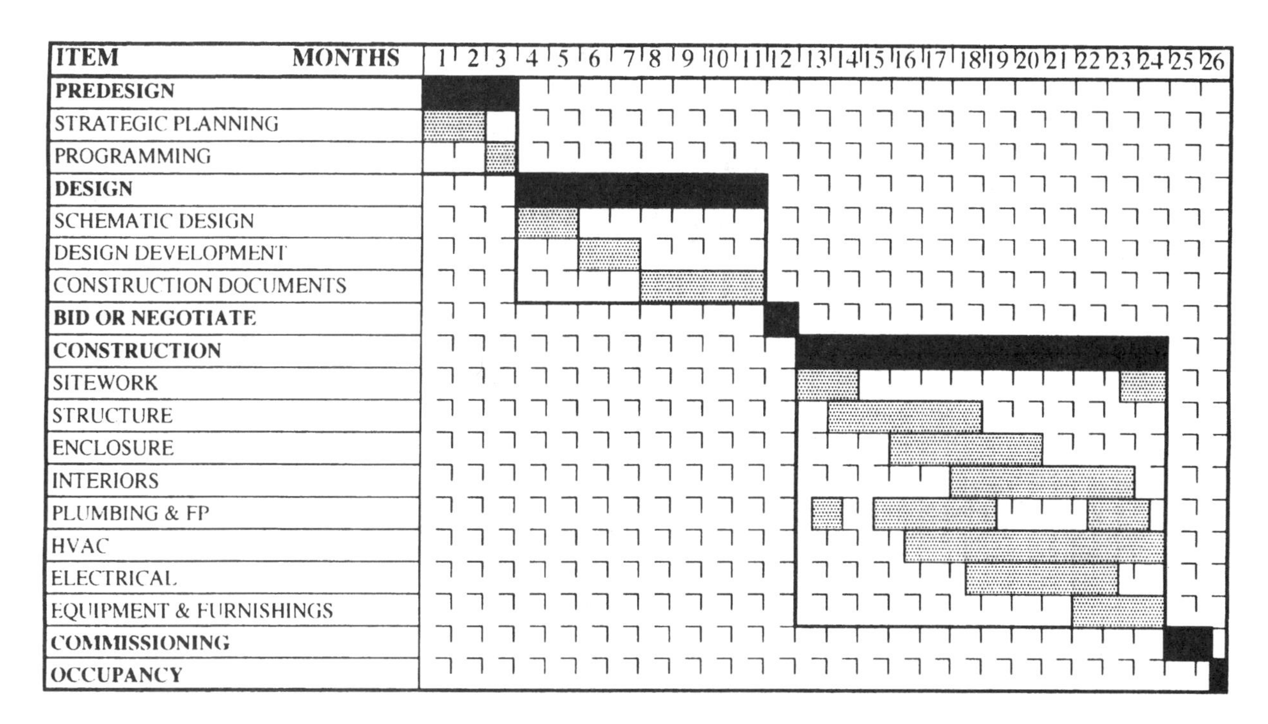

FIGURE 3 Typical project schedule.

with the support capabilities. One should not design and install systems that cannot be maintained for budgetary or staffing reasons. Organizational experience with certain products, availability of supplies and parts, staff training, and maintenance budgeting should be evaluated before the design phase.

Energy Savings

With high electrical power use, bright lighting, and 100% exhaust systems operating 24 h/day, laboratories are expensive to operate. Selection of highly sophisticated, energy-efficient equipment is one means of long-term cost reduction. There are some laboratories in which the systems require such major modifications after the initial installation that energy and cost savings are lost. Other laboratories do not maintain the pressure relationships and safe ventilation patterns that formed the original reason for trying to lower energy use. Still others require more specialized maintenance personnel than are normally found in a laboratory facility.

All efforts should be made to provide energy savings that come with good basic common-sense design before deciding to develop costly and complex systems for energy reduction. Room volume should be minimized when room volume is the driving force in energy consumption. Appropriate insulation and solar barriers should be provided to reduce heat loss or gain in conditioned spaces. Natural light should be used where possible to augment artificial lighting.

Preplanning can provide a solid foundation for the development of the design of the facility, reducing the potential for delays and cost increases that occur with changes in scope or approach to the project. At the end of preplanning, the entire design team should understand the goals and limitations of the project.

DESIGN: DEVELOPING SAFE OPERATIONAL SYSTEMS

The design team develops a space validation package for each room in the facility composed of the following:

- Review of applicable codes and regulations
- Development and analysis of hazardous material use data
- Development and analysis of program data
- Code classification
- Design recommendations and review
- Development of room data sheets for recording requirements including the casework and fixed equipment data, equipment listing and data, and functional layout and diagram
- Proximity relationship diagram to other rooms

This provides a comprehensive analysis and record of requirements that will produce the following benefits:

- Rooms will meet codes and guidelines.

Equipment will fit and have the proper services.

The requirements and layout of each room will be understood.

Hazardous materials will be identified and appropriate safeguards planned.

Quality control reviews will verify that all requirements are met.

Laboratory Safety Concerns

Biomedical laboratories should be designed to address safety concerns inherent or anticipated in such facilities. The potential for spread of contamination from the laboratories to other areas throughout the building needs to be minimized. The relationship of air handling within the building is critical. (There should be no air movement from laboratories, containment, or animal care facilities to other spaces.)

Engineering controls can minimize hazards within the laboratory. Correct airflow and primary containment equipment allow investigators to perform their operations safely and may reduce contamination. For example, BSCs and chemical fume hoods should be located away from doorways and air supply vents. Handwashing sinks should be installed near exits. Space should be provided for storage of supplies, collection of wastes, and break rooms.

Methods for handling exhaust air, waste, and hazardous by-products should be selected to properly disperse or dispose of these wastes into the environment and to provide adequate protection for the worker and the community. Systems should be easy to maintain and safe for support personnel.

Primary and Secondary Barriers

Primary barriers are specialized items designed for capture or containment of chemical or biological agents. BSCs, chemical fume hoods, and animal cage dump stations are examples of larger primary barriers. Trunnion centrifuge cups, aerosol-containing blenders and high-speed mixers, and related devices are examples of smaller primary barriers.

Safe laboratories use an appropriate layered approach, combining primary and secondary barriers to provide for personnel and environmental occupational safety and health.

Secondary barriers are facility-related design features that separate laboratory from nonlaboratory areas or from the outside. Many of these barriers are physical in nature (e.g., walls, doors), and others result from mechanical devices such as air handling systems.

Further information on primary barriers can be found in Chapter 11.

Architecture and Engineering

Architectural and engineering features in a laboratory provide a secondary barrier system for protection that complements the primary barrier. These systems are designed to move the hazard away from the laboratory workers, restrict the hazard to a specific area, treat the hazard if necessary, and allow for easy cleanup in a confined area.

Protocol or Practices

The primary tool for laboratory safety involves the practices and procedures used by the laboratory worker. The effectiveness of the laboratory design will be greatly increased if these protocols for use, cleanup, and maintenance of the facility are available to the design team during the early parts of the project. Care should be taken to use a realistic approach to the development of protocols that will affect the laboratory design. It is as much a mistake to develop a laboratory to too stringent a protocol, as it is to develop it to too lax a protocol. Design teams also need to consider the practices and procedures used by the laboratorians to ensure that necessary support facilities are provided. Conversely, a realistic approach needs to be taken relative to final practices so that engineering designs are not too complex. A protocol is only effective if it is followed. If the building makes it difficult to follow protocol or is designed around unnecessary procedures, it is unlikely that the protocol will be followed. Further information on biosafety practices is found in Chapter 13, as well as Chapters 3 and 4.

Hazards

The main hazards that occur in biomedical laboratories that the design team must address are

Chemical: flammables, carcinogens, toxins, compressed gases

Biological: known etiologic agents, materials that may contain etiologic agents

Radioactive: radionuclides, equipment that produces ionizing radiation

Physical: laser, magnetic, high voltage, UV, high noise

How these hazards are to be manipulated is vitally important to laboratory design. Therefore, the risk of each hazard must be individually assessed. Appropriate measures must be taken for proper storage, handling, and disposal of all chemicals and biological and radioactive material. The threat to life and property from these hazards has caused the development of codes, regulations, and guidelines governing facilities and practices to minimize potential problems. Physical barriers, interlocked room access devices, and noise abatement strategies must be addressed.

Regulations, Codes, and Guidelines

Regulations, codes, and guidelines in the design of laboratories fall into two main areas: building and life safety codes that are adopted and administered on the local level; and laboratory safety codes that may be local, state, federal, or private association in nature. Determine the edition of code adopted by the authority having jurisdiction governing your project, as requirements may differ greatly between various editions. In some instances, local jurisdictions have adopted even more restrictive codes than state or federal regulations.

Building, Fire, and Life Safety

Local zoning and building codes often govern size, occupancy, construction, fire protection, and other requirements such as handicap accessibility or environmental assessment for the facility. These are often administrated by the local building and fire departments. The main life safety code in use nationally is *NFPA 101 Life Safety Code.* This National Fire Protection Association (NFPA) code provides facility requirements for protecting life safety and property in the event of a fire. It applies to all facility types and includes the requirements for safe design and operation.

Laboratory Safety Codes, Regulations, and Guidelines

Because of increased scrutiny from the federal government, research funding agencies, facility accreditation organizations, and animal rights groups, the design of your facility needs to conform to a variety of regulations and guidelines. The following are the main national regulations and guidelines that generally govern the design of laboratory facilities:

CDC/NIH Biosafety in Microbiological and Biomedical Laboratories: This guideline from the CDC and the NIH recommends minimum facility and operational requirements for laboratories working with biological hazards. It includes separate animal biosafety facility criteria (17).

NFPA 45 Fire Protection for Laboratories Using Chemicals: This NFPA code provides facility requirements for laboratories using chemicals for research purposes, including university facilities. It includes the requirements for safe design and operation (9).

NFPA 99 Fire Protection for Health-Related Laboratories: This NFPA code provides facility requirements for laboratories using chemicals in health institutions including clinical laboratories. It includes the requirements for safe design and operation (8).

OSHA 29 CFR 1910 Safety Standards: This OSHA regulation provides minimum laboratory facility requirements for personnel protection (18–21).

OSHA 29 CFR 1990 Safety Standards for Carcinogens: This Occupational Safety and Health Administration regulation provides minimum facility requirements to limit occupational exposure to carcinogens (21).

USDA 9CFR Parts 1,2,3 Animal Welfare: This regulation of the Department of Agriculture defines the requirements for licensing, registration, identification, records, facilities, health, and husbandry for all animals covered by the Animal Welfare Act. Specific facility and operating standards are defined for various species of animals (13).

DHHS 85-33 Guide for Care and Use of Laboratory Animals: This guideline developed by the Department of Health and Human Services is the basic guideline to be used for laboratory animal research carried out through grants from the Department of Health and Human Services (15).

NIH Guidelines for Laboratory Use of Chemical Carcinogens: This NIH guideline provides recommendations for dealing with chemical carcinogens for programs with NIH funding including specific facility requirements.

NCI Safety Standards for Research Involving Oncogenic Viruses: This National Cancer Institute (NCI) standard provides minimum facility requirements for NIH/NCI funded grants involving oncogenic viruses (7).

NCI Design Criteria for Viral Oncology Research Laboratories: This NCI guideline provides recommendations for facilities doing research into viral oncology (6).

51 FR 16598 NIH Guidelines for Research Involving Recombinant DNA: This NIH guideline provides recommendations on facility design for laboratories involved in recombinant DNA research (16).

Many other codes, guidelines, and design criteria may apply to a facility of this type.

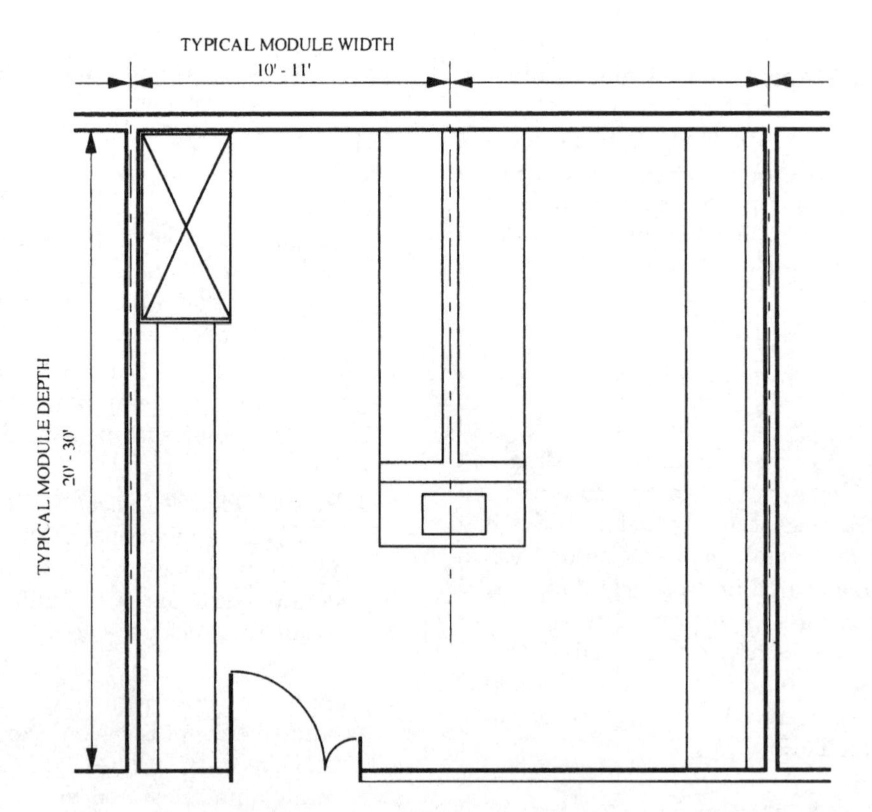

FIGURE 4 Typical two-module laboratory.

Components

Offices or Conference Administration

An administrative area, physically separated from all hazardous aspects of laboratory work, should be planned near the main entry to each building or floor. This area provides administrative support for the laboratories and serves to control access to the laboratory area.

Scientific staff offices should be positioned as close as possible to the occupant's main nonoffice work space. Research faculty offices typically range from 100 to 160 ft^2; a detailed layout should be developed to ensure adequate space is provided for bookshelves, computers, desk, filing, and guest chairs. Each laboratory or office should ideally have a window to permit the occupant to look either into the corridor, the laboratory, or the outside.

Corridors

A fire-rated corridor, providing two means of exit from any point, should service the laboratory block. Safety showers and spill control centers should be located in this corridor. If corridor widths are constructed to meet minimum fire code and equipment access requirements, it is impossible for the corridors to be used as auxiliary laboratory work spaces, storage areas, or offices and break areas. If the corridors are wider than minimum requirements, these problems will inevitably occur.

Basic Research Laboratories

A standard laboratory module should be developed for flexibility (11). Laboratory space can be analyzed by the efficiency of workstations, generally measured by the footage of usable bench and equipment space available. Typical laboratory modules vary in width from 9 ft 6 in. to 11 ft 6 in. (Fig. 4). Although the most frequently used widths are 10 to 11 ft, the increase from 10 to 11 ft can add 5% to the cost of the facility (Table 2). Module depths generally range from 20 to 30 ft. A simple masking tape mockup can provide insight for this critical decision.

Access and egress. To allow clearance for access of most equipment expected in a laboratory, laboratory doors should be a minimum of 3 ft 6 in. in width for single doors, or be a pair of doors 4 ft 0 in. in width using an active leaf of 3 ft 0 in. in width and an inactive leaf of 1 ft 0 in. in width. All active leafs should have closers. The preferred door is the 4 ft 0 in. pair, which allows easy personnel access while still allowing equipment access. Most doors should open out of the laboratory, but doors opening into an exit corridor should be in a recessed pocket.

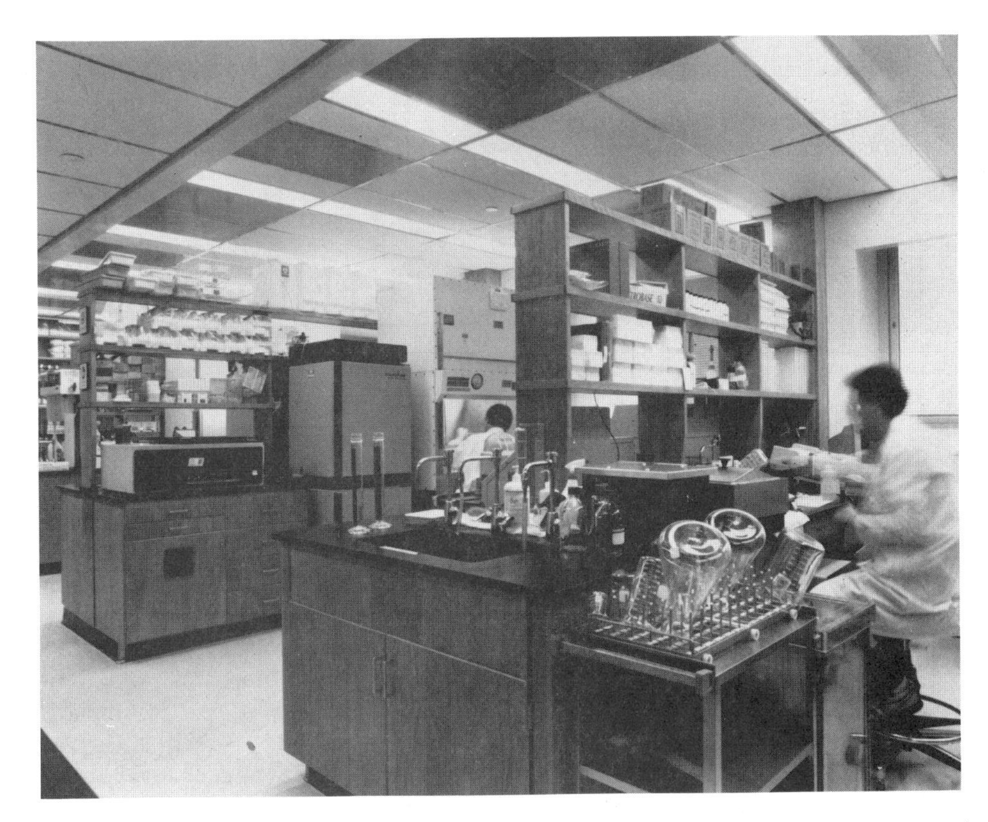

FIGURE 5 BSL 2 laboratory at the Aaron Diamond AIDS Research Center for the City of New York. (Photo by Elliott Kaufman, courtesy of Lord, Aeck & Sargent, Inc., Atlanta, Georgia.)

Doors and windows in fire-rated walls should also be fire rated as required by code.

Ventilation and finishes. Air should be supplied to the laboratory near any dedicated desk space, and the exhaust should be taken at the other end of the laboratory near potential hazards. Mechanical components requiring service should be located where possible in ceilings outside the laboratory areas, to reduce noise and service access in the laboratories. Flooring in basic laboratories can be sheet vinyl or vinyl composition tile with standard rubber or vinyl base. Walls can be enamel paint. Ceilings in these labs should be lay-in acoustic tile to reduce noise and provide access to system components above the ceiling (Fig. 5).

Basic Clinical Laboratories

Clinical laboratories should be set up to handle the volume of specimens anticipated in a manner that allows the specimens to flow from collection and receiving to processing, analysis, reporting, and storage in a logical manner. Much clinical laboratory work is now automated, and care must be taken in the design to provide space for the equipment to operate and be maintained. Record keeping positions should be located at appropriate positions along the work flow. Although some components of clinical laboratories may be modular in nature, each individual area should be designed with its unique requirements in mind. Much has been written about clinical laboratories, and we refer you to *Medical Laboratory Planning and Design* (5), NFPA 45 (9), and other literature to discuss laboratory sizing and detailed design; however, Fig. 6 represents the basic functions and relationships in a typical clinical laboratory.

Specialized Laboratories

Animal care areas. The main considerations and components in the design of animal facilities outlined by the *Guide for Care and Use of Laboratory Animals* (15) are

Separation of animal facilities from personnel areas

Separation of species

Isolation of individual projects (when required by protocol)

Areas to receive, quarantine, and isolate animals

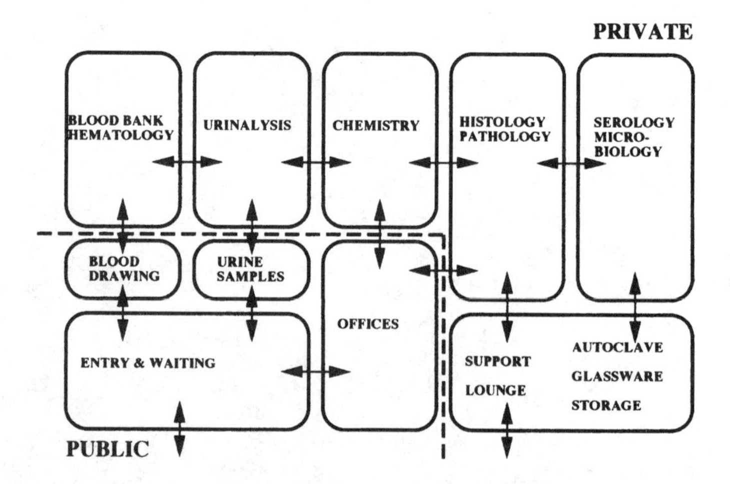

FIGURE 6 Basic functions and relationships in a typical clinical laboratory.

Areas to house animals

Specialized laboratories or individual areas contiguous with or near animal housing areas for such activities as surgery, intensive care, necropsy, radiography, preparation of special diets, experimental manipulation, treatment, and diagnostic laboratory procedures

Containment facilities or equipment, if hazardous biological, physical, or chemical agents are to be used

Receiving and storage areas for food, bedding, pharmaceuticals and biologics, and supplies

Space for the administration, supervision, and direction of the facility

Showers, sinks, lockers, and toilets for personnel

An area separate from animal rooms for eating, drinking, and applying cosmetics

An area for washing and sterilizing equipment and supplies and, depending on the volume of work, machines for washing cages, bottles, glassware, racks, and waste cans; a utility sink; an autoclave for equipment, food, and bedding; and separate areas for holding soiled equipment and materials separate from clean ones

An area for repairing cages and equipment

An area to store wastes before incineration or removal

Additional specific requirements for construction can be found in the *Guide for Care and Use of Laboratory Animals* (15).

Traffic flow patterns are generally the determining factor in the layout of animal facilities, with consideration given to controlled access, functional conveniences, environmental control, easy movement of cages, waste, and personnel. Facilities are designed with standard corridors, clean-dirty corridor systems, or barrier systems depending on the species housed, the functional needs of the facility, and the type of housing unit selected. Isolation of animals either for biohazard containment or animal protection (such as severe combined immunodeficient mice) can be handled at the cage level with isolation cages, at the rack level with airflow modules, or at the room level with filtered, pressurized airflow in cubicle rooms.

Easy access and movement of cages and racks into and out of the cage washing areas and storage space are important for a smoothly functioning facility. Storage space for cages becomes a critical factor when a variety of animals, using differing cage types, are held in the same facility. Consideration should be given to standardizing the animal care module for each major species grouping. Standardization provides uniform air requirements and consistency of airflow in modules and allows for maintenance of the room HVAC systems. The size of room should be based on rack layout, flexibility desired, and type of animal housed. Rooms that are over- or undersized are inefficient and waste energy. Material selection for rooms, flooring, and walls is important for accreditation and ongoing Department of Agriculture inspections (13).

Containment facilities. The layout requirements for containment facilities vary according to size and purpose (Fig. 7 and 8). An airlock should have adequate space if used for gowning and ungowning, along with space for storage and disposal of gowns, masks, and gloves. Handwashing facilities must be provided and shower facilities should be considered, depending on the nature of the hazard. (Showers are not a requirement of BSL 3.)

Space within the containment facility should be sufficient to handle the specific work that is

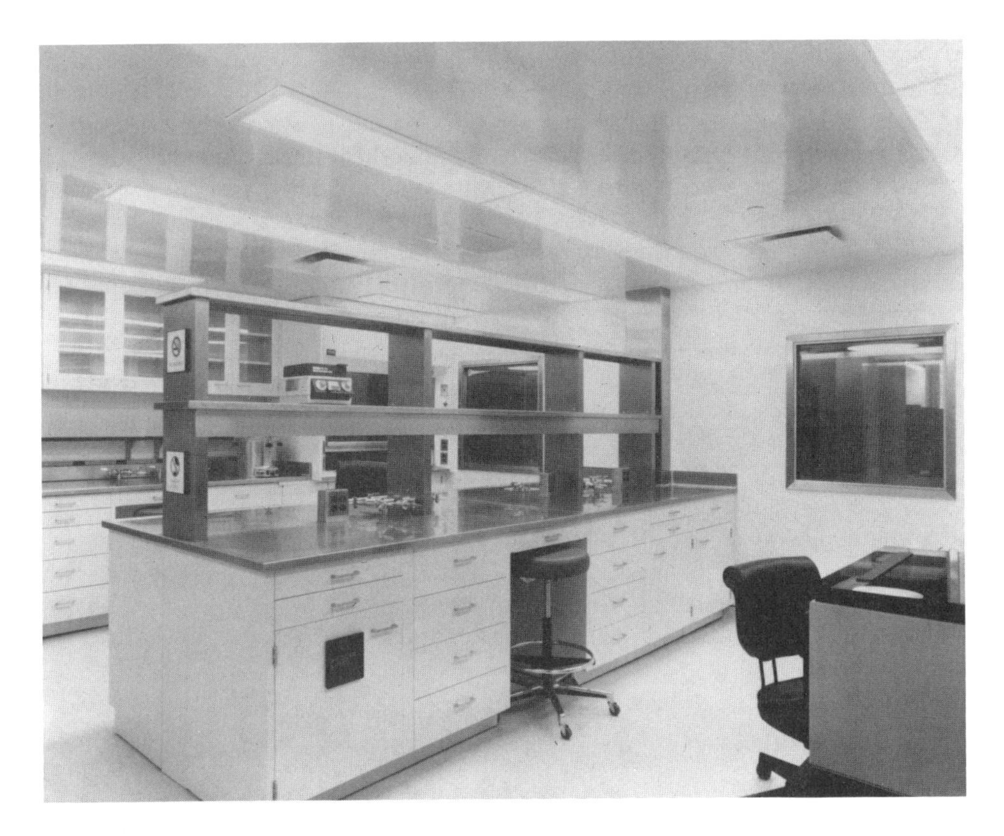

FIGURE 7 BSL 3 work area in the Aaron Diamond AIDS Research Center for the City of New York. (Photo by Elliott Kaufman, courtesy of Lord, Aeck & Sargent, Inc., Atlanta, Georgia.)

FIGURE 8 BSL 3 module in the Aaron Diamond AIDS Research Center for the City of New York. (Photo by Elliott Kaufman, courtesy of Lord, Aeck & Sargent, Inc., Atlanta, Georgia.)

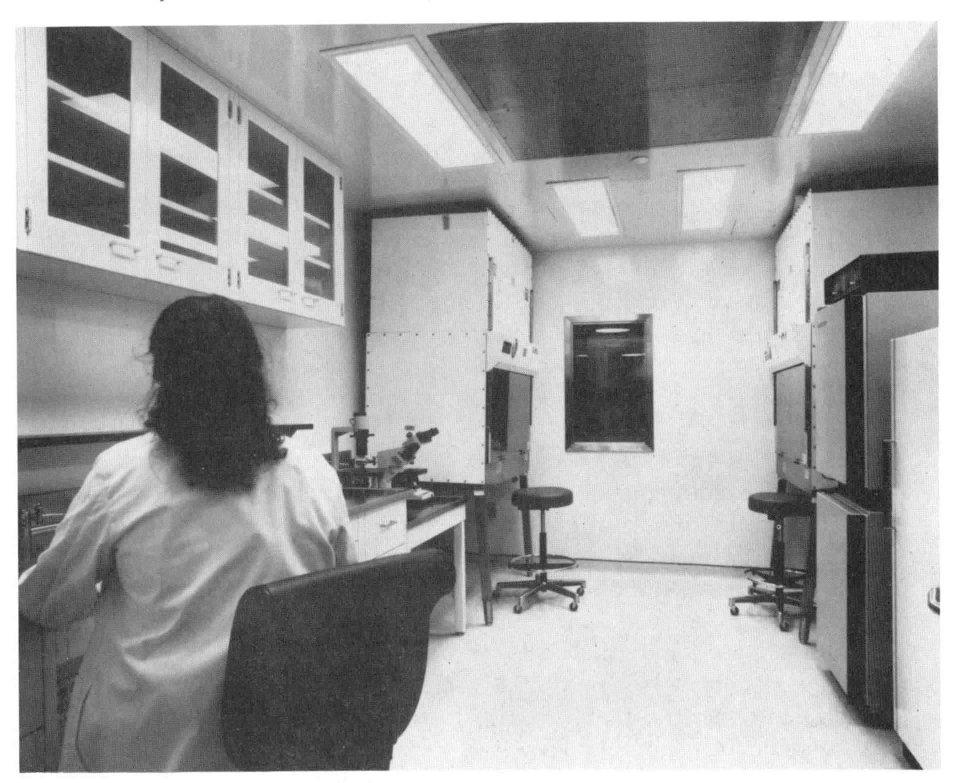

planned. Containment laboratories should have troweled epoxy or seamless sheet vinyl floors with integral coved base. Walls and ceilings should be coated with epoxy paint. Care should be taken in placing mechanical components requiring access above these ceilings to minimize service access in the containment laboratory and access panels in the ceiling. Standard lay-in acoustical tile ceilings are not acceptable in BSL 3 areas.

Other specialized laboratories such as isotope rooms for working with higher-level radioisotopes, radio-frequency enclosures to minimize the effects of electrical "noise" on sensitive equipment, tissue culture rooms, and facilities for specialized equipment such as magnetic resonance imagers, lasers, and electron microscopes are often part of biomedical facilities. These rooms must be designed to the specific needs and criteria of the programs and equipment to be housed.

Support

Most laboratories require support rooms. Large, expensive pieces of equipment (such as ultracentrifuges and scintillation counters) shared between laboratory programs require shared equipment rooms. These rooms also allow for noise-generating equipment to be housed away from the working laboratory. Separate rooms for freezers and ultra-low-temperature freezers are recommended to minimize noise in laboratories and to adequately address the heat generated by the operation of this equipment.

Environmental rooms ranging from –20°C (freezers) to 4°C (cold rooms) to 37°C (warm rooms) are found in many laboratory facilities. These rooms may be used for storage or work. Working rooms must be ventilated. Environmental rooms are complex systems that must be designed to the specific project requirements. Their costs will vary greatly according to the accuracy and degree of temperature and humidity required. Environmental rooms also require ramped access if they are not inset into the structure.

Appropriate facilities need to be provided for preparation of culture media. An assessment of media preparation and storage requirements should be made . This work can produce unpleasant odors and potentially allergenic dusts; thus, it is advisable to provide a well-ventilated exhausted room.

Space for washing and drying some glassware is required in the laboratory. Space for ice machines, dry ice storage, and for autoclaving, which produces heat and noise, should be provided in suitably ventilated support rooms outside the laboratory. Provision for ventilation above the doors of autoclaves and glassware washing and drying equipment helps remove heat, humidity, and odors from the space. This equipment also requires adequate service space to minimize maintenance costs. Particular attention should be paid to flooring and floor drains in these areas to minimize damage caused by leaks from equipment malfunction.

Photography and autoradiography. Dark rooms should be provided for both autoradiographic and photographic work. These rooms often have automatic film processors that require special lighting, drainage, and ventilation. Much imaging is being performed with computerized image analyzers and image generators, which also requires specialized services, along with space for preparation of graphics and slides.

Accessibility

Many states and the federal government have adopted facility design standards to allow access and use by disabled persons. Wheelchair turn radii, required clearances at doors, handicap accessible workstations, and removal of obstructions are examples of types of requirements in these codes. These items have a large impact on size and layout of laboratory areas, providing adequate aisle width between benches, access to eyewash stations, and emergency showers. Retrofits are expensive; however, routine provision is not required unless handicapped personnel are to be accommodated in a laboratory.

Systems

Heating, Ventilation, and Air Conditioning

Historically, laboratories have had fairly simple static HVAC systems. A certain amount of air was supplied to the room and a proportional amount was exhausted (constant volume systems) (Fig. 9). The temperature of the room was adjusted by varying the temperature of the air coming into the room (reheat systems). Pressure relationships were maintained by the proportion of the supply air exhausted. Once balanced, these systems, with few moving parts, were simple to maintain. Most problems came from dynamic changes in the system such as static pressure change caused by supply, exhaust, or ducted biosafety cabinet filter loading because these simple systems could not automatically respond to such changes in conditions.

The past decade has seen a rise in the use of variable volume systems (Fig. 10) that respond to dynamic changes in the laboratory environment.

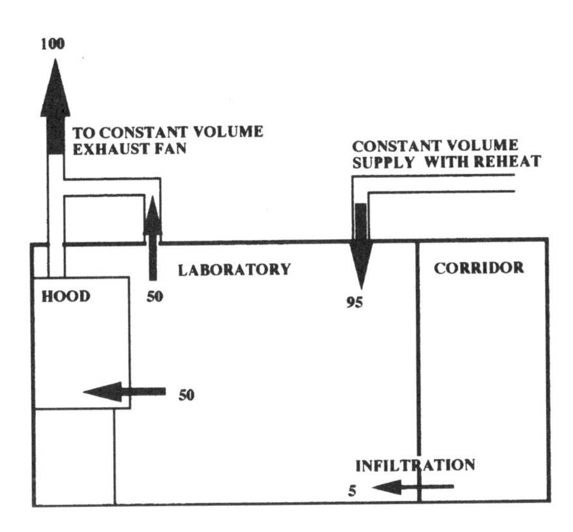

FIGURE 9 Section through laboratory with constant volume system.

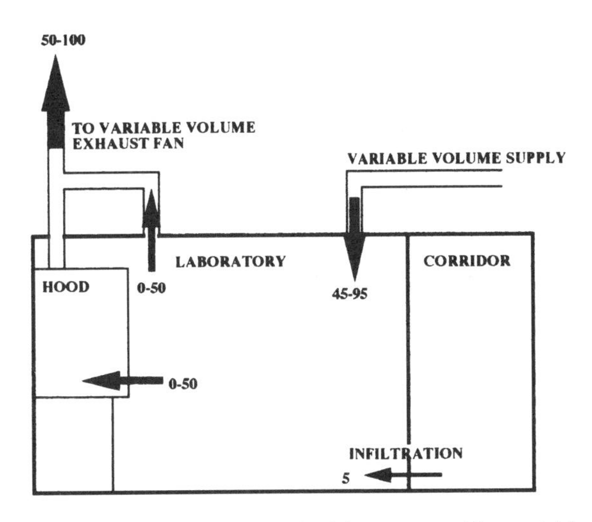

FIGURE 10 Section through laboratory with variable volume system.

These systems control temperatures in laboratories by supplying varying amounts of air to match the heat load produced in the space. Instead of a fixed damper, these systems use modulating dampers, or air valves, to control the amount of air supplied. Room pressure relationships are maintained by modulating dampers or air valves in the exhaust system, varying the amount of air exhausted in proportion to the air being supplied. These valves and their controls add an additional level of design, construction coordination, and maintenance to the system; however, they can better account for the dynamic conditions noted above.

Although the individual components of the system have remarkable accuracy, when these individual components become interrelated, the design, construction, and operations of such systems are difficult to control. When the airflow changes or a door opens in one laboratory, it may affect other laboratories or the entire system, and cumulative effects are difficult to predict. These systems must be well thought out, properly calibrated, commissioned, and maintained to work properly.

The HVAC system is the most critical system in a laboratory to ensure worker comfort and safety. Early planning for this system is essential to allow proper sizing and placement of mechanical rooms with their large air intake louvers and integration of the exhaust stacks into the design of the building. This is particularly critical in a renovation of, or conversion to, a laboratory where floor-to-floor heights begin to dictate system concepts. In a new building, floor-to-floor heights and location of chases for continuation of ducts through floors should be considered early.

The integration of architecture and engineering in a laboratory is critical to the cost, schedule, and quality control of a project. Module width for laboratories is critical where minimum air changes in laboratories is a governing factor in air handling system design. For example, an 11-ft module adds an additional 10% to the air handling system requirements compared with a 10-ft module. This additional requirement has an effect on the air handling systems, supply and exhaust duct sizes, chiller size, the cost of structural and architectural systems (Table 2), and operating costs during the life of the building.

Ceiling height (even the decision to put a ceiling in the laboratories) must be decided early, as both decisions affect the volume of the room. When air changes are the governing factor, the lack of a ceiling can add about 33% to the size and costs of running the HVAC system. Floor-to-floor height is critical for operating HVAC systems. A floor to below-beam height of 12 to 13 ft is usually sufficient to allow duct systems to be installed with minimum offsets when properly planned. Offsets increase the resistance to airflow in the duct and waste energy. In existing buildings, where minimal floor-to-floor height is available, the entire system may have to be planned around possible duct routing (Fig. 11).

Room size can be established early to maximize the match between the HVAC system and the room volume and to eliminate duplication of systems such as fume hood exhaust and room exhaust.

Ventilation and airflow is a critical factor in minimizing spread of aerosolized microorganisms, thereby protecting animals and personnel. Air movement is necessary for controlling odors in an animal facility. The air supply and exhaust systems

TABLE 2 Dimensional and cost comparison between 10- and 11-ft modules for a typical five-story building

	10-ft module	11-ft module
Dimension		
Height (ft)	75	75
Length (ft)	200	220
Width (ft)	100	100
Square footage	100,000	110,000
Item		
Superstructure	$ 3,720,000	$ 4,054,800
Exterior enclosure	2,360,000	2,515,760
Interior construction	2,300,000	2,415,000
Equipment furnishings	3,160,000	3,160,000
Plumbing/fire protection	1,840,000	1,913,600
HVAC	4,560,000	4,970,400
Electrical	2,060,000	2,142,400
Total	$20,000,000	$21,171,960
Increase in cost		5.9%

should be independent and provide 100% outside air, totally exhausted with no recirculation. Systems should be designed for a high degree of reliability in providing constant temperature, air change rate, and humidity in the rooms. These requirements will vary with species of animals housed. The air pressure relationships between dirty areas and clean areas should be maintained at all times.

The air intake for the units should be on the building face, and the exhaust air should be discharged on the roof level. Air intakes must not pick up vehicle fumes from loading dock areas or discharges from the building exhausts.

Coordination and integration of the architectural treatment of the space with the HVAC systems are essential parts of design due to the large quantities of supply and exhaust air that must be moved.

Controls. There are two main types of HVAC system controls used in laboratories: pneumatic controls that use compressed air, and direct digital controls that use electronic signals to monitor and modulate systems. Combinations of the above systems are also used. The important factors to consider are reliability, repeatability, and accuracy. More complex HVAC systems require more complex controls; direct digital controls should be considered. These controls can provide sophisticated monitoring and diagnostics; however, a highly trained maintenance staff is required to operate them.

Design parameters. Design parameters for HVAC systems are directional airflow, ventilation rates (air changes), pressure relationships, temperature, and humidity control (2). Fundamental guidelines are presented here, but conditions in each laboratory are different and each should be engineered to their specific requirements.

Most research facilities at BSL 2 or above are designed with no recirculation of air between laboratories. This minimizes the potential for spread of hazardous fumes or bioaerosols and reduces the chance of cross-contamination between laboratories. Air from the laboratories is exhausted directly to the outside. Air is usually introduced into the laboratory near the entry or desk working areas and sweeps the laboratory before being exhausted near the area of hazards. This reduces the area of the laboratory that becomes exposed to hazardous aerosols and is another good reason to put desk areas at the front of the laboratory. Care should be taken in the types of diffusers used to minimize drafts and cross-currents that might have an effect on experiments or upset the balance of safety devices. This is particularly important when air is introduced into the area of fume hoods or BSCs. Laminar-flow air supply diffusers, which are large perforate panels that allow air to ooze through, can be provided to minimize the velocity of air entering laboratories, thus reducing drafts that can affect hoods, cabinets, and laboratory work.

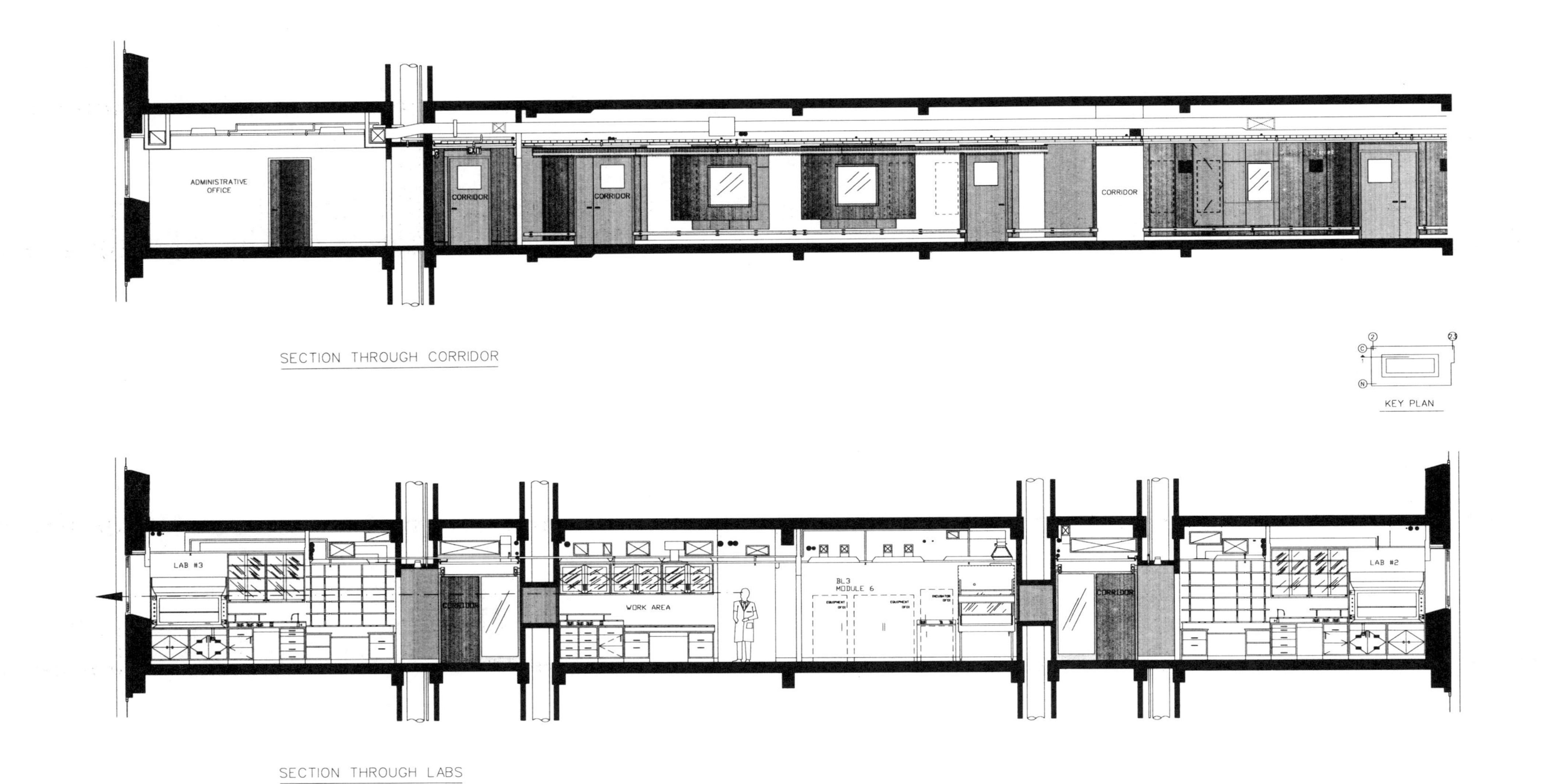

FIGURE 11 Section through Aaron Diamond AIDS Research Center for the City of New York showing space constraints on systems imposed by an existing building. (Courtesy Lord, Aeck & Sargent, Inc., Atlanta, Georgia.)

Air change rates vary from laboratory to laboratory, depending on specific needs, types of HVAC systems, number of hoods per laboratory, and cooling requirements of rooms. Typical laboratories might have 10 air changes per hour, with containment and animal facilities having 15 or more air changes per hour.

The rooms are then balanced so that air will flow from the corridor (or lower-hazard room) into the higher-hazard room, eventually being exhausted from the building (9). To have air moving from areas of low hazard to areas of higher hazard, the higher-hazard room is at a "negative" pressure relative to the lower-hazard room. Pressure relationships between rooms vary from 5 to 15%, depending on specific system requirements.

Airflow at low pressure differential is difficult to control, particularly with variable volume systems whose accuracy may be ±5%, which might allow in the worst case the room pressures being reversed. With higher differentials, doors become difficult to open or are snatched open, depending on their direction of swing, and air infiltrates into the room from any unsealed opening such as electrical outlet boxes, door frames, and windows. The mechanical system design and level of architectural integrity of the laboratory must be carefully balanced. High negative pressures can also cause high infiltration of outside air through window joints and other locations that can bring in spores, pollen, fungi, and other pollutants. This may disturb experiments, particularly in laboratories working with this type of specimen, or contribute to contamination of tissue cultures or personnel problems (allergies).

Temperature in a biomedical research laboratory should be cool enough to provide comfort for workers wearing personal protective equipment, such as laboratory coats, Tyvek suits, and gloves. Comfort levels are difficult to maintain unless sufficient cooling is provided to overcome the heat generated by equipment used in the laboratory. Humidity control is important to minimize condensation problems caused by high humidity or static electricity caused by low humidity.

Acoustical Considerations

Acoustical considerations in laboratories include vibration which can affect sensitive equipment such as balances, microscopes, microtomes, and electron microscopes, and noise that can be detrimental to the health and comfort of the occupants. Most vibration in laboratory buildings comes from either the ground under the building or from moving mechanical equipment in the building, such as chillers, fans, and air handling units. Proper design of these systems minimizes vibration transmission and local dampening of equipment (e.g., balance tables) and makes vibration a minimal problem in most laboratories. Noise, however, is a constant battle. Equipment noise and the noise inherently associated with large volumes of air movement are more difficult to control. Fume hoods, freezers, BSCs, centrifuges, blenders, and vacuum pumps produce noise. Where possible, items such as freezers and centrifuges that produce noise constantly or for long periods of time should be located in separate rooms that are seldom occupied. Fume hood noise can be minimized by proper system design. Although most laboratories can meet OSHA noise level requirements of 85 dB over an 8-h period, they may not provide auditory comfort to the occupants. Sessler and Hoover (12) suggested that noise levels should not exceed 45 to 55 dB in research laboratories where telephone communication and creative thinking are done; however, these levels may be difficult to achieve in a laboratory and a more realistic criteria may be 50 to 60 dB. Consideration of noise early in the design process will reduce potential problems.

Plumbing

Sanitary drainage, laboratory drainage, and vent systems. Laboratories often have two types of drainage systems: sanitary and laboratory waste. The sanitary drainage handles liquid waste from water closets, lavatories, drinking fountains, and other nonlaboratory sources. This system may also be used for the disposal of human clinical specimens (e.g., blood, urine). The laboratory waste system is acid-resistant and may serve sinks in laboratories and hoods that could be contaminated with chemicals due to improper disposal or spills. Acid dilution tanks are often provided with such laboratory waste systems. Although regulations minimize the amount of waste that can be legally put into drainage systems, spills at fume hoods can occur. Care must be taken to ensure that any wastes put into drainage systems meet local codes. Vent piping must be taken up through the laboratory to relieve sewer gases.

Domestic and laboratory water supply systems. Separate systems should be provided for potable and laboratory water in laboratory facilities. Backflow preventers (vacuum breakers) are required on the laboratory water side to prevent contamination to potable water systems.

Eyewash stations should be provided at main laboratory sinks. Safety showers should be placed in corridors where required by code and ideally in each laboratory that contains a chemical fume hood. The safety showers should each have an eyewash associated with them. Considerations and your local codes may govern whether these are on the laboratory or domestic water system and if the water should be tempered.

Vacuum and compressed air systems. Central systems should be provided for vacuum and compressed air. The vacuum system from biosafety areas should have a disinfecting trap and HEPA filter in line before entering the vacuum lines, to keep the lines and collection systems from being contaminated and to keep biohazards from being discharged to the atmosphere. In-line air filters should be provided where necessary, to keep the compressed air system from becoming contaminated or from contaminating equipment or experiments. Alternatively, the use of oil-free compressors can alleviate the problem.

Laboratory gases. Natural gas, carbon dioxide, and cryogenic and other specialty gases are normally found in laboratory facilities. High-volume-use gases are normally piped through central systems and low-volume-use gases are supplied from in-laboratory cylinders. It is advantageous to keep cylinders out of sensitive areas, such as tissue culture rooms, to minimize contamination from traffic entering the room. It is also prudent to provide outside access to BSL 3 laboratories to reduce the potential exposure of support personnel. Sufficient storage areas and restraining devices must be provided for gas cylinders, both empty and full. All cylinders in use or in storage must be restrained. Also, some hazardous specialty gases such as hydrogen fluoride may warrant additional safety support such as showers or leak detection systems.

Fire Protection

Fire extinguishers. Fire extinguishers must be provided at locations required by the life safety codes and should be installed at other hazardous locations. Type ABC extinguishers are usually provided; other types (e.g., carbon dioxide foam, halon) should be provided for specialized needs.

Sprinklering. The facility may not have to be equipped with automatic sprinklers to meet the current NFPA code, but consideration should be given to sprinklering the facility while it is being built. Sprinklering provides the following benefits in the event of fire:

Increases life safety
Minimizes loss of experimental data
Reduces possibility of breach of containment
Minimizes loss of use of your unique facility

Other fire protection systems. Special electronic equipment or high-volume chemical transfer areas may require other than water sprinkler systems. Halon, carbon dioxide foam, and other systems are available for specialized needs but are not generally used in laboratories.

Spill control and fire stations. Spill control and fire stations should be set up in areas near hazards for quick access and response in the event of an emergency. Neutralization, absorption, and disinfectant materials to handle the hazards in the laboratory should be stored along with fire blankets, extinguishers, and first-aid supplies. Limited access laboratories (BSL 3) should be stocked with sufficient emergency response materials.

Electrical

Normal power supply and distribution. Laboratories have become high-volume power users, and use of electricity in laboratories is likely to increase. Normal power should be generously supplied to all laboratories, with many circuits feeding each laboratory. Also, particular equipment with unusual power requirements should be identified during the design phase so that appropriate power can be supplied. Additional empty conduits and junction boxes from the laboratory panels should be installed to allow easy wiring of future laboratory equipment. All levels of the laboratory power system, from the transformers to the individual laboratory panels, should allow for future growth.

Emergency power. Consider installation of an emergency generator to serve the life safety equipment, air handling, exhaust, biosafety systems, freezers, and incubators in the event of a loss of normal power. Establish in advance the critical equipment to be powered by the emergency generator to provide adequately for the power required.

Lighting. Lighting in laboratories should be evenly distributed and task-oriented. Lighting should be placed over benches in a manner to minimize glare and shadows and should be of sufficient brightness to illuminate the workers as well as the work surfaces of the laboratory.

Instrument ground. Laboratory equipment and computers are sensitive to "electrical noise" from building systems and motors that may be distributed along the building grounding system. Care should be taken to minimize this problem by providing system separation between electronic noise producers and equipment that might be affected by the noise. Grounding systems can be developed to reduce this problem.

Fire alarm. A fire alarm system should be provided to give early warning of fires or other life safety problems. The requirements for these systems are usually governed by the local codes. Visual systems should be provided for the hearing impaired and for those in high-noise areas or limited access laboratories (BSL 3) who cannot hear the building system.

Access control, security, and monitoring. Controlled access to the facility should be considered. As noted previously, all parties have an interest in knowing that the facility is operating as designed. We suggest that a computerized monitoring system be installed to provide continuous monitoring of the conditions in the facility. Several systems have been developed and programmed for laboratory and animal facilities. Options for monitoring include

- Card key access control to the facility
- Operation of critical equipment (e.g., HVAC, emergency generator, BSCs, fume hoods, autoclaves, cold rooms, freezers)
- Airflow, air changes, and pressurization
- Local environmental conditions (temperature, humidity, lighting)
- Automatic 24-h notification of alarm systems

This system could be monitored in the facility, in the administrative area, or in the maintenance area. Some of the potential benefits are

- Advanced warning of systems malfunction
- Record of entries and exits to the facility, including records of improper entry attempts
- Monitoring the condition of incubators and freezers to prevent potential loss of research specimens
- Record of space conditions as backup for experimental result validation
- Higher level of public comfort with the facility

Communications

Telephone and data connections should be an integral part of the planning of all laboratory spaces. Cable trays should be provided above the ceiling for communication and computer wiring. This wiring should extend to all laboratory areas and a central point for connection to building-wide systems (existing or projected). Space should also be provided for data system racks and modems to connect to outside networks.

Laboratory Information Management Systems

Computer use is now integral to laboratory work. Laboratories use computer systems for accounting, data analysis, data acquisition, equipment operation, HVAC systems operation, personal use, quality control, and systems monitoring. Stand-alone and networked applications are being developed into laboratory information management systems, allowing for collection and analysis of data from a variety of laboratory sources. Most new laboratory equipment is computerized to some degree with data output into networks. Currently, much of this equipment has a PC-type computer and printer dedicated to it. A typical laboratory may have three or four computer printer combinations dedicated for equipment use. Standards are being developed for systems to communicate through a network allowing one computer to handle multiple functions in each laboratory. This could reduce the need for up to 6 to 10 lf of bench space per laboratory.

Audiovisual Systems

Conference and seminar spaces should be provided with projection and speaking rostrum capabilities. Projectors, projection screens, VCR players, and monitors now form an integral part of laboratory research and teaching programs. Lighting level control and location are important considerations in audio-visual planning. Photography, slide making, and graphic arts are also important to the operation of some laboratories. The need for such equipment and services must be considered in design.

Systems Distribution and Layering

A plan must be developed to provide a clear method for distribution of HVAC, plumbing, and electrical systems to the facility to allow for ease of operation. Supply and exhaust ducts, pipes, conduit, and cable tray must be routed to minimize turns (which can reduce efficiency), crossings (which reduce ceiling heights or raise structure), and access locations (which may require access panels or drains). This routing must allow access for maintenance and repair. As systems account for most costs in a laboratory, the distribution scheme can waste or save a lot of money. Evaluation should be made as to the best point to switch from main to branch distribution. Again, common sense should come into play as systems should be located and developed to minimize the extent of the most expensive components.

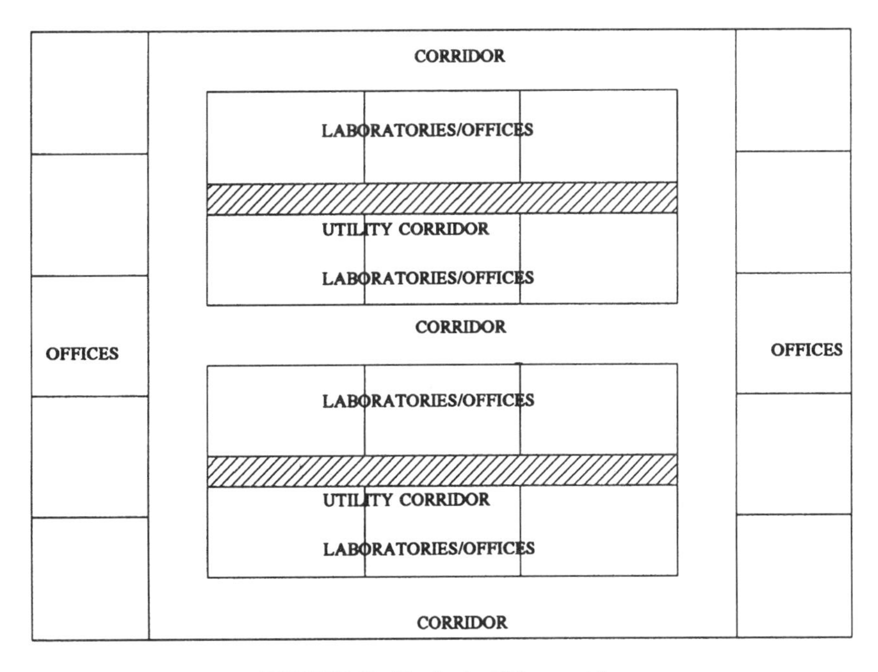

FIGURE 12 Typical utility corridors.

Gross Building

Gross distribution brings the main systems into the building and to the floors. In 1979, five basic distribution schemes for laboratory buildings were published by the NCI Research Facilities Branch (7). The advantages and disadvantages of each are addressed according to the service needs and costs.

Utility corridors. A utility corridor 8 to 12 ft wide is placed between each bank of laboratories (Fig. 12). Ducts and piping are brought to the floor in vertical shafts, then piping is distributed horizontally in the utility shaft branching off into each laboratory. This scheme provides easy access for service and a great deal of flexibility. The large amount of floor space devoted to service makes this scheme

FIGURE 13 Typical multiple interior shafts.

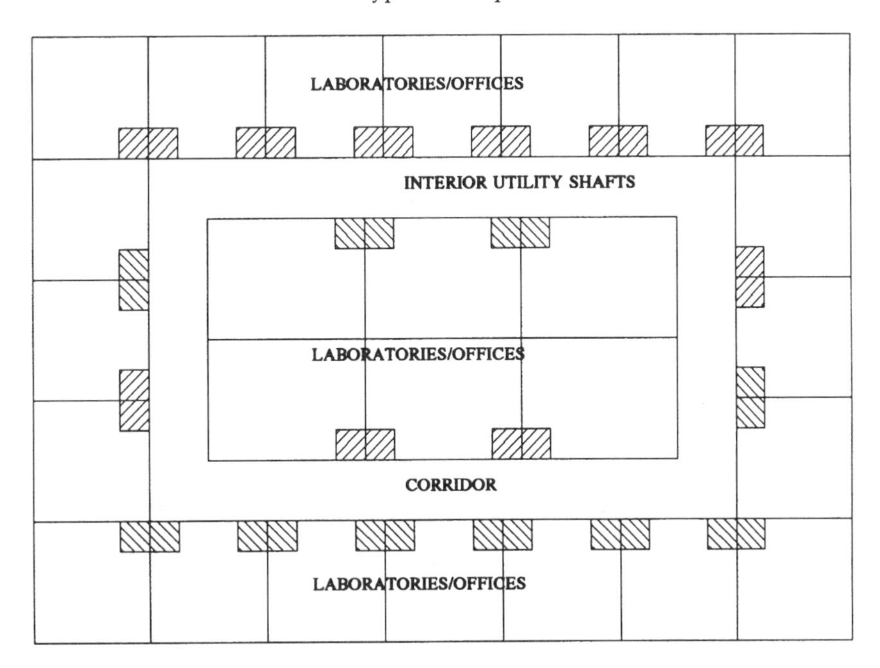

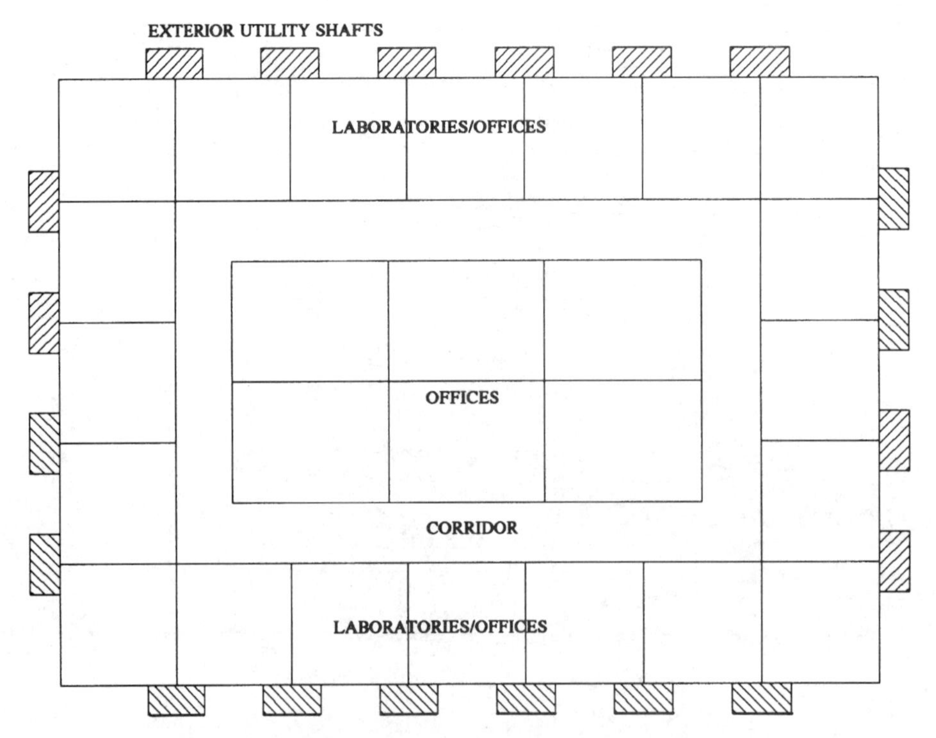

FIGURE 14 Typical multiple exterior shafts.

FIGURE 15 Typical corridor ceiling distribution.

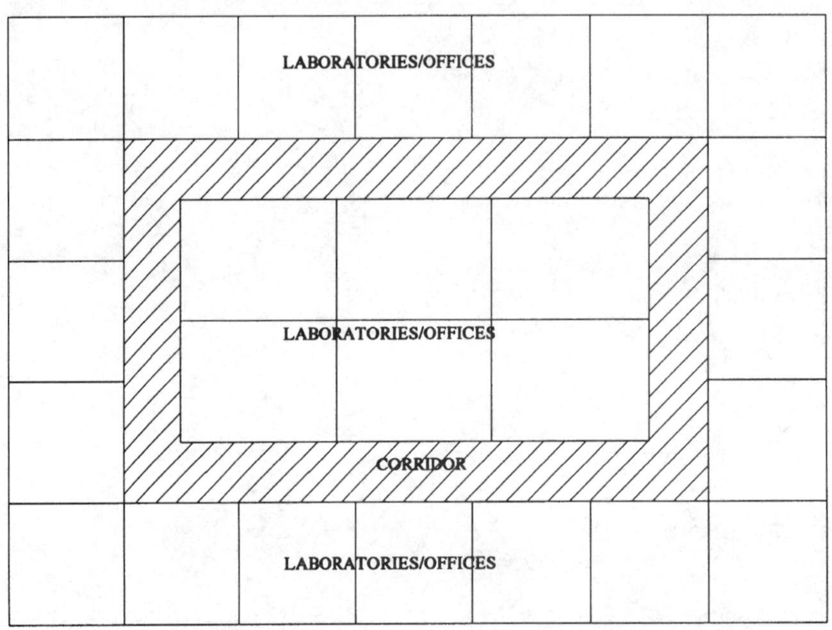

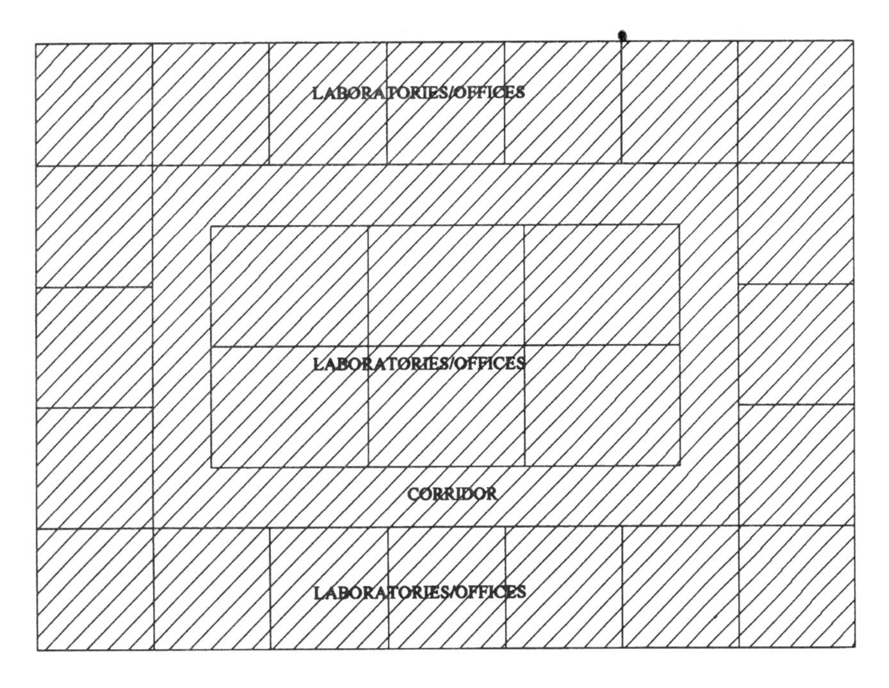

FIGURE 16 Typical interstitial space distribution.

relatively less efficient in net-to-gross square footage ratios.

Multiple interior shafts. Ducts and piping are brought to the floor through vertical chases located on the corridor sides of the laboratories at each or alternate laboratories (Fig. 13). Services are distributed horizontally into the laboratories. This scheme takes a large amount of floor space and decreases the flexibility of the laboratories. Access to service or modify the chases is moderately difficult.

Multiple exterior shafts. Ducts and piping are brought to the floor through vertical chases located on the exterior of the building at each or alternate laboratories (Fig. 14). Services are then distributed horizontally into the laboratories from the shafts. Although taking minimal floor space and providing internal flexibility, the vertical chases are difficult to access for service or modifications.

Corridor ceiling distributions. Ducts and piping are brought to the floor in a few main shafts, and services are distributed out to the floor in the ceiling of the corridor and distributed into the laboratories horizontally (Fig. 15). This scheme minimizes floor space devoted to services but must be extremely well coordinated to allow access to ducts and piping for service or maintenance.

Interstitial spaces. The distribution concept of interstitial space is similar to the corridor ceiling distribution scheme; however, a service floor is inserted above each laboratory floor (Fig. 16 and 17). This allows services to be distributed down into laboratories from above the laboratories. This scheme provides a great deal of access for service and modifications, is very flexible, and minimizes laboratory floor space required for service. However, its high initial costs do not make it cost-effective except in unusual circumstances.

In-Laboratory Local Systems Distribution and Layering

Because of the many systems serving laboratory and containment facilities, building systems must be integrated with the architecture of the facility. As noted above, building systems are distributed above, below, and to the floor to serve laboratories and office spaces. The following describes typical system ideas regarding locations and layers to provide a coordinated laboratory support system:

Below floor: plumbing sanitary drain and acid waste and vacuum piping.

Laboratory floor: on-floor or "through" systems include shafts along corridors or exterior walls, ducts and vents from spaces below through to roof, roof drain conductors, elevator and stairs, plumbing risers, electrical risers.

On-floor systems that penetrate the floor: include plumbing sanitary drain, vacuum, and acid waste piping, floor drains, condensate drains,

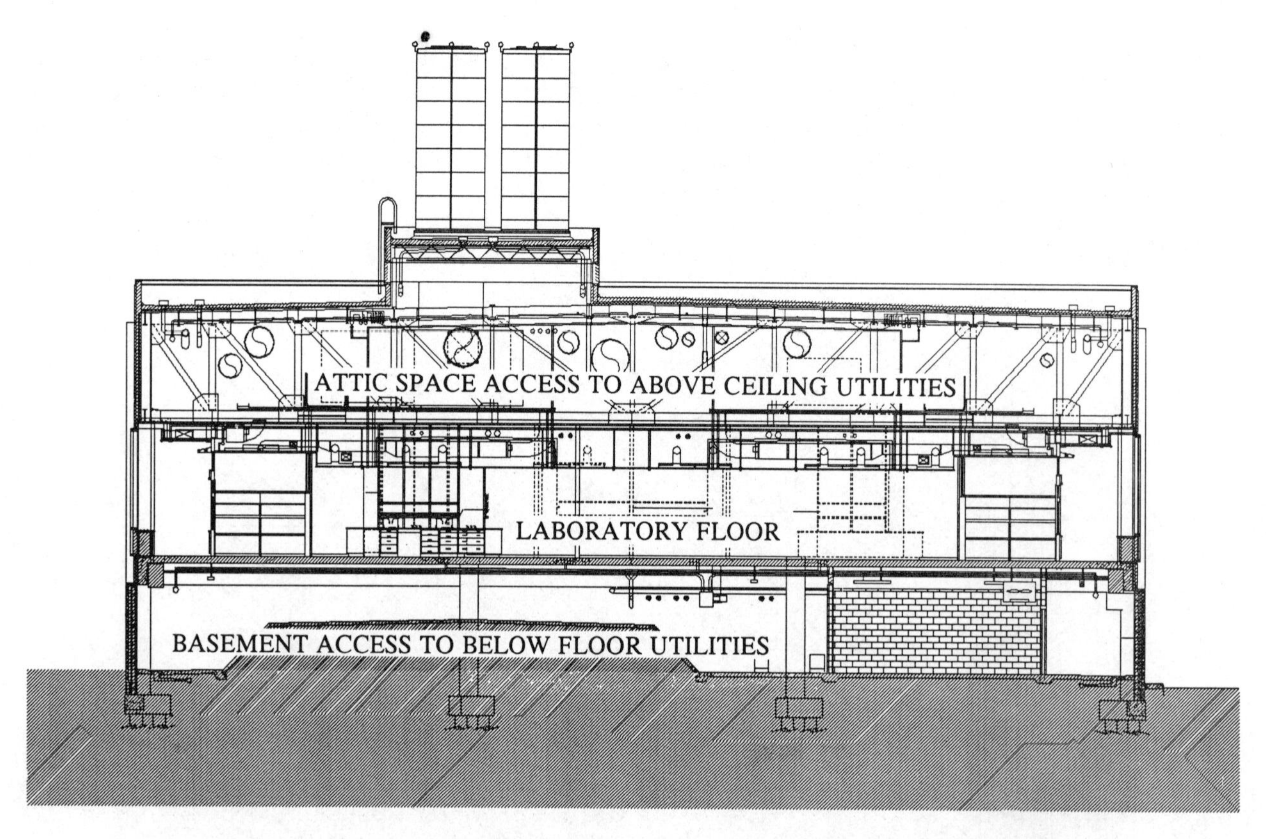

FIGURE 17 Section through planned Laboratory for Toxicology and Parasitic Diseases at the CDC showing interstitial attic space and crawl space for future flexibility. (Courtesy of Lord, Aeck & Sargent, Inc., Atlanta, Georgia.)

and electrical main buss feeder up to laboratory power panels.

Ceiling-mounted systems or penetrations: include lighting fixtures, HVAC diffusers, exhaust grilles, hoods, speaker systems, security, alarm, smoke detectors, fire protection systems (sprinklers), emergency shower, vent piping, fume hood exhaust ducts, piped systems distribution.

Wall-mounted systems: include fire alarm pull stations and horns, laboratory power panels, power outlets, wire mold raceways, telephone or data outlets, telephones, fire extinguishers, hoses, valves, blankets and cabinets, spill kits, safety kits, laboratory monitoring systems, water treatment valves and systems, emergency power off switches, thermostats, card readers or security systems, light switches, and cold room or autoclave control panels.

"On-bench" systems: include laboratory gases, data or computer outlets, telephone outlets, power outlets or wire mold, cup sinks, sinks, pure water system secondary polishing units and associated piping, laboratory gas system monitoring valves, unistrut support racks, reagent shelving, and eyewashes.

Above-ceiling or interstitial space: includes telecommunication cabling distribution in cable trays, lighting system and flexible power cabling for lighting, HVAC, security, and other control system cabling in cable trays, HVAC duct distribution for supply and exhaust, terminals and boxes, laboratory gases, deionized water and distribution loop, and laboratory water supply and return, laboratory air, laboratory gas intermediate pressure regulators, chilled water, hot water, and steam or condensate supply and return piping, laboratory condenser water system piping, room air exhaust duct, fume hood exhaust ducts, plumbing vents, and ceiling or lighting support grid.

Equipment Selection

Discussions on equipment selection and operation can be found in Chapters 11 and 13. The discussion below relates equipment to design considerations.

Risk Assessment

Laboratory equipment used with biological, radioactive, or chemical hazards can pose risks if releases of these agents occur during operation. Containment requirements must be assessed before laboratory design begins. Equipment placement is also critical to proper safe operation (e.g., BSCs and fume hoods should be located at the back of laboratories, away from doors, air supply diffusers, drafts, and movement of personnel).

Casework

Casework type, construction, adaptability, and cost are important decisions in the design of any laboratory. Initial costs versus long-term cost must be considered, and casework tops and fixtures in relation to utility services are key components of design (Table 3).

Tops are available in several materials; those most commonly used in biomedical laboratories are stainless steel (easily cleanable, high solvent-resistant), epoxy resin (impact-resistant, high solvent, and acid-resistant), and plastic laminate (moderately chemical-resistant).

Cabinets for flammable and acid storage should be provided in each laboratory where these chemicals are used and stored.

Self-supporting (flexible) casework systems should be considered in laboratories where a high degree of change is anticipated. Clinical laboratories are one of the most successful users of flexible casework.

Benches should be designed with the appropriate mixture of cupboards and drawers for the use intended. An appropriate number of kneespace workstations should be provided, to allow comfortable sit-down areas for working and prevent them from becoming storage areas. Dry areas should be provided for working with paper and computers. The most common mistake in casework design is putting too much casework into the laboratory during design. Floor space should be left for unanticipated large equipment. Casework can always be added; it is seldom removed.

Fume Hoods

Fume hoods are enclosures designed to maximize capture of chemical vapors and fumes and to move them away from the worker and into the building exhaust system. Most fume hoods use the building's exhaust fans to move air through them. Preferred design has the building fans at the end of the run of exhaust duct (usually on the roof) to place the exhaust system under negative pressure, thereby minimizing the potential of contaminants escaping from the system while in the facility. However, this ductwork is contaminated; service workers must be properly trained before opening ducts for maintenance.

There are many types of fume hoods: general-purpose (for work with solvents, acids, low-level isotopes), isotope (for work with higher-level isotopes), and specialty hoods (e.g., distillation, walk-in, or perchloric acid hoods to be used with special chemicals or experiments). Recommended face velocities for fume hoods of various types and conditions can be found in *Laboratory Fume Hood Standards* (4). However, the only true test of a fume hood's effec-

TABLE 3 Comparison of casework types

Cabinet construction	Cost range	Comments
Wood	Low	Good general purpose
		Difficult to judge construction quality
		Moderately adaptable
		Corrosion resistant
Plastic laminate	Low to moderate	Good general purpose
		Difficult to judge construction quality
		Moderately adaptable
Metal	Moderate	Good general purpose
		Easily cleanable
		Moderately adaptable
		Subject to corrosion
Molded plastic/metal (flexible systems)	High	Lower storage capacity
		High adaptability
		Lower weight-carrying ability

tiveness is the ability of an actual fume hood, in use, to reduce the hazard quality (below the threshold limit value for the hazard at the worst-case conditions of face velocity) in its actual operating location.

Fume hoods should be located away from drafts, convection currents, and passageways to minimize airflow disruption from these sources. Fume hoods should be placed at the back of the laboratory to minimize the possibility of trapping occupants in the event of an accident or fire.

Several types of fume hoods have been developed to minimize energy usage in laboratories. These can be effective in reducing operating costs in laboratories, if heat loads can be overcome and air changes are acceptable, with exhaust levels below that which would normally be required to operate the fume hood.

Auxiliary air hoods have a significant percentage of their exhaust air requirements met by introducing unconditioned air just above the hood. The energy savings come from a reduction in the heating and cooling of air. In extreme climates, comfort problems for the hood users develop if this air introduced is not sufficiently tempered. Hot, humid air or cold drafts make work at the hoods uncomfortable. Also, the introduction of auxiliary air adds noise to the hood installation.

Variable volume hoods create energy savings by reducing the exhaust air requirements to match the position of the sash, thereby reducing the air exhausted to the minimum required for safe operating conditions. Their advantage is in a constant face velocity over the entire range of sash positions, which minimizes drafts in hoods that might affect experiments. Disadvantages include relatively high initial cost (often offset by operational savings) and complexity of operation.

Another option for energy savings includes reducing the exhaust to provide the required face velocity at a sash position less than the full open position and providing a sash stop to limit operations to the appropriate opening.

Face velocity alarms on hoods can be a helpful tool in commissioning the facility as well as important safety indicators.

Biological Safety Cabinets (see Chapter 11)

BSCs are either ducted or nonducted, depending on type and use. Unlike fume hoods, BSCs have internal fans to provide a balanced airflow and near-sterile work environment in the cabinet. Some BSCs do require external fans for exhaust. Fume hoods are designed to provide personal protection (by drawing air away from the worker) and environmental protection (filtration or dilution to the outside). BSCs are equipped with HEPA filters and are designed to provide personnel, product, and environmental protection by careful airflow balance, keeping contaminated air away from the product and biohazards in the cabinet with filtration of air leaving the cabinet.

Nonducted BSCs referred to as class II, type A (10) (which exhaust cabinet air back into the laboratory through HEPA filters) can be located easily in most laboratories. Like fume hoods, they should be kept away from drafts, convection currents, diffusers, and traffic paths (Fig. 18). Sufficient ceiling height is

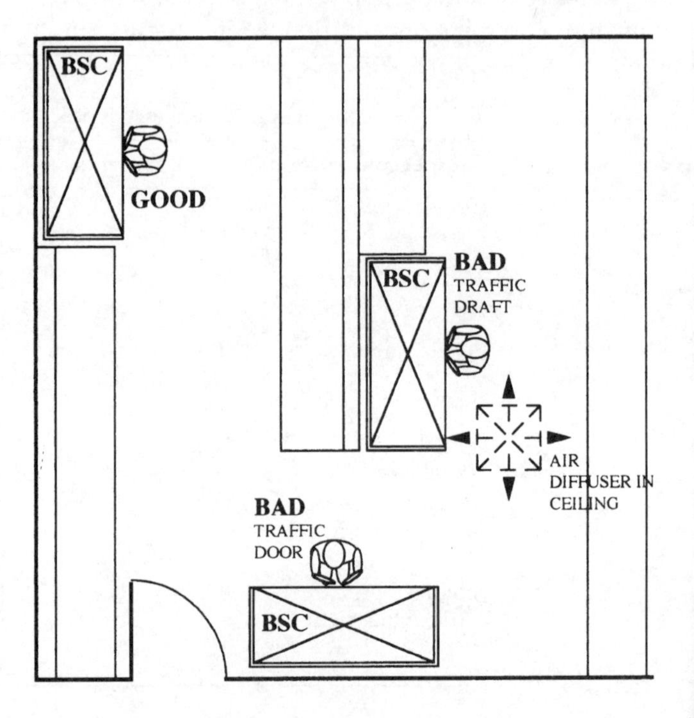

FIGURE 18 Evaluation of proposed biological safety cabinet locations.

required to allow free exhaust from the top of the cabinet and to provide space for cabinet certification or HEPA filter change. BSCs are heat generators, and sufficient cooling should be provided to overcome the load.

Ducted BSCs (exhaust air is first passed through a HEPA filter) have the added concern of being tied into building exhaust systems. The cabinet's fan provides exhaust into the system. The building exhaust system must provide the balanced exhaust to take the air out of the building and yet not affect cabinet performance, which depends on a delicate airflow balance. This is a difficult design issue. When hard-ducted BSCs are used with variable volume and ganged exhaust systems, the exhaust systems must work with minimum variations because all variations will be reflected in cabinet performance.

Thimble connections are exhaust systems that allow the air from the cabinet to move into the exhaust duct without direct connection, thereby reducing the system's effect on cabinet performance. The thimbles must be designed correctly to make sure that the BSC operation is not compromised if the exhaust system malfunctions.

Autoclaves, Glassware, and Ice Machines

The following should be considered during design:

Provision of steam autoclaves for the preparation of materials to be brought into the facility, sterilization of instruments, and decontamination of infectious wastes.

Air capture vents above autoclaves for heat, steam, and odor removal.

Ethylene oxide autoclaving and aeration for heat-sensitive items. (Ethylene oxide requires stringent environmental safeguards to minimize occupational exposure [18].)

Evaluation of steam contamination and effect on products for use in the laboratory. If required, pure or clean steam sterilization can be provided to minimize the impact of steam contamination on tissue cultures, virus production, and other processes. This can add considerable expense to the cost of an autoclave facility.

Adding cross-contamination seals and filtration on the autoclaves for BSL 3 and 4 laboratories where double-ended autoclaves are installed in walls separating clean and dirty areas.

Prevacuum autoclaves may be advisable to ensure steam penetration and decontamination of the load but are expensive to purchase.

Provision of glassware washers and dryers.

Ice machines for flaked ice.

Laboratory Water Systems

Reagent-grade water can be provided by central or point-of-use systems by many methods depending on quantity and quality required. Distillation, deionization, reverse osmosis, and ultrafiltration are typical methods that can provide up to type I reagent-grade water from tap water; however, pretreatment is generally recommended to increase system life.

Refrigerators or Freezers

Refrigerators, mechanical freezers, and ultra-low-temperature mechanical or liquid nitrogen freezers are used for specimen storage in most biomedical laboratory facilities. Voltage, ambient temperature requirements, brownout protection, back-up power, heat release, noise, and system alarms are all considerations for proper design of specimen storage facilities. Explosion-proof freezers and refrigerators may be needed for storage of volatile or flammable materials and must be appropriately designed and labeled. In some laboratories, sparking devices may be removed from refrigerators, by facility maintenance personnel, to provide adequately for such safety needs.

Centrifuges (see also Chapter 11)

Heat release is a factor in space design for centrifuges as medium-speed, high-speed, and ultra-speed centrifuges generate from 5,000 to 25,000 Btu/h when running and at times can run up to 36 h continuously. Some centrifuges can be water-cooled to avoid releasing heat into the room.

Centrifuges that are used with biohazards and draw a vacuum in the chamber need to have the vacuum HEPA-filtered. Containment apparatus such as trunnion centrifuge cups and aerosol covers should be used to minimize aerosol generation of hazards; they can be loaded and unloaded in a BSC. Centrifuges can be placed or built into a BSC if the risk assessment requires additional containment.

Consideration should be given to placing the centrifuges in a separate room to minimize personnel exposure from an incident such as a rotor failure.

Vermin or Rodent Control

Vermin and rodents can be a nuisance and hazard in biomedical facilities, particularly animal and con-

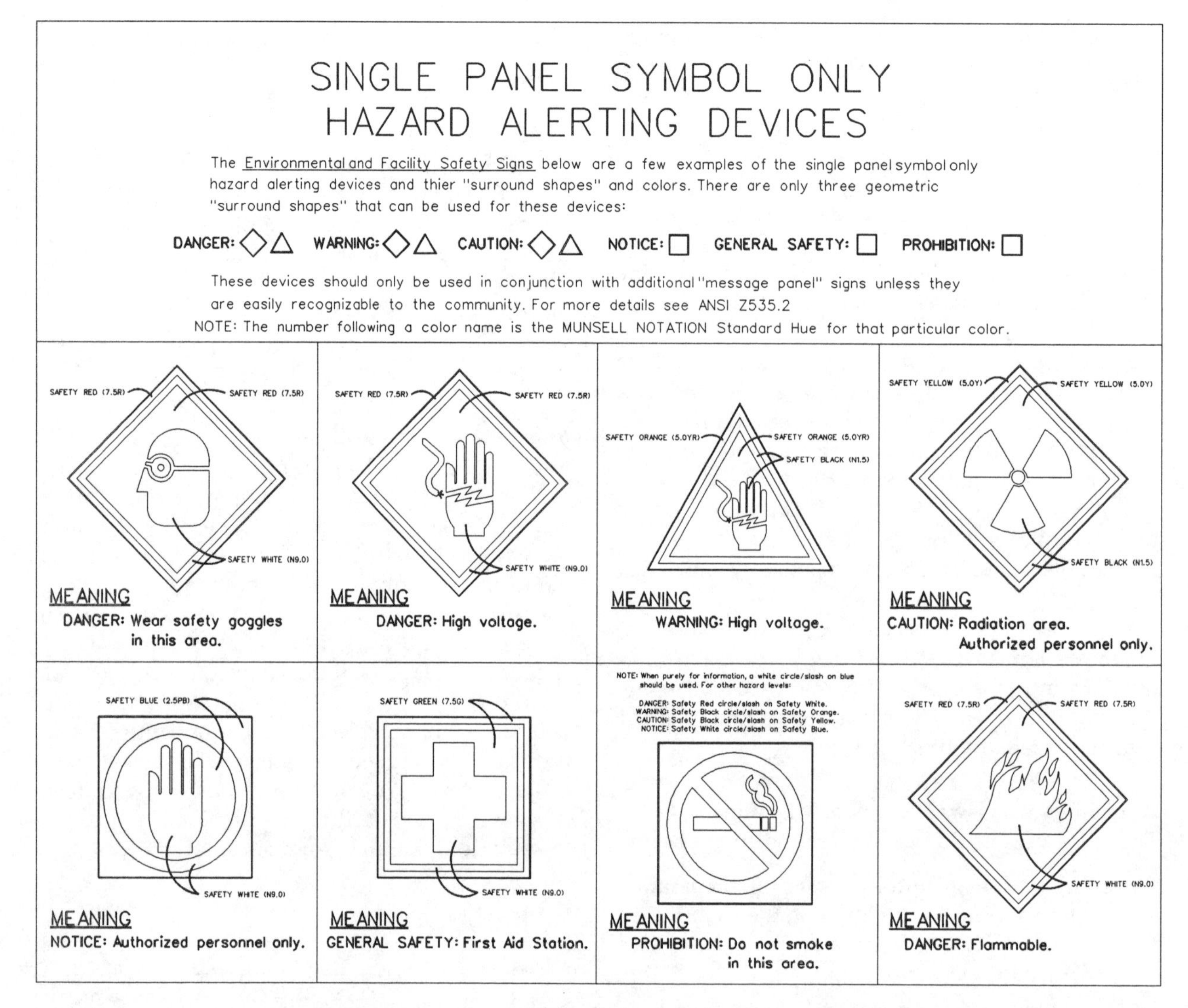

FIGURE 19 Examples of single-panel hazard alerting devices.

tainment facilities. These pests can spread contamination throughout the facility or could cross-breed with laboratory animals that escape. Procedures in design, construction, and operation can minimize such occurrences. All penetrations, small holes, openings, and cracks should be sealed well. Materials and systems should be selected to minimize areas for vermin to hide. Storage of shipping cartons and corrugated boxes should be minimized, and should be on pallets off the floor. Continuous pest control treatment of the facility should be provided, especially during construction. Protocols that ban food in laboratories should be enforced. Waste must be removed promptly and completely. Adequate receiving areas, storage, and break rooms should be located away from laboratories to minimize these problems.

Waste Handling and Removal

Space for waste storage, biomedical waste storage, radioactive waste storage, chemical waste storage, flammable or solvent storage, cold rooms or freezers for animal carcass storage, and full and empty gas cylinder storage should be an integral part of the facility. Provisions for the handling and storage of these wastes must be made during design. There is an increasing emphasis on waste management by institutions and regulatory agencies. Space must be provided if chemical recycling programs are to be carried out in your facility.

Space for general waste, biological waste, and noninfectious animal bedding should be provided along with appropriate pick-up facilities. If not properly handled, this waste can create odor and

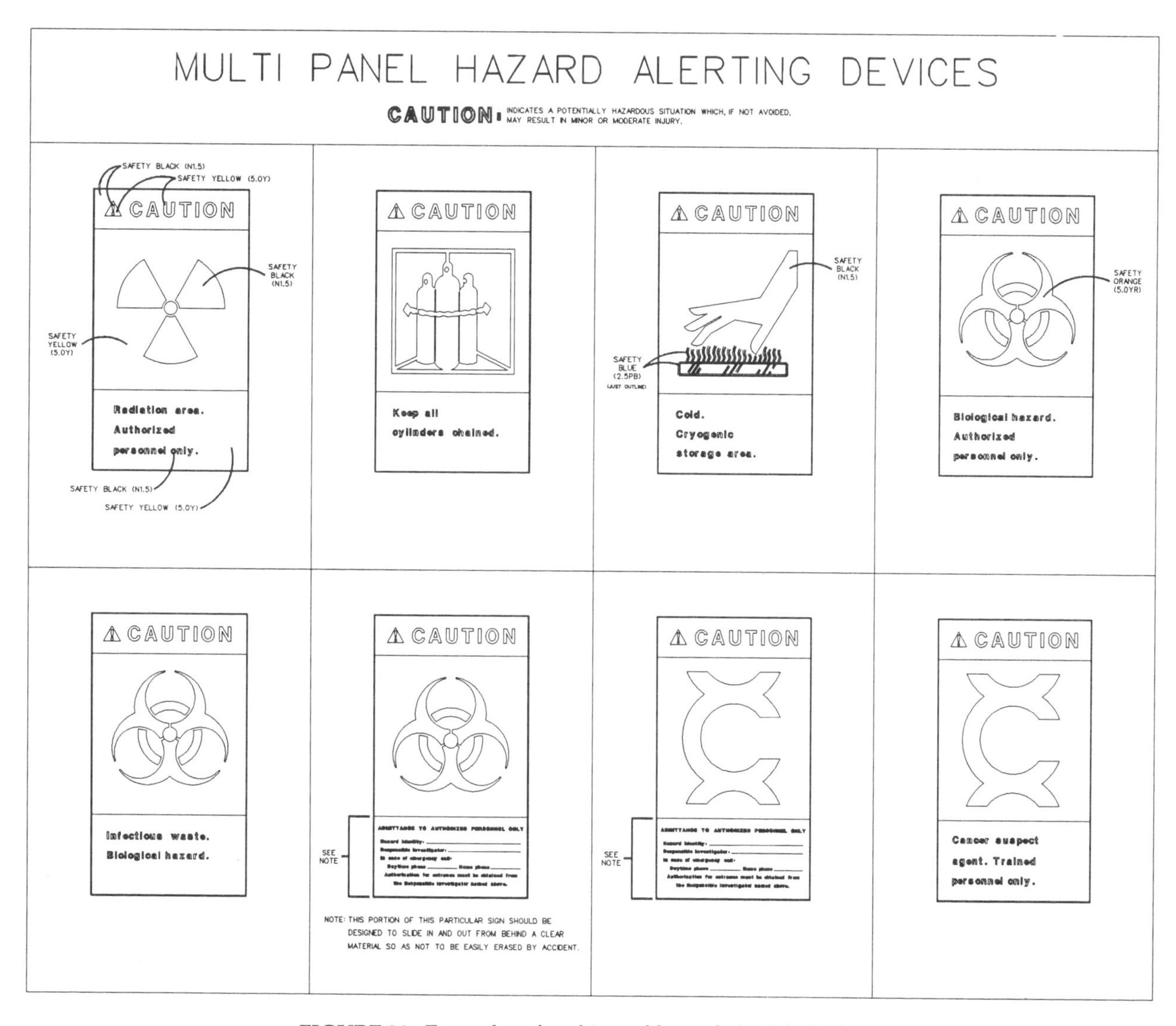

FIGURE 20 Examples of multipanel hazard alerting devices.

pest problems. Some state or local areas allow only limited holding times for such materials, according to refrigeration capabilities, and so on.

Space should be provided for radioactive waste to be held in the laboratory, collected and transported to a storage area for and removal by the appropriately authorized waste handler. Freezer space for holding radioactive animal carcasses should be provided. An incinerator should be considered for biological or pathological waste treatment if volume and local regulations warrant the expense.

Decontamination

Consideration of how your facility will be decontaminated should be a part of the design. Provisions to allow separation of rooms from supply and exhaust systems should be made. Openings into the room should be sealed or be capable of being sealed depending on the level of hazard (14). Spill control, clean-up, and decontamination centers should be provided in the laboratory or with easy access to all hazardous laboratories.

Signage and Information Systems

Five levels of signage are to be considered for a laboratory project: general directional, informational, life safety, hazard identification, and system identification.

General directional signage should direct persons to various locations in the facility. It should also include notification of access restrictions to laboratory areas.

Informational signage should include room numbers and names, operating hours, and general restrictions such as no smoking. Many buildings have such a sign at the main entrance because the entire building is "smoke-free."

Life safety signage deals with warnings and movement of occupants in the event of an emergency. Exit signs, fire stair identification, fire exit routes, and emergency equipment identification (e.g., spill control centers, fire extinguishers and blankets, and emergency showers and eyewashes) are examples of this type of signage.

Hazard identification signage has four categories. "Notice" states a policy related to safety of personnel or protection of property but is not for use with a physical hazard. "Caution" indicates a potentially hazardous situation that, if not avoided, may result in minor or moderate injury. "Warning" indicates a potentially hazardous situation that, if not avoided, will result in death or serious injury. "Danger" indicates an imminently hazardous situation that, if not avoided, will result in death or serious injury. There are specific sizes, shapes, colors, messages, and lettering requirements for each type (Fig. 19 and 20). Some of the more common hazards found in laboratories that should be identified are biohazards, radiation hazards, no smoking, laser light, chemical hazards, explosive or flammable liquids, prohibition of mouth-pipetting, cryogenic hazards, compressed gas storage, noise hazards, and UV light.

Specific requirements can be found in NFPA 30 and 45; ANSI Z53.1, A13.1, A35.1, and Z535; OSHA 29 CFR 1910.96, 1910.97, 1910.106, 1910.144, 1910.145, 1910.1200, and 1910.1450; and the CDC/NIH guidelines.

Signs for system identification, including flow direction of all pipes and valves, are critical to allow quick identification and response to laboratory emergencies as well as ensure ease of maintenance. This also allows positive identification and notification for system shutdowns.

In addition to signage, notice and bulletin boards, display areas, and computerized information systems can facilitate the exchange of information in laboratory facilities.

SUMMARY

Many considerations and decisions are involved in the design of a biomedical research facility. The laboratory users, administrators, and facility designers must provide adequate information to allow correct decisions to be made at timely points in the design process. In the end, this will minimize lost time and added costs that occur when issues have to be revisited. Good communication between all parties is a key to the successful design of a biomedical laboratory facility. Each laboratory is unique; no method of laboratory design can provide a finished laboratory without adequate time, effort, and thought. The most successful facilities with provide for their unique requirements as well as the common elements required in every laboratory. Such a laboratory will meet the functional requirements for specific projects and also be adaptable for future projects.

References

1. **American Institute of Architects, Committee on Architecture for Health.** 1987. *Guidelines for construction and equipment of hospital and medical facilities.* American Institute of Architects Press, Washington, D.C.
2. **American Society of Heating, Refrigerating and Air-Conditioning Engineers Technical Committee.** 1991. Industrial and special air conditioning and ventilation, laboratories, p. 14.1–14.17. *Applications. ASHRAE Handbook.* American Society of Heating, Refrigerating and Air-Conditioning Engineers, Inc., New York.
3. **Browner, A. B.** 1990. *Laboratory renovation vs. new construction: how to decide—how much to pay.* Technical Paper No. 90.04. Hamilton Industries, Two Rivers, Wis.
4. **Chamberlin, R. I., and J. E. Leahy.** 1978. *Laboratory fume hood standards, recommended for the U.S. Environmental Protection Agency.* Contract No. 68-01-4661. Environmental Protection Agency, Washington, D.C.
5. **College of American Pathologists.** 1985. *Medical laboratory planning and design.* College of American Pathologists, Skokie, Ill.
6. **Dolan, D. C.** 1981. Design for biomedical research facilities: architectural features of biomedical design, p. 75–86. *In* D. G. Fox (ed.), *Design of biomedical research facilities: proceedings of a cancer research safety symposium, 1979 October 18–19.* NIH Publication No. 81-2305. Frederick Cancer Research Center, National Institutes of Health, Bethesda, Md.
7. **National Cancer Institute, Research Facilities Branch.** 1979. *Design and construction guide for cancer research facilities.* National Institutes of Health, Bethesda, Md.
8. **National Fire Protection Association.** 1990. *NFPA 99 health care facilities.* National Fire Protection Association, Quincy, Mass.
9. **National Fire Protection Association.** 1991. *NFPA 45 fire protection for laboratories using chemicals.* National Fire Protection Association, Quincy, Mass.
10. **National Sanitation Foundation.** 1992. *Standard 49, class II (laminar flow) biohazard cabinetry.* National Sanitation Foundation, Ann Arbor, Mich.
11. **Ruys, T. (ed.).** 1990. *Handbook of facilities planning,* vol. 1. *Laboratory facilities.* Van Nostrand Reinhold, New York.
12. **Sessler, S. M., and R. M. Hoover.** 1983. *Laboratory fume hood noise, heating piping and air conditioning.* Penton/PC Reinhold, Cleveland, Ohio.
13. **U.S. Department of Agriculture.** 1992. *Animal welfare.* 9 CFR Parts 1,2,3 Animal and Plant Health Inspection Service, U.S. Government Printing Office, Washington, D.C.

14. **U.S. Department of Health and Human Services.** 1979. *Laboratory safety monograph.* National Institutes of Health, Bethesda, Md.
15. **U.S. Department of Health and Human Services.** 1985. *Guide for the care and use of laboratory animals.* HHS Publication No. 85-23. U.S. Government Printing Office, Washington, D.C.
16. **U.S. Department of Health and Human Services.** 1986. *Guidelines for research involving recombinant DNA molecules. Fed. Regist.* **51**(88):29,800–29,814.
17. **U.S. Department of Health and Human Services.** 1993. *Biosafety in microbiological and biomedical laboratories.* HHS Publication No. (CDC) 93-8395. U.S. Government Printing Office, Washington, D.C.
18. **U.S. Department of Labor.** 1990. *Ethelyne oxide.* 29 CFR Part 1910.1047. Occupational Safety and Health Administration, U.S. Government Printing Office, Washington, D.C.
19. **U.S Department of Labor.** 1990. *Occupational exposures to hazardous chemicals in laboratories, final rule.* 29 CFR Part 1910.1450. Occupational Safety and Health Administration, U.S. Government Printing Office, Washington, D.C.
20. **U.S. Department of Labor.** 1991. *Bloodborne pathogens, final rule.* 29 CFR Part 1910.1030, Occupational Safety and Health Administration, U.S. Government Printing Office, Washington, D.C.
21. **U.S. Department of Labor.** 1991. *Identification, classification, and regulation of potential occupational carcinogens.* 29 CFR Part 1990. Occupational Safety and Health Administration, U.S. Government Printing Office, Washington, D.C.
22. **West, D. L., and M. A. Chatigny.** 1986. Design of microbiological and biomedical research facilities, p. 124–137. *In* B. M. Miller (ed.), *Laboratory safety: principles and practices.* American Society for Microbiology, Washington, D.C.

Laboratory Biosafety Practices

DIANE O. FLEMING

13

INTRODUCTION

Good laboratory practices can prevent exposure to hazardous agents. Prudent biosafety practices are based on a need to protect the worker, coworkers, and the local community from infection and to protect the work, product, and the environment from contamination. Although employers may be interested in following safety guidelines to protect their employees, regulatory requirements tend to drive compliance activities.

Certain laboratory safety practices are mandated by federal regulations. A formal program for radiation safety is required by the Nuclear Radiation Commission. There are chemical safety requirements under the Occupational Safety and Health Administration (OSHA) laboratory standard 29 CFR 1910.1450 (13) and the Hazardous Waste Operations and Emergency Response Standard, 29 CFR 1910.120q. The OSHA blood-borne pathogen standard, 29 CFR 1910.1030, driven by health care union demands for protection from hepatitis B and AIDS, introduced biological agents into the realm of regulated materials (14). The Coordinated Framework for Regulation of Biotechnology (38) added the Environmental Protection Agency, the U.S. Department of Agriculture, and the Food and Drug Administration to the list of federal agencies, including the Centers for Disease Control and Prevention (CDC) and the National Institutes of Health (NIH), that provide oversight and information on risk assessment for work with biohazardous agents. Certain biological and chemical agents associated with terrorist activities or biological warfare require oversight from federal defense agencies as well (37).

When the biohazardous agents or materials are identified and assessed for risk, biosafety practices can be developed to provide employee protection and to manage the risk to the community and also to product integrity. In developing appropriate barriers to exposure and infection, a biosafety program must take into account the large variety of potential pathogens encountered in clinical microbiology, the volume of potential pathogens that can be used in research and production activities, the ability of microbes to adapt to new environments, and the potential for work procedures to bypass normal routes of exposure and modes of entry into the employee.

Publications are available to assist in developing an appropriate system for laboratory safety management (3, 23, 28, 43). Guidelines for risk assessment of biohazardous agents are available from government agencies (35, 44, 45, 46) in publications of professional associations (1, 10, 16, 17, 20–22, 24, 30–34) and from industrial and academic sources where they have been modified for internal use (25, 29, 41, 42).

Information on the hazard in use and the names of those at risk of exposure can be provided to workplace safety professionals for evaluation of worker protection methods (25) and safety training (42) and to employee health personnel for inclusion in preventive and postexposure treatment programs (39). Although this chapter emphasizes worker protection in biomedical laboratories, the use of good biosafety practices can provide safety and security for the worker and the community.

Regulatory Requirements and Associated Biosafety Guidelines

Regulatory requirements to protect employees from most biohazardous agents are implied in the "general duty" clause of the Occupational Safety and Health Act of 1970, which requires the employer to provide a workplace free of recognized hazards(12). Workplace biosafety is now mandated by federal regulations for work with certain human tissues and body fluids, the hepatitis B virus, and the human immunodeficiency virus. The OSHA final standard 1910.1030 for protection of workers from blood-borne pathogens is the first OSHA standard to regulate the use of infectious materials (14). Compliance requires the use of controls to prevent infection associated with exposure to blood and certain body fluids. The requirements include an employee Exposure Control Plan, specific training, provision of hepatitis B vaccine, and a postexposure follow-up. In this standard, risk management is based on the practice of "universal precautions" derived from the category-specific blood and body fluid precautions and the disease-specific hepatitis precautions found in hospital infection control guidelines published by the CDC (7, 8, 18, 19). Further information on hepatitis, HIV-1, and other blood-borne pathogens may be found in Chapters 3 and 4 of this book.

In the United States, laboratory biosafety guidelines developed by the CDC and the NIH describe four biosafety "containment" levels (BSL 1 to 4), which correspond well with four risk groups described by the World Health Organization (WHO) for international use (47). The CDC/NIH guidelines also take into account the host resistance to the agent and the relative risks of different work environments where procedures involve the agent (44). The CDC/NIH guidelines include lists of general and special biosafety practices, which are reproduced in their entirety as Appendix I.

Under these guidelines, clinical specimens of human blood and body fluids are handled at BSL 2 due to the risk of hepatitis B infection. Compliance with these CDC/NIH guidelines is now required under the OSHA blood-borne pathogen standard (14), which is provided as Appendix II in this book.

TABLE 1 Prudent biosafety practices[a]

Barrier precaution	Exposure route(s)[b] blocked
Do not mouth-pipette	A, I, C
Manipulate infectious fluids carefully	A, C
Restrict the use of needles and syringes	P, A
Use protective laboratory clothing	C
Wash hands	C
Decontaminate work surfaces	C
Do not eat, drink, store foods, or smoke in the laboratory	I, C

[a]Based on reference 10.

[b]A, airborne; C, contact (skin, mucous membranes); I, ingestion; P, percutaneous.

Specific etiologic agents may be handled under higher or lower levels of containment as recommended in the agent summary statements of the CDC/NIH guidelines (44). Industrial large-scale biosafety guidelines developed and adopted by the NIH and published in Appendix K of the recombinant DNA guidelines (35, 36) are actually applicable to any large-scale work with microbial agents, especially those that pose a potential risk to human health (2).

Good biosafety practices can be selected or developed to allow the trained, healthy, adult worker to manage the risk of work with a biohazardous agent by using barriers that prevent exposure and infection without actually removing the biohazard itself (Table 1 and Appendix A). For example, a tagging system can be devised to notify coworkers and repair staff that equipment has been decontaminated. The removal of all potential hazards can be documented by a tag-out system, in which a separate section of a form is sent to appropriate parties (see sample tag, Fig. 1). Such practices, combined with appropriate containment devices and facility design, can be used to reduce the risk of exposure of the worker to a level that is low enough to prevent the transmission of an infectious disease.

Written documentation of the risk estimate and risk management protocols are to be placed in an infection control plan or biosafety manual. Regulatory requirements for the prevention of occupationally acquired blood-borne infectious diseases now mandate such documentation in an Exposure Control Plan (14).

Selection of containment practices requires a knowledge of the biohazardous materials in use,

WORK AUTHORIZATION/SAFETY TAG CLEARANCE

REQUESTOR ______________ PHONE NO. ______________ DATE ________

ITEM/EQUIPMENT ______________________ WORK ORDER NO. ________

BUILDING __________ ROOM __________ Write problem/comments on back of the tag.

CHECK SERVICE REQUIRED: ______ Repair ______ Manufacturer service ______ Surplus/discard

CHECK AND IDENTIFY (attach list of hazardous materials that contaminated this item/equipment):
______ Biological ______ Chemical ______ Radiological

DECONTAMINATED BY ______________ Supervisor ______________

____

CLEARED BY SAFETY ______________ Date ________

Send White—Safety Yellow—Facilities Engineering Tag—With equipment

FIGURE 1 Equipment tag for safety.

knowledge or assumption of the health status of the worker, and a thorough understanding of the practices, procedures, and volume amounts of biohazardous agents or materials used in the process. The increased severity (risk group) rating is reflected by the requirement for a higher containment level, from BSL 1 to 4, for work with a biohazardous agent. The risk assessment is further quantified by a job safety analysis, in which each task involving the use of a biohazard is carefully reviewed to identify the best work practices and equipment to be used (see also Chapter 17). The importance of evaluating the epidemiologic triad (agent-host-environment) rather than the use of a simple agent risk group classification in establishing a containment level has been emphasized by the National Research Council Committee on Hazardous Biological Substances in the Laboratory (10), the CDC, and the NIH in their biosafety guidelines (44).

The work practices recommended in the United States come from the CDC/NIH guidelines *Biosafety in Microbiological and Biomedical Laboratories* (44), the NIH *Guidelines for Research Involving Recombinant DNA Molecules* (35), the CDC guidelines for hospital infection control (7, 8, 18, 19), the National Committee on Clinical Laboratory Standards (33, 34), and the prudent practices recommended by the National Academy of Sciences (NAS) (10). They must also include the OSHA standard on blood-borne pathogens (14).

The WHO (47) and the Organization of Economic and Cooperative Development (26, 27, 40) have attempted to lay the framework for a standard code of practice. The European Economic Community has provided a list of the WHO-recommended risk groups for biohazardous agents to accompany a comprehensive directive on the protection of workers from exposure to biological agents in the workplace (15, 39). Their concerns for worker safety harmonize well with our own precautionary measures, differing only in the classification of a few endemic agents. The European Economic Community has also published several directives relating to the contained used or planned release of genetically modified organisms. There is a need for recognition in the United States that the large-scale guidelines now recorded in Appendix K of the recombinant DNA guidelines (35, 36) also apply to nonrecombinant work with biohazardous agents (2, 11).

HAZARD MANAGEMENT

Biosafety Containment Levels

Background

In the 1970s, etiologic agents were classified by the CDC on the basis of hazard, and a few directions were given for safe laboratory practices (5) based upon earlier studies (46). That classification, originally compiled for the safe transport of specimens to the CDC, still exists in slightly modified form in many government documents. Because of many taxonomic changes, and the recognition that all species of organisms may contain strains that are more or less virulent or even avirulent, the lists currently in use need to be revised. For appropriate identification of the category to be used in packaging, labeling, and shipping etiologic agents, the lists provided by the CDC in 1974 (5) and in the 1980 revision of 42 CFR Part 72 (6) are no longer relevant. The shipper must perform the risk assessment.

Published Guidelines of Containment Levels

The biosafety guidelines commonly used in the United States for containment of work with biohazardous agents are those recommended by the CDC, the NIH, and the NAS (10, 35, 44, 45). Other guidelines found in the literature are usually interpretations of the recommendations of these agencies (14, 24, 33, 34).

NIH: Guidelines for research with recombinant DNA. The CDC Classification of Etiologic Agents on the Basis of Hazard (5) was slightly modified and incorporated into Appendix B of the *NIH Guidelines for the Use of Recombinant DNA* (35). This classification implied to the user that all members of the groups, species, and strains on that list were pathogenic. The original CDC classification system, on which the NIH Appendix B was based, provided criteria and "points to consider" in estimating the degree of hazard by noting that it depended on the etiologic agent and the nature and kind of use. These points were not included in the NIH guideline. Instead, the authors acknowledged the guidelines as not able to take into account all existing and anticipated information on special procedures. Users were encouraged to *recommend changes in the guidelines* rather than being empowered by the NIH to do a thorough risk assessment. Both the NIH guidelines on recombinant DNA and the CDC/NIH guidelines evaluate the risk from the "product" rather than the "process" of the work itself, as recommended by the NAS below (11). The basic biosafety requirements for the work are the same.

National Academy of Sciences. 1. *Biosafety in the Laboratory: Prudent Practices for the Handling and Disposal of Infectious Materials* (1989) (10): To avoid exposure to infectious agents, the National Research Council, Committee on Hazardous Biological Substances in the Laboratory, recommended seven basic prudent biosafety practices. These prudent practices provide barriers against the known routes of exposure for most diseases and are recommended for work with any biohazardous agents. The basic principles and the barriers they provide are outlined in Table 1. The NAS recommendations are to be supplemented with additional practices, equipment, and facility design as the severity of the hazard increases. The practices recommended by the NAS, when accompanied by recommendations for facility design and containment equipment are basically the same as the CDC and the NIH biosafety containment levels.

2. *Introduction of Recombinant DNA-Engineered Organisms into the Environment: Key Issues* (1987) (11): The NAS has concluded that there is no evidence of any unique hazard posed by recombinant DNA techniques. The risks associated with recombinant DNA are the same in kind as those associated with unmodified organisms or organisms modified by other means. The NAS recommended that risk assessments be based on the nature of the organism and the environment into which it is introduced and not on the method by which it was produced. This recommendation, for risk assessment of the "product" rather than the "process," was also accepted by the Office of Science and Technology Policy.

Given this conclusion, the redundancy in guidelines, one for recombinant work and the other for work with human pathogens, could be eliminated. Guidelines for a single code of practice for protection of the worker from exposure to biohazardous agents are appropriate and have already been accepted in Europe (39, 40).

CDC/NIH guidelines for microbiological and biomedical research laboratories (44). In the mid-1980s, the CDC and the NIH developed agent summary statements that provided recommendations for specific containment levels for etiologic agents of human disease that pose a risk for workers in biomedical and microbiological laboratories. These BSLs were not derived directly from a simple classification of the pathogenicity of the biohazardous agent because such a qualitative risk assessment would have been too static for use in worker protection. The requirements of BSL 1 to 4 are provided in the CDC/NIH guidelines (44), which are reprinted as Appendix I of this book.

In using the CDC/NIH guidelines, the laboratory director is to take into account interactions of the virulence of the agent, immune status of the host, and the hazards of the procedure. The four containment levels described are to be used to establish requirements for work practices, as well as equipment and containment facilities. Under these guidelines, the laboratory director is also responsible for the appropriate risk assessment of agents not included in the Agent Summary Statements and for the use of appropriate practices, containment equipment, and facilities for work with the agent. The laboratory director is allowed some flexibility in resolving the safety requirements of a specific work environment but is expected to be familiar with the subject of risk assessment regarding biological agents or seek the advice of one who has such expertise.

Although risk evaluation is the responsibility of the laboratory director, no method is suggested for determining the virulence of the agent. The CDC/NIH guidelines do not account for the use of most agents that have not been reported as a laboratory-acquired infection (e.g., *Haemophilus influenzae* type b) unless they pose a serious hazard to a healthy adult as a laboratory-acquired infection. Information on human communicable diseases (4) and the relative risk of isolates from human clinical specimens (22) is available in the literature.

The workplace can create unnatural situations that increase the risk of employee exposure and infection. Information regarding industrial-scale use and the potential pathogenicity for plants and animals is not currently available as Agent Summary Statements. The NIH recombinant DNA guidelines (35, 36) provide assistance in Appendix K for developing specific practices for large-scale work. Regulations and guidelines from the Department of Agriculture and the Environmental Protection Agency are available to enable the user to evaluate plant and animal pathogens on the basis of known risk levels (38). A mechanism for a timely review of the existing data on agent pathogenicity for humans, plants, and animals is needed to facilitate user compliance.

Criteria for the Use of Containment Levels for Human Pathogens

Biohazard risk management is attained through the practices, facilities, and equipment of the biosafety containment levels. Biosafety practices are an important part of a program to manage the risk of exposure to potentially infectious agents. The general agent descriptions and applicable work environments described by the CDC and the NIH for each of the BSLs are listed below as compiled directly from the NIH guidelines (35, 36) and those from the CDC (5) and the CDC/NIH (44). Some of the specific guidelines are presented in Tables 2 and 3. The guidelines are translated into barriers to specific routes of exposure in Appendix A; some specific suggestions for safe pipetting are provided in Appendix B.

The standard practices of the CDC/NIH guidelines are required at all four BSLs. Also, special practices are added for the higher levels of containment.

Biosafety Level 1. BSL 1 is used for work involving well-characterized strains of viable microorganisms not known to cause disease in healthy adult humans and of minimal potential hazard to laboratory personnel and the environment. This level is appropriate for high school and undergraduate college teaching or training laboratories. No special competence is required, although training in the specific procedures is to be provided, and supervision is to be by a scientist with general training in microbiology or a related science. The laboratory is not necessarily separated from general building traffic patterns, and work is generally conducted on the open bench using standard microbiological practices. Examples of organisms used under these conditions are *Bacillus subtilis*, *Naegleria gruberi*, and canine hepatitis virus. Much of the recombinant DNA work with *Escherichia coli* K-12 and *Saccharomyces cerevisiae* has been approved at BSL 1.

BSL-1LS. For large-scale work (scaleup or production), BSL-1LS is used for those agents handled at BSL 1 in small scale. Microbial agents that have been safely used for large-scale industrial production for many years may qualify for good large-scale practices status (e.g., *Lactobacillus casei*, *Penicillium camembertii*, *S. cerevisiae*, *Cephalosporium acremonium*, *Bacillus thuringiensis*, *Rhizobium meliloti*). These microbes are used in the production of beer, wine, bread, and cheese. The criteria for good large-scale practices was originally developed for the European Community through the Organization for Economic Cooperation and Development as good industrial large-scale practices (17, 26, 27, 40) and was slightly revised by the NIH for acceptance and use in the United States (36). These large-scale practices are appropriate for work with nonpathogenic agents, not just those developed through recombinant DNA techniques.

Biosafety Level 2. BSL-2 is used for work with many moderate risk agents present in the community (indigenous) and associated with human disease of varying degrees of severity. Agents are usually of moderate potential hazard to personnel and the environment. BSL 2 is appropriate for clinical, diagnostic, teaching, and other research facilities in which work is performed by individuals with a level of competency equal to or greater than one would expect in a college department of microbiology. Users must be trained in good microbiological

TABLE 2 Laboratory and large-scale BSL criteria[a]

Biosafety level criteria	Laboratory scale				Large scale			
	BSL 1	BSL 2	BSL 3	BSL 4	GLSP	BSL-1LS	BSL-2LS	BSL-3LS
Procedures								
Implement institutional safety code	+	+	+	+	+	+	+	+
Written instructions and training	+	+	+	+	+	+	+	+
Biosafety manual	−	−	+	+	−	−	+	+
Good occupational hygiene	+	+	+	+	+	+	+	+
Compulsory shower out	−	−	−	+	−	−	−	−
Provision and use of appropriate PPE	+/−	+	+	+	+	+	+	+
Good microbiological techniques	−	+	+	+	−	+	+	+
Surfaces disinfected daily or after spills	+	+	+	+	+	+	+	+
Mouth-pipetting prohibited	+	+	+	+	+	+	+	+
Eating, drinking, smoking prohibited	+	+	+	+	+	+	+	+
Unnecessary use of sharps (hypodermics) restricted	−	+	+	+	−	+	+	+
Bench-top work prohibited	−	+/−	+	+	−	+/−	+	+
Internal accident reporting	+	+	+	+	+	+	+	+
Medical surveillance	−	+/−	+	+	−	+	+	+
Operation and equipment								
Biohazard sign	−	+/−	+	+	−	−	+	+
Restricted access	−	+/−	+	+	−	+	+	+
No children under 12 allowed (none under 15 at BSL 4)	−	+	+	+	−	+	+	+
No animals allowed except as part of experiment or process	−	+	+	+	+	+	+	+
No plants allowed except as part of experiment or process	−	−	+	+	+	+	+	+
Control aerosols	−	+	+	+	+/−	+	+	+
Laboratory doors closed when agent is in use	+/−	+	+	+	+/−	+	+	+
Insect and rodent control program	+/−	+	+	+	+	+	+	+
Special equipment (BSCs) required	−	+/−	+	+	−	+	+	+
Vacuum line protected by overflow flask (F) and/or filter (f)	F	F/Ff	Ff	Ff	F	Ff	Ff	Ff
Equipment and facilities								
Handwashing facility	+/−	+	+	+	+	+	+	+
Changing facility	−	−	+/−	+	−	−	+	+
Special engineering design	−	−	+	+	−	+	+	+
Monolithic construction	−	−	−	+	−	−	−	+
Emergency shower facility	−	−	−	+	−	−	−	+
Laboratory separated from general traffic	−	−	+	+	−	−	+	+
Airlock	−	−	−	+	−	−	−	+
Effluent from sinks and showers collected and treated	−	−	−	+	−	−	−	−
Effluents from process decontaminated	−	−	−	+	−	+	+	+
Solid waste decontaminated or packaged	+/−	+	+	+	−	+/−	+	+
Liquid waste decontaminated	−	+	+	+	−	+	+	+
Site negative pressure controlled	−	−	+	+	−	−	−	+
HEPA filters in air ducts	−	−	−	+	−	−	−	+
Area hermetically sealable for decontamination	−	−	+	+	−	−	−	+
Site designed to contain large losses	−	−	−	+	−	+	+	+

[a]+, Required; −, not required; +/−, required when appropriate; GLSP, Good Large Scale Practices.

TABLE 3 Large-scale biosafety criteria[a,b]

Biosafety level criteria	GLSP	BSL-1LS	BSL-2LS	BSL-3LS
Work with viable microorganisms should take place in closed systems that minimize (m) or prevent (p) aerosol release	–	m	pe	pe
Aerosol release is minimized (m) or prevented (p) by treatment of exhaust air or gas from closed systems	–	m	p	p
Aerosols are to be controlled during sampling from closed systems	mp	me	pe	pe
Aerosols are to be controlled during addition of materials to closed systems	mp	me	pe	pe
Aerosols are to be controlled during removal of materials, products, and effluents from closed systems	mp	me	pe	pe
Penetrations of closed systems are designed to minimize or prevent release of agent	–	m	p	p
Closed system has monitoring and sensing devices to monitor and validate integrity	–	+	–	+
Closed system is kept at low pressure	–	–	+	+
Closed system is not opened until sterility of fluid with viable organisms is validated	–	+	+	+
Closed system is permanently identified for records	–	+	+	+

[a]Based on reference 36.
[b]m, Minimize; me, minimize by engineering; mp, minimize by practices; p, prevent; pe, prevent by engineering; +, required; –, not required.

techniques to allow the use of these agents on the open bench when the potential for aerosol production is low. Laboratory personnel must have specific training in handling pathogenic agents and must be directed by competent scientists. Users must be trained in proper use of biological safety cabinets (BSC) or other appropriate primary containment equipment when the risk of aerosol production is high, as when centrifuging, grinding, homogenizing, blending, vigorous shaking or mixing, sonic disruption, opening containers with increased internal pressure, inoculating animals intranasally, harvesting infected tissues from animals or eggs, or harvesting human cells from tissues using a cell separator. Access to the laboratory is limited when work is in progress. Primary hazards to workers include accidental inoculation, exposure of nonintact skin or mucous membranes, or ingestion. Examples of organisms used under BSL 2 conditions are hepatitis B virus, *Salmonella enteritidis*, *Neisseria meningitidis* (9), and *Toxoplasma gondii*.

BSL-2LS. In vitro work with 10 liters or more of culture using agents requiring BSL 2 in small scale requires BSL-2LS containment. A detailed description of the containment requirements for BSL-2LS can be found in Appendix K-III of the recombinant DNA guidelines (35, 36). Requirements are also outlined in Tables 2 and 3.

Biosafety Level 3. BSL 3 is used for work with indigenous or exotic agents in which the potential for infection by aerosols is real and the disease may have serious or lethal consequences. Indigenous and exotic agents vary by country and within regions of some countries; thus there must be some flexibility for assignment of containment levels. BSL 3 is appropriate for certain clinical diagnostic work when tuberculosis or brucellosis is suspected and also for special teaching and research situations that require the handling of such agents. Partial containment equipment, such as class I or class II BSCs (44), is used for all manipulations of infectious material at BSL 3. Worker competency must equal or exceed that of college-level microbiology and include *special* training in handling these potentially lethal human pathogens and infectious materials. Supervisors must be competent scientists who are trained and experienced in working with these agents.

The most important route of exposure of workers to BSL 3 hazards is by inhalation. The extra personal protective clothing required at BSL 3 can serve as a reminder of the hazard level and promote an awareness that reduces incidents involving ingestion or accidental autoinoculation. Special engineering design criteria for BSL 3 laboratories involve the access zone, sealed penetrations, and directional airflow to protect the employee from respiratory exposure. Examples of organisms used under BSL 3 conditions are *Mycobacterium tuberculosis*, *Brucella suis*, St. Louis encephalitis virus, Borna virus (an exotic agent when used in the United States), and *Coxiella burnetti*.

BSL-3LS. Although agents requiring this level of containment are rarely used in production, the detailed requirements for BSL-3LS are to be found in Appendix K-IV of the NIH recombinant DNA guidelines (35, 36) and are outlined in Tables 2 and 3.

Biosafety Level 4. BSL 4 is used for work with dangerous and exotic agents that pose a high individual risk of an aerosol-transmitted laboratory infection that could result in a life-threatening disease. Such agents have a low infectious dose and usually pose a danger for the community due to person-to-person spread. BSL 4 containment is appropriate for all manipulations of potentially infectious diagnostic materials and isolates. These precautions are also appropriate when handling animals that are naturally or experimentally infected with agents in WHO risk group IV. Maximum containment equipment (e.g., a class III BSC) or partial containment equipment (class II BSC) in combination with a full-body, air-supplied, positive-pressure personnel suit is used for all procedures and activities. Examples of organisms used under these conditions are agents of viral hemorrhagic fevers, Lassa fever, Machupo, Ebola, and other filoviruses, arenaviruses, and certain arboviruses.

The highest hazard to laboratory or animal care personnel working with agents requiring such extreme caution and containment is respiratory exposure to infectious aerosols. Mucous membrane exposure to infectious droplets and accidental parenteral inoculation are expected to play a reduced role in laboratory-acquired infections at this level when appropriate personal protective equipment (PPE) and engineering controls are in place. Worker competency must equal or exceed that of college-level microbiology with specific, thorough training in handling extremely hazardous infectious agents. They must understand the function of the primary and secondary containment equipment and facility design. Supervisors must be competent scientists trained and experienced in working with such agents.

Laboratory access is strictly controlled. The facility is either separated from other buildings or completely isolated from other areas. A separate facility operations manual is required. The maximum containment facility has special design and engineering features that prevent dissemination of microbes to the environment. The requirements for laboratory-scale BSL 4 are described in reference 44 and are outlined in Table 2. Appendix K of the NIH recombinant guidelines does not include a description of requirements for BSL-4LS. The requirements for large-scale production at this level are to be determined on a case-by-case basis if and when they are requested.

Animal biosafety. Other special biosafety precautions described in the CDC/NIH guidelines are for use of naturally or experimentally infected animals. Separate animal BSLs are described for use in containment from BSL 1 through BSL 4. Animals are restricted from the laboratory unless they are a part of the experiment at BSL 2 and higher, and decorative plants are restricted from use at BSL 3 and higher. Animals and plants harbor their own microbial flora that could contaminate the work or even infect the worker; it is prudent to prohibit their use in all microbiology laboratories.

Interpreting Guidelines to Protect Employees from Exposure to Biohazardous Agents

To manage the risk of working with biohazards and to actually reduce the risk of exposure, one must identify and interpret the recommended practices that provide barriers to block the route(s) of entry into the worker. Barriers to direct or indirect contact with infectious agents, to percutaneous inoculation, and to ingestion of the agent or material are important at all BSLs (see Appendix A). Although barriers to infection by inhalation are required starting with BSL 2, sensitive individuals may require protection from allergens even at BSL 1. Ventilation requirements and access restrictions are important barriers to exposure. Prudent practices as barriers to block host entry are listed in Table 1. The reader is also referred to an extensive history of pipetting and pipetting accidents found in the first edition of this book. The summary of the safe use of pipettes and pipetting aids is important enough to be repeated here as Appendix B to this chapter.

Using the CDC/NIH guidelines, decisions on containment levels for work at BSL 1 through BSL 3 at small and large scale can be made at the local level. The expertise of the biosafety committee and/or biosafety officer is needed for a quantitative risk assessment of the proposed use of pathogenic agents and infectious materials. A professional biosafety consultant may be needed for small facilities. An excellent resource book, *Control of Communicable Diseases of Man* (4), describes the epidemiology of specific diseases as well as general methods for control in the community. Other valuable references include the tables of allowable exposure limits for chemical and physical agents, published by the American Conference on Industrial Hygienists. There are some plans to add biohazard limits to these tables in the future (1).

Risk assessment for agents to be used at large scale, especially for industrial production, should begin early in the research and development process. Information on pathogenicity should be known when the organism is to be scaled-up for production, because special large-scale containment practices and facilities are too costly to be provided if they are not actually needed.

Emergency Response Requirements

Selection and Training of First Responders

A biosafety liaison from the work area or from a safety department should be designated in advance and trained to respond to biohazardous incidents. Spill cleanup is actually a scaleup of routine decontamination procedures. Laboratory personnel, trained in microbiology and the use of aseptic technique, are expected to be trained and capable of cleaning up their own spills at BSL 1 through BSL 4. Current OSHA requirements under 29 CFR 1910.120(q) for worker safety in emergency response situations may require training for those involved in laboratory spill cleanup.

Laboratory personnel should be encouraged to take training in first-aid and cardiopulmonary resuscitation. They are considered "Good Samaritans" under the OSHA blood-borne pathogen standard (14) unless they are designated as responders as a part of their job. These workers should be thoroughly trained in the health and safety aspects of the laboratory (42).

If a registration procedure for biohazards is in place, a data base can be kept to indicate special hazards of concern, locations of use and storage, emergency response personnel available, and the responsible supervisor. This information can expedite the response in an emergency situation and make the timely identification of the hazard possible. The information posted on the door to the laboratory and stored in the data base should be updated as changes occur.

Equipment and Materials Required for Spill Cleanup

Examples of spill cleanup procedures may be found in the literature (44). Although it is important to have general guidelines for spill cleanup, the laboratory spill procedure must address the specific hazards used (23). An example of a generic protocol, with consideration for the inclusion of site-specific requirements, can be found in Fig. 2. Further information on disinfection, sterilization, and decontamination can be found in Chapter 14.

ASSESSMENT OF THE BIOSAFETY PRACTICES

Using Audits to Identify Problem Areas

Self-audits of the required safety practices provide a measure for compliance achievement. The criteria for the designated BSL can be used for the critical elements of the audit. A sample audit form is included as Fig. 3.

Regular safety audits may be carried out quarterly or semiannually by designated safety specialists accompanied by the laboratory supervisor and a facilities management representative. Deficiencies can be pointed out during the inspection. Later, a written report, with suggestions for corrective actions, can be sent to the laboratory supervisor. The supervisor is to report progress on remediation to the safety specialist within a fixed time period determined by mutual agreement.

The supervisor is to notify the safety specialist of any significant changes of biohazards in use and any change in personnel. The inspection program serves as a reminder to keep the biohazard data base up to date, triggering a review of the safety of the facility and the safety practices associated with the work.

Support of higher administration is absolutely essential for an audit to have the desired effects of improving employee safety as well as institutional compliance with federal, state, and local requirements. If some laboratories are designated as "off-limits" due to internal politics, the audit program is rendered useless and the credibility of safety personnel is lost.

Annual Biosafety Registration Review

An annual renewal of a biohazard registration procedure also helps to remind each responsible supervisor to review the work in progress and keep the information updated. A pathogen registration program provides a mechanism for such an update.

Incident and Accident Statistics

Some institutions judge the status of the safety program by a statistical review of changes in OSHA recordable incidents. The emphasis on small statistical changes in such figures does not usually highlight deficiencies in a biosafety program. The positive changes brought about through education and training in preventive methods can be measured by specific outcome audits (i.e., the effect of training and safety equipment on needlestick incidents). Trends in reports on such injuries show the cost-benefit of the changes. Observations of increases in such injuries, or the reporting of "sentinel

MEDICAL EMERGENCY

Immediate procedures:

- Remain calm.
- Initiate lifesaving measures if required.
- Call for emergency response.
- Do not move Injured person unless there Is danger of further harm.
- Keep Injured person warm.

MAJOR INCIDENT

Immediate procedures:

- Attend to injured or contaminated persons and remove them from exposure.
- Alert people to evacuate the area.
- Call for emergency response:
  - FIRE ____________
  - RADIATION SPILL ____________
  - CHEMICAL SPILL ____________
  - BIOLOGICAL SPILL ____________
  - PERSONAL INJURY ____________
- Close doors to affected area.
- Have person knowledgeable of incident and laboratory assist emergency personnel.

BIOLOGICAL SPILL

Notes and precautions: Biological spills outside biological safety cabinets will generate aerosols that can be dispersed in the air throughout the laboratory. These spills are very serious if they involve microorganisms that require BSL 3 containment, since most of these agents have the potential for transmitting disease by infectious aerosols. To reduce the risk of inhalation exposure in such an incident, occupants should hold their breath and leave the laboratory **immediately.** The laboratory **should not** be reentered to decontaminate and clean up the spill for at least 30 min. During this time the aerosol will be removed from the laboratory by the exhaust air ventilation system. Appropriate protective equipment is particularly important in decontaminating spills involving microorganisms that require either BSL 2 or BSL 3 containment. This equipment includes lab coat with long sleeves, back-fastening gown or jumpsuit, disposable gloves, disposable shoe covers, and safety goggles and mask or full-face shield. Use of this equipment will prevent contact with contaminated surfaces and protect eyes and mucous membranes from exposure to splattered materials.

Spill involving a microorganlsm requiring BSL 1 containment:

- Wear disposable gloves
- Soak paper towels in disinfectant and place over spill area.
- Place towels in plastic bag for disposal.
- Clean spill area with fresh towels soaked in disinfectant.

Spill involving a microorganlsm requiring BSL 2 containment:

- Alert people in immediate area of spill.
- Put on protective equipment.
- Cover spill with paper towels or other absorbent materials.
- Carefully pour a freshly prepared 1 in 10 dilution of household bleach around the edges of the spill and then into the spill.
- Avoid splashing.
- Allow a 20-min contact period.
- Use paper towels to wipe up the spill, working from the edges into the center.
- Clean spill area with fresh towels soaked in disinfectant.
- Place towels in a plastic bag and decontaminate in an autoclave.

Spill involving a microorganlsm requiring BSL 3 containment:

- Attend to Injured or contaminated persons and remove them from exposure.
- Alert people in the laboratory to evacuate.
- Close doors to affected area.
- Call biological spill emergency response number.
- Have person knowledgeable of incident and laboratory assist emergency personnel.

Additional site instructions:

FIGURE 2 Excerpts from the *Emergency Guide* distributed by the Howard Hughes Medical Institutes. Reprinted with permission of Dr. Emmett Barkley, Director of Safety, Howard Hughes Medical Institutes, Chevy Chase, Md.

Laboratory director:
Name:
Location:
Telephone no.:

	Yes	No	N/A	Comments
Laboratory facilities				
Benches				
Wiped down daily?				
Absorbent cover?				
Wiped down after spills?				
Special containment areas marked?				
Windows				
Do they open?				
Are screens in place?				
Handwashing				
Is a sink available?				
Soap, paper towels in place?				
Access				
Is access restricted?				
Is janitor access allowed?				
Is a door sign posted?				
Are doors kept closed?				
Equipment				
Is required equipment located outside the room?				
Are there appropriate labels on equipment containers where potentially infectious material is stored?				
Identify equipment and location:				
Shaker				
Centrifuge				
Sonicator				
Other				
Vacuum system				
Central				
Pump				
Aspirator				
None				
Vacuum shield in place?				
Pest control				
Have insects, rodents been observed?				
Reported?				
Work practices				
Pipettors				
Available?				
Used?				

FIGURE 3 Biosafety audit form for use in determining compliance with the CDC/NIH guidelines at BSL 1 to 3, including work with recombinant DNA. Modified from a form prepared by the Office of Biosafety, Harvard University, L. Harding and G. Casper.

	Yes	No	N/A	Comments
Sharps				
Needles used?				
Disposed of appropriately?				
Spills				
Are significant spills reported to PI?				
Are spill kits (paper towels and disinfectant) available?				
Can personnel handle spills of viable organisms?				
Centrifuge				
Is potentially infectious material spun?				
Are safety cups or rotors with O-rings used?				
Biosafety cabinet				
Used when infectious aerosols expected?				
Used for all manipulations of infectious materials?				
Protective clothing				
Available?				
Laboratory coat?				
Gown?				
Gloves?				
Other?				
Worn?				
Personal practices				
Handwashing				
Do personnel wash hands after work?				
Food, etc.				
Eating, drinking in laboratory?				
Smoking?				
Serum storage				
Collected?				
Stored appropriately?				

Decontamination

Disinfectants	Dilution available
Hypochlorite (chlorine bleach)	1%
Iodophor-detergent	2%
Ethanol	70%
Phenol (Lysol)	3%
Other	___

Decontamination of biologic wastes: For each of the following wastes indicate the decontamination method:

	Disinfectant	Dilution	Time	Steam	Container	Time
Disposable plastics (flasks, dishes)						
Pipettes (disposable, recyclable)						
Other recyclables						
Liquids						
Other						

Are autoclaving procedures verifed? Yes No

If yes, explain how

FIGURE 3 *Continued*

Training of personnel
Documented safety training
Documented hazard notification

Surveyor
Name
Title Date

FIGURE 3 *Continued*

events" or near misses that could have had serious consequences, highlight the need for intervention efforts. Biosafety practices must be tailored to the work and assessed for efficacy.

In summary, the tools for selecting the appropriate biosafety practices for work with biohazardous agents are currently available in the literature. They should only be used by a trained and competent laboratory director. The assistance of an expert in biohazard safety assessment should be obtained when information on quantitative risk assessment and containment requirements is needed.

APPENDIX A: BARRIERS TO BLOCK ROUTES OF TRANSMISSION

Barriers to Exposure by Known Routes of Transmission

The following are examples of barriers that block the known routes of transmission of infection. They can be used as an adjunct to the biosafety containment information from the CDC, the NIH, and the NAS.

1. *Barriers to prevent transmission by direct and indirect contact:* Procedures to prevent contact of infectious agents or contaminated materials with eyes, nose, mouth, mucous membranes, and/or nonintact or abraded skin include
   - A. Keeping potentially contaminated hands and materials (i.e., pencils) away from mouth and nonintact skin.
   - B. Bandaging or using an occlusive dressing on nonintact/abraded skin and wearing gloves when handling infectious agents or contaminated materials.
   - C. Using robotics or tools in place of hands; using automated bacterial culturing devices to replace the multiple blind sampling previously performed by hand.
   - D. Selecting and wearing PPE appropriate for the task, with the understanding that these are not completely impenetrable barriers.
     1) Hand protection: Latex, vinyl, or other appropriate protective gloves are to be worn when handling infectious agents or materials.
     2) Eye protection: Safety glasses are worn to prevent direct splashing into eyes and/or contact of contaminated hands with eyes. The extra protection of goggles or face shields is needed with certain tasks involving infectious liquids (e.g., homogenizing, blending, centrifuging) in which the potential for splashing and splattering exists.
     3) Body protection: Gowns and aprons are used to keep street clothes from contamination (BSL 2); jumpsuits or wrap-around gowns with ribbed cuffs are used in containment areas starting with BSL 3.
     4) Respiratory protection: Using a surgical mask to prevent contact of contaminated hands or materials or splashes with nose and mouth. Using a particulate respirator, a half-faced or full-faced respirator with HEPA cartridge for work with infectious aerosols when appropriate.
   - E. Removing the Hazardous Agent to Prevent Infection.
     1) Substitute nonpathogens for pathogens whenever possible, especially in teaching laboratories and in the development of processes for scaleup and production.
     2) Remove contamination from hands by vigorous handwashing with a mild soap for 10 full seconds for good frictional removal of transient materials. Antiseptic soaps should be reserved for those situations in which an extra level of cleanliness is required, as in survival surgery. Hands should be washed on leaving the laboratory, after gloves are removed, or more often if obviously soiled. If infectious agents are used for a series of tasks, hands should be washed when the batch or job is done.
     3) Decontaminate work surfaces before and after work and after the spill of a biohazardous agent. Prepare a written spill plan and train workers to use it in advance of a spill. A sample plan is illustrated in Fig. 2. Use appropriate physical methods (wet or dry heat, UV light, gamma radiation, filtration) or chemical methods (liquid or gas) to decontaminate equipment and wastes (see Chapter 14).
     4) Add disinfectant to water baths according to manufacturers' instructions to prevent contamination of the water bath and the outside of partially submerged containers.

F. Providing warning of hazards and advice about special requirements.
   1) Identify materials that need to be decontaminated; separate contaminated items from reusable or disposable items before reprocessing or disposal. Prepare a decontamination procedure as a section of the technical biosafety manual.
   2) Devise a tagging system to notify coworkers and repair staff that equipment has been decontaminated. Document the removal of all potential hazards by a tag-out system in which documentation is sent to appropriate parties (see sample tag, Fig. 1).

G. Restricting access as a barrier to direct and indirect exposure.

2. *Barriers to prevent percutaneous exposures:* Suggestions for protection from autoinoculation with infectious agents include but are not limited to the following:
   A. Eliminate the use of sharp objects, such as glass Pasteur pipettes, lancets, and needles to prevent percutaneous exposure; use a blunt or flexible cannula whenever possible.
   B. Use needles with self-storing sheaths.
   C. Keep sharp objects in view and limit use to one open needle at a time whenever possible.
   D. Use appropriate gloves to prevent cuts and skin exposure; look for new products with puncture-resistant features.
   E. Use puncture-resistant containers for the disposal of sharp objects, especially hypodermic needles.
   F. Handle animals with care to prevent scratches and bites.
3. *Barriers to prevent exposure by ingestion:* To prevent the ingestion of infectious agents or materials, workers should
   A. Use automatic pipetting aides (with filters at BSL 2 and above); do *not* mouth-pipette.
   B. Prohibit smoking, eating, drinking, and applying cosmetics in laboratories.
   C. Keep hands and contaminated items away from the mouth.
   D. Protect the mouth from splash and splatter. Use a face shield or wear a face mask if appropriate.
4. *Barriers to prevent exposure by inhalation:* Important physical barriers and engineering controls include
   A. Aerosol containment procedures to prevent, control, or decrease aerosol generation or dissemination:
      1) Use of absorbent materials on table tops to collect splashes and drips.
      2) Use of disinfectants or antiseptics to kill microbes in waste collection containers.
      3) Careful manipulation of equipment when pipetting the last drop in a pipette, removing needle from rubber diaphragm on vial, opening centrifuge tube, using a hot loop to take a sample from a liquid culture, opening a Waring blender, and homogenizing.
      4) Selection of appropriate PPE to protect user from splashes and splatter. This would include head, face, and body protection as mentioned in 1D above.
   B. Engineering controls for reduction of biohazardous aerosols
      1) Laminar-flow class I and class II BSCs provide some level of protection from aerosols. A class I cabinet can be used with a front-panel insert that has glove ports. *Note:* Clean-air benches are not containment devices and pose additional hazards.
      2) Centrifuge safety equipment: safety cups, rotors with covers, and O-rings protect against release of aerosols.
      3) Puncture-resistant needle disposal containers. Detailed requirements are mandated in the OSHA blood-borne pathogen standard (14).
      4) Containment homogenizers are commercially available and designed for use on the open bench. They are meant to be opened in a fume hood or BSC for filling or for emptying after use.
   C. Ventilation requirements are necessary to protect workers from biohazards. Specific requirements for BSL 3 and 4 may be found in reference 44, as reproduced in Appendix I of this book.

APPENDIX B: PIPETTING SAFELY

The elimination of pipetting hazards will provide increased safety with operational costs limited to the purchase price of pipetting devices. The safety regulations of many research and clinical laboratories still recommend rather than require the use of pipetting aids when pipetting infectious or hazardous materials. The NIH guidelines for research with recombinant DNA molecules prohibit mouth-pipetting as does the CDC/NIH *Biosafety in Microbiological and Biomedical Research Laboratories.*

Recommendations for the Safe Use of Pipettes and Pipetting Aids

1. Never suction or pipette by mouth; always use some type of a pipetting aid when pipetting infectious materials with a designated risk requiring BSL 3 facilities or operations or agents with a BSL 2 designation when protection from contamination is imperative. Preferably, all activities should be confined to a ventilated BSC or other acceptable primary barrier. Pipetting of toxic chemicals should be performed in a chemical fume hood or, when product protection is necessary, in a class II-type BSL 2 BSC. Mouth-pipetting should be prohibited even with mouth-pipetting devices that use an entire hydrophobic membrane filter not requiring fingers to touch the mouthpiece. This reusable pipetting device requires storage on the bench or other location between usage, which can result in contamination of the end piece that inserts into the mouth.

2. Infectious or toxic materials should never be forcefully expelled from a pipette. Mark-to-mark pipettes are preferable to other types, because they do not require expulsion of the last drop.

3. Infectious or toxic fluids should never be mixed by bubbling air from a pipette through the fluid.

4. Infectious or toxic material should not be mixed by alternate suction and expulsion through a pipette.

5. Discharge from pipettes should be as close as possible to the fluid or agar level, and the contents should be

allowed to run down the wall of the tube or bottle whenever possible, not dropped from a height.

6. Pipettes used for transferring infectious or toxic materials should always be plugged with cotton, even when safety pipetting aids are used.

7. Avoid accidentally dropping infectious or toxic material from the pipette onto the work surface. Place a disinfectant dampened towel or other absorbent material on the work surface, and autoclave the towel before discard or reuse. Disposable absorbent paper bonded to polyethylene is a suitable cover for the work surface and provides a waterproof lining.

8. Contaminated pipettes should be placed horizontally in a pan or tray containing enough suitable disinfectant, such as a solution of hypochlorite, to allow complete immersion of the pipettes. Pipettes should not be placed vertically in a cylinder that, because of its height, must be placed on the floor outside the cabinet. Removing contaminated pipettes from the BSC and placing them vertically in a cylinder provides an opportunity for dripping from the pipette onto the floor, or the rim of the cylinder, thereby creating an aerosol, and the top of the pipettes often protrude above the level of the disinfectant. An aerosol may also be produced by displacement of contaminated air from inside an unplugged pipette during vertical insertion into a cylinder. However, a tray containing the pipettes within the BSC can be covered and then removed and placed directly in the steam autoclave.

9. Discard pans for used pipettes are to be housed within a BSC.

10. The pan and used pipettes should be autoclaved as a unit and replaced by a clean pan with fresh disinfectant.

References

1. **American Conference of Governmental Industrial Hygienists, Inc.** 1991. *Threshold limit values and biological exposure indices, 1991–1992.* Cincinnati, Ohio.
2. **Andrup, L., B. Nielsen, and S. Kolvraa.** 1990. Biosafety considerations in industries with production methods based on the use of recombinant DNA. *Scand. J. Work Environ. Health.* **16:**85–95.
3. **Balows, A., W. J. Hausler, K. L. Herrmann, H. D. Isenberg, and H. J. Shadomy (ed.).** 1991. *Manual of clinical microbiology,* 5th ed. American Society for Microbiology, Washington, D.C.
4. **Benenson, A. S. (ed.).** 1990. *Control of communicable diseases in man,* 15th ed. American Public Health Association, Washington, D.C.
5. **Center for Disease Control, Office of Biosafety.** 1974. *Classification of etiologic agents on the basis of hazard,* 4th ed. U.S. Department of Health, Education and Welfare, Public Health Service, Atlanta.
6. **Centers for Disease Control.** 1980. *Title 42 code of federal regulations (42 CFR) part 72.2 transportation of diagnostic specimens, biological products and other materials; part 72.3 transportation of materials containing certain etiologic agents: minimum packaging requirements.* Superintendent of Documents, Government Printing Office, Washington, D.C.
7. **Centers for Disease Control.** 1981. Guidelines on infection control. Guidelines for hospital environmental control. *Infect. Control* 2:131–146.
8. **Centers for Disease Control.** 1982. Guidelines on infection control. Guidelines for hospital environmental control(cont.). *Infect. Control* **3:**52–60.
9. **Centers for Disease Control.** 1991. Laboratory-acquired meningococcemia—California and Massachusetts. *Morbid. Mortal. Weekly Rep.* **40:**46–47, 55.
10. **Committee on Hazardous Biological Substances in the Laboratory, National Research Council.** 1989. *Biosafety in the laboratory: prudent practices for the handling of infectious materials.* National Academy Press, Washington, D.C.
11. **Committee on the Introduction of Genetically Engineered Organisms into the Environment, National Research Council.** 1987. *Introduction of recombinant DNA-engineered organisms into the environment: key issues.* National Academy Press, Washington, D.C.
12. **Department of Labor, Occupational Safety and Health Administration.** 1970. Public Law 91–596. December 29, 1970. CFR Title 29. Part 1904.8 Section 5 (a). General Duty Clause.
13. **Department of Labor, Occupational Safety and Health Administration.** 1987. CFR Title 29. Part 1910.1450. Occupational exposure to hazardous chemicals in laboratories. 52 FR 31877, Aug. 24, 1987, as amended at 52 FR 45080, Dec. 4, 1987; 53 FR 15035, Apr. 27, 1988; 54 FR 24334, June 7, 1989; 54 FR 6888, Feb. 15, 1989.
14. **Department of Labor, Occupational Safety and Health Administration.** 1991. CFR Title 29. Part 1910.1030 Protection from bloodborne pathogens *Fed. Regist.* **56**(235)**:**64175–64182.
15. **European Economic Community.** 1991. EN Document No. 4645/1/91 EN. Draft proposal for a council directive, amending directive 90/679/EEC.
16. **Findlay, B., and S. Falkow.** 1989. Common themes in microbial pathogenicity. *Microbiol. Rev.* **53**(2)**:**210–230.
17. **Frommer, W., B. Ager, L. Archer, G. Brunius, C. H. Collins, R. Donikian, C. Frontali, S. Hamp, E. H. Houwink, M. T. Kuenzi, P. Kramer, H. Lagast, S. Lund, J. L. Mahler, F. Normand-Plessier, K. Sargeant, G. Tuijnenburg Muijs, S. P. Vranch, and R. G. Werner.** 1989. Safe biotechnology. III. Safety precautions for handling microorganisms of different risk classes. *Appl. Microbiol. Biotechnol.* **30:**541–552.
18. **Garner, J. S., and M. S. Favero.** 1986. CDC guideline for handwashing and hospital environmental control. *Am. J. Infect. Control* **7:**231–243.
19. **Garner, J. S., and B. P. Simmons.** 1983. CDC guideline for isolation precautions in hospitals. *Am. J. Infect. Control* **4:**245–325.
20. **Groschel, D. H. M., K. G. Dwork, R. P. Wenzel, and L. W. Scheibel.** 1986. Laboratory accidents with infectious agents, p. 261–266. *In* B. M. Miller, D. H. M. Groschel, J. H. Richardson, D. Vesley, J. R. Songer, R. D. Housewright, and W. E. Barkley (ed.), *Laboratory safety: principles and practices.* American Society for Microbiology. Washington, D.C.
21. **Hartree, E., and V. Booth (ed.).** 1977. *Safety in biological laboratories.* Biochemical Society Special Publication No. 5. The Biochemical Society. London.
22. **Isenberg, H. D., and R. F. D'Amato.** 1991. Indigeneous and pathogenic microorganisms of humans, p. 2–14. *In* A. Balows, W. J. Hausler, K. L. Herrmann, H. D. Isenberg, and H. J. Shadomy (ed.), *Manual of clinical microbiology,* 5th ed. American Society for Microbiology, Washington, D.C.
23. **Kent, P. T., and G. P. Kubica.** 1985. Safety in the laboratory, p. 5–10. *In Public health mycobacteriology. A guide*

for the level III laboratory. Department of Health and Human Services, Public Health Service, Atlanta.
24. **Kruse, R. H., W. H. Puckett, and J. H. Richardson.** 1991. Biological safety cabinetry. *Clin. Microbiol. Rev.* **4:**207–241.
25. **Kuehne, R. W.** 1989. Personal protection and hygiene, p. 209–226. *In* D. Liberman and J. G. Gordon (ed.), *Biohazards management handbook.* Marcel Dekker, Inc., New York.
26. **Kuenzi, M., F. Assi, A. Chmiel, C. H. Collins, M. Donikian, J. B. Dominguez, L. Financsek, L. M. Fogarty, W. Frommer, F. Hasko, J. Hovland, E. H. Houwink, J. L. Mahler, A. Sandkvist, K. Sargeant, C. Sloover, and G. Tuijnenburg Muijs.** 1985. Safe biotechnology. General considerations. A report prepared by the Safety in Biotechnology Working Party of the European Federation of Biotechnology. *Appl. Microbiol. Biotechnol.* **21:**1–6.
27. **Kuenzi, M., L. Archer, F. Assi, G. Brunius, A. Chmiel, C. H. Collins, T. Deak, R. Donikian, I. Financsek, L. M. Fogarty, W. Frommer, S. Hamp, E. H. Houwink, J. Hovland, H. Lagast, J. L. Mahler, K. Sargeant, G. Tuijnenburg Muijs, S. P. Vranch, J. G. Oostendorp, and A. Treur.** 1987. Safe biotechnology 2. The classification of microorganisms causing diseases in plants. *Appl. Microbiol. Biotechnol.* **27:**105.
28. **Liberman, D., and J. G. Gordon (ed.).** 1989. *Biohazards management handbook.* Marcel Dekker, Inc., New York.
29. **Liberman, D., and L. Harding.** 1989. Biosafety: the research/diagnostic laboratory perspective, p. 151–170. *In* D. Liberman and J. G. Gordon (ed.), *Biohazards management handbook.* Marcel Dekker, Inc., New York.
30. **Lynch, P., M. J. Cummings, P. L. Roberts, M. J. Herriott, B. Yates, and W. E. Stamm.** 1990. Implementing and evaluating a system of generic infection precautions: body substance isolation. *AJIC* **18:**1–12.
31. **Lynch, P., M. M. Jackson, M. J. Cummings, and W. E. Stamm.** 1987. Rethinking the role of isolation practices in the prevention of nosocomial infections. *Ann. Intern. Med.* **107:**243–246.
32. **Miller, B. M., D. H. M. Groschel, J. H. Richardson, D. Vesley, J. R. Songer, R. D. Housewright, and W. E. Barkley (ed.).** 1986. *Laboratory safety: principles and practices.* American Society for Microbiology, Washington, D.C.
33. **National Committee for Clinical Laboratory Standards.** 1991. *M29-T2 protection of laboratory workers from infectious disease transmitted by blood, body fluids and tissue, tentative guideline,* 2nd ed. National Committee for Clinical Laboratory Standards, Villanova, Pa.
34. **National Committee for Clinical Laboratory Standards.** 1991. *I17-P protection of laboratory workers from instrument biohazards, proposed guideline.* National Committee for Clinical Laboratory Standards, Villanova, Pa.
35. **National Institutes of Health.** 1986. Guidelines for research involving recombinant DNA molecules. 51 FR 16958, May 7, 1986, as amended at 52 FR 31848, Aug. 24, 1987; 53 FR 28819, July 29, 1988; 53 FR 43410, Oct. 26, 1988; 54 FR 10508, March 13, 1989; 55 FR 7438, March 1, 1990; 55 FR 37565, Sept. 12, 1990.
36. **National Institutes of Health.** 1991. Action under the guidelines, NIH guidelines for research involving recombinant DNA molecules. *Fed. Regist.* **56**(138):33174–33183.
37. **Nettleman, M. D.** 1991. Biological warfare and infection control. *Infect. Control Hosp. Epidemiol.* **12**(6):368–372.
38. **Office of Science and Technology Policy.** 1986. Coordinated framework for the regulation of biotechnology. *Fed. Regist.* **51:**23302–23393.
39. **Official Journal of the European Communities.** 1990. Directive 90/679/EEC on the protection of workers from risks related to exposure to biological agents at work. No. 374: 1–12, 31.12.1990.
40. **Organization for Economic Cooperation and Development.** 1986. *Recombinant DNA safety considerations.* Organization for Economic Cooperation and Development, Paris.
41. **Richardson, J. H., E. Schoenfeld, J. Tulis, and W. Wagner. (ed.).** 1986. Proceedings of the 1985 Institute on Critical Issues in Health Laboratory Practice: Safety Management in the Public Health Laboratory. The DuPont Co., Wilmington, Del.
42. **Schwartz, C. (ed.).** 1988. *Health and safety training in the biotechnology laboratory: a practical guidebook.* Occupational and Environmental Health Center at the Cambridge Hospital, Cambridge, Mass.
43. **Songer, J. R.** 1986. Laboratory safety program organization, p. 1–9. *In* B. M. Miller, D.H. M. Groschel, J. H. Richardson, D. Vesley, J. R. Songer, R. D. Housewright, and W. E. Barkley (ed.), *Laboratory safety: principles and practices.* American Society for Microbiology, Washington, D.C.
44. **U.S. Public Health Service, Centers for Disease Control, and National Institutes of Health.** 1993. *Biosafety in microbiological and biomedical laboratories,* 3rd ed. HHS Publication No. (CDC)93–8395. U.S. Department of Health and Human Services, Washington, D.C.
45. **U.S. Public Health Service, National Institutes of Health.** 1974. *National Cancer Institute safety standards for research involving oncogenic viruses.* Publication No. (NIH) 75–790. U.S. Department of Health Education and Welfare, Washington, D.C.
46. **Wedum, A. G.** 1964. Laboratory safety in research with infectious aerosols. *Public Health Rep.* **79**(7):619–633.
47. **World Health Organization.** 1993. *Laboratory biosafety Manual,* 2nd ed. World Health Organization, Geneva.

Decontamination, Sterilization, Disinfection, and Antisepsis

DONALD VESLEY AND JAMES L. LAUER

14

INTRODUCTION

The decontamination of cultures and items contaminated by biohazardous agents is a vital step toward protection of laboratory workers from infectious disease. The decontamination process is also necessary to prevent release of such agents into the community at large. Those releases that have been reported primarily involve person-to-person rather than environmental transmission (4). Conversely, sterilization of media and equipment is vital to the integrity of any microbiological manipulation as a component of standard quality control practices. Microbiological safety processes of decontamination and sterilization are taken for granted by most laboratory workers. Considerable confusion remains over terminology, specific techniques to be used, and the effectiveness of a given microbiological process for its intended purpose.

The emergence of human immunodeficiency virus as a major pathogen in the 1980s raised great concerns in those laboratories working with that virus. Fortunately, early evidence published by Martin et al. (22) indicated that human immunodeficiency virus is very susceptible to virtually all common germicides used in the laboratory. No heroic germicidal measures were necessary, and emphasis was rightly placed on protecting workers from large-scale exposure to concentrated virus. Further information on human immunodeficiency virus may be found in Chapter 4.

In this chapter, workable terminology and practical recommendations are proposed for use in various laboratory situations. Publications on the theoretical aspects of microbiological action and the intricacies of specific microbicidal agents are available in the literature (5). This chapter focuses on practical aspects of routine microbicidal tasks in the microbiology laboratory and the categories or objectives of these tasks. Pitfalls, limitations, and hazards associated with the use of microbicidal agents are identified along with suggested procedures for many specific tasks.

TERMINOLOGY OF MICROBICIDAL ACTION

Microbicidal terminology must be defined. Sterility implies the killing of all living agents. The dilemma of testing for sterility is discussed by Kelsey in *The Myth of Surgical Sterility* (17). Sterility testing is impractical due to its destructive nature, the large number of samples needed to detect even a relatively high percentage of nonsterile items, and the ever-present possibility of false-positives due to laboratory contamination. The alternative, of basing destruction cycles on experimentally determined D values (the time required for a 90% reduction in number of surviving microbes under specific test conditions), also has pitfalls because D values are also subject to variation. This criterion has the built-in difficulty of requiring assignment of a probability of sterility caused by the logarithmic decline in survivors, which implies an infinite time to reach zero, the only accepted definition of sterility. Brown and Gilbert (6) proposed abandonment of the term *sterility* in favor of *safe for its intended use,* provided that an "acceptable" probability of sterility can be demonstrated. Campbell (7) proposed that this probability be expressed in terms of a "microbiological safety index," a number representing the logarithm of the reciprocal of the theoretical number of survivors. Thus, an autoclave pack with an assumed initial bioload of 10^6 resistant spores, subjected to a cycle transversing 12 logs (e.g., a 24-min cycle based on a D value of 2 min), would end up with 10^{-6} survivors, for a microbiological safety index of 6. It is obvious that this concept also depends on assumptions that cannot be verified easily. However, in practice, autoclave sterilization cycles probably grossly overestimate the initial spore load and the D value, thus creating a large margin of safety if the steam does, in fact, penetrate to all organisms in or on the item to be sterilized.

The difficulties inherent in assessing sterility are compounded when the objective is decontamination for safety purposes. Decontamination is usually defined as the destruction or removal of microorganisms to some lower level, but not necessarily total destruction. Logically, the safety objective can only be met by a sterilization process when a high-risk etiologic agent (Biosafety Level [BSL] 3 or 4) is present, but it may be met by a lesser microbicidal treatment for lower-risk agents. Thus, decontamination may variably be achieved by the application of sterilization, disinfection, or antisepsis treatments as the situation dictates. The situation is further complicated by the nature of the material to be decontaminated, which often includes large quantities of organic matter such as microbiological media. Decontamination then requires increasing cycle time or increasing the concentration of the microbicidal agent. Verification of decontamination under these conditions is more difficult than for conventional quality control sterilization processes carried out on "clean" items. Under such circumstances, chemical and/or biological indicators can be used to determine the assumption of sterility (15). Chemical indicators can confirm exposure of the indicator to minimum time and temperature conditions. Recent technical advances have produced indicators that integrate time and temperature progression in a steam sterilizer cycle, providing confirmation of theoretically adequate microbicidal conditions at a given chamber location.

The biological indicator comes closest to a legitimate measure of microbial spore kill because a negative result actually requires the total destruction (or at least nonrecoverability) of a given number of resistant spores (*Bacillus stearothermophilus* for steam and *Bacillus subtilis* subsp. *globigii* for ethylene oxide or dry heat). Even biological indicators, however, have limitations (e.g., variable resistance levels and a variable number of spores inoculated on the strip). Also, biological indicators do not compensate for the high organic loads inherent in decontamination and thus are of questionable value for those tasks. However, they remain the most popular indicators of total microbial destruction. In using either chemical or biological indicators, the location of the indicator for meaningful interpretation must be at a point in the container and the chamber that is most resistant to steam penetration.

Obviously, a biological indicator (spores) cannot be used to determine the effectiveness of nonsporicidal chemical germicides. The success of such compounds can only be assumed if the manufacturer's recommendation for dilution and contact time are followed. The Environmental Protection Agency registration of such products requires that the Association of Official Analytical Chemists use-dilution test be performed and satisfactorily passed. Attempting to quantify the degree of destruction achieved under given circumstances (e.g., microbiological sampling of surfaces subjected to germicidal treatment) may be useful in the evaluation of specific treatments but is generally not cost-effective for routine monitoring. Because we do not have consistently reliable methods for confirming the degree of destruction actually achieved in day-to-day microbiological activities, the objectives of decontamination can be reached by selecting empirically acceptable techniques and applying them diligently.

Decontamination

Decontamination is routinely required in microbiological laboratories. The laboratory environment is subjected to contamination with both infectious and noninfectious agents, often in the presence of extraneous organic matter (e.g., culture media). For this reason, greater concentrations of the decontaminating agent and longer contact times are usually recommended for decontamination treatments than for sterilization or disinfection of clean items or media. The steam (121 °C) cycle for sterilization is 20 to 30 min, but that for decontamination is 1 h; disinfection techniques call for 100 ppm of chlorine, but 10,000 ppm is recommended for decontamination (emergency spill cleanup). Decontamination requires the application of microbicidal steam, gas, solid (granular), or liquid chemical agents in situations in which microbes may be protected from contact by extraneous matter. It is not possible to assume total destruction of either spores or vegetative microbes under such conditions.

Sterilization, disinfection, and antisepsis are all forms of decontamination. Liquid cultures are decontaminated before being discharged to the sewer; instruments are decontaminated (sometimes by both a liquid chemical soak and an autoclave cycle) before being washed and sterilized for reuse. Even routine floor cleaning can be considered decontamination of a contaminated surface that is treated with a germicide mixed with a cleaning agent (detergent-disinfectant) to make it suitable for further use.

The objective of decontamination is to render a contaminated material safe for further handling, whether it is an instrument to be washed and sterilized for reuse or a material to be disposed of as waste. For high-risk agents, total destruction may be required for adequate safety. Materials for reprocessing or disposal are treated to kill the hazardous agent(s) before removal from the laboratory under BSL 3 and 4 containment.

Sterilization

The term *sterilization* is reserved for

1. Processing of clean, prewrapped items in steam sterilizers or ethylene oxide autoclaves using cycles that consistently produce negative results with chemical and biological indicators. It is assumed that the packaging and wrapping will permit penetration of the steam or gas to all areas of interest and then will protect against post-treatment recontamination.
2. Liquids, such as microbiological culture media, sterilized in steam autoclaves to ensure the integrity of the experiment.
3. Items decontaminated by steam or gas with increased cycle times to compensate for heavy organic loads.

Disinfection

The term *disinfection* implies the use of antimicrobial agents on inanimate objects (e.g., work surfaces, equipment items). The purpose of disinfection is to destroy all nonsporeforming organisms that could pose a potential hazard to humans or compromise the integrity of the experiment. Thus, disinfection may be carried out for safety and/or quality control purposes. The term *pasteurization* is used when heat treatment of a liquid or a subboiling wet-heat treatment is applied to an object for the destruction of nonsporeforming organisms.

Antisepsis

The term *antisepsis* refers to the application of a liquid antimicrobial chemical to living tissue (human or animal). The objective is to prevent sepsis, either by destroying potentially infectious organisms or by inhibiting their growth and multiplication. No sporicidal activity is implied. This category includes activities such as the swabbing of an injection site on a research animal or handwashing with an antiseptic agent. With handwashing, the objective includes preventing the spread of infectious or contaminating agents for safety and quality control.

Sterilants, Disinfectants, and Antisepsis

Words such as *sterilant, disinfectant,* or *antiseptic* should be used to denote products or processes capable of achieving sterilization, disinfection, or antisepsis, respectively, under ideal conditions.

PRECAUTIONS, LIMITATIONS, AND TOXIC HAZARDS OF MICROBICIDAL AGENTS

The capacity for DNA disruption that makes antimicrobial agents useful may also make them dangerous to personnel as well as disruptive to research microbes. The more obvious of these undesirable properties can be mitigated by common-sense precautions, but some hazards are more subtle and require greater attention to potential dangers. For example, the hazard of burns from escaping steam can be overcome by good autoclave design and maintenance, together with thorough training of personnel. The risk of exposure to ethylene oxide remains controversial. In 1984, the Occupational Safety and Health Administration (OSHA) standard was reduced to 1.0 ppm (time weighted average), with an action level of 0.5 ppm. These levels replaced a previous standard of 50 ppm and are near the limit, which can accurately be measured by currently available monitoring instruments. The National Institute for Occupational Safety and Health has published a comprehensive review of ethylene oxide hazards (14). More recently, glutaraldehyde

TABLE 1 Human hazards associated with selected antimicrobial agents

Antimicrobial agent	Indicated hazard	Reference
Physical agents		
Steam (autoclave)	Burns from escaping steam or hot liquids; cuts from exploding bottles	Many incidents reported
	Aerosols and chemical vapors from improper vacuum exhaust	Barbeito and Brookey (1)
UV light	Corneal and skin burns from direct or deflected light	Many incidents reported
Gases		
Ethylene oxide gas	Eye and respiratory irritant, skin desiccant, mutagen, potential carcinogen	Glaser (14)
Formaldehyde gas	Highly irritating; toxicity and hypersensitivity	National Institute for Occupational Safety and Health (24)
Liquid chemicals		
Alcohol (isopropyl)	Acute toxicity	Freireich et al. (12)
	Contact dermatitis	Fregert et al. (11)
Chlorine	Gaseous form highly toxic; liquid Cl_2 not toxic at active dilutions	CDC-MMWR (9)
Glutaraldehyde	Contact dermatitis	Sanderson and Cronin (28)
Hexachlorophene (bisphenol)	Acute neurotoxin	Mullick (23)
Iodine (iodophors)	Skin irritation	Zinner et al. (32)
Phenols	Occupational leukoderma	Bentley-Phillips (3)
	Depigmentation	Odom and Stein (25)
	Idiopathic neonatal hyperbilirubinemia	CDC (8)
Quaternary ammonium compounds	Minor contact dermatitis	Shmunes and Levy (29)
Anesthetic		
Chloroform	Weak carcinogen (animal models only)	Mansuy et al. (21)

has emerged as a distinct occupational health threat with an OSHA-imposed ceiling limit of 0.2 ppm.

The literature is replete with reports of acute toxic effects for virtually every class of antimicrobiological agent used in laboratories. Chronic effects have been difficult to document, and the projection of mutagenic, carcinogenic, or teratogenic consequences remains speculative. Effects on animals have been better documented than those on humans, because animal models are used for toxicity evaluations. The complication for microbiology laboratories is obvious, however, when the effect of microbicidal agents on research animals becomes an important quality control consideration.

Table 1 summarizes some of the harmful effects reported for the various types of microbicidal agents used in laboratories. A few representative literature citations are included.

In many instances, toxic hazards are related to overuse of microbicidal agents at concentrations well above the recommended use-dilution. In fact, deaths from hexachlorophene and the neonatal hyperbilirubinemia problem are examples of that type of careless mistake (8). There is a variety of precautions related to microbicidal treatments that stem from theoretical knowledge of potential hazards. Barbeito and Brookey (1) described the potential for aerosols to escape from a vacuum autoclave with air exhausted before the steam cycle. Methods are described to ensure that autoclave design will not allow such escape. Another theoretical potential for hazards exists in the practice of autoclaving pans of items that are soaked in a chlorine solution, because it is possible for poisonous gaseous chlorine to be generated during autoclaving. Generally, the high organic load in such a situation will serve to neutralize the chlorine, but a worthwhile precaution would be to add sodium thiosulfate (1 ml of 5% sodium thiosulfate per ml of 5% hypochlorite ion). This would also help to prevent corrosive action of

TABLE 2 Physical properties of commonly used laboratory plastics relative to decontamination or sterilization[a]

Plastic	Maximum use temperature (°C)	Transparency	Suitability for:			
			Microwave	Steam autoclave	ETO[b] or formaldehyde gas	Dry heat
Low-density polyethylene	80	Translucent	Yes	No	Yes	No
High-density polyethylene	120	Translucent	No	—[c]	Yes	No
Polypropylene	135	Translucent	Marginal	Yes	Yes	No
Nylon (polyamide)	90	Translucent	No	No	Yes	No
Polymethylpentene	175	Clear	Yes	Yes	Yes	No
Teflon PFR (perfluoroalkoxy)	205	Translucent	Yes	Yes	Yes	Yes
Teflon TFE (tetra-fluoroethylene)	121	Opaque	—	Yes	Yes	No
Polycarbonate	135	Clear	Marginal	Yes[d]	Yes	No
Polyvinyl chloride	70[e]	Clear	Yes	No[e]	Yes	Yes
Polysulfone	165	Clear	Yes	Yes	Yes	Yes

[a]Used by permission of Nalge Co., Division of Sybron Corp., Rochester, N.Y.
[b]ETO, ethylene oxide.
[c]Can be autoclaved at 120°C for 20 min if containers are empty and uncovered.
[d]Mechanical strength may be reduced by autoclaving.
[e]Polyvinyl chloride in tubing can withstand 121°C.

chlorine on susceptible autoclave components. Chlorine gas toxicity has also been reported from mixing household bleach with a phosphoric acid cleaner (9). Another precaution is to limit the quantity of ethanol used in wiping down laminar-flow cabinets. The explosion potential of ethanol should restrict that quantity to 400 ml for a 4-ft cabinet and 600 ml for a 6-ft cabinet. These quantities are significantly above the amount needed for surface decontamination but refer to total quantities contained in the cabinet.

The damaging effect of microbicidal treatments on containers and other surfaces is a legitimate concern. Table 2 summarizes the characteristics of various plastics relative to steam, ethylene oxide, and dry-heat treatments. Table 3 indicates the relative resistance of these plastics to various categories of liquid chemicals. Centrifuge rotors and accessories

TABLE 3 Resistance of commonly used laboratory plastics to a variety of substances[a,b]

Plastic	Resistance to:						
	Weak acids	Strong acids	Alcohols	Aldehydes	Bases	Halogenated hydrocarbons	Strong oxidizing agents
Low-density polyethylene	E	E	E	G	E	N	F
High-density polyethylene	E	E	E	G	E	F	F
Polypropylene	E	E	E	G	E	F	F
Nylon (polyamide)	F	N	N	F	F	G	N
Polymethylpentene	E	E	E	G	E	N	F
Teflon—PFA (perfluoroalkoxy) or TFE (tetrafluoroethylene)	E	E	E	E	E	E	E
Polycarbonate	E	N	G	F	N	N	N
Polyvinyl chloride (bottles)	E	E	E	N	E	N	G
Polysulfone	E	G	G	F	E	N	G

[a]Used by permission of Nalge Co., Division of Sybron Corp., Rochester, N.Y.
[b]E, No damage after 30 days of constant exposure (may involve years); F, some effect after 7 days of constant exposure; G, little or no damge after 30 days; N, not recommended for continuous use.

are particularly susceptible to damage from chemicals that may be used for decontamination purposes. Manufacturer's instructions should always be followed for such practices. Generally, titanium rotors are more resistant than aluminum, and stainless steel tubes and caps are more resistant than most plastics. Many rotors, including O-rings and gaskets, can be autoclaved at 121°C for 1 h without damage (2).

The concentrations of microbicidal agents recommended for decontamination tasks are frequently significantly higher than the conventional use-dilution recommended by the manufacturer. Therefore, precautions are all the more important when considering the risk versus benefit for a particular activity.

GENERAL CONSIDERATIONS

Some issues may be overlooked in a busy laboratory. The first of these has to do with the general laboratory clutter that occurs due to poor housekeeping and an increase in the use of disposable materials that are stored in the laboratory. The result is a working environment incompatible with decontamination needs. A requirement, even at the basic laboratory level, is for daily wipe-down of bench tops with a suitable disinfectant. In fact, even this task is frequently ignored or is complicated by clutter. Because the correction of this situation requires an attitude change on the part of the worker, it can be difficult to implement despite its technical simplicity. The reality of laboratory safety depends on individual willingness and desire to comply rather than on technical considerations. The procedures for emergency cleanup become extremely cumbersome in the presence of clutter. Examples of obstacles to efficient cleanup include venetian blinds on windows and cardboard boxes stored on the floor. Thus, limiting the materials kept in the laboratory to those items specifically used in the current work will expedite decontamination tasks and, incidentally, contribute to quality control and to fire safety.

Another general consideration involves the separation and clear identification of items to be decontaminated. A "dirty" cart clearly labeled "To Be Autoclaved" is recommended for all laboratories where infectious disease agents are handled. The diligence and training of workers is essential to ensure that appropriate items are promptly placed on the cart. Autoclave cycles must be scheduled with sufficient frequency to avoid overloading of the cart and the consequent misplacement of contaminated items. Also, separate containers must be provided for reusable and disposable items to avoid problems with melting plastic.

RECOMMENDED PROTOCOLS

A list of various decontaminants, their effectiveness against different microbial groups, their important characteristics, and their most appropriate application in research and clinical laboratories is given in Table 4. However, for the effective use of these decontaminants, specific protocols must be followed.

Emergency Spill Protocols

It is recommended that every laboratory prepare a specific protocol to be followed in the event of a biohazardous spill. The protocol should include the assignment of responsibility for each step in the spill response. In most instances, spill cleanup should be assigned to the person who perpetrated the accident. Some institutions may choose to assign the biosafety officer to this task, particularly when high-risk or even moderate-risk agents are involved.

To make the cleanup protocol effective, all materials designated in the written procedure must be readily available and functional on a continuous basis. At a minimum, the materials available for emergency cleanup should include suitable disinfectants, paper towels, forceps, an autoclavable squeegee and dust pan, sponges, and autoclavable plastic bags. All laboratory personnel should be familiar with the procedures and should have immediate access to the necessary equipment. Periodic emergency drills should be scheduled to ascertain that this state of readiness indeed exists.

In a properly designed and equipped laboratory, spill cleanup will be enhanced by easily cleanable surfaces and properly installed equipment (either built-in or free-standing to permit cleaning access under and around all components). Minimizing unnecessary storage of supplies and removing items and equipment not needed in the day's activities will serve to simplify cleanup should a spill occur. Accessories such as venetian blinds or decorations should be minimized for the same reason.

Specific protocols should be developed for various potential emergency spill situations. The following common accidents are discussed in this section: (i) spill in a laminar-flow biological safety cabinet; (ii) spill on a laboratory bench or floor (distinction will also be made for containment versus noncontainment laboratories); and (iii) spill outside a laboratory (such as during transport of biohazardous agents).

Spill in a Laminar-Flow Biological Safety Cabinet
If a spill should occur within a laminar-flow biological safety cabinet, the cabinet should continue to operate, as it is designed specifically to contain aerosols and all exhaust air is filtered. Cleanup should be initiated as soon as possible, using a suitable disinfectant. A germicidal detergent such as a phenolic or iodophor should be used in wiping all reachable cabinet surfaces. If the cabinet incorporates a catch basin beneath the work surface, this should be flooded with the disinfectant. Alcohol is not recommended due to the explosive potential of large quantities.

At least 20 min of contact time with the germicide should be allowed before items are removed from the cabinet. All items within the cabinet should be packaged for transport to the autoclave or, for items that cannot be autoclaved, such as intact containers of viable cultures to be retained, wiped carefully with the disinfectant. The operator's gloves and long-sleeved gown should be considered contaminated and should be placed in a container for autoclaving after the cleanup process is completed. The cabinet should be run for 10 min or so after this process and before activity is resumed.

This procedure is recommended for all spills of moderate- or high-risk agents, but may be modified for low-risk agents (i.e., conventional cabinet wipedown may be used without flooding of the catch basin if a small quantity of an agent being handled under BSL 2 is spilled).

After spill cleanup, a decision must be made concerning the need for formaldehyde decontamination of the cabinet, including filters. In general, this should be performed for any high-risk agent (although BSL 4 usually implies the use of class III cabinets) and for a major spill of a moderate-risk agent.

The procedure recommended by the National Institutes of Health for formaldehyde decontamination should be followed (see Protocols for Laminar-Flow Biological Safety Cabinets below). This procedure uses either flake paraformaldehyde, at 0.3 g/ft^3, or 37% formalin. The humidity in the cabinet should be raised (by vaporizing water) to 70% or more, and the cabinet must be sealed with plastic sheets and tape to contain the formaldehyde gas. If the cabinet does not exhaust directly to the outdoors, a flexible hose connection must be used to remove the gas after decontamination. During the decontamination process, the exhaust must be sealed to achieve adequate contact time (1 to 3 h is recommended). After this period, the exhaust damper is opened and the cabinet is turned on to exhaust all residual formaldehyde. It is recommended that the cabinet be run at least overnight before normal activity is resumed. Formaldehyde is highly irritating and toxic, with an OSHA-established permissible exposure level of 2 ppm. Recent development of a vaporized hydrogen peroxide protocol for safety cabinet decontamination gives promise for replacing this cumbersome formaldehyde protocol.

Spills in a Laboratory
The procedure recommended assumes a spill of a moderate-risk agent and therefore also assumes that the agent is being handled in a containment laboratory with negative pressure (BSL 3). If that is the case, the initial response should be to warn others in the room (or suite of rooms) to leave as quickly as possible. The worker involved in the spill should remove contaminated clothing (in the containment laboratory, a wrap-around garment is worn) and place it in an autoclavable plastic bag for decontamination immediately on leaving the laboratory.

If the laboratory is not under negative pressure, it may be necessary to begin cleanup immediately after others leave. The advantage of the negative pressure is that it permits a suggested 30-min waiting period for any aerosol resulting from the spill to settle or be cleared by the ventilation system before cleanup is begun. During the waiting period, the cleanup materials can be assembled, and the person who will carry out the procedure can dress in suitable attire. For moderate-risk agents, disposable booties, a long-sleeved gown, single-use mask, and rubber gloves are suggested. If a high-risk agent is involved, a jumpsuit with tight-fitting wrists and a full-face respirator with high-efficiency particulate air filter cartridge should be considered.

The cleanup process in the laboratory should use a concentrated germicide, such as 5% bleach (50,000 ppm Cl_2) although a 1:10 dilution of household bleach is recommended by the Centers for Disease Control and Prevention (CDC) for blood spills. The germicide should be poured around the edges of the spill to avoid further aerosolization, or alternatively, paper towels soaked in the germicide can be placed over the spill area. The extent of the spill area must be determined to define the scope of the cleanup activity. It is at this point that the value of having removed extraneous items will be apparent, as all objects within the spill area must be decontaminated. Approximately 20 min of contact time should be allowed to ensure adequate germicidal action. All spill materials are then gathered with the autoclavable squeegee and dust pan and transferred to an autoclavable plastic bag, which may itself be put in

TABLE 4 Decontaminants and their use in research and clinical laboratories[a]

Decontaminant	Use parameters: Concn. of active ingredients	Temp. (°C)	Relative humidity (%)	Contact time (min)	Effective against: Vegetative bacteria	Lipo viruses	Tubercle bacilli	Hydrophilic viruses	Bacterial spores	Important characteristics: Inactivated by organic matter	Residual	Corrosive	Skin irritant	Eye irritant	Respiratory irritant	Toxic (absorbed or ingested)
Autoclave																
15 lb/in^2	Saturated steam	121		50–90	+	+	+	+	+							
27 lb/in^2	Saturated steam	132		10–20	+	+	+	+	+							
Dry-heat oven		160–180		180–240	+	+	+	+	+							
Incinerator	Heat	649–926		1–60+	+	+	+	+	+							
UV radiation (253.7 nm)	40 μW/cm^2			10–30	+		+	±		+[c]			+	+		
Ethylene oxide	400–800 mg/L	35–60	30–60	105–240	+	+	+	+	+		+		+	+	+	+
Paraformaldehyde (gas)	0.3 g/ft^3	>23	>60	60–180	+	+	+	+	+		+			+	+	+
Vaporized hydrogen peroxide	2.4 mg/L	4–50	<30	8–60	+	+	+	+	+			+	+	+	+	
Quaternary ammonium compounds	0.1–2%			10–30	+	+				+						+
Phenolic compounds	0.2–3%			10–30	+	+	+	±		±	+	+	+	+	±	+
Chlorine compounds	0.01–5%			10–30	+	+	+	+	±	+	±	+	+	+	+	+
Iodophor compounds	0.47%			10–30	+	+	+	±		+	+	+	+	+		+
Alcohol (ethyl or isopropyl)	70–85%			10–30	+	+	+	±		+				+		+
Formaldehyde (liquid)	4–8%			10–30	+	+	+	+	±		+		+	+	+	+
Glutaraldehyde	2%			10–600	+	+	+	+	+		+		+	+	+	+
Hydrogen peroxide (liquid)	6%			10–600	+	+	+	+	+		+	+	+	+		

[a]+, Very positive response; ± less positive response. A blank denotes a negative response or not applicable.
[b]Contact manufacturer of instruments.
[c]Soil and other materials are not penetrated by UV radiation.
[d]The pungent and irritating characteristics of formaldehyde preclude its use for biohazard spills.

a stainless steel bucket or pan. Larger items that can be autoclaved may be separately bagged for transport to the autoclave. A cart with side rails, to prevent materials from sliding off, should be used to transport the contaminated items by a direct route, avoiding populated areas, to the autoclave. It is for this reason that a steam autoclave should be located within a containment laboratory or very nearby. The specific precautions and parameters for use of the steam autoclave will be described in a later section. They are also outlined in Chapter 17.

Spills outside the Laboratory: During Transport

If a biohazardous agent is spilled during transport outside a laboratory, the main difference from the previous procedure would be to initiate the cleanup immediately, as in a noncontainment laboratory. Because it would already be too late to prevent aerosolization with its unpredictable consequences, the main emphasis should be on preventing spills during transport. All cultures that must be removed from the laboratory for incubation, refrigeration, or any other purpose should be placed in an unbreakable container that would prevent escape of any liquid or aerosol if it should be dropped. One-gallon or half-gallon paint pails are an inexpensive means of accomplishing this objective. A biohazard symbol should be affixed to containers used for this purpose. Protective plastic jug carriers are also commercially available.

Applications																					
Work surface maintenance	Floor maintenance	Biohazard spill, floor surfaces	Safety cabinet surface maintenance	Safety cabinet biohazard spill	Safety cabinet total decontamination	Large area and air systems decontamination	Water baths	Contaminated liquid—discard	Infectious laboratory waste	Contaminated glassware	Contaminated instruments	Equipment surfaces	Equipment total decontamination	Floor drains	Animal injection site	Contaminated animal bedding	Infected animal carcass	Microbial transfer loop	Books, paper, shoes	Electrical instruments	Lensed instruments[b]
								+	+	+	±					+	±				
								+	+	+	±					+	±				
										±	±					±					
									+							+	+	+			
			±																		
										±	±		±						+	+	
					+	+							+							+	
					+								+								
+	+	±	+	±			±			+		+		+							
+	+	+		+			+	+	+	+	+	+		+							
		+		+			±			+	+	+		+							
+		±	+	±			+			+		+									
+		±	+				±					+			+						
		±[d]						±	±	±	±	±									
										+	+	+									±
										+	+	+									+

Combination Spills (Biohazardous and Radioactive Materials)

In the event that a spill of biohazardous materials should also involve radioactive substances, the cleanup procedure becomes somewhat more complex. The institution's radiation protection officer should be consulted before cleanup begins to determine what, if any, modification may have to be instituted. The most likely type of radiation hazard in a microbiological laboratory will be either ^{14}C or ^{3}H, neither of which presents an external hazard. The spill should be surveyed to determine whether there is a need to protect against hand and body radiation exposures from more energetic beta and gamma emitters. Under the joint guidance of the biosafety officer and the radiation protection officer, the cleanup can proceed along the lines previously specified for laboratory biohazard spills. The radiation protection officer should evaluate the potential hazard of radioactivity release before the steam autoclave is used. Again, usually ^{14}C and ^{3}H can be autoclaved without hazard. However, ^{125}I may be of sufficient activity to preclude that step. In that case, the biohazardous agent may have to be inactivated by use of a compatible liquid chemical germicide before the shipment is packaged as a radioactive waste.

Do not use chlorine (Cl_2) as a liquid germicide if there is the potential for release of iodine (I_2) through the chemical reaction, as determined by the radiation protection officer. Either formalin or glutaraldehyde would be a suitable substitute.

After the area of the spill is cleaned up, a final radiation survey should be made. If the radiation level exceeds twice the background level, radiation decontamination is needed and should be carried out under the supervision of the radiation protection officer, using an approved detergent for radiation cleanup.

Handling of Infectious Waste

Since the publication of the first edition of this book there has been a great deal of attention given to the handling, treatment, and disposal of infectious wastes in laboratories and in medical facilities in general. Cultures and stocks of biological agents have invariably been included in the definition of infectious wastes. Deliberately infected research animals and sharps of all kinds have also been included. Thus, microbiological laboratories generate significant quantities of infectious wastes, which must now be handled under strict regulation, usually under a plethora of different state regulations. Uniform federal regulations for infectious waste have not been adopted (as of early 1994), although the potential for such regulation exists. For most laboratories, cultures and stocks slated for discard have routinely been decontaminated in steam autoclaves (per CDC/NIH guidelines) even before the recent publicity. Procedures for that task are described in this section. Infected research animals are generally not suitable for autoclaving or any of the alternative infectious waste treatment methodologies currently being advocated other than through incineration. Infected animals should not be processed through rendering plants. Thus, incineration remains the method of choice for most infectious and pathologic wastes.

Sharps are generally agreed to be the most hazardous (from an occupational health perspective) of the items in the infectious waste stream, particularly with current emphasis on blood-borne pathogens. Deposition of sharps (without recapping or clipping of needles) into puncture-resistant containers is the first and most important step in the handling process. Subsequent treatment and disposal can be by any of several means, either on or off site. An effective decontamination treatment (e.g., autoclaving, microwave, incineration) is considered to be a desirable option for sharps disposal after sharps are made unrecognizable through a shredding or incineration process. The key is to minimize personnel exposure (by leaving the sharps in the puncture-resistant containers) through the decontamination and destruction steps, whereupon the final residue can safely be subjected to final disposal directly into a landfill or as a component of incinerator ash into a landfill.

Protocols for Decontaminating Infectious Waste

All laboratories handling moderate- to high-risk agents should have a steam autoclave within a restricted area that serves several laboratories. For laboratories handling low-risk agents, an autoclave should also be available, preferably on the same floor and in the general vicinity of the laboratory.

Infectious laboratory waste (e.g., petri dishes, culture tubes, animal bedding, contaminated liquid) and contaminated instruments and glassware can be decontaminated in either a gravity displacement (most common type) or a prevacuum autoclave operating at 121 to 132°C. However, the processing time necessary to achieve decontamination will depend on specific loading factors. These factors include the type of waste container used (metal versus polypropylene), the use of autoclavable waste bags, the amount of water added to the waste, and the weight of the waste load (19).

Figures 1 through 5 show the average time-temperature profiles in waste relative to these waste loading factors. Figure 1 shows that when 1 liter of water was added to waste in a steel container and to waste in an autoclavable waste bag in the steel container, the temperature of the waste increased rapidly with processing time, reaching 120 to 121°C at 50 min. When water was not added to the steel container or to the autoclavable bag in the steel container, the temperature increase was significantly slower. Note that the worst time-temperature profile occurred when water was not added to the waste in the autoclavable bag.

Figure 2 depicts the average temperature of waste being processed in polypropylene containers. The data indicate that the addition of 1 liter of water to the container and to the autoclavable bag in the container did not significantly increase the temperature in the waste as compared with the trials with no additional water. Also, note that the time-temperature profiles of the waste in polypropylene containers are much lower than those in steel containers (Fig. 1).

The effect of the amount of water added to the waste in an autoclavable bag that is processed in a steel container is illustrated in Fig. 3. Note that 1 liter of water was better than 100 ml of water, which was better than no additional water.

Figure 4 shows that the average time-temperature profile was depressed (after 24 min of processing) by about 3 to 8°C when the amount of waste being

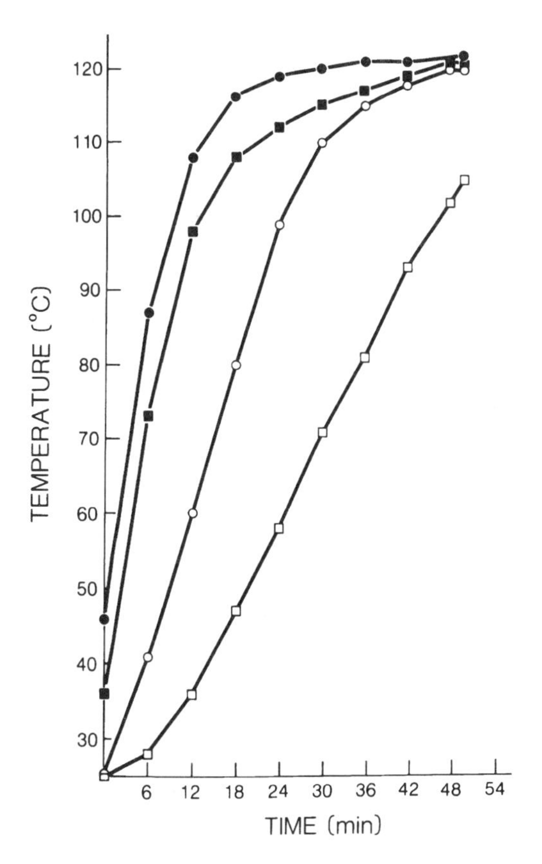

FIGURE 1 Average time-temperature profiles of waste loads (1,750 g) consisting of petri dishes (100 by 15 mm) containing agar. The loads were autoclaved in a steel container with (●) or without (○) 1 liter of water and in a steel container with an autoclavable plastic waste bag with (■) or without (□) 1 liter of water. Averages are based on three separate trials conducted in a gravity displacement autoclave that reached 121 to 122°C within 3 min after the cycle was initiated.

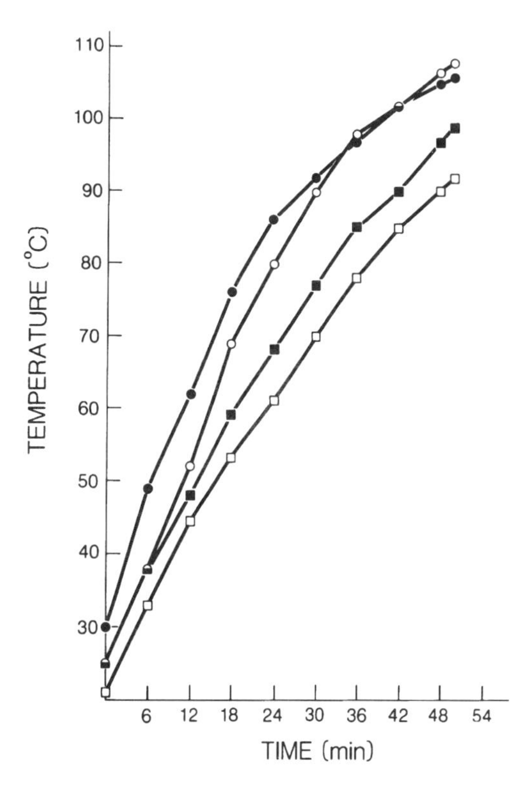

FIGURE 2 Average time-temperature profiles of waste loads (1,750 g) consisting of petri dishes (100 by 15 mm) containing agar. The loads were autoclaved in a polypropylene container with (●) or without (○) 1 liter of water and in a polypropylene container with an autoclavable plastic waste bag with (■) or without (□) 1 liter of water. Based on three separate trials conducted in a gravity displacement autoclave that reached 121 to 122°C within 3 min after the cycle was initiated.

processed was doubled. Figure 5 depicts the average time-temperature profiles when 1 liter and 500 ml of water were added to 10 lb of waste in an autoclavable bag, which was then processed in a steel pan. Note that the time-temperature profile was higher when 1 liter of water was used. These results are in contrast with the interpretation by Rutala et al. (27) that the addition of water does not significantly affect the average time-temperature profile of waste.

The data in these figures are based on studies using a gravity displacement autoclave. An inherent problem with this type of autoclave is that air entrapment will occur in upright containers. The temperature within an air pocket is much lower than that of the saturated steam surrounding it. Thus, materials in an air pocket will take longer to reach an adequate decontamination temperature (5) (see Chapter 17 for an outline of the safe operation of a gravity displacement autoclave). A prevacuum autoclave removes the air from the chamber and upright containers by pulling a vacuum before the saturated steam enters. This resolves the air entrapment problem, except when the prevacuum autoclave is operated on the liquid cycle. In the liquid cycle, air must be removed by gravity displacement because the mechanical vacuum is bypassed (liquid in containers could be forcefully drawn from the containers if a vacuum were applied).

We suggest that each laboratory review its present protocol to ensure that the processing time used for decontamination produces a temperature in waste of at least 115°C for 20 min. This may necessitate a total processing time of 60 to 90 min or more, depending on the loading conditions used. Also, extreme caution must be used when adding water to infectious waste so that aerosols containing infectious agents are not generated. Water, if added, should be trickled gently down the sides of the con-

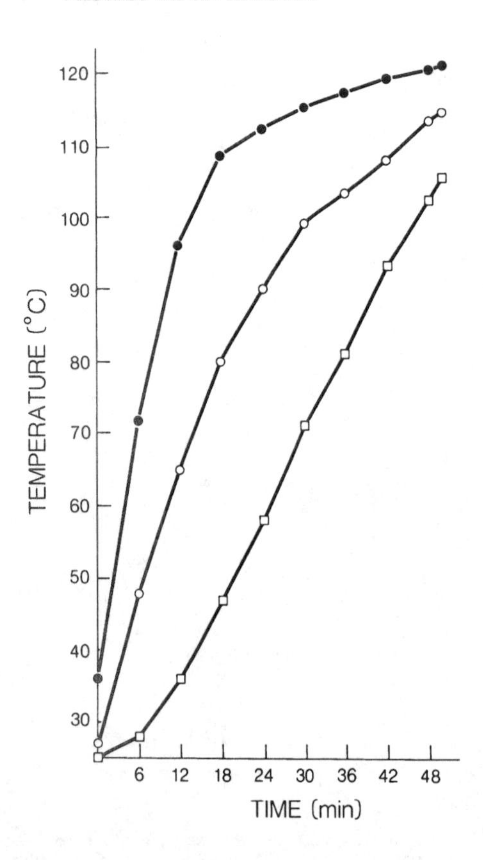

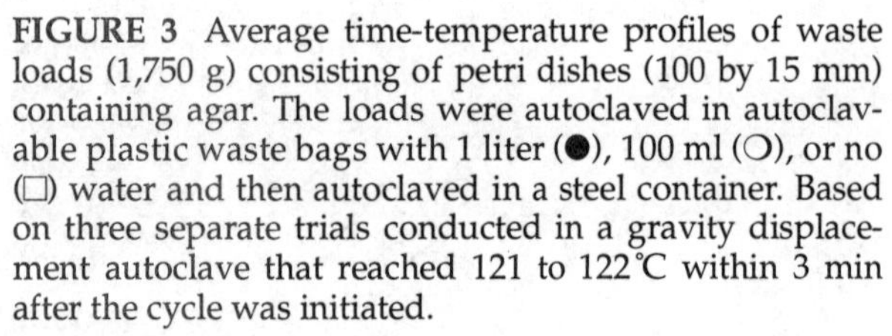
FIGURE 3 Average time-temperature profiles of waste loads (1,750 g) consisting of petri dishes (100 by 15 mm) containing agar. The loads were autoclaved in autoclavable plastic waste bags with 1 liter (●), 100 ml (○), or no (□) water and then autoclaved in a steel container. Based on three separate trials conducted in a gravity displacement autoclave that reached 121 to 122°C within 3 min after the cycle was initiated.

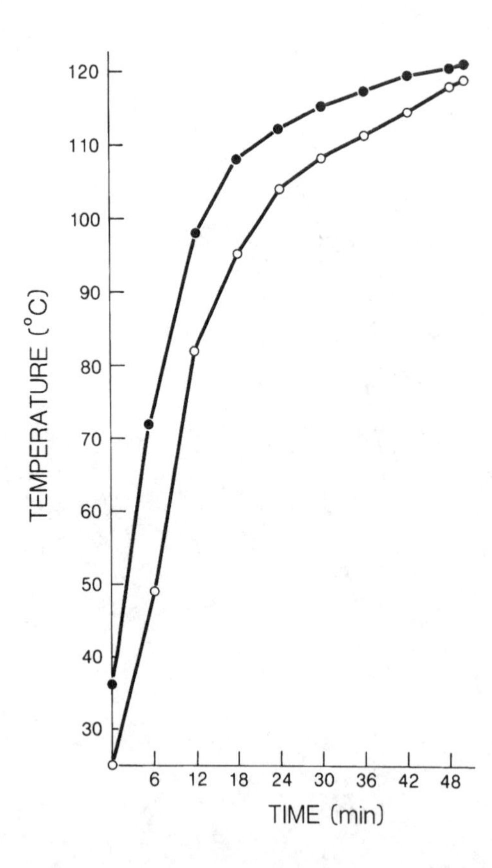

FIGURE 4 Average time-temperature profiles of waste loads 1,570 g (●) or 3,500 g (○) consisting of petri dishes (100 by 15 mm) containing agar. The loads were placed in autoclavable plastic waste bags with 1 liter of water and then autoclaved in a steel container. Based on three separate trials conducted in a gravity displacement autoclave that reached 121 to 122°C within 3 min after the cycle was initiated.

tainer rather than poured in directly, and any items added to the container after the water should be handled gently to avoid splashing.

Incineration can also be used to destroy infectious laboratory waste, contaminated animal bedding (bedding waste from animals infected with pathogenic agents), infected animal carcasses, and human pathologic waste. However, as for any other processes, specific design and operating criteria must be adhered to (5). Inadequate or incomplete incineration may well be accompanied by inadequate decontamination. These criteria will not be covered here but should be reviewed by persons responsible for operating the incinerator.

Incineration is the process of choice for the decontamination and destruction of infected and noninfected animal carcasses and for human pathologic waste (e.g., limbs, organs). It is recommended that such waste be placed in a designated labeled waste bag, preferably within a metal or reinforced cardboard container (e.g., "Infectious Waste for Incineration—DO NOT PLACE SHARP OBJECTS, NORMAL WASTE, ETC., IN THIS BAG"). This labeling is important to designate the contents of the bag and to alert custodial personnel to use greater care in handling it. The waste bag should be of durable material that is not easily ripped and can hold small amounts of liquid (e.g., 3-mm polyethylene bags, plastic-lined paper bags). Laboratory staff should then transport this bag to a secured, refrigerated storage area and place it in a covered metal container. Custodial personnel trained in the handling of infectious waste should transport the covered metal containers to the incinerator or to the loading dock where they will be picked up and delivered to the incinerator.

The emptied metal container should be decontaminated and washed at a site next to either the incinerator or the loading dock. The metal containers should be cleaned with a disinfectant-detergent, such as a phenolic or iodophor compound, followed by steam. An alternative to the metal containers

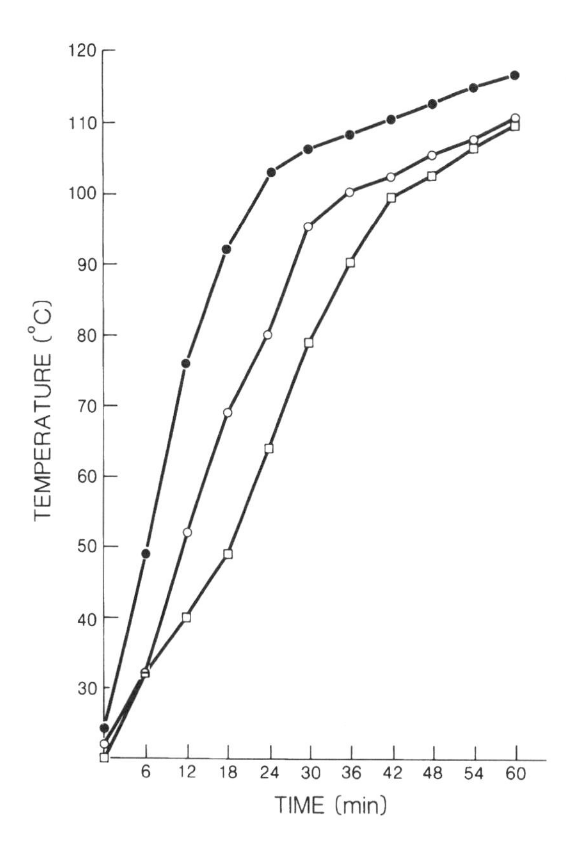

FIGURE 5 Average time-temperature profiles of waste loads (4.54 kg) consisting of petri dishes (100 by 15 mm) containing agar. The loads were placed in autoclavable plastic waste bags with 1 liter (●), 500 ml (○), or no (□) water and then autoclaved in a steel tray (53 by 36 by 13 cm). Based on three separate trials conducted in a gravity displacement autoclave that reached 121 to 122°C within 3 min after the cycle was initiated.

would be to use leakproof cardboard boxes or drums. These could be placed directly in the incinerator.

Rendering of noninfected animal carcasses and infected animal carcasses containing low-risk agents is also a potentially adequate decontamination-disposal process. Rendering plants may not accept small animals such as rats or mice. Also, one should contact the state agency responsible for regulating rendering plants to ascertain the requirements that must be complied with, especially as related to infected animal carcasses.

The decontamination or sterilization of glassware, instruments, and other materials can also be achieved by dry heat. Dry-heat ovens are used for glassware, instruments, and anhydrous materials such as oils, greases, and powder. The longer time necessary for sterilization is because dry heat is less effective in destroying microbes than is moist heat (steam) (5). However, for the sterilization of anhydrous materials and closed containers, dry heat is recommended over steam autoclaving. The reason for this is that the moisture component of saturated steam at 121 to 132°C, which is responsible for the rapid destruction of microbes, will not penetrate anhydrous materials and closed containers. Thus, the highest dry-heat temperature that materials and container materials will reach in an autoclave is 121 to 132°C. Obviously, the highest temperature reached in a dry-heat oven will depend on the temperature setting (i.e., 160 to 180°C). The contact time given in Table 4 takes into account the lag time for certain materials to reach the temperature in the oven and may or may not be applicable to all situations. Good examples of the times required to decontaminate various materials are presented by Reddish (26).

Contaminated glassware and instruments, regardless of the microbial agent present, can be decontaminated with a 2% alkaline glutaraldehyde solution. This method is a possible alternative for items that cannot tolerate steam autoclaving. Glutaraldehyde is commercially available as a 2% alkaline solution and should be used undiluted. Its use is mainly restricted to items that can be submerged and soaked in a covered container of the solution. The soak or contact time will depend on several factors, including the contaminating microbe. Items contaminated with large concentrations of bacterial spores should be soaked for 10 to 12 h. Lightly contaminated items with less-resistant microbes can be soaked for 10 to 30 min. The lumen in various instruments and equipment can also be decontaminated with this solution. If it is used in this manner, great care must be taken to ensure that the solution has reached all areas in the lumen. Also, if decontaminating a lensed instrument, make sure that the components of the instrument are compatible with glutaraldehyde. After decontamination, the item must be rinsed with sterile water to remove all residuals. Glutaraldehyde has been implicated as an occupational hazard due to sensitivity problems developing in workers using it as a high-level disinfectant in endoscopy units. OSHA has lowered the ceiling exposure limit to 0.2 ppm (16). Hydrogen peroxide (6%) has recently been suggested as an alternative to glutaraldehyde because of the severe sensitization problems, but caution must be exercised due to potential damage to instruments from this product.

Formaldehyde can also be used for much the same applications as glutaraldehyde, with a few exceptions. An 8 to 10% solution of formaldehyde applied for several hours is necessary to decontaminate items containing bacterial spores and some strains of atypical mycobacteria. Formaldehyde vapors from the solution are very pungent and irritating, making it extremely difficult to work with this decontaminant. Iodine, phenolic, and chlorine solutions are usually not recommended to be used in this manner because they are too corrosive, are rapidly inactivated by organic matter, or are not sporicidal. However, a chlorine or iodine solution (0.5 to 1.0%) may be used if the items are glass or plastic, the organic load is low, and a sporicidal activity is not needed. Phenolic solutions are not recommended because some phenolics corrode plastics and very high concentrations are needed to inactivate hydrophilic viruses.

Floor Care Protocols

Many chemical disinfectant-detergent compounds are commercially available for the general maintenance of laboratory and animal care area floors. Selection of a suitable compound depends on the microbicidal effectiveness needed, the corrosiveness of the compound, cost, and its residual and irritating properties.

Phenolic-detergent and quaternary ammonium compounds are the most commonly used decontaminants for floor maintenance. Phenolic-detergent-based compounds have a broad microbicidal spectrum and are less affected by organic matter than are the quaternary ammonium-based compounds. However, phenolic-detergent compounds are more toxic, and many can be absorbed through intact skin. Both types of compounds are sold as a concentrated solution and must be diluted per manufacturer's directions (1 to 3 oz [ca. 30 to 89 ml] per 128 oz [ca. 3,785 ml] of water) before use.

A two-bucket protocol is recommended for wet mopping of laboratory floors. The principle of this protocol is that fresh disinfectant-detergent solution is always applied to the floor from one bucket, and all spent solution removed from the floor is wrung from the mop and collected in the second bucket. In practice, the mop is saturated with the disinfectant-detergent solution from the first bucket, very slightly wrung into the second bucket, and applied to the floor using a side-to-side motion while slowly rotating the mop head. An area of 100 to 150 ft^2 is covered in this manner. Then after allowing at least a 5-min contact period, the solution is removed with a wrung-out mop. This procedure is continued until the total floor area is treated.

An alternative protocol would be to flood the floor area with the disinfectant-detergent solution and then, after 5 min, either pick it up with a wet vacuum or push it down a floor drain with a squeegee. This protocol requires that floors be completely sealed or of monolithic construction so that liquid cannot leak to adjacent areas. This protocol is more suitable to animal care areas where there is an absence of permanently placed benches and fixed equipment. The disinfectant-detergent solution is applied at a rate of about 1 gal/144 ft^2. If tank or wall sprayers are used to apply the solution, the setting on the nozzle should be adjusted so that the solution flows freely (not a spray) and is applied close to the floor.

When wet mopping, a freshly laundered cotton mop head should be used daily. Before a used mop head is sent to the laundry, it should be soaked in a freshly prepared disinfectant-detergent solution for 30 min or autoclaved. For laboratories handling moderate- to high-risk agents, the mop head should be autoclaved at 121°C for 20 to 30 min.

Sweeping and vacuum cleaning for general floor maintenance should not be allowed in laboratories handling infectious agents. Such cleaning techniques will readily disperse microbes present into the air.

Protocol for Floor Drains

If a floor drain(s) is present in a laboratory, a sufficient amount of liquid must be present in the trap to prevent the backup of sewer gases. Also, floor drains can serve as a reservoir for microbial agents. Thus, a disinfectant solution should be poured into the drain each month. A half-gallon (ca. 1.9 liter) of either a 1:100 dilution of household bleach (chlorine) or a phenolic-detergent solution is recommended.

Protocol for Water Baths

The addition of a disinfectant to the water in baths is recommended to resolve two potential concerns. First, inapparent contamination of the water during the incubation of pathogenic agents in the bath is not unlikely. Such contamination could lead to the transmission of this agent to susceptible staff. Second, water, including distilled, deionized, and tap water, usually contains nonpathogenic bacteria. Thus, laboratory experiments requiring a water bath could become contaminated with these bacteria, causing erroneous results. Because of this, it is recommended that either 0.1 oz (ca. 2.9 ml) of bleach (chlorine) or 1 oz (ca. 29 ml) of phenolic detergent be added to each gallon (3.8 liter) of water used in a bath. The phenolic disinfectant is preferred over chlorine because it is more stable, less corrosive, and less inactivated by organic matter. Propylene glycol has been used effectively as an alternative in cold-water baths, and if the bath can be turned up to temperatures of 90°C and more, that should be done as a decontaminating process for about 30 min every week.

Protocols for Laminar-Flow Biological Safety Cabinets

Surfaces in the work area of laminar-flow biological safety cabinets should be wiped down with a germicide both before and after each use. Before work is started, the cabinet should be operated (fan on) for 5 to 10 min. The surface of the work area is then decontaminated by wiping the area with a clean cloth soaked with a 70 to 85% solution of ethanol. If moderate-risk agents or hepatitis B virus have been handled in the cabinet, a wipe-down with a 2% iodophor solution, followed by alcohol to remove the iodine, would be more appropriate. After the completion of work, the same wipe-down procedure should be used while the cabinet is operating. Gloves and a gown should be worn during the wipe-down procedure. The cloth used should be placed in a container that will be subsequently autoclaved. The gloves and gown worn during the work may continue to be worn while cleaning up afterward, but after the final wipe-down, they should also be autoclaved.

UV radiation produced by germicidal lamps is also used within cabinets to decontaminate the work surfaces. UV radiation is effective against most microbes. Its effective use, however, requires an understanding of its abilities and limitations. The 253.7-nm wavelength emitted by a germicidal lamp has limited penetrating power. It is, thus, primarily effective against unprotected microbes on exposed surfaces or in the air (i.e., will not penetrate soil or dust). The intensity or destructive power decreases by the square of the distance from the lamp (ca. $I = 1/d^2$). Thus, exposure time is always related to distance. The intensity of the lamp will also decrease with time. Therefore, it should be checked with a UV meter at least yearly. The intensity from the lamp is drastically affected by the accumulation of dust and dirt on it. Therefore, unless the lamp is cleaned weekly or biweekly, the intensity from it may be inadequate. (Cleaning involves turning off the UV light and wiping off the surface of the bulb with alcohol.) Eye protection against direct or indirect (reflected from surfaces or clothing) UV exposure is required. The UV lamp should *never* be on while an operator is working in the cabinet. On the basis of these points, UV cannot be recommended as a decontaminant unless the lamp is properly maintained (checked yearly and cleaned often) and UV safety glasses are worn when there is a potential for exposure.

The decontamination of the total cabinet, which includes the work surfaces, the supply and exhaust filters, the surfaces of the air plenums, and the fan unit, is best achieved by exposing these areas to formaldehyde gas. Such decontamination is recommended (i) before maintenance work on cabinet; (ii) before filter changes; (iii) before the cabinet is moved; (iv) before performance tests that require access to the sealed portion of the cabinet; (v) after a gross spill of infectious material in the cabinet; or (vi) before the cabinet activity is changed from work with moderate- or high-risk infectious materials to work with noninfectious materials (31).

Caution: **Because of the potential for exposure to biohazardous agents as well as the hazardous chemicals (formaldehyde) used, decontamination of biological safety cabinets should only be performed by trained personnel who are monitored for potential formaldehyde exposure and enrolled in a respiratory protection program.**

Formaldehyde gas has a very pungent and irritating odor and is a toxic substance having a permissible exposure limit (maximum amount a worker can be exposed to, averaged over an 8-h workday) of 2 ppm. Repeated exposure to formaldehyde is known to produce a hypersensitive condition in certain individuals. Because of this and the potential for exposure to formaldehyde during cabinet decontamination, it is essential that a self-contained breathing apparatus or an industrial-type gas mask is immediately available. Also, it is strongly recommended that formaldehyde gas decontamination be performed only by individuals with sufficient experience and training.

Formaldehyde gas is most easily generated by heat-accelerated depolymerization of flake paraformaldehyde, which can be purchased commercially. The recommended protocol for use of this decontaminant is as follows:

1. Calculate the total cabinet volume in cubic feet by multiplying in feet the height, width, and depth of the cabinet.
2. The amount of paraformaldehyde to be used is determined by multiplying the volume of the cabinet in cubic feet by 0.3 g of paraformaldehyde. This amount will provide an equivalent concentration of 0.8% by weight or 10,000 ppm by volume in air. *Caution:* Be careful not to use a greater amount of paraformaldehyde than is required. Concentrations of formaldehyde over 8% in air are explosive! Also, an excess amount can cause the formaldehyde to polymerize on surfaces as a white powder, which

will necessitate a longer aeration period before the cabinet can be used.

3. Weigh out the amount of paraformaldehyde calculated and place in a plastic bag until it is needed. Flake paraformaldehyde is recommended because it has less tendency to splatter when heated.

4. Assemble the following equipment and materials:

Electric frying pan with a thermostatic control that can be set at 450°F

Polyethylene film (thickness, 3 to 4 mm)

Duct tape

Heavy-duty electrical extension cords (two)

Hot plate

Beaker of water (500 ml)

Temperature- and humidity-indicating devices

A sign labeled "Keep Out—decontamination with formaldehyde gas in process"

Safety glasses, face shield, or goggles

Industrial canister-type gas mask or a self-contained breather apparatus

Coveralls and gloves

Knife or scissors

5. Setup of equipment and materials and preparation of cabinet and exhaust air system

Attach the sign "Keep Out—etc." to cabinet.

Put on coveralls, eye protection, and gloves.

Spread the preweighed paraformaldehyde evenly in the bottom of the frying pan, which is preset at 450°F, and place the pan in the cabinet. Then attach the electrical cord of the pan to an extension cord and extend it out of the front of the cabinet. *Caution:* Do not connect the extension cord to the electric supply at this time.

Place the hot plate in the cabinet and put the beaker of water on it. Attach the plate's electric cord to the second extension cord and extend it out the front of the cabinet. Do not connect the electric supply at this time.

Place temperature- and humidity-indicating device(s) in cabinet so that the indicators can be seen through the front glass portion of the cabinet.

If the cabinet exhaust air port is not connected directly to the building's exhaust ventilation system, attach the flexible hose to the cabinet's exhaust port (seal and tape around this connection). Then extend the hose to a room air exhaust grill or an exterior window. Seal the opening at the end of the hose with plastic film and tape. *Caution:* If the room exhaust air is recirculated within the building, the hose must be extended to a window. Also, if the volume of air exhausted through the room grill is less than the volume of air being exhausted by the cabinet, extend hose to window.

If the cabinet exhaust air port is connected directly to the building's exhaust ventilation system, close the exhaust damper. *Caution:* The building's exhaust ventilation system should not in this case be connected to the building's air recirculation system. However, check to make sure that the air is not recirculated. If it is, extend hose to window and seal as in the step above.

Now tape all cracks and seams of the cabinet to make it as gas tight as possible.

Then, using the plastic film and the duct tape, seal the work opening of the cabinet. Tape around the protruding electrical cords should produce a seal that is gastight.

Plug in the extension cord attached to the hot plate until the temperature is about 74°F and the relative humidity is between 60 and 80%.

Now plug in the electric fry pan and allow it to heat until about half the paraformaldehyde has depolymerized (about 10 min). Then turn on the cabinet fan for 3 to 5 s to disseminate the gas to inaccessible areas.

Depolymerize the remaining paraformaldehyde and then turn on the fan again for 3 to 5 s. Allow the cabinet to stand for a minimum of 1 and preferably 3 h.

After 1 to 3 h of exposure, begin the ventilation process by removing the film from the flexible hose or by opening the exhaust damper. Then, slit the plastic covering the front opening of the cabinet and immediately turn on the cabinet fan.

Ventilate the cabinet overnight or for about 16 h to remove formaldehyde gas and polymer that has adsorbed to surface areas and filters.

6. Now work (e.g., changing filters) can be performed. Large areas, air handling systems, laboratory equipment, and electrical instruments can also be decontaminated with formaldehyde gas. The protocol for equipment and instrument decontamination would be similar to that used in cabinet decontamination, except that large polyethylene bags made by heat-sealing polyethylene film would be used to enclose the equipment or instruments. Also, an exhaust fan and duct would have to be attached to the bag to disseminate the gas throughout the bag

and then to exhaust the gas from the bag to the outside atmosphere.

Recently, vaporized hydrogen peroxide has been proposed as a safer alternative to both formaldehyde and ethylene oxide for gaseous decontamination tasks in laboratories. Commercial applications are being developed that could result in less-cumbersome procedures than those associated with formaldehyde and ethylene oxide. The breakdown products of hydrogen peroxide are oxygen and water, but it is a strong oxidizing agent and product compatibility will have to be demonstrated (18).

Protocol for Work Surface Maintenance

Research and clinical laboratories that deal with microorganisms or clinical specimens from animals and human beings should decontaminate work surfaces. It is recommended that work surfaces be decontaminated each morning before starting work, before restarting work after a break, and at the end of the work day.

A quaternary ammonium compound diluted according to the manufacturer's instructions, or a 70 to 85% alcohol solution, is suitable for laboratories dealing with ordinary noninfectious microbes. These solutions could be used in other laboratories dealing with specific low-risk microbes as long as these germicides are known to be effective against the agent(s). For laboratories working at BSL 3 or 4, agents should not be handled on the open bench except under extenuating circumstances. In that case, a phenolic or iodine solution should be used, followed by an alcohol wipe (alcohol will remove the residual phenolic compound that can be adsorbed through intact skin and will remove the sticky iodine-detergent residual left by iodophors); manufacturer's directions should be followed in preparing these solutions. The disinfectant solution should be applied to the surface with a clean cloth or paper towel dampened with the solution (spray or pour on cloth). Light equipment and materials should be moved so that the entire surface is dampened. Allow at least 5 min of contact before wiping the surface with a different cloth or paper towel that has been dampened with alcohol. The cloth or towel should be discarded in a container that will be autoclaved or incinerated. The individual applying the solution should wear a glove on the hand that is in contact with the cloth or towel.

The same procedure is followed at the end of the day and after breaks. It is also a good practice to wipe down all vertical surfaces, at least on a monthly basis, with a disinfectant.

Handwashing Protocol

Handwashing is an extremely important procedure for preventing exposure to and dissemination of infectious agents. Hand contamination with transient microbes (pathogenic and saprophytic microbes that are not normal components of the resident skin flora) can readily occur during manipulation of specimens, equipment, and supplies and during contact with work surfaces (13, 20). Unless microbial contamination is routinely removed, exposure via contact with mucous membranes, inoculation through skin, or ingestion becomes inevitable.

A nonantiseptic soap can be used for handwashing in laboratories that do not handle human and animal specimens and in laboratories handling low-risk or noninfectious agents. The use of an antiseptic handwashing compound is recommended for laboratories using high-risk infectious agents or those involved in sterile technique. The antiseptic compound used should be effective against most microbes and should not cause excessive skin dryness, cracking, or dermatitis on repeated use. Also, compounds that can be absorbed through intact skin and cause systemic poisoning should never be used.

Two compounds that are widely used as antiseptic handwashing agents are chlorhexidine gluconate and povidone-iodine (10). An alternative would be to use a nonantiseptic soap followed by a rinse with 70% ethanol. Liquid dispensers rather than bar soap are recommended for all handwashing products.

Laboratory personnel should wash their hands

- When coming on duty
- On leaving the laboratory for whatever reason
- When hands are obviously soiled
- Before and after completion of a task in a biological safety cabinet, even if gloves are worn
- Before contact around one's face or mouth
- On completion of duty

A protocol for handwashing is as follows:

1. Turn on faucets and wet hands with tepid water.
2. Dispense nonantiseptic soap or antiseptic compound into a cupped hand.
3. Spread soap or compound around both hands and between fingers. If needed, add a little more water to facilitate spread and lathering.
4. Wash hands for about 10 s. Vigorously rub both sides of hands starting from a few inches above the wrist, extending downward between the fingers and around and under the fingernails.
5. Rinse thoroughly under the tepid running water. Rinsing should start above the wrist area and proceed to the tips of the fingers. *Note:* If faucets are

not knee- or foot-operated, do not turn off water (touch faucet handles) yet.

6. Dry hands thoroughly with paper towel(s). If faucets are hand-operated, turn them off now, using a dry paper towel to protect clean hands.

Handwashing sinks should be of adequate size and conveniently located throughout the laboratory and at each entry or exit door to encourage handwashing. The faucet should be operated by a knee- or foot-control device, especially for laboratories handling materials containing moderate- to high-risk microbial agents. Dispensers for handwashing compound and for paper towels should be available and maintained at each of the handwashing sinks.

GLOSSARY

Antisepsis: Application of a chemical to living tissue for the purpose of preventing infection.

Decontamination: Destruction or removal of microorganisms to some lower level, not necessarily zero. (Also applies to removal or neutralization of toxic agents and, in microbiology laboratories, implies microbicidal action for safety purposes.)

Disinfection: Chemical or physical treatment that destroys most vegetative microbes (or viruses), but not spores, in or on inanimate substances.

Dunk tank: Device containing liquid germicide, used for the passage of materials into or out of a gastight chamber without the transfer of microorganisms.

***D* value:** Time required, in a given treatment, to reduce the number of surviving organisms by one logarithm (90%).

Microbiological safety index: Logarithm of the reciprocal of number of survivors after a sterilization treatment.

Pasteurization: Heat treatment of a liquid medium to destroy the most resistant vegetative organisms of interest.

Sanitization: Reduction of microbial load on an inanimate surface to a "safe" public health level.

Sterilization: Unequivocal total destruction of all living organisms. In practice, sterility is very difficult to determine. Therefore, the term is applied to a very low probability (e.g., usually 10^{-6}) that even one organism has survived the process.

Thermal death time: Time required to achieve sterility under a given set of circumstances.

References

1. **Barbeito, M. S., and E. A. Brookey.** 1976. Microbiological hazard from the exhaustion of a high-vacuum sterilizer. *Appl. Environ. Microbiol.* **32:**671–678.
2. **Beckman Instruments.** 1980. Rotors and tubes for preparation ultracentrifuges: an operator's manual LR-1M-7. Spinco Division of Beckman Instruments, Palo Alto, Calif.
3. **Bentley-Phillips, B.** 1974. Occupational leucoderma following misuse of a disinfectant. *S. Afr. Med. J.* **48:**810.
4. **Blaser, M. J., and J. P. Lofgren.** 1981. Fatal salmonellosis originating in a clinical microbiological laboratory. *J. Clin. Microbiol.* **13:**855–858.
5. **Block, S. (ed.).** 1991. *Disinfection, sterilization and preservation,* 4th ed. Lea & Febiger, Philadelphia.
6. **Brown, M. R. W., and P. Gilbert.** 1977. Increasing the probability of sterility of medicinal products. *J. Pharm. Pharmaceut.* **29:**517–523.
7. **Campbell, R. W.** 1980. Sterile is a sterile word (the microbiological safety index). *Radiat. Phys. Chem.* **15:**121–124.
8. **Centers for Disease Control.** 1975. Neonatal hyperbilirubinemia. *Epidemiol. Notes Rep.* **24:**35.
9. **Centers for Disease Control.** 1991. Chlorine gas toxicity from mixture of bleach with other cleaning products—California. *Morbid. Mortal. Weekly Rep.* **40:**619–627.
10. **Favero, M. S.** 1982. Iodine—champagne in a tin cup. *Infect. Control.* **3:**30–32.
11. **Fregert, S., Ø. Growth, N. Hjorth, B. Magnusson, H. Korsman, and P. Ovium.** 1969. Alcohol dermatitis. *Acta Dermatol. Venerol.* **49:**493–497.
12. **Freireich, A. W., J. J. Cinque, and G. Zanthasky.** 1967. Hemodialysis for isopropanol poisoning. *N. Engl. J. Med.* **277:**699–700.
13. **Garner, J. S., and M. S. Favero.** 1985. *Guidelines for handwashing and hospital environmental control.* Centers for Disease Control, Atlanta.
14. **Glaser, A. R.** 1977. *Special occupational hazard review with control recommendation for the use of ethylene oxide as a sterilant in medical facilities.* Department of Health and Human Services Publication No. 77–200. U.S. Government Printing Office, Washington, D.C.
15. **Health Industry Manufacturers Association.** 1978. *Biological and chemical indicators.* Medical Device Sterilization Monograph Series. Health Industry Manufacturers Association, Washington, D.C.
16. **Jachuck, S. J., and C. L. Bound.** 1989. Occupational hazard in hospital staff exposed to 2% glutaraldehyde in an endoscopy unit. *J. Occup. Med.* **39:**69–71.
17. **Kelsey, J. C.** 1972. The myth of surgical sterility. *Lancet* **ii:**1301–1303.
18. **Klapes, N. A., and D. Vesley.** 1990. Vapor-phase hydrogen peroxide as a surface decontaminant and sterilant. *Appl. Environ. Microbiol.* **56**(2):503–506.
19. **Lauer, J. L., D. R. Battles, and D. Vesley.** 1982. Decontaminating infectious laboratory waste by autoclaving. *Appl. Environ. Microbiol.* **44:**690–694.
20. **Lauer, J. L., N. A. VanDrunen, J. W. Washburn, and H. H. Balfour, Jr.** 1979. Transmission of hepatitis B virus in clinical laboratory areas. *J. Infect. Dis.* **140:**513–516.
21. **Mansuy, D., P. Beaune, T. Crestell, M. Lange, and J. P. LeRoux.** 1977. Evidence for phosgene formation during liver microsomal oxidation of chloroform. *Biochem. Biophys. Res. Commun.* **79:**513–517.
22. **Martin, L. S., J. S. McDougal, and L. S. Loskoski.** 1985. Disinfection and inactivation of the human T

lymphotropic virus type III/lymphadenopathy-associated virus. *J. Infect. Dis.* **152:**400–403.

23. **Mullick, F. G.** 1972. Hexachlorophene toxicity—human experience at the Armed Forces Institute of Pathology. *Pediatrics* **51:**395–399.
24. **National Institute for Occupational Safety and Health.** 1976. *Criteria for a recommended standard-occupational exposure to formaldehyde.* Department of Health and Human Services Publication No. 76-142. U.S. Government Printing Office, Washington, D.C.
25. **Odom, R. B., and K. M. Stein.** 1973. Depigmentation caused by a phenolic detergent-germicide. *Arch. Dermatol.* **108:**848–852.
26. **Reddish, G. F. (ed.).** 1975. *Antiseptics, disinfectants, fungicides, and chemical and physical sterilization.* Lea & Febiger, Philadelphia.
27. **Rutala, W. A., M. M. Stiegel, and F. A. Smith, Jr.** 1982. Decontamination of laboratory microbiological waste by steam sterilization. *Appl. Environ. Microbiol.* **43:**1311–1316.
28. **Sanderson, K. E., and D. Cronin.** 1968. Glutaraldehyde and contact dermatitis. *Br. J. Med.* **3:**802–805.
29. **Shmunes, E., and E. J. Levy.** 1972. Quaternary ammonium compound contact dermatitis from a deodorant. *Arch. Dermatol.* **105:**91–95.
30. **Steere, A. C., and G. F. Mallison.** 1975. Handwashing practices for the prevention of nosocomial infections. *Ann. Intern. Med.* **83:**683–690.
31. **U.S. Department of Health and Human Services.** 1979. *Laboratory safety monograph: a supplement to National Institutes of Health guidelines for recombinant DNA research.* National Institutes of Health, Bethesda, Md.
32. **Zinner, D. D., J. M. Jablon, and M. S. Saslow.** 1961. Bactericidal properties of povidone-iodine and its effectiveness as an oral antiseptic. *Oral Surg. Oral Med. Oral Pathol.* **14:**1377–1382.

Packaging and Shipping Biological Materials

JOHN W. McVICAR AND JANE SUEN

15

INTRODUCTION

Several national and international regulations and guidelines apply to the packaging and shipping of biological materials that pose a potential threat to humans or animals. It is the intent of all these rules to protect the public from accidental direct contact with such materials by ensuring to the greatest practical extent that packages will arrive at their destinations with the contents confined to the inside. The shipper, however, will also benefit from following these rules because material that is packed properly is more likely to reach its destination in usable condition and because liability, which can result from leaking packages, is less likely to be incurred. Simply stated, the packaging requirements are based on three principles: Pack the material so that the container that holds it is not likely to break or leak during shipment; provide packaging that ensures that if the container holding the material should break or leak, the contents will not reach the outside of the package; and affix appropriate identification labels.

Persons anticipating shipment of potentially hazardous biological agents must have appropriate personnel trained in the requirements for the safe shipment of such material. Training may be obtained through shippers, couriers, and package manufacturers.

The requirements for a particular institution should be incorporated into policies and procedures established by and responsibly accepted by management. These policies and procedures should be reviewed and updated at least annually, due to the frequent changes in these regulations. Training should be provided so that everyone understands what is required for shipping and receiving biohazardous materials at that specific location.

This chapter attempts to provide an overview of applicable regulations and to guide the reader to the appropriate references for more definitive answers. The intent is not to provide all the information that will be needed for a particular shipment, because the regulations are constantly evolving. The regulations describe the minimum requirements at any point in time; carriers may have additional requirements. It is important to know how the regulations work and why and to be able to get the information when you need it. The Department of Transportation (DOT), Research and Special Programs Administration, determines how hazardous materials, which include biohazardous materials, can be transported in the United States. The U.S. Public Health Service (PHS), through the Centers for Disease Control and Prevention (CDC), determines what biohazardous agents or materials should be regulated. The carrier selected by each institution should be contacted to provide specific instructions for packaging and labeling.

REGULATIONS

Applicable Regulations

Three federal regulatory agencies specify the minimum requirements for the packing and shipping of biological materials: the PHS, the U.S. Postal Service (PS), and the U.S. DOT. The United Nations (UN) publishes recommendations for packing and shipping such materials, and the International Civil Aeronautics Organization (ICAO) and the International Air Transport Association (IATA) publish regulations based on the UN recommendations. Fortunately for the shipper of hazardous biological materials, all these various regulations are directed to the end result previously stated (i.e., delivery of the material without breakage or leakage). However, the requirements for the various regulations are similar enough that in complying with one regulation, the *Dangerous Goods Regulations* (DGR), the shipper has basically complied with almost all of them.

Applicable regulations include the following:

- U.S. Public Health Service, 42 CFR Part 72, *Interstate Shipment of Etiologic Agents*
- U.S. Postal Service, 39 CFR Part 111, *Mailability of Etiologic Agents, Mailability of Sharps and Other Medical Devices,* and Publication 52, *Acceptance of Hazardous, Restricted, or Perishable Matter*
- U.S. Department of Transportation, 49 CFR Parts 171–180 and amendments
- United Nations, *Recommendations of the Committee of Experts on the Transportation of Dangerous Goods*
- International Civil Aviation Organization, *Technical Instructions for the Safe Transport of Dangerous Goods by Air*
- International Air Transport Association, *Dangerous Goods Regulations*
- U.S. Department of Labor, Occupational Safety and Health Administration, 29 CFR Part 1910.1030, *Bloodborne Pathogens*

Although changes are proposed in the transport regulations for the United States, there is usually a prolonged period before actual changes in these regulations occur. For example, the PHS regulation, *Interstate Shipment of Etiologic Agents,* went into effect on August 20, 1980. In 1990, the PHS proposed revisions to 42 CFR Part 72, which were published in the *Federal Register* (Vol. 55, No. 42, March 2, 1990, p. 7678). As we go to press in 1994, the final version of this revision has yet to be adopted. At the present time, under DOT *Hazardous Materials Regulations,* infectious substances may be shipped using one of three different sets of regulations: the DOT hazardous materials regulations; the revision for infectious substances, based on the sixth edition of the UN recommendations, the effective date for which has been postponed several times and is now October 1, 1994; or the DGR.

International regulations apply not only to shipments to other countries but, increasingly, to shipments within the United States. The Air Transport Association of America recommends that its members, which include most of the North American air carriers, implement the DGR; therefore, they may be applied to almost any shipment that moves by air. Federal Express, perhaps the largest mover of infectious materials, is now using the DGR as a standard. The DOT regulations (49 CFR Parts 171 to 180) are being brought more in line with the international principles exemplified by the DGR with each revision.

Scope of Regulations

The regulations do not dictate how a shipment may be made but give the minimum acceptable standards for a shipment. Shippers of biological materials must first determine what, if any, regulations apply to their shipment. In general, the regulations specify acceptable packaging and required shipping methods and documentation for any material that contains or can reasonably be presumed to contain an etiologic agent. The addition of a few simple steps should make the shipment acceptable to all the regulations. For example, the packaging standard for infectious substances is written in a different form in the different regulations. Careful review shows that the DGR can be used to cover almost all of them. Shippers using private commercial carriers must seek guidance from them. Usually more than one regulation will apply to a given shipment, and the one with the most stringent requirements will have to be followed.

Obtaining the Regulations

The current editions of the *Codes of Federal Regulations* can be obtained from the Superintendent of Documents, U.S. Government Printing Office, Washington, D.C. 20402. Inquiries about PS publications may be directed to U.S. Postal Service, 475 L'Enfant Plaza, Washington, D.C. 20260-5365. DGR can be obtained from Publications Assistant, International Air Transport Association, 2000 Peel Street, Montreal, Quebec, Canada H3A 2R4.

CLASSES OF REGULATED MATERIALS

The CDC (PHS) determines the hazard classes of etiologic agents and biohazardous materials; their expertise is focused on what should be regulated. The DOT focuses on how these materials should be regulated in transport. The materials regulated by the PHS are defined as biological products and diagnostic specimens that are reasonably believed to contain an etiologic agent, material containing etiologic agents, and special etiologic agents. The PS regulates biological products and clinical specimens not reasonably believed to contain an etiologic agent and also etiologic agent preparations. The DOT regulates infectious substances affecting humans and infectious substances affecting animals only and does not currently regulate biological products and diagnostic specimens that do not contain an infectious substance (deferring to PHS regulations). The DOT regulation considers "the terms 'infectious substances' and 'etiologic agent' are synonymous." The Occupational Safety and Health Administration (OSHA) standard on protection of workers from blood-borne infection is applicable to the shipment of blood and certain body fluids and to specific agents for hepatitis B and AIDS. IATA also regulates human and animal infectious substances but does not regulate biological products unless they contain, or are reasonably believed to contain, an infectious substance.

Clearly, any material known or presumed to contain an etiologic agent (infectious substance) is regulated by these agencies, and in addition, the PS also requires leakproof packaging for all clinical specimens that move through the mails. There are similarities in these definitions but also some differences in the requirements for the main classes, biological products, clinical or diagnostic specimens, etiologic agents, and special etiologic agents, that must be understood and followed. The best way to do this is by consultation with the carrier.

Biological Products

The federal regulations define biological products as products that are produced under the following regulations and that may be shipped in interstate traffic: 9 CFR 102—licensed veterinary biological products; 9 CFR 103—biological products for experimental treatment of animals; 9 CFR 104—imported biological products; 21 CFR 312—investigational new drug application; and 21 CFR 600 to 680—biologics. As previously stated, the PHS only regulates biological products that are "reasonably believed to contain an etiologic agent"; the PS does regulate certain biological products such as the live poliovirus vaccine. Both the DOT and IATA take the position that biological products are exempt from regulation unless they contain or are reasonably believed to contain an infectious substance. In that case, they are considered as hazardous or as dangerous goods and handled accordingly. One must also consider any requirements under the OSHA blood-borne pathogen standard for handling biological products derived from certain blood or body fluids.

The IATA defines biological products in the DGR as materials that meet one of the following criteria:

—finished biological products for human or veterinary use manufactured in accordance with the requirements of national public health authorities and moving under special approval or license from such authorities;
—finished biological products shipped prior to licensing for development or investigational purposes for use in humans or animals;
—finished biological products for experimental treatment of animals, and which are manufactured in compliance with the requirements of national public health authorities.

The IATA also covers unfinished biological products prepared in accordance with procedures of specialized governmental agencies. Live animal and human vaccines are considered biological products and not infectious substances.

Diagnostic Specimens

Diagnostic specimens are defined by the PHS, DOT, and IATA as any human or animal material including, but not limited to, excreta, secreta, blood and its components, tissue, and tissue fluids being shipped for purposes of diagnosis; the IATA specifically excludes live infected animals at the present time. Although the PHS does not require special packaging or compliance with the hazard communication requirements for an infectious substance, most diagnostic specimens from a human immunodeficiency virus patient would be considered as infectious substances under the DOT regulations. The PS substantially broadens this class of material to include essentially all clinical material sent through the mails (see next section). As with biological products, neither the DOT nor IATA considers diagnostic specimens to be hazardous or dangerous goods unless they are known or are reasonably believed to contain an infectious substance. The IATA does require that diagnostic specimens be specially packaged to not be considered dangerous goods. Again, any requirements for handling certain blood or body fluid specimens under the OSHA blood-borne pathogen standard must be taken into account.

Clinical Specimens

This class of materials is defined in the PS regulation as "any human or animal material, including but not limited to, excreta, secreta, blood and its components, tissue and tissue fluids, but excluding animal materials, such as leather goods and poultry eggs that are produced commercially." Such a definition substantially broadens this class of materials because it now includes all clinical specimens and not just those shipped for diagnostic purposes as specified in the other regulations. The PS regulation includes specific packaging instructions for "a clinical specimen that is not reasonably believed to contain an etiologic agent, such as a urine and blood specimen used in drug testing programs or for insurance purposes," leaving the assumption that those that do contain etiologic agents will be treated like an etiologic agent preparation. Compliance with the requirements for handling blood and body fluids under the OSHA blood-borne pathogen standard is also necessary.

Etiologic Agents (Infectious Substances)

Biological agents that cause disease in humans or animals are referred to as etiologic agents or infectious substances. The PHS regulation defines an etiologic agent as a "viable microorganism or its toxin which causes, or may cause, human disease" and includes a list of agents to be regulated. The PS regulation excludes "viable" from the definition and includes agents that exclusively affect animals as well as humans. The DOT and IATA refer to etiologic agents as "infectious substances" but define them almost the same as the PHS does at the present time.

The PHS considers all materials that contain the etiologic agents listed in 42 CFR Part 72.3, except for biological products and diagnostic specimens, as a separate class that is to be carefully tracked. The PS regulation defines an etiologic agent preparation as "a culture or suspension of an etiologic agent" and includes "purified or partially purified spores or toxins that are themselves etiologic agents." The DOT regulation defines an infectious substance like the PHS defines an etiologic agent and refers to the PHS list; however, the DOT adds agents that cause disease in animals and also "any other agent that causes or may cause severe, disabling or fatal disease." The IATA definition is also like that of the PHS but includes agents that are "known, or suspected, to cause disease in animals or humans."

Special Etiologic Agents

The PHS regulation requires that shippers of certain etiologic agents use "registered mail or an equivalent system" that provides for notification of receipt to the sender immediately on delivery, and when notice of delivery is not received by the sender "within 5 days following anticipated delivery of the package," the sender must notify the Director, Centers for Disease Control and Prevention, 1600 Clifton Road, Atlanta, Ga. 30333, (404) 633-5313. These special etiologic agents are currently listed in 42 CFR Part 72.3 (f). Proposed changes to this regulation as 42 CFR Part 72.5 (f) will provide a single list of agents that will require confirmation of receipt; they are expected to be adopted in 1994.

PACKAGING, LABELING, AND SHIPPING REGULATED MATERIALS

Once the decision is made as to whether the material to be shipped is regulated and the federal or international regulations that apply are determined, the remainder is an exercise in packaging, labeling, marking, and documentation according to that regulation. It is of interest to remind the reader that a package marked as per the DGR is acceptable to everyone; IATA's Declaration of Dangerous Goods is acceptable to everyone; and IATA/ICAO's Packing Instruction 602 is acceptable by all regulations. It is for this reason that the DGR are increasingly used by couriers in the United States.

Volume Limitations

The various regulations limit the amount of material allowed to be included in a single package based on several criteria. These limitations will not be listed here because they are subject to frequent changes. The reader is encouraged to obtain a current edition of the applicable regulations.

Packaging

Two general packaging methods are specified in all the regulations. They are referred to here as basic leakproof packaging, of which there are three variations, and triple packaging. Obviously, triple packaging will suffice for any of the basic packaging requirements, but it is prohibitively expensive as a method for shipping biological products and diagnostic or clinical specimens.

Basic Leakproof Packaging I

Basic leakproof packaging I is specified (PHS) as that which will "withstand leakage of contents, shocks, pressure changes, and other conditions incident to ordinary handling in transportation."

Basic Leakproof Packaging II

In the PS regulation, clinical specimens and biological products

> must be packaged in a securely sealed primary container(s) with sufficient absorbent material to take up the contents in case of leakage, and in an outer shipping container with secondary leakproof material so that if there should be leakage of the primary container during shipment, the contents will not escape from the outer shipping container. Shock resistant material shall be used to withstand conditions incident to ordinary handling in transit, including but not limited to shock and pressure changes.

Basic Leakproof Packaging III

When clinical specimens and biological products exceed 50 ml per package, the PS requires that the outer shipping container be constructed of fiberboard or other material that complies with standards published by the DOT for packaging of infectious substances. The December 31, 1991, revision of 49 CFR sets forth standards for packaging material in Part 173.196 and test standards for such packaging in Part 178.609.

Triple Packaging

This type of packaging is described in all three federal regulations and in the DGR. The material is placed in a primary container that is watertight and securely sealed. The PS regulation requires that "sufficient outage (space for liquid expansion) must be provided so that the primary container will not be liquid full at 130°F (55°C)." The primary container(s) is then enclosed in a durable, watertight secondary container. More than one primary container may be enclosed in a single secondary container subject to volume limitations. The PS requires that primary containers be surrounded with sufficient shock-absorbent material "to prevent breakage during ordinary handling while in transit," and the DOT requires that they be "wrapped individually." PHS only requires such shock-absorbent material when the total volume of material exceeds 50 ml per package. All regulators require that the space between the primary and secondary containers must contain sufficient absorbent materials to absorb the entire contents of the primary container in case of breakage or leakage. Each set of primary and secondary containers is then enclosed in an outer shipping container. The PHS specifies that the outer shipping container be constructed of "corrugated fiberboard, cardboard, wood, or material of equivalent strength." The PS requires that the outer shipping container be constructed of fiberboard or other material that complies with DOT standards in 49 CFR 173.387(b). The DOT has revised its regulation since the publication of the PS regulation, and the cited DOT requirements are now found in 49 CFR 173.196 and 49 CFR 178.609.

The PS goes on to require that if the package were subjected to the environmental and test conditions prescribed by the DOT, "there would be no release of the contents to the environment, and no significant reduction in the effectiveness of the packaging."

The IATA requires "strong outer packaging" and recommends fiberboard, wood, metal, or sturdy plastic. A special UN number indicates that the packaging has passed the required tests.

The DOT and IATA also require that the outer shipping container and/or the package itself be capable of withstanding specified durability tests described in 49 CFR 178.609 and DGR 10.5, respectively. Application of the various packaging requirements should be discussed with the carrier. The reader should obtain a copy of the current edition of the applicable regulations.

Labeling

The PHS and DOT regulations and the DGR describe and illustrate a label to be affixed to the outer shipping container of packages containing regulated materials; the PS regulation specifies the PHS label (see Fig. 3). The various labeling requirements may require more than one label to be used.

The diamond-shaped label illustrated in Fig. 1 is the one specified by the IATA. The DOT label is essentially identical except that it specifies notification of the CDC in case of damage or leakage as does the PHS label.

Documentation

The DOT regulation requires shipping documents to accompany packages shipped by air, but no specific form is mentioned (49 CFR 175.30). The IATA, which regulates air shipments, specifies the use of a document called the Shippers Declaration for Dangerous Goods (Fig. 2) for all packages containing regulated

FIGURE 1 International label for packaging containing infectious substances or etiologic agents (IATA DGR 7.3.17). DOT label is similar. The PS requires the international label on air shipments of etiologic agents (DMM 124.386). Color is black on white.

materials (i.e., those with an infectious substances label). Instructions for filling out these forms are found in section 8 of the DGR.

Special Packaging Requirements

In addition to the general requirements presented above, each regulating agency has some special requirements. The shipper must determine those that apply. Assistance is usually provided by the carrier.

Coolant Material

Each regulatory agency has specific requirements for coolant materials:

Ice: The DOT specifies that the packaging must be leakproof if ice is used.

Dry ice (carbon dioxide, solid): Both the PHS and DOT specify that dry ice be placed outside the secondary container and that the secondary container be secured by some arrangement of shock-absorbent material or interior supports so that it will not become loose when the dry ice sublimates. The PS and DOT specify that the outer packaging be such that it will allow the release of carbon dioxide gas. The DOT specifies testing for packaging of material that will use dry ice to ensure the stability of the secondary container after the dry ice is gone. The PS requires that all packages for air transport and containing more than 5 lb of dry ice must be labeled as such and accompanied by a Shippers Declaration for Dangerous Goods (Fig. 2). This requirement is waived if: "(1) The weight of the dry ice in the package does not exceed 5 pounds and the net weight of the dry ice is marked on the package; (2) The dry ice is a refrigerant for a material being used for diagnostic or treatment purposes, e.g. frozen medical specimens, and the material is so marked on the package; and (3) The package is marked 'Carbon Dioxide' or 'Dry Ice.'" The IATA has requirements similar to those of the PS.

Liquid nitrogen: The DOT requires that packaging be able to withstand very low temperatures and that "requirements for shipment of liquid nitrogen must also be observed."

Damaged Packages

Persons handling any package of biological material that bears either an etiologic agents label (Fig. 3) or a DOT infectious substances label (Fig. 1) are required to notify the Director, Centers for Disease Control and Prevention, 1600 Clifton Avenue, Atlanta, Ga. 30333, (404) 633-5313 and the sender if the package is damaged or leaking. The IATA requires inspection of packages of dangerous goods "upon unloading from the aircraft" and, if evidence of damage or leakage is found, inspection of the stowage area and that "any hazardous contamination [be] removed." The IATA goes on to state in the DGR that

> If any person responsible for the carriage or opening of packages containing infectious substances becomes aware of damage to or leakage from such package, that person must:
>
> avoid handling the package or keep handling to a minimum,
> inspect adjacent packages for contamination and put aside any that may have been contaminated,
> inform the appropriate public health authority or veterinary authority, and provide information on any other countries of transit where persons may have been exposed to danger, and
> notify the consignor and/or the consignee.

Variations in Requirements

The PHS (42 CFR Part 72.5) provides for variations from the regulatory requirements under the following circumstances:

> The Director, Centers for Disease Control, may approve variations from the requirements of this section if, upon review and evaluation it is found that

(Provide at least two copies to the airline.)

SHIPPER'S DECLARATION FOR DANGEROUS GOODS

Shipper

Air Waybill No.

Page of Pages

Shipper's Reference Number
(optional)

Consignee

Two completed and signed copies of this Declaration must be handed to the operator

WARNING

Failure to comply in all respects with the applicable Dangerous Goods Regulations may be in breach of the applicable law, subject to legal penalties. This Declaration must not, in any circumstances, be completed and/or signed by a consolidator, a forwarder or an IATA cargo agent.

TRANSPORT DETAILS

This shipment is within the limitations prescribed for: *(delete non-applicable)*

PASSENGER AND CARGO AIRCRAFT	CARGO AIRCRAFT ONLY

Airport of Departure

Airport of Destination:

Shipment type: *(delete non-applicable)*

NON-RADIOACTIVE	RADIOACTIVE

NATURE AND QUANTITY OF DANGEROUS GOODS

Dangerous Goods Identification				Quantity and type of packing	Packing Inst.	Authorization
Proper Shipping Name	Class or Division	UN or ID No.	Subsidiary Risk			

Additional Handling Information

I hereby declare that the contents of this consignment are fully and accurately described above by proper shipping name and are classified, packed, marked and labelled, and are in all respects in the proper condition for transport by air according to the applicable International and National Government Regulations.

Name/Title of Signatory

Place and Date

Signature
(see warning above)

STYLE F83 LABELMASTER, CHICAGO, IL 60646
312/478-0900

FIGURE 2 Shippers Declaration for Dangerous Goods required by the IATA and PS to accompany air shipments (IATA DGR section 8 and DMM 124.386). Required information is found in DGR 4.2 and 49 CFR 172.102. Form is black on white with a red and white border.

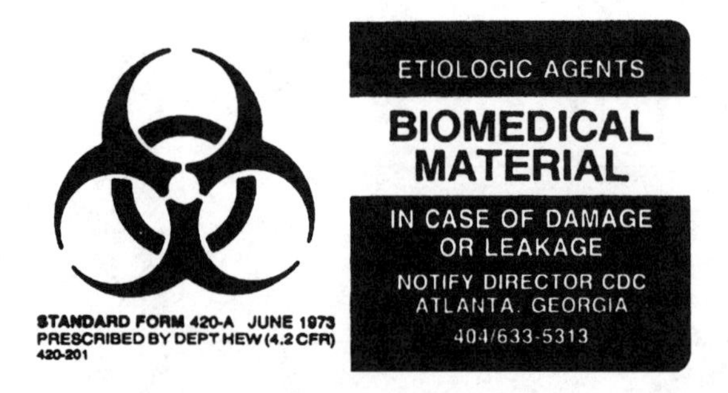

FIGURE 3 PHS label currently required on packages containing etiologic agents. Color is red on white, and details of size and type are in 42 CFR 72.3(d).

> such variations provide protection at least equivalent to that provided by compliance with the requirements specified in this section and such findings are made a matter of the official record.

Packaging Materials

The following discussion of primary containers and secondary packaging (from 49 CFR Part 173.196) is provided as an example of the variety of packaging requirements.

For lyophilized substances, the DOT specifies "flame-sealed glass ampoules or rubber-stoppered glass vials fitted with metal seals" as primary containers.

For liquids or solids shipped at ambient temperature or higher, the DOT specifies "primary receptacles . . . of glass, metal or plastic" with a "positive means of ensuring a leakproof seal, such as heat seal, skirted stopper or metal crimp seal. . . . if screw caps are used, they must be reinforced with adhesive tape."

For substances shipped in liquid nitrogen,

> plastic primary receptacles capable of withstanding very low temperatures must be used. Secondary packaging must also withstand very low temperatures and in most cases will need to be fitted over individual primary receptacles. . . . Whatever the intended temperature of shipment, the primary receptacle and secondary packaging . . . must be capable of withstanding, without leakage, an internal pressure which produces a pressure differential of not less than 95 kPa (14 psi) and temperatures in the range of –40°C to +55°C (–40°F to 131°F).

Testing of Packaging

Performance tests for packaging are described in detail in 49 CFR 178.609 for the DOT and section 10.5 and Packing Instruction 650 (for general diagnostic specimens) in the DGR. Samples of packaging are subjected to free-fall and puncture tests, usually with some preconditioning such as being soaked in water, and examined to see if the contents are protected. DOT and IATA tests require that "there must be no leakage from the primary receptacle(s) which should remain protected by absorbent material in the secondary packaging." The PS also requires testing as specified by the DOT but requires that "there would be no release of the contents to the environment, and no significant reduction in the effectiveness of the packaging." The shipper may rely on the firm supplying the packaging material to certify that it has satisfied the various requirements.

Package Size

The DOT requires that packages "consigned as freight" must be at least 100 mm (3.9 in.) in the smallest overall dimensions.

Packing List

The DOT requires that "an itemized list of contents must be enclosed between the secondary packaging and the outer packaging."

Although the myriad of current regulations still needs to be simplified, the requirements are slowly becoming more realistic and achievable. The determination of hazard class for shipping and the selection of appropriate packaging containers, materials, and labels can be done. Organizations cannot use the excuse that the regulations are too confusing to justify their failure to comply.

Chemical Safety in the Microbiology Laboratory

FRANK S. LISELLA AND SCOTT W. THOMASTON

16

BACKGROUND

Since 1987, regulatory actions by the Occupational Safety and Health Administration (OSHA) have underscored the necessity for proper safety programs involving the use of chemicals in laboratories. The promulgation of the standard *Occupational Exposures to Hazardous Chemicals in Laboratories* in 1990 was tacit recognition by OSHA that laboratories are unique environments with regard to the use of hazardous chemicals (14). This standard (29 CFR 1910.1450) requires employers engaged in the laboratory use of hazardous chemicals to meet certain requirements. This performance-oriented standard, when coupled with extant guidelines and codes such as those prepared by the National Fire Protection Association (NFPA) (7–9), provides a strong foundation for the development of credible chemical safety programs. This chapter addresses the legal requirements and presents the main features that should be included in a laboratory chemical safety program.

INTRODUCTION

The management of hazardous chemicals in biomedical laboratories represents a unique challenge because of the diversity of the work and the variety of chemicals used. Virtually every type of solvent, acid, base, or mixture may be used in microbiology laboratories. Many of these chemicals are carcinogens, mutagens, or teratogens, which require special handling (see also Chapter 10). Laboratorians must make certain that these chemicals are properly used, stored, and disposed of after use. Further, the appropriate engineering controls must be in place to ensure that the proper ventilation, including fume hoods, is used to prevent or minimize chemical exposures of employees.

Although the availability of the proper equipment is important, employees must have a positive attitude toward safety. Workers can tend to disregard the hazardous properties of chemicals with which they are familiar, particularly if the chemicals do not cause an obvious health problem after short-term exposure. Unfortunately, it often takes a major injury to an employee for coworkers to realize that adherence to safety practices might have prevented the exposure or minimized the outcome. This chapter attempts to identify the safety elements required in the use of chemicals, the type of physical facilities provided, and the attitude of the user.

LABORATORY ADMINISTRATION

Safety and health activities are the responsibility of each employee. Laboratory administrators are responsible for the development of a detailed policy statement that also emphasizes employee responsibility regarding use of chemicals. This chemical safety policy should clearly state management's safety objectives and the responsibilities of management and workers. The policy should be distributed and explained to each employee before being posted in a prominent location.

The role of a full-time or collateral duty safety coordinator or, in the absence of such an individual, a safety committee must be established in advance. (The organization of a safety program is addressed in more detail in Chapter 17.) A committee on chemical safety should make recommendations to management regarding restrictions required in the purchase or use of certain chemicals. Action items from the chemical safety committee meetings should be shared with line management and with laboratorians as appropriate.

PROCUREMENT OF CHEMICALS

The procurement of laboratory chemicals should be carefully monitored to prevent the unnecessary purchase of hazardous chemicals. Monitoring is also performed to ensure that toxic chemicals are identified and handled appropriately and that the least toxic material is ordered. A chemical inventory should be kept to ensure that an acceptable means exists for the handling and disposal of the chemical and any hazardous by-products resulting from its use. Before procurement, the hazardous properties of a chemical should be reviewed, and a containment level should be identified. Although there are several options available, the simplest is for knowledgeable safety personnel to review requisitions. If this is not practical, procurement personnel should be alerted so they can obtain clarification from investigators with regard to the purchase of chemicals that may require special safety precautions. Included in this category would be materials such as ultratoxins (diisopropylfluorophosphate), material that may form explosive peroxides (ethers), chemicals that may become explosive if dehydrated (picric acid), carcinogens, or other dangerous compounds. The safety committee could develop a list of "restricted" materials to serve as a flag for further review. A computer program could automate the system to identify these restricted chemicals at the time the order is placed.

The following guidelines can be applied by those who plan to purchase hazardous chemicals:

1. Obtain only that quantity of material that will be used within a reasonable length of time as appropriate for the required tasks.
2. Substitute nonhazardous chemicals for hazardous chemicals when possible.
3. Exchange unused portions of surplus stock chemicals with other laboratories to prevent unnecessary purchases of the same chemical.
4. Plan procedures and experiments carefully to minimize or prevent the use of hazardous chemicals.
5. Develop procedures to ensure that all incoming chemicals are accompanied by a material safety data sheet (MSDS).
6. When appropriate, use procedures that require microscale quantities of chemicals.

USE OF CHEMICALS

The use of certain hazardous chemicals requires oversight by the safety coordinator and/or safety committee. A protocol should be prepared for each such chemical, with relevant information about the chemical and the conditions under which it can be used. The laboratory worker has a right to know the hazards of the materials to be used as well as the safety practices required or recommended. The OSHA standard *Occupational Exposure to Hazardous Chemicals in Laboratories* (14) now requires laboratories to have a chemical hygiene plan (CHP) in place (see Appendix III in this volume). The CHP must include the standard operating procedures to be followed when working with hazardous chemicals. The CHP must also include

- Criteria for the determination of hazards
- Control measures to prevent or minimize exposures
- Employee training and information
- Policies and procedures for medical consultation and evaluation
- Record keeping
- Labeling and availability of MSDSs

The CHP outlined in Appendix A of the OSHA laboratory standard (14) is reprinted as Appendix III in this book. The OSHA requirements are based on the recommendations of the National Academy of Sciences. Readers are encouraged to review the original recommendations for additional background on this subject (10, 11, 13). An MSDS should be reviewed before any work with a hazardous chemical. The MSDS provides information on the physical characteristics of the compound, toxicologic data, spill clean-up information, and disposal practices and should thus be kept in a location accessible to all users. If the MSDS does not arrive with the chemical product, the manufacturer or supplier should be contacted immediately. Work with the chemical should not begin until the actual MSDS has been reviewed. If the chemical has been synthesized in the laboratory, the hazard assessment and the containment requirements are the responsibility of the laboratory director.

MSDSs will specifically include

- Chemical identity as it appears on the label
- Chemical name and common name
- Physical and chemical characteristics
- Signs and symptoms of exposure
- Routes of entry
- Exposure limits
- Carcinogenic potential
- Safe handling procedures
- Spill clean-up procedures
- Emergency first-aid

Container labels must include

- Common name
- Company name and address
- Signal words (Danger, Warning, Caution)
- Principal hazard (physical and health)
- Precautionary measures
- First-aid instructions
- Proper handling and storage
- Special instructions

PERSONAL PROTECTIVE EQUIPMENT

Personal protective equipment should be used if appropriate engineering practices cannot prevent exposure to hazardous materials. The proper use of personal protective equipment such as eye protection, gloves, and other protective clothing, in conjunction with good laboratory practices, can reduce the risk or prevent employee exposure to hazardous laboratory chemicals.

Eye Protection

Eye protection should be used in all areas where chemicals are used or stored. Users of contact lenses should be cautious in areas where gases and vapors may be present. These gases can become concentrated under the lenses and cause eye damage. The use of contact lenses is prohibited in some laboratories.

Safety glasses and face shields should be composed of hardened glass or plastic and comply with the American National Standards Institute standard, Z87.1979. Side shields that attach to regular safety glasses provide some protection, but they do not protect the eyes and face from major splashes. Goggles of the impact-protection type are intended for wear where there is a danger of splashing chemicals and flying debris. Goggles offer little face and neck protection. Full face shields are recommended for potential exposures to such materials; they should be worn in conjunction with goggles or safety glasses when maximum protection from flying particles and toxic materials is required. An eyewash fountain is required to gently flush out ocular contaminants. Eyewash fountains should be readily accessible, marked as to their location and tested at least weekly by running water to clear the pipes. Eyewash fountains should be kept free of obstructions that could block access in an emergency.

Gloves

Hands are more likely to contact chemicals than any other part of the body. Gloves made of appropriate materials can effectively protect the hands from exposure if they are worn during routine handling of chemicals. Selection of protective gloves is based on the chemical hazard and the tasks involved. Some glove fabrics are very resistant to chemicals. The glove fabric must have an acceptably slow breakthrough time and permeation rate for the chemical of interest. Breakthrough time (in minutes) between chemical contact and chemical penetration should be as prolonged as possible to also delay permeation, the rate at which a chemical passes through after breakthrough has been achieved. Knowledge of the characteristics of glove fabrics (e.g., natural rubber, nitrile, butyl rubber, polyvinyl chloride) and the type of use is necessary for the selection of appropriate gloves. For example, leather gloves may be appropriate to use while removing broken glass, but a solvent or corrosive material could readily pass through the leather and onto the skin. Special gloves are available for handling hot substances or cryogenic liquids, such as liquid nitrogen or dry ice, which can instantly freeze living tissue. These gloves are usually multiple layers of fabric that have been treated to make them water-resistant. Gloves for handling hot substances, such as materials removed from hot autoclaves, are typically composed of cotton, wool, or synthetic fibers, although asbestos gloves may still be in use in some laboratories. Asbestos is a suspect human carcinogen, thus gloves containing asbestos should be removed from service and disposed of properly.

Protective Clothing

A variety of protective clothing is commercially available to meet most needs. Certain clothing should not be worn in the laboratory (e.g., no open-toed shoes or shorts). A laboratory coat should routinely be worn when handling chemicals, although it provides only minimal protection of street clothing and skin. Typically, these laboratory coats are composed of cotton and/or synthetic fibers and do not resist penetration by liquid chemicals. When a high level of protection is required, it is prudent to use clothing composed of materials that will better resist chemical penetration. Tyvek gowns or nitrile aprons can prevent skin contamination. The user should be familiar with the advantages and limitations of the protective garments before use.

Emergency Equipment

A safety shower is needed in an emergency (i.e., to remove hazardous chemicals from the skin), thus the location of safety showers should be well marked. Such safety devices should be tested periodically and remain accessible.

Respiratory Protection

Respiratory protection is mandated in situations in which the type or amount of chemicals causes contamination above the exposure control limit. Ventilation devices such as conventional fume hoods or elephant trunks may not remove enough of the contaminant, and the worker may need to use a respirator. The respiratory protection program mandated by OSHA requires that the wearer be trained in the use and limitations of a respirator. Testing must be performed to ensure an adequate fit. If respirators are used, the respiratory protection program must be documented in accordance with OSHA regulations. Whenever possible, engineering design should be used to preclude the need for respiratory protection.

Before a laboratory employee uses any type of protective equipment, he or she should be trained in the capabilities and limitations of the equipment. Information is available from the manufacturer or the distributor.

PROTECTIVE ENGINEERING DESIGN

OSHA prefers that hazards, including exposure to hazardous chemicals, be prevented by proper engineering techniques. Appropriate laboratory ventilation and proper use of fume hoods can significantly reduce exposures.

Ventilation

Laboratory ventilation should be configured to maximize dilution and prevent contamination of employee breathing zones and adjacent areas. Ventilation should provide several air changes per hour. Exhaust air from a laboratory using hazardous chemicals should not be circulated to other areas or recirculated into the laboratory. The laboratory should be maintained under negative pressure relative to corridors and adjacent rooms to prevent the release of contaminants. General ventilation practices provide a degree of protection, but if toxic or volatile chemicals are used, employee protection must be provided by a fume hood.

Chemical Fume Hoods

Fume hoods can provide protection from many chemicals when installed, maintained, and used properly. The proper type of hood must be selected to protect the employee from the chemicals used. Fume hoods should be located away from doorways and ventilation registers to prevent disturbance from air currents. Optimally, fume hoods have a face velocity of approximately 100 lfpm (80 to 120 linear feet per minute). Airflow is critical for containment and should be checked periodically and adjusted as necessary. Fume hood ducting should remain under negative pressure and should not be integrated with other ventilation ducts. If the face velocity is too low, protection of the worker may be lost. However, if the airflow is too high, turbulence created inside the hood may allow vapors to escape into the room. Hoods should be left on at all times unless the general laboratory ventilation is adequate to prevent the accumulation of toxic materials or flammable vapors in the room.

Fume hoods should be checked periodically (every 6 to 12 months, depending on the use) to ensure that proper airflows and containment capabilities are being met. If ultratoxins or other extremely hazardous chemicals are used, it may be prudent to check the hood before each use. A small anemometer can be used for this purpose. Readings can be verified by a more sophisticated instrument. Airflow should be checked with the sash height in the operating position (usually 12 to 18 in.), which provides upper body and eye protection. The airflow reading, the date the unit was checked, and sash position at 100 lfm should be posted on the side frame of the hood. Fume hoods that do not function within the ideal range should be removed from service for repair by a qualified technician.

Ductless fume hoods that exhaust into the laboratory may seem like an attractive alternative to ducted hoods when costs are of concern, but these hoods must be used with caution. The carbon bed filters used to remove contaminants will desorb the chemicals over time and reintroduce the contaminants into the laboratory (3). Ductless hoods should not be used with flammable or toxic substances.

CHEMICAL STORAGE

The space and equipment for the proper storage of chemicals should be factored into the design of biomedical laboratories. Chemicals must be stored according to chemical compatibility (e.g., flammables, oxidizers, corrosives, poisons) (1, 10, 12, 16). A centralized stockroom should be accessible to employees to minimize the storage of chemicals in individual laboratories. Laboratory design should be based on the class and amount of solvents to be used. Flammable solvent bulk purchase (drums or 5-gal cans) is being replaced by purchases of 4-liter or smaller containers. Solvents should be stored either in flammable liquid storage cabinets or inside a stockroom with special electrical fixtures, proper ventilation, berms or dikes, fire suppression systems, and other fire prevention devices. NFPA standards 45, 30, and 99 (7–9) provide information for developing proper storage guidelines. When a centralized storage facility is used, the area should have easy access to a safety shower and eyewash station. Ventilation at the floor and ceiling level is necessary to prevent accumulation of vapors; none of this air should be recirculated.

Fire extinguishers of the B-C type should be provided adjacent to the room, and all employees should be familiar with locations of alarm pull stations. Fire extinguishers are classed as follows:

- Class A—paper or wood. Extinguish with water.
- Class B—flammable liquid or gases. Extinguish with carbon dioxide, dry chemical, or foam
- Class C—combustion caused by electrically energized circuits. Extinguish with carbon dioxide or dry powder
- Class D—metal fires. Extinguish with graphite or sodium chloride

Flammable liquids are classed as follows:

- Class IA—flash point <73°F and boiling point <100°F (*n*-pentane)
- Class IB—flash point <73°F and boiling point ≥100°F (acetone or gasoline)
- Class IC—flash point 73 to 100°F (*n*-butyl acetate, xylene)

Combustible liquids are classed as follows:

- Class II—flash point 100 to 140°F (kerosene)
- Class IIIA—flash point 140 to 200°F (phenol)
- Class IIIB—flash point >200°F (ethylene glycol)

Chemicals should be stored appropriately in chemical storage cabinets or ventilated storage areas, not in fume hoods or on laboratory benches. Chemical compatibility groups must be identified before chemical storage. Most modern laboratory casework provides an area beneath the fume hood for storage of acids and/or flammable solvents. Small quantities of flammable liquids should be stored in NFPA-approved flammable liquid storage cabinets. These cabinets should be sized so that the total storage capacity does not exceed the NFPA requirement for the laboratory area. The NFPA standards, NFPA 45 for laboratories (8) and NFPA 99 (9) for health care facilities, provide guidance on this matter.

COMPRESSED GASES

Compressed gas cylinders should be handled with care during delivery, while in storage and use, and on removal from the laboratory. Cylinders should always be secured in an upright position so that they cannot fall over, and when not in use, the safety cover should be in place to protect the valve. If a cylinder is dropped in such a manner that the valve is broken, the cylinder can become a projectile with the potential to cause injury or death. Compressed gas cylinders should be transported on a specially designed cart and should never be dropped, rolled, slid, or placed in contact with sharp objects.

Gas cylinders should be stored in an area with adequate ventilation to prevent the buildup of vapors in the event of a leak. The temperature should not be allowed to exceed 125°F (51.7°C). Gas cylinders should be stored by compatibility groups, and all fittings and piping must be compatible with the gas to be used. Oxygen should be stored away from other gases and never be allowed to come in contact with grease, oil, or other organic materials that can become flammable or explosive. Although oxygen is not flammable, it is a powerful oxidizer and accelerates combustion. Compressed gas cylinders should be labeled as to their hazards and contents.

Centralizing compressed gases at a manifold near the point of delivery avoids the need for service personnel or laboratory employees to deliver full cylinders and remove empty cylinders from crowded or congested laboratory spaces. The manifold area should be well ventilated and protected from weather extremes. A fire control system should be appropriate for those flammable gases stored in the manifold area. The system should be tested for leaks after each cylinder is changed by using a soap bubble test or other acceptable method. A *Handbook of Compressed Gases* is available as a resource on this subject (2).

LABORATORY REFRIGERATORS

When laboratory refrigerators or freezers are used for the storage of chemicals, there are several important issues to be considered. Only heat-labile chemicals, reagents, or supplies requiring near- or subfreezing temperatures should be stored in refrigerators or freezers. If flammable chemicals are to be stored under refrigeration, only explosion-proof or laboratory-safe units that meet NFPA 45 (8) requirements are to be used. Users should be aware that although NFPA-approved refrigerators are free from ignition sources within the closed refrigeration chamber, most laboratories have other ignition sources. When refrigerator doors are opened, heavier-than-air vapors that accumulate in the unventilated gasketed chamber are released and may be ignited by operating equipment. Solvents are frequently stored inappropriately in refrigerators or freezers, where they can be a fire hazard. Food and drinks are not to be stored in laboratory refrigerators, especially those used for hazardous chemicals, because of the potential for ingestion of contaminated food.

REACTIVE CHEMICALS

Laboratorians must be aware of the inherent dangers of certain chemicals that can mix improperly and cause explosions (1). Other reactions are highly exothermic and may cause splattering. Some chemicals are unstable and can become extremely dangerous when they crystallize or dry out due to neglect or extended storage. Sodium azide, allowed to come in contact with metals, forms explosive compounds. Pouring sodium azide solutions down the drain can allow the formation of explosive lead or copper azides from contact with the plumbing fixtures. Perchloric acid, if heated, should only be used in a perchloric acid hood because of the formation of explosive metal perchlorates. Perchloric acid may also react explosively with organic compounds, including acetic acid (6). Concentrated acids or bases can generate large amounts of heat when mixed with water, subsequently causing splatter and/or the release of steam. The laboratorian should also be aware that compounds, such as cyanides or sulfates, will release toxic fumes when in contact with acids (5). Some chemicals can become explosive if neglected or stored beyond expiration dates. For example, picric acid is a commonly used reagent that becomes shock-sensitive if allowed to become dehydrated (<10% water) and is a more powerful explosive than TNT. Picric acid should not be allowed to come in contact with metals or concrete. Other compounds such as 2,4-dinitrophenylhydrazine and nitrated aromatics are also explosive under certain conditions.

Ethyl and isopropyl ether are among those that form explosive peroxides on exposure to air or light. Great care should always be observed when handling ethers because of their tendency to form peroxides and their exceptional flammability. These peroxides can be set off by heat or friction and will cause ignition of the ether. For this reason, ethyl ether should not be stored for more than 12 months and isopropyl ether should not be stored more than 3 months. Laboratory workers should not handle chemicals suspected of being explosive due to age or neglect. The laboratory supervisor should be alerted to contact a munitions expert, or a disposal firm trained in the handling of explosive materials, for removal of the chemical.

HAZARDOUS WASTE DISPOSAL

When a chemical becomes contaminated or is no longer needed or useful, it must be removed from the laboratory. Disposal of chemicals must be in accordance with federal, state, and local laws and regulations. Improper disposal is not only dangerous but can result in fines and imprisonment. At the federal level, chemical waste disposal is regulated by the Resource Conservation and Recovery Act (RCRA) passed by Congress in 1976. Subsequent legislation that amended RCRA, entitled *The Hazardous and Solid Waste Amendments,* was passed in 1984 and again in 1986 (3a). This is the regulatory foundation on which the disposal of hazardous chemicals is based. Both of these laws are enforced by the U.S. Environmental Protection Agency. Many state laws more stringent than the RCRA are in effect. Readers are urged to consult the applicable laws and regulations in their jurisdiction.

Chemicals may be considered "wastes" if they have no economic value to the generator. Wastes are considered to be hazardous if they have the characteristic of being ignitable, corrosive, or toxic or are listed as hazardous waste (15). It is important that laboratorians understand the characteristics of the chemical waste generated within their own facilities, so that incompatible materials can be properly segregated before disposal. Waste chemicals should be

TABLE 1 Suggested scheme for chemical categorization

Category	Chemical type	Hazard class
Group A	Acids (separate inorganic acids from organic acids; perchloric acid should be separated from other acids)	Corrosive
Group B	Bases	Corrosive
Group C	Cyanides and sulfides (incompatible with acids)	Poison
Group D	Flammable liquids	Flammable
Group E	Flammable solids	Flammable solid
Group F	Acutely toxic (Environmental Protection Agency "P" list or low median lethal dose)	Poison
Group G	Inorganic oxidizer	Oxidizer
Group H	Organic oxidizer	Organic peroxide
Group I	Water reactives	Dangerous when wet

properly stored, by chemical compatibility groups, until they are removed from the laboratory by qualified personnel. A waste disposal contractor or chemical hygiene officer can consolidate the compatible solvents such as ether, alcohol, and xylene and "lab pack" solid materials. It is incumbent on the generator to minimize the amount of waste from laboratory operations (4). Chemicals in unopened containers should be made available for use by other laboratory personnel.

The selection of a hazardous waste disposal contractor should also take into consideration the potential for participation in fuels blending programs. Typically, compatible organic solvents are mixed with various industrial sludges, such as paint waste, to make a product that can be used for fuel at industrial facilities such as cement kilns. Solvents handled in this manner can be disposed of at a fraction of the cost of conventional incineration.

TRANSPORT

Chemicals should be packaged to prevent spills when moved from one laboratory to another, from stockroom to laboratory, or to a central staging area in preparation for disposal. Chemicals should be categorized by compatibility type for appropriate storage or packing for disposal. Table 1 shows a suggested scheme for chemical categorization.

These categories are *generally* compatible within themselves; however, the possibility of some specific incompatibility cannot be overlooked. It is a good idea to retain the original shipping container if chemicals are routinely transported in a facility. This ensures that the chemicals will be placed in a safe container for transport. If this is not possible, special acid and solvent carriers should be made available for transporting chemicals inside a facility. When multiple containers are to be transported and the original shipping containers are not available, the chemicals should be placed in a sturdy box and padded with a material such as vermiculite that will prevent breakage, absorb liquids, and not react with most chemicals. (Hydrofluoric acid is incompatible with vermiculite.) Bottles should be shipped upright and the box sealed with shipping tape. An arrow indicating which end is to be kept up must be drawn on the side of the container. After the chemicals are packaged, the box should be marked with contents, the name and telephone number of the responsible person, and the appropriate label for the hazard class (e.g., oxidizer, corrosive, poison).

A qualified person from the generating facility needs to be available to take care of the legal requirements and record keeping associated with disposal of hazardous waste. Such items as the uniform hazardous waste manifest, the Department of Transportation requirements for shipping hazardous waste, land bans, and biennial reports must be properly handled. The RCRA is designed to make generators of hazardous waste responsible for chemicals until they are suitably destroyed.

The laboratory of today is a complex environment with many potential hazards. The implementation of the chemical safety recommendations and requirements described herein can provide a significant level of protection to the laboratory worker.

References

1. **Bretherick, L.** 1990. *Handbook of reactive chemical hazards,* 4th ed. Butterworths, Stoneham, Mass.
2. **Compressed Gas Association, Inc.** 1990. *Handbook of compressed gases,* 3rd ed. Van Nostrand Reinhold, New York.
3. **Keimig, S., N. Esmen, S. Erdal, and E. Sansone.** 1991. Solvent desorption from carbon beds in ducted and

non-ducted laboratory fume hoods. *Appl. Occupational Environ. Hyg.* **6**(7):592–597.

3a. **Lisella, F. S.** 1994. *The VNR dictionary of environmental health and safety.* Van Nostrand Reinhold, Inc., New York.

4. **Lunn, G., and E. B. Sansone.** 1990. *Destruction of hazardous chemicals in the laboratory.* John Wiley & Sons, Inc., New York.

5. **Meyer, E.** 1989. *Chemistry of hazardous materials,* 2nd ed. Prentice-Hall, Inc., Englewood Cliffs, N.J.

6. **Myer, R.** 1987. *Explosives,* 3rd ed. VCH Weinham, New York.

7. **National Fire Protection Association.** 1986. *Fire protection handbook,* 16th ed. National Fire Protection Association, Quincy, Mass.

8. **National Fire Protection Association.** 1986. *Fire protection standard for laboratories using chemicals.* NFPA-45. National Fire Protection Association, Quincy, Mass.

9. **National Fire Protection Association.** 1990. *Health care facilities handbook.* 3rd ed. NFPA-99. National Fire Protection Association, Quincy, Mass.

10. **National Research Council. Committee on Hazardous Substances.** 1981. *Prudent practices for handling hazardous chemicals in laboratories.* National Academy Press, Washington, D.C.

11. **National Research Council. Committee on Hazardous Substances in the Laboratory.** 1983. *Prudent practices for disposal of chemicals from laboratories.* National Academy Press, Washington, D.C.

12. **Sax, I. N., and R. Lewis, Sr.** 1989. *Dangerous properties of industrial materials,* 7th ed. Van Nostrand Reinhold, New York.

13. **U.S. Department of Labor. Occupational Safety and Health Administration.** 1983. *Hazard communication standard.* 48 FR 53280 as amended and revised (Code of Federal Regulations, 29 CFR Part 1910, Subpart Z, Section 1910.1200).

14. **U.S. Department of Labor. Occupational Safety and Health Administration.** 1990. *Occupational exposure to hazardous chemicals in laboratories.* 55 FR 3327–3335 (Code of Federal Regulations, 29 CFR Part 1910, Subpart Z, Section 1910.1450).

15. **U.S. Environmental Protection Agency.** 1980. *Hazardous waste management systems.* (Code of Federal Regulations, 40 CFR Parts 260–263).

16. **Young, J. A.** 1987. *Improving safety in the chemical laboratory: a practical guide.* John Wiley & Sons, Inc., New York.

Safety Program Management

IV

Laboratory Safety Management and the Assessment of Risk

JOSEPH R. SONGER

17

LABORATORY SAFETY PROGRAM ORGANIZATION

Properly conceived, the organization of a laboratory safety program permits laboratory management to share and to assign responsibility for accident prevention and to ensure adherence to safety standards among the staff. The safety program is not something imposed on the organization; rather, it must be built into every technique and operation within the laboratory. It must be an integral part of each laboratory function. Although there were no specific sections of the Occupational Safety and Health Act of 1970 that dealt exclusively with laboratories, the provisions of Occupational Safety and Health Administration (OSHA) general industry standards codified in 29 CFR Part 1910, Subpart Z (13) have been applied. In 1990 specific regulations on laboratories were enacted under the title *Occupational Exposures to Hazardous Chemicals in Laboratories*, 29 CFR Part 1910 (14).

Under these new regulations, laboratory management is required to formulate and implement a chemical hygiene plan (CHP). The CHP must outline specific work practices and procedures that are necessary to ensure that the employees are protected from health hazards associated with hazardous chemicals with which they work. The regulation also states that "it is not the intention of this standard to dictate the approach that the employer may find effective in meeting the objectives of the CHP or the manner in which it is implemented." The standard specifies certain elements that must be addressed but leaves the details to the employer's discretion. (Appendix III of this book outlines the elements of a CHP.)

Generally, when this standard applies, it supersedes the provisions of all other standards in 29 CFR Part 1910, Subpart Z, except in specific instances identified by this standard. For laboratories covered by this standard, the obligation to maintain employee exposures at or below the permissible exposure limits specified in 29 CFR Part 1910, Subpart Z, is retained.

During the comment period on this standard, recommendations were received that OSHA include measures to protect laboratory workers from additional hazards such as biological, radiological, and physical hazards such as fire and explosion. Rather than jeopardize the enactment of this standard by expanding it to include other hazards, it was approved as written.

The drafters of the regulations further commented that OSHA recognizes that laboratory employees may be exposed to potential hazards that are not addressed by this standard. However, because the initial emphasis for a separate laboratory standard was directed toward the inappropriateness of OSHA's health standards for laboratory work, the record is not completely developed regarding other hazards facing laboratory personnel. Although this standard exempts laboratories from most provisions of Subpart Z, other subparts of 29 CFR Part 1910 that address physical hazards remain in effect for laboratories. For example, laboratories and other general industry employers must comply with Subpart H, which pertains to hazardous materials and includes regulations for compressed gases and flammable and combustible liquids, and Subpart C—Occupational Health and Environmental Control, which contains regulations for noise exposure and radiation.

The only OSHA regulation that addresses biological hazards is entitled *Occupational Exposure to Bloodborne Pathogens* 56 CFR, 235:64004-182, 1991 (5). Laboratory workers are included under its coverage. OSHA further comments that "several guidelines are available which make recommendations pertaining to biosafety for laboratories. For example, the Centers for Disease Control and the National Institutes of Health have jointly published 'Biosafety in Microbiological and Biomedical Laboratories.' In addition, the National Committee for Clinical Laboratory Standards has issued a proposed guideline entitled 'Protection of Laboratory Workers from Infectious Diseases Transmitted by Blood, Body Fluids and Tissue'" (8a).

There is a sense in which the guidelines developed in *Biosafety in Microbiological and Biomedical Laboratories* serve as prestandards. Many institutions use them as though they were standards. Official standards will likely be developed around these guidelines. *Biosafety in the Microbiological and Biomedical Laboratories* appears in its entirety in Appendix I of this book. Assessment of risk by cross-referencing agents and activities is a logical and straightforward approach. If proper values are entered, a valid assessment is made.

HAZARDS OF CODIFYING RISKS

The main purpose for codifying risk with accompanying regulations is to form a basis on which to build safe control procedures. A problem we encounter when we attempt to codify risks with their accompanying control regulations is that safety is a very inexact science and the interacting systems of agents and activities are dynamic. They resist our attempts to place them in neat little packages. Another factor we must consider is that new technologies intrinsically—not accidentally—generate uncertainties and ignorance, which necessarily lead to unanticipated side effects that we characterize as risks. Regulations, guidelines, and directives are essential elements in a safety program; however, safety programs will be grossly inadequate if we rely on regulations alone. There are shortcomings to regulations. They form the essential framework for a safety program but are quite inadequate alone.

Regulations are inadequate for at least two reasons: (i) because it is impossible to write enough regulations to cover every possible situation, and (ii) because even if we could write enough regulations, we would be hard pressed in an emergency to recall the specific regulation to follow. There is also a sense in which codification of risks along with prescribed ritual antidotes may stifle further search for the best safety procedure. Once a practice has been established, especially for a long period of time, it goes unchallenged. Risk management should include a continual search for better control methods. As we look at these codes and guidelines, we should continually attempt to translate them into work practices, keeping in mind that they are not intended to cover every detail of activity but need to be translated to fit each immediate risk situation. One of the most difficult attitudes to overcome in the laboratory is that a well-qualified, highly educated researcher will automatically be well qualified to judge biohazard risks. Risk assessment, for the most part, is not covered in the university curriculum.

Rather than a code based on a long list of dos and don'ts, I believe biosafety is best served by the approach taken in the drafting of the OSHA standard *Occupational Exposure to Hazardous Chemicals in Laboratories* (14). With only general directions and a nonmandatory program outline in Appendix A, the responsibility for developing a CHP is laid on each laboratory. Instead of attempting to anticipate all the risk situations that might occur in a biomedical laboratory, safety will be best served by general guidelines that are results oriented. (See Appendix III.)

JUDGING THE ACCEPTABILITY OF RISKS

Assessment of risk is an essential element of safety. A thing is considered safe if the risks associated with it are judged to be acceptable. Although measuring risk is an objective but probabilistic exercise, judging the acceptability of that risk (judging safety) involves both personal and social value judgments. Lowrance (8) pointed out that it is false to expect that scientists can measure whether something is safe. Scientists are prepared principally to measure risk. Although the level of a risk might remain the same, the acceptability of that risk is constantly subject to change. The courts are continually recalibrating the yardstick of social values (8).

Safety is not measured; risks are measured. Only when those risks are weighted in a value judgment system can the degree of safety be judged. As previously indicated, a thing is safe if its attending risks are judged to be acceptable. Judging the acceptability of risk is a normative process. We must avoid any suggestion that safety is an intrinsic, absolute, measurable property of things. Lowrance (8) provided some criteria for judging the acceptability of risks.

Reasonableness is by far the most commonly cited principle in safety judgments; this is often referred to as the "rule of reason." The problem is to decide what is reasonable. Tort liability charges that might occur in the laboratory are most likely to be the result of negligence. Negligence in the eyes of the law may be defined as conduct that falls below a standard of care established by law to protect others against an unreasonable risk of harm (19). If the standard of care has not been specifically established by statute, the actions or inactions of an individual will be measured against what a hypothetical reasonably prudent individual would have done under the same circumstances.

Several factors enter into the determination of reasonableness, as follows.

1. Custom of usage (or prevailing professional practice): Many laboratory activities have been performed for years and are generally recognized as safe; if a practice has been in common professional use it must be okay, because any adverse effects would have become evident, and a thing approved by custom is safer than one not tested at all. However, long-time usage does not automatically mean that an activity is safer. A prime example is mouth-pipetting, which was commonly practiced for many years.

2. Best available practice, highest practicable protection, and lowest practicable exposure: In the laboratory, we should seek to find the best available procedure with the highest level of protection and lowest level of exposure practicable. A series of publications by the National Research Council dealing with prudent practices for handling hazardous chemicals and infectious materials are available (9–11).

3. Degree of necessity or benefit: Are the benefits worth the risk? Sometimes it is difficult to justify the risks taken in the laboratory on the basis of the benefits derived. Therefore, we should ask ourselves, is the procedure really necessary? Is there an alternate, safer way?

4. No detectable adverse effects: This is a weak criterion because it is an admission of uncertainty or ignorance.

5. Toxicologically insignificant levels: "There are no safe substances, only safe doses" is a maxim in common usage today. This approach could be criticized as being quite arbitrary. What is judged a safe dose today might not be judged safe tomorrow.

Codification of risks is an important part of a safety program. Regulations, guidelines, and directives are very important in the implementation of safety in the laboratory; however, we must not rely on regulations alone. Using the framework provided by regulations and guidelines, we must measure risks and make safety judgments on a day-to-day basis. It is important to remember that safety is not measured. We can only measure risks and make safety judgments regarding the acceptability of those risks.

Assumed Responsibility

The attitude of top management toward safety is reflected in the attitudes of supervisors. Similarly, the worker's attitude usually reflects that of the supervisor. If top management is not genuinely interested in accident prevention, no one else is likely to be.

Details for carrying out a laboratory accident prevention program may be assigned, but responsibility for the basic policy cannot be merely delegated to a safety officer or safety committee. The laboratory director is responsible for overall safe performance and must constantly review this performance, which is best performed through lines of supervision.

Declaration of Policy

Managing the safety function in the laboratory must start with a written safety policy. Petersen (15) stated that a safety policy should (i) affirm a long-range purpose; (ii) commit management at all levels to reaffirm and reinforce this purpose in daily decisions; and (iii) indicate the scope left for discretion and decision by lower-level management.

Safety is a line responsibility, but the line organization will accept this responsibility only when management assigns it. It is important that the safety policy be written to ensure that there will be no confusion concerning direction and assignment of responsibility.

The safety policy must be put in writing and each employee made aware of it. It should be a part of the laboratory's operational manual. The policy statement should include not only the intent but also the scope of activities covered. Responsibility and accountability should be clearly stated. If a staff safety officer is appointed, his or her role should be clearly stated. The policy statement may be conceived and written by the safety officer, but it should be signed by the laboratory director.

ASSIGNMENT OF RESPONSIBILITY

Although the laboratory director has the ultimate responsibility for safety, he or she delegates authority for safe operation down through all levels of management.

Management Officials

Middle levels of management should play an active role in the safety program. Managers of divisions within a laboratory should provide personal leadership in implementing the laboratory safety policy. They should interpret the policy as it applies to their division and give it their full support. A casual or indifferent approach to safety on the part of middle management will be reflected in an indifferent attitude among supervisors and laboratory workers.

Safety Officer

Any laboratory with more than just a few employees should have a safety officer. It is important that management definitely assign full staff responsibility for safety activities to one responsible individual. The safety officer will advise top management, help in administering the safety policy, and ensure the continuity of the safety program. The duties of the safety officer usually include

1. Planning, administering, and making necessary changes in the accident prevention program
2. Reporting to the laboratory director on a regular basis on the status of safety in the laboratory
3. Acting as an advisor on all matters of safety to all levels of management, laboratory supervisors, and other departments such as purchasing, engineering, and personnel
4. Maintaining accident records, investigating accidents, obtaining supervisor's accident reports, and checking corrective action
5. Conducting or coordinating supervisor safety training program; advising and assisting supervisors in employee safety training
6. Developing and coordinating a medical surveillance program
7. Making personal safety inspections and supervising inspections by the safety staff and special employee committees for the purpose of discovering and correcting unsafe conditions or practices in the laboratory
8. Reviewing and approving designs of new equipment and facilities
9. Supervising fire prevention activities
10. Making sure that the laboratory is in compliance with OSHA requirements and state or local laws and ordinances
11. Maintaining outside professional contacts to exchange safety information relevant to laboratory operations

Supervisors

The first-line supervisor is a key person in any safety program because he or she is in constant contact with the employees. The laboratory supervisor has the following responsibilities:

1. Establish work methods in the laboratory
2. Give job instructions to employees
3. Make the job assignments
4. Supervise the work of employees under him or her
5. Be responsible for maintenance of equipment and laboratory facilities

All these responsibilities have safety overtones. Establishing work methods that are well understood and consistently followed is essential to an orderly and safe operation. Many laboratory accidents have been reported to result from unsafe methods or procedures.

Giving job instruction, with emphasis on safety aspects of the job, will help eliminate one of the most frequent causes of accidents, lack of knowledge. A technique called job safety analysis (JSA) is frequently used in industry to emphasize safety in the job assignment (12). This technique can be used in planning new jobs or elucidating hazards associated with established jobs. JSA is covered in detail later in this chapter.

Assigning people to jobs is closely related to job instruction. Whenever a laboratory supervisor makes a work assignment, safety as well as good job performance requires that he or she be sure the worker is qualified to do the job and thoroughly understands the work method.

Supervising people in the laboratory is necessary even after a safe work method has been established. People deviate from established safe practices, and injuries result. It is the supervisor's responsibility to make sure that his or her subordinates are working safely.

Maintaining equipment and laboratory facilities in safe condition is the supervisor's responsibility. Accidents can result from unsafe equipment, a disorderly workplace, or makeshift arrangements. The supervisor who keeps the laboratory and equipment in top condition helps prevent accidents as well as improves efficiency.

Safety Committees

A laboratory of 20 or more employees should have a safety committee representing all levels of employment. The committee can be chaired by any member; however, the safety officer should keep records and coordinate the committee's activities. Meetings should be held once each month, on laboratory time, to discuss safety problems and programs. Minutes should be taken. The safety committee will review accidents and take corrective action. They will also review safety suggestions and return their favorable or unfavorable response to the originators of the suggestions.

When a safety committee is formed, certain policies and procedures should be set forth in writing, including at least, (i) the scope of the committee's activities; (ii) the extent of the committee's authority; and (iii) points of procedure for meetings, including frequency, time, place, attendance requirements, the order of business, and records to be kept.

Committee membership should rotate at intervals long enough so that each member has time to make a positive contribution and to gain personal experience and interest and yet short enough so that the opportunity to serve can be passed around.

Safety inspection committees should understand that their job is essentially a helpful and constructive one. The laboratory supervisor should accompany the committee on its inspection of his or her area. The supervisor can supply information needed by the committee and should be kept informed of its actions. In making inspections, the committee should watch particularly for unsafe practices and report them immediately to the supervisor. The supervisor should then be allowed the opportunity to correct them.

The inspection committee should comply with all regulations. If laboratory coats, safety goggles, rubber gloves, or respirators are required for workers, committee members should wear them also.

Under no circumstances does the committee or its members interfere with the work of employees or with the conditions of the laboratory nor usurp any of the supervisor's authority.

EMPLOYEES' RESPONSIBILITIES FOR THEIR OWN SAFETY

Regardless of how well a laboratory is engineered for safety, the employees' safety depends most on their own conduct. It is impossible to oversee everything a worker does in a laboratory, so in addition to providing direct supervision, it is necessary to influence the voluntary acts of workers by education and motivation. Management can state in its policy that safe performance from employees is desirable; it cannot, however, force the employees to perform safely. Employees decide for themselves whether they will work, how hard, and how safely. Their decisions are shaped by their attitudes toward themselves, their environment, their supervisor, the laboratory, and the job, as well as by their own knowledge and skills. Group attitudes also affect safety performance. Although group attitudes are more obvious in factory settings among assembly line workers than in the usual laboratory setting, they nevertheless do exist in the laboratory. Once an activity becomes an accepted norm among a group of laboratory workers, it is performed without question. Essentially, the group sets and keeps its own safety standards, regardless of what management's standards are. Employees might even abide by the letter of the safety directives but still not perform safely. By building the staff into strong groups of mature and competent people, management is more likely to be able to motivate them to perform safely.

Employee selection is very important in building mature competent groups. Petersen (15) pointed out that government today has a large amount of control over this process, and because of this, some industries have almost abdicated control in this area.

Training is a powerful influence and motivator for safety in the laboratory, just as it is in many other areas. With training, management gives the employee two things—knowledge and skill; if it is successful, three are returned—knowledge, skill, and motivation.

Job Safety Analysis

A technique used in industry that can be used to implement safety in the laboratory is JSA. This technique, if properly applied, can also be the basis of compliance with the OSHA standards, *Occupational Exposure to Hazardous Chemicals in Laboratories* (14) and *Occupational Exposure to Bloodborne Pathogens* (Appendix II). This technique can be used in planning new jobs or for elucidating hazards associated with established jobs. Using a work sheet, the job safety analyst (i) selects the job; (ii) breaks the job down; (iii) identifies hazards and potential accidents; and (iv) develops solutions.

Such jobs as operating laboratory equipment or titrating a virus are appropriate selections for job safety analysis; large, complex jobs as well as simple, one-step jobs should be avoided. In the job breakdown, the analyst should be especially alert for potential accidents. For example, what is the potential for aerosol generation that could result in respiratory exposure? for a puncture injury from a hypodermic needle or an animal bite? for respiratory or contact exposure with toxic chemicals? Methods of control should be sought for each potential hazard.

In the laboratory, JSA should be performed by the supervisor, with the technician or other subordinate participating. This process helps the supervisor better understand the jobs he or she supervises; the employee's understanding is also increased, improving his or her safety knowledge and attitude.

A common job in the laboratory that can be used to illustrate JSA is operating a sterilizer. Improper use of sterilizers has resulted in laboratory infections, cross-contamination, steam burns, and in rare cases, loss of life. The analysis sheet (Fig. 1) is divided into three columns entitled "What to Do," "How to Do It," and "Key Points." In column 1, the steps of the job are listed. Column 2 provides instructions for doing the job, and column 3 lists key points to be kept in mind.

JSA not only provides an excellent opportunity for training in job performance and safety, but it also provides a record of that training.

JSA has been used extensively in industry. Some industrial plants have prepared in excess of 30,000 such analyses. This technique, if applied to laboratory procedures, could help pinpoint potential hazards and enable management to eliminate or protect against them.

Safety Promotion Programs

It is difficult to maintain a high level of interest and awareness of safety over extended periods of time. Because of this, an ongoing safety awareness program should be instituted. Even if the laboratory has been engineered for safety, laboratory procedures and equipment have been made as safe as possible, and supervisors have trained their workers thoroughly and continue to enforce safe work procedures, it is necessary to maintain interest in safety. Accident prevention basically depends on the desire of people to work safely.

Many hazardous conditions and unsafe acts cannot be anticipated, so each employee must frequently use his or her own imagination, common sense, and self-discipline to protect him- or herself. A program that will stimulate safety awareness will help prevent accidents.

The safety officer should be a well-informed specialist. His or her role is to coordinate the safety awareness program and supply the ideas and inspiration, while enlisting the wholehearted support of management, supervisors, and employees.

The laboratory supervisor is the key figure in any program to create and maintain interest in safety because he or she is responsible for translating management's policies into action and for promoting safety activities directly among the employees. The supervisor's attitude toward safety is a significant force in the success, not only of specific promotional activities but also of the entire safety program, for his or her views will generally be reflected by the employees in the laboratory. The supervisor who is sincere and enthusiastic about safety can be more effective than the safety officer in maintaining interest. Supervisors sometimes resist safety promotional ideas; in that case, it is the safety officer's job to sell the programs to the supervisor.

Supervisors can motivate by example. Wearing safety glasses and other protective equipment when they are needed is one of the best ways a supervisor can promote safety.

The laboratory supervisor should maintain a high level of safety in the laboratory at all times. He or she should not have to suddenly adopt a "get tough" approach to enforce safety rules. He or she should be consistent, firm, and fair. If the workers sense that the supervisor either cannot recognize unsafe conditions and unsafe acts or is indifferent to them, they, too, will become lax.

Safety films with a laboratory orientation are good safety motivators. Films on microbiological hazards, chemical hazards, the hazard of compressed gases, and many other subjects are available.

Regularly scheduled "stand-up" safety meetings are also good safety motivators. Subjects and materials for these regular meetings can be suggested by the safety officer, but the meetings themselves should be conducted by the supervisor. These meetings should be short and timely and oriented toward the work at hand.

Information on safety off the job ought to be distributed to employees. Interestingly, most of the time lost from work caused by accidents is from accidents that occur outside of work.

Safety is always a timely subject for inclusion in the laboratory newsletter or publication; the use of cartoons can be very effective.

Well-planned and properly conducted safety contests can also be effective. One successful contest is "Safety Suggestion of the Month." Employees are encouraged to place suggestions on safety in well-located suggestion boxes, with the monthly winner receiving, for example, a month's parking in a choice "reserved" parking stall. This has the dual

LABORATORY JOB SAFETY ANALYSIS WORK SHEET
JOB: OPERATING A STERILIZER

WHAT TO DO (Steps in sequence)	HOW TO DO IT (Instructions)	KEY POINTS (Safety always a key point)
1. Prepare for start-up	Remove plug screen from bottom of chamber and clean	If not clean, will interfere with free flow of steam
	Replace chart in controller	Chart is record to verify sterilization cycle
2. Heat jacket	With operating handle at "off" position, open steam supply valve	Do not attempt to sterilize until jacket gauge shows 15 to 17 lb of pressure
3. Arrange load	Place flat packs of supplies on edge. If several tiers, place alternative tiers crosswise	To ensure adequate steam circulation
	Loading containers of liquid:	
	—Do not mix loads of liquids with other supplies	
	—Use only vented closures	Sealed bottles may explode
	—Use only type 1 borosilicate (Pyrex) glass bottles	Stress of pressure and temperature may rupture ordinary glass
	—Use sterilizer slow exhaust cycle only	Fast exhaust causes rapid boiling within the bottles with loss of fluids. Do not place flammable chemicals or chemicals that are unstable at high temperatures in the sterilizer
	Moisten loads of clothing or other fabrics	Dry fabrics remove moisture from steam, causing superheating, which chars the fabric
4. Close autoclave door	If equipped with "quick throw" handle, move "quick throw" handle clockwise until arms are positioned radially in slots	
	Continue tightening door by turning hand wheel clockwise	
5. Sterilize	To sterilize all materials:	
	—Open "steam to chamber valve" and "chamber drain valve"	
	—Close "air filter valve," "vacuum valve," and "condenser exhaust bypass valve"	
	—Turn timer to desired exposure period	Time required for sterilization varies with load: surgical packs, 30 min; solutions (aqueous) in Pyrex flasks, 30 min; chicken eggs in 10-gal GI cans, 6 h
	—Turn selector to appropriate position: "slow exh" for liquids; "fast exh" if drying not required; "fast exh and dry" for surgical packs, wrapped supplies, etc.	
	—Turn operating handle (clockwise only) to "Ster"	
6. Open sterilizer	Turn hand wheel counterclockwise. Move "quick throw" handle counterclockwise and open door	
When load is completely processed, sterile light will come on and alarm will sound	At end of a liquid cycle, open sterilizer door no more than 0.5 in. Wait 10 min before unloading sterilizer	Rapid boiling will occur if door is opened too soon

FIGURE 1 JSA for operating a sterilizer.

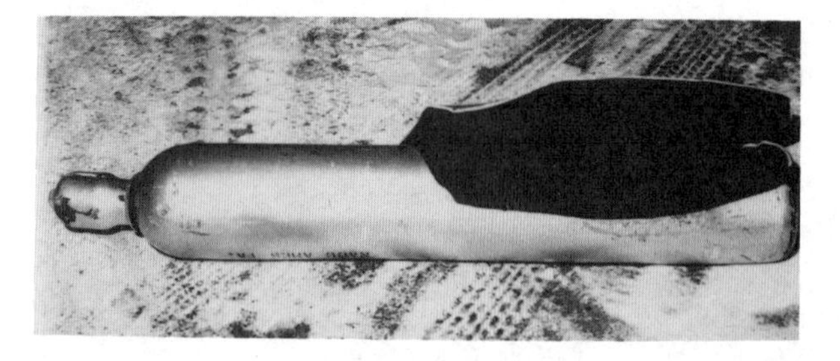

FIGURE 2 This cylinder, which exploded during storage, was thrown through the roof of the building and landed 330 ft from the site of the explosion. The explosion occurred at night, and no one was injured. Had it been delivered on schedule, it would have exploded in a laboratory. Near misses can be effective communicators of safety.

benefit of stimulating employees to think about safety and bringing out safety problems of which supervisors might not have been aware.

Safety bulletin boards by themselves are not very effective; however, in a well-rounded safety program, they can be useful as information centers. Timely safety posters are of value if they are part of a total program. Safety-related displays and exhibits can also be effective, especially if they relate directly to the laboratory. Accidents or "near misses" can be very effectively used to call attention to hazards. Figure 2 presents an example. A carbon dioxide cylinder scheduled for delivery to a research laboratory was overfilled and was fitted with an improper safety device. When it exploded in a storage area, it was hurled through the roof and landed 330 ft from the site. Had the cylinder been delivered on schedule, the explosion would have occurred in a laboratory. A graphic illustration of this type is worth a thousand words.

It is important in all motivational efforts to convey the concept that safety is not something a worker puts on like a hat but should be an integral part of his or her life. There are no absolutes in safety, and to some extent, risk is the essence of life. The objective is to be aware of and to minimize the risks faced daily in the laboratory and elsewhere.

ESTABLISHING RESPONSIBILITY FOR MAINTAINING SAFE WORKING CONDITIONS IN THE LABORATORY

Maintaining safe working conditions in the laboratory is everyone's responsibility; however, some key positions play an important role.

Inspectors

Inspectors play an important part in maintaining safety in the laboratory. It is, at times, necessary that many people assume the role of inspector. The inspector may be a safety professional trained to look for specifically hazardous equipment and activity, or a person from any level of management, including the laboratory director who makes a casual inspection to support the safety program. The first-line supervisor is one of the most important inspectors in the laboratory and is a key figure in the maintenance of safe working conditions. Practically all the supervisor's time in the laboratory is spent in constant contact with the workers. The supervisor should be thoroughly familiar with all hazards that may develop and should be constantly on the alert to correct unsafe conditions and practices.

Maintenance engineers should make frequent trips through the laboratory to assure that proper ventilation and air balance are maintained and that mechanical equipment is functioning properly. At times very sophisticated equipment and systems are installed in laboratories with no provision for their inspection, evaluation, and maintenance. Such equipment as sterilizers, laminar flow biological safety cabinets, and fume hoods need to be monitored on a regular schedule. Biological safety cabinets should be certified at least once each year. If they are heavily used or operated in an area where dust is a problem, they should be certified every 6 months.

Employees who are constantly on the alert can be of great value in preventing accidents. Any safety problems should be reported immediately to the laboratory supervisor.

Safety committees often function in an inspection capacity. They make periodic or intermittent inspections of laboratories, which serve to uncover hazards as well as call attention to safety. Safety committees may also investigate accidents.

Paperwork for inspectors should be kept to a minimum. Checklists will speed up the process, but space should be allowed for listing those problems not covered in the checklist.

Purchasing

Purchasing departments should have excellent liaison with the safety department. Safety standards should be developed to be used as guidelines when equipment is purchased for the laboratory. These purchasing requirements could be computerized to flag unacceptable equipment or materials during the procurement process. Refrigerators, for example, which are to be used for holding flammable solvents, should be explosion-safe. If purchase orders for refrigerators are routed through the safety department, proper equipment can be selected and proper labeling can be performed. Many hazardous chemicals can now be purchased in plastic-coated glass bottles. Orders for hazardous chemicals should be reviewed by the safety department to ensure that this safety feature is requested. Safety should always be of greater consideration than cost.

Engineering

Engineering design plays an important role in accident prevention. The goal of engineering should be to design laboratory environments and equipment and to set up job procedures so that laboratory workers' exposure to infection or injury will be eliminated or minimized. Human factors engineering is a scientific approach to matching humans with their work environment to maximize safety and production while minimizing effort and risk. Counter heights, location of equipment, door locations, visual displays, operating knobs, and many other factors influence how efficiently and for how long a laboratory worker can satisfactorily perform a job. Fatigue and strain often contribute to accidents. Cumulative trauma disorders can also result from repetitive motions such as using some automatic pipettors.

Supervisors

Supervisors, as has been previously stated, and laboratory instructors play a key role in maintaining safe working conditions in the laboratory. The supervisor should be satisfied that all equipment is operating safely and that laboratory procedures are outlined and followed in a safe manner.

Legal Liability

The Resource Conservation Recovery Act of 1976 was enacted by Congress for the purpose of providing technical and financial assistance for the development of management plans and facilities for the recovery of energy and other resources from discarded materials and for the safe disposal of discarded materials and to regulate the management of hazardous waste. Standards governing the generation, storage, transport, and disposal of hazardous waste were put into effect November 19, 1980. The full impact of these standards has yet to be realized. Compliance with these laws is primarily the responsibility of management.

Legal liability in the laboratory can be incurred in several ways; however, for this discussion, tort liability will be of primary interest. Most people are only vaguely aware of what lawyers call torts. This is largely because no one has satisfactorily defined a tort. A person is far more likely to be the victim of a tort, or to commit a tort, than a crime. A tort is a civil wrong against an individual. A crime is an offense against the public at large or the state (1). A tort is an act that violates one's private or personal rights. If a tort is committed against you, it is entirely up to you to seek relief by suing the offender in the civil courts. Costs for bringing such suits are borne by you, and all damages are paid to you. Probably torts is the most social of all the areas of the law, in that it affects more people in their day-to-day living than do contracts, criminal law, bankruptcy, or any of the other legal fields. More than 75% of all the litigation in the United States is for torts (6).

To recover damages for a tort, one must prove either that the act was committed with deliberate intent or that it was the result of negligence (2). Intent is an essential element in such torts as libel or trespass.

Torts that might occur in the laboratory are for the most part likely to be the result of negligence. Negligence in the eyes of the law may be defined as conduct that falls below a standard of care established by law to protect others against an unreasonable risk of harm (2). If the standard of care has not been specifically established by statute, the actions or inactions of an individual will be measured against what a hypothetical reasonably prudent individual would have done under the same circumstances (4).

One important aspect of conduct of the hypothetical reasonable person is anticipation. A supervisor must be able to anticipate the common ordinary events and, in some cases, even the extraordinary.

Joyce (7) points out that there are two main sources of any legal standard or requirement: (i) the requirement may be imposed by a statute or government regulation that is ultimately derived from a legislature, whether federal or state; or (ii) it may be contained in the common law, that is, the law established by judges in actual courtroom cases.

The law of negligence comes primarily from the common law. Court opinions will provide guidance about most of the legal requirements placed on teachers and laboratory supervisors. In a sense, the courts are continually recalibrating the yardstick of social values (8).

Legal liability will also differ significantly between academic and nonacademic laboratories. Academic laboratories in institutions of learning are more likely places for torts to occur than are industrial or governmental laboratories. Schmitz and Davies (16) pointed out that educational institutions undertaking to provide laboratory courses have a duty to provide an outline of reasonably safe experiments. Due care should be exercised to exclude dangerous experiments with limited educational benefits.

Laboratories must be equipped with necessary safety devices such as proper ventilation and fume hoods, operable fire fighting equipment, safe fire escapes, and adequate access for first-aid facilities. Laboratory safety rules should be established, and instructors are expected to enforce them. Schools have a duty to furnish qualified and responsible instructors educated in science and having a knowledge of the specific dangers involved in handling hazardous materials. Laboratory instructors have an active duty to maintain proper supervision commensurate with the potential dangers surrounding the prescribed experiments. To protect their instructors, several chemistry departments have resorted to giving tests on safety before beginning the course. One such department required that the students get a perfect score on the examination before laboratory work could be started. The examination could be taken any number of times but had to be signed (17).

Laboratory procedures manuals cannot be relied on to instruct students sufficiently. Specific warnings describing latent dangers of each experiment must be conveyed by the instructor (16). Corbridge and Clay (3) pointed out that the duty to warn also applies to the writer of a textbook that sets out laboratory experiments for students. The California Supreme Court has said that a jury may find a writer of a laboratory manual negligent for failing to include a warning that an iron mortar is not to be used when grinding a certain compound. The court also held that the teacher may also be held liable for such a failure to warn if he or she knew that an iron mortar might be used by the student conducting the experiment and he or she knew that the use of an iron mortar could cause injury (19). Wyatt and Wright (20) published a paper entitled "How Safe Are Microbiology Texts?" in which they were quite critical of several textbooks. Laboratory techniques and procedures recommended in these books were replete with hazards. They concluded that writers and publishers of textbooks have a responsibility to eliminate hazardous procedures or at least to provide adequate warning.

"Contributory negligence" is sometimes a factor in a liability case. Contributory negligence simply means that the plaintiff was negligent in helping to cause his or her own injury. The modern trend in the law is for the jury to compare the plaintiff's neglect with the defendant's neglect and reduce the plaintiff's recovery by the percentage of neglect attributed to him or her (3).

The Sovereign Immunity Doctrine is based on the concept that any governmental operation could do no wrong and therefore could not be sued without its consent. The U.S. government and the governments of many states have passed laws to permit such suits to be brought against them. Some states have laws that permit schools to buy general liability insurance and waive immunity up to the amount of the policy.

The Sovereign Immunity Doctrine varies from state to state. For example, in 1974 a suit charging negligence was brought against a chemistry professor who taught at a state college in Pennsylvania. The courts held that mere negligence is not sufficient to place liability on a public official. Something more, such as malice or recklessness, must be present to overcome the governmental immunity doctrine of Pennsylvania. This governmental immunity is rarely found to extend to professors. In Ohio, the Attorney General issued an opinion that governmental immunity may extend as far as the dean of a college in a state university (18).

In some states, statutes exist that require indemnification of the teachers by the school board. When the teacher is found liable for injuries, the board must "save the teacher harmless" by paying the damages. This shifts the financial hardship from the teacher to the board.

A teacher may insure him- or herself against a recovery based on his or her simple negligence. Some education associations carry insurance that either may cover members as a benefit of membership or may be purchased from the association.

In tort liability cases, damages are established by the jury. In general, when physical injury has occurred, a plaintiff seeks to recover for medical expenses, both past and future, for income lost since the injury, for the impairment of earning capacity, which will lead to income not gained in the future, and for pain and suffering. These are known as compensatory damages. All juries can do is make a guess at what amount would be fair and reasonable compensation.

Punitive damages may also be sought by the injured party. The purpose of these damages is to punish the wrongdoer and to make him or her an example to others who might think of acting wrongly (3).

Industrial Laboratories

Schmitz and Davies (16) pointed out that, in contrast to a student, the industrial laboratory supervisor is confronted with a master servant status; and to recover for injuries related to research employment, master-servant laws must be considered.

Under Workmen's Compensation laws, the employer is strictly liable for injuries irrespective of the employer's negligence and disregarding common law defenses. All states of the Union, as well as the federal government, have adopted Workmen's Compensation programs, administered by Workmen's Compensation agencies or by the courts.

Several states limit the payments for injuries suffered in certain occupations that are recognized as particularly hazardous. Some states are also restrictive when it comes to occupational diseases. These states list the disorders that entitle a worker to compensation. If a person is not eligible for Workmen's Compensation, it is his or her right to bring a suit for damages against his or her employer or any other person who may be responsible for an injury. However, if the worker is entitled to Workmen's Compensation, the state laws usually prohibit filing a suit.

Safe Equipment and Facilities

The employer is charged with a duty to provide safe equipment and instruments and must exercise reasonable care in keeping them in safe operating condition. Liability may arise for failure to provide adequate safety equipment such as safety goggles and respirators.

The employer has a duty to warn of known latent dangers and dangers existing that are chargeable to the employer's knowledge. The duty to warn is particularly applicable to young and inexperienced employees. Thus, a technician may rely on his or her supervisor's assurance that no danger exists, and the employer is liable if, in fact, an unsafe working condition does exist. By the very nature of laboratory work, a worker must assume some risk; however, he or she is not expected to assume the risks of perils arising from the employer's negligence. The principle is well established that the cost of doing business includes providing medical care and compensation for employees injured in the course of business (16).

SUMMARY

In summary, the development of a safety program is a management responsibility. Implementation and enforcement of the safety program are line responsibilities, and the most important person in the system is the first-line supervisor. Ultimately, however, each individual is responsible for his or her own safety. Although management desires safe performance, the employees ultimately decide if they will work safely. Employees' decisions are shaped by their attitudes toward themselves, their supervisors, the laboratory, and their entire situation. An effective safety program, including training and other motivational efforts, will help develop proper attitudes toward safety.

Safe working conditions are the responsibility of many disciplines. Persons with inspection responsibilities, purchasing agents, engineers, and first-line supervisors all have responsibility for ensuring that safety is given due consideration in the selection of equipment and maintenance of work environments.

Legal liability in the laboratory can be incurred in several ways; tort liability is the most likely. Torts that occur in the laboratory are, for the most part, likely to be the result of negligence. Supervisors must be informed about the potential hazards in their work environments and must convey this knowledge to workers.

References

1. **Anonymous.** 1977. *You and the law,* p. 84. The Reader's Digest Association, Inc., Pleasantville, N.Y.
2. **Anonymous.** 1981. Legal liability, p. 17. *In* J. A. Gerlovich and G. Downs (ed.), *Better science through safety.* Iowa State University Press, Ames.
3. **Corbridge, J. N., Jr., and A. R. Clay.** 1979. Laboratory teachers and legal liability. *Phys. Teacher* **17:**449–454.
4. **Department of Health and Human Services.** 1986. Guidelines for research involving recombinant DNA molecules. Appendix K. *Fed. Regist.* **51**(88).
5. **Department of Labor, Occupational Safety Administration.** 1991. Occupational exposure to bloodborne pathogens: final rule. *Fed. Regist.* **56** (235):64004–64182.
6. **Evans, John E., Jr.** 1971. Legal liability for laboratory accidents, p. 64–66. *In* N. V. Steere (ed.), *Handbook of laboratory safety.* Chemical Rubber Co., Cleveland, Ohio.
7. **Joyce, E. M.** 1978. Law and the laboratory. *Sci. Teacher* **45:**23–25.
8. **Lowrance, W. W.** 1976. *Of acceptable risk.* Wm. Kaufmann, Inc., Los Altos, Calif.

8a. **National Committee for Clinical Laboratory Standards.** 1991. Protection of laboratory workers from infectious disease transmitted by blood, body fluids and tissue, 2nd ed. Tentative guidelines. National Committee for Clinical Laboratory Standards, Villanova, Pa.

9. **National Research Council, Committee on Hazardous Substances in the Laboratory.** 1981. *Prudent practices for handling hazardous substances in laboratories.* National Academy Press, Washington, D.C.
10. **National Research Council, Committee on Hazardous Substances in the Laboratory.** 1983. *Prudent practices for the disposal of chemicals from laboratories.* National Academy Press, Washington, D.C.
11. **National Research Council, Committee on Hazardous Substances in the Laboratory.** 1989. *Prudent practices for the handling and disposal of infectious materials.* National Academy Press, Washington, D.C.
12. **National Safety Council.** 1974. *Accident prevention manual for industrial operations,* 7th ed. National Safety Council, Chicago, Ill.

13. **Occupational Safety and Health Administration.** 1983. 29 CFR 1910, Subpart 2.
14. **Occupational Safety and Health Administration.** 1990. *Occupational exposure to hazardous chemicals in laboratories.* 29 CFR 1910 1450. FRSS 3300–3335, Jan. 31, 1990.
15. **Petersen, D.** 1978. *Techniques of safety management,* 2nd ed. McGraw-Hill, Inc., New York.
16. **Schmitz, T. M., and R. K. Davies.** 1967. Laboratory accident liability: academic and industrial. *J. Chem. Educ.* **44:**A654–A659.
17. **Scott, R. B., Jr., and A. S. Hazari.** 1978. Liability in the academic chemistry laboratory. *J. Chem. Educ.* **55:** A196–A198.
18. **Sweeney, T. L.** 1977. The personal liability of chemical educators. *J. Chem. Educ.* **54:**134–138.
19. **U.S. Department of Health, Education, and Welfare.** 1977. Legal aspects of classroom safety, p. 250. *In Safety in the school science laboratory.* National Institute for Occupational Safety and Health, Cincinnati, Ohio.
20. **Wyatt, H. V., and K. A. Wright.** 1974. How safe are microbiology texts? *J. Biol. Educ.* **8:**216–218.

Behavioral Factors in Safety Training

ROBYN R. M. GERSHON AND BARBARA G. ZIRKIN

18

INTRODUCTION

Today's laboratory workers are faced with the difficult task of working safely amid a complex array of potentially hazardous materials. Fortunately, the dramatic changes in laboratory science over the past several years have been accompanied to some extent by similar advances in the field of laboratory occupational health and safety. These advances have been driven, in part, by rapidly increasing federal and state regulatory activities in this area.

There are several well-defined approaches to minimizing hazardous exposure in the laboratory. These include facility design, engineering controls, administrative controls, and work practice controls. This chapter focuses on safety training programs that are under administrative control. Although training and educational programs are an important example of the administrative control of hazards, other administrative controls include the development of standard operating practices and improvements in the overall safety program and safety climate of the laboratory to encourage safe work practices.

The development and implementation of effective training and educational programs is often problematic, because it is difficult to measure the effectiveness of such programs. Although it would appear that they are useful in improving overall compliance with safe work practices and with motivating employees to adhere to these practices, very few studies have demonstrated this effectiveness, especially in the long term. Yet, the demand for health and safety training programs has never been greater. New Occupational Safety and Health Administration (OSHA) standards such as the blood-borne pathogens standard (23) and the laboratory standard and new recombinant DNA guidelines have a large effect on the laboratory. The evolution of new laboratory procedures, practices, and materials has also led to dramatic changes in safety-related techniques, equipment, learning tools, regulatory requirements, and training needs. These changes have resulted in the need to update procedures continually and to retrain and re-educate laboratory personnel, administrators, and health and safety specialists.

Increased demand for training programs has also come from the employees themselves. Employee concern and fear regarding novel pathogens such as the human immunodeficiency virus (HIV) and multiple drug-resistant tuberculosis has increased laboratory employee interest and involvement in health and safety issues in general and in blood-borne pathogens in particular.

To meet the needs successfully of both the employee and the employer, training and educational programs should be designed and implemented only after thorough planning and development. Training programs are costly in terms of lost work time and concomitant training expenses. Thus, it is important to maximize the effectiveness of these programs as much as feasible. This chapter provides a framework for developing and implementing training programs that will meet the needs of both the employer and the employee with a minimum disruption to the workplace and with a maximum benefit to both employee and employer.

SAFETY TRAINING OBJECTIVES

The objectives of providing specialized health and safety training to laboratory workers include the following: (i) to provide information regarding potential workplace hazards; (ii) to instruct workers in the safe handling of equipment, supplies, cultures, and laboratory chemicals; (iii) to provide information on emergency procedures including fire safety and evacuation; (iv) to advise on the required safety precautions under OSHA, College of American Pathologists, Joint Committee on Accreditation of Healthcare Organizations, Environmental Protection Agency, and other state and federal regulatory agencies; (v) to familiarize employees with the safety policies and procedures of the institution; and finally, and probably most important, (vi) to motivate workers to practice safe work procedures.

The ultimate goal of safety training programs is to reduce occupational exposures, accidents, near accident incidents, injuries, and illnesses. Improved safety climate may also result in enhanced performance and productivity. In turn, absenteeism and turnover may be reduced, as will workers' compensation rates. Employer liability may also be reduced. Because there is some evidence that training programs enhance the overall safety climate of the institution, there may be an additional benefit of improving employee-employer relations by having such programs. These objectives are summarized in Table 1.

An important assumption made regarding safety training programs is that the information will somehow be translated into the desired response or behavior. Unfortunately, it has been shown that knowledge is necessary but not sufficient for behavior change or behavior adoption. In the next section of this chapter, preventive behavioral theories are briefly discussed to highlight the limitations of safety training programs.

TABLE 1 Objectives of training and educational programs

1. Provide information to improve knowledge
2. Demonstrate safe work techniques
3. Provide instruction on emergency response
4. Provide information on regulatory controls
5. Provide information and yearly updates on instructional safety policies and procedures
6. Motivate workers to work safely

PREVENTIVE BEHAVIOR THEORIES

Several psychosocial theories have been developed to help predict and understand preventive health behaviors in the face of threat. These theories have several determinants in common; those found to be particularly relevant in the development of protective behavior are as follows: (i) perceived susceptibility; (ii) perceived severity of the health threat; (iii) perceived effectiveness of the precaution; (iv) perceived cost of prevention action; and (v) enabling (environmental) factors.

For example, for workers to take precautionary action, such as wearing personal protective equipment (PPE), they must (i) perceive that they personally may be at risk of exposure; (ii) perceive that the exposure may result in severe health threats; (iii) believe that the precautionary action will, in fact, protect them; and (iv) believe that the trouble it takes to protect themselves is worth the effort. Finally, various enabling factors, such as the availability of gloves, are important in the determination of preventive behavior. Some examples of these enabling factors include "structural" characteristics of the workers' environment, such as the accessibility and provision of PPE, administrative support of safe work practices (e.g., "safety climate"), and support of coworkers and colleagues. Even workers with the best of intentions in using PPE will be unable to do so if the equipment is simply not available.

Four preventive behavior models are particularly useful in understanding the limits of training programs in terms of safety behavior. These are the theory of precaution adoption (31, 32), the theory of reasoned action (4, 12), the health belief model (5–7), and DeJoy's behavioral diagnostic model (9). Each of these is briefly discussed and examined in the light of its influence on the conceptual basis for training program design and implementation.

Precaution Adoption Model

Weinstein's precaution adoption model (31, 32) describes a continuum of movement or development through five distinct behavior stages, culminating in the adoption of precautionary behavior. The five stages are (i) awareness of the risk; (ii) awareness of the risk to others; (iii) awareness of the risk to self; (iv) intention to protect oneself; and (v) preventive behavior. Weinstein emphasizes the third stage, self-perception of risk, as being the most critical and the one most difficult to achieve.

Key points regarding these various stages of development are (i) people at different stages of the process behave differently; and (ii) different interventions and information are thus needed at each stage to allow for a smooth transition to the next stage. In this model, to understand the barriers to precaution adoption, it is necessary to determine the individuals' stage of development to facilitate the design of specific targeted interventions that serve to encourage the individual to progress to the next stage. An example of each stage is given below to indicate how the model might operate with respect to laboratory workers and the threat of occupational exposure to HIV-1.

Stage 1. Has heard of the hazard

Example: Workers have heard about the risk of occupational exposure and infection with HIV-1

Determinants: Communications about the hazard; experience with the hazard

Stage 2. Believes in significant likelihood for others

Example: Workers believe that occupational exposure and infection with HIV-1 has happened to other workers

Determinants: Stage 1 determinants plus credibility and clarity of communications

Stage 3. Acknowledges personal susceptibility

Example: Workers believe that they are at risk of occupational exposure and infection with HIV-1

Determinants: Stage 1 and 2 determinants plus information about peers' risk; personalized risk factor information

Stage 4. Decides to take precaution

Example: Workers intend to adopt universal precautions

Determinants: Stage 1 to 3 determinants plus beliefs on seriousness of the threat; beliefs on personal severity; salience; beliefs about the effectiveness of precautions; beliefs about barriers to precautions; influence of subjective norm

Stage 5. Takes precaution

Example: Workers practice strict adherence to universal precautions

Determinants: Stage 1 to 4 determinants plus cost-benefit analysis of preventive behavior

Theory of Reasoned Action

The theory of reasoned action, developed by Ajzen and Fishbein, attempts to integrate attitudes and behavioral intention with health behaviors (2, 4, 12, 13). The four main assumptions in this theory are (i) humans are rational beings; (ii) information is used to arrive at a decision; (iii) most behaviors are under volitional control; and (iv) intentions are usually followed by the behavior. According to this theory, the immediate determinant of a given behavior is the intention to perform that behavior. Intentions, in turn, are a function of two other basic determinants: (i) attitudes, and (ii) influence of subjective norms (and significant others) (3). Thus, only attitudes and the subjective norms can directly influence one's intentions. All other variables can only act indirectly on behavior (by directly affecting attitudes or by affecting attitudes and/or expectations of the subjective norm group).

According to this theory then, any attempt at persuasion should be directed toward changing either the key target beliefs or the impact of the subjective norm. Again, the development of effective interventions is dependent on knowing the beliefs that are held concerning any given risk. Fishbein and Ajzen (13) pointed out that the bigger the discrepancy between the belief and the preventive health message, the less likely it is that persuasion will occur. However, the persuasiveness of a communication may be facilitated by several factors (e.g., trainer, choice of media, audience factors) (20). If the facilitation is great enough, it can overwhelm even large discrepancies.

A modified version of the theory of reasoned action, developed by Ajzen and Madden, is called the theory of planned action (1). Behavioral controls are introduced into the model suggesting that information alone is not sufficient to reduce risk-taking behavior but that long-term social reinforcement (e.g., support groups) coupled with environmental factors (e.g., material resources, such as sharps containers, disposable gloves, respirators) is necessary for long-lasting behavior change (22). In the updated model, the subjective norms serve to generate new group values and behavioral expectations and to reinforce and support newly adopted behaviors (e.g., seeing a coworker wash hands serves as a reminder to wash one's hands).

Health Belief Model

The health belief model, developed by social psychologists, attempts to help predict, explain, and understand health-related preventive behaviors (5–7, 17, 19, 25). In this model, benefits are perceived to result from specific preventive actions, and decisions are made based on individual perceptions of vulnerability. The dimensions of the health belief model are as follows: (i) perceived susceptibility; (ii) perceived severity; (iii) perceived benefits; and (iv) perceived barriers (7). The degree to which a person believes that the risk exists and has the perception that a particular behavior, or set of behaviors, will reduce that risk (benefit-to-cost ratio) will determine the extent of preventive behavior. The stimulus to trigger this decision-making process has been referred to as "cues to action." Certain modifiers, such as salience or experience, may serve as either internal or external cues to action. The process is also stimulated by several "enabling" factors, as, for example, sociodemographic variables and environmental (resource) variables (5). The model explains many different health behaviors quite well, up to a point, and has recently been used to understand compliance with preventive recommendations for community-acquired AIDS (21). Although the health belief model has had previous success in predicting responses to persuasive health educational programs, psychologists have been less hopeful about its role in reducing the risk of AIDS.

Several unique characteristics of HIV/AIDS make it difficult to predict behaviors using this model. First, perception of personal susceptibility may be difficult to achieve because of strong social disapproval related to the disease and because many people are so afraid of AIDS that they cannot even consider the possibility that they are at risk for infection (24). Second, the preventive behavior that is recommended for AIDS can only prevent new infection and cannot, at this time, cure infection; thus the benefits of the behavior may be seen as limited. Finally, the costs of the protective behavior are high, because behavior change affecting powerful and basic urges (e.g., sex and drug addiction) are very difficult to change. Psychologists have concluded that when the health threat has no cure and also requires costly preventive behavior, the health belief model is not effective in predicting health behaviors (21). If the perceived benefits of the recommended behavior do not outweigh the cost, people will not even be motivated to attempt, let alone maintain, such behavior.

DeJoy's Model

Another theoretical model has been adapted from DeJoy's behavioral diagnostic model. It is based on the *PRECEDE* framework: Predisposing, Reinforcing, and Enabling Causes in Educational Diagnosis and Evaluation (9, 15). This adapted model integrates both worker-centered determinants and organizational determinants, thus emphasizing the interconnectedness between the individual and the environment. In this model, the preventive behavior elicited by the threat of exposure to a hazard such as blood-borne pathogens is the final step in a series of events. In other words, there is a "chain of causation" (11).

Barriers to Safe Work Practices

Psychosocial determinants that have been found to be important in the adoption of preventive health behaviors include knowledge, perception of risk, fear, subjective norm, and sociodemographics. These determinants are often referred to as "worker-centered" (i.e., they are characteristics specific to each individual employee). Each of these is briefly reviewed with respect to the adoption of safe work practices.

Lack of Knowledge

Simply knowing about the recommended work practices may not result in adoption by laboratory workers. Although knowledge is thought to be a critical factor in the adoption of preventive health behavior, recent findings suggest that it is only the first step in a longer chain of events leading to behavior change (32). Thus, it is expected that knowledge is necessary but not sufficient in the adoption of recommended work practices.

Biased Risk Perception

Self-perception of risk is the second key determinant in the adoption of precautionary behavior. Perception of risk may be affected by several factors, including individual personality characteristics, psychosocial input from others, characteristics of the risk itself, and communication of the risk (18, 27–29). Biases may also result because (i) low-risk probabilities are difficult to judge; (ii) strong initial perceptions or strongly held beliefs are difficult to change; (iii) risks that are seen as involuntary are much less likely to be acceptable; and (iv) individuals tend to see themselves as uniquely invulnerable to many risks. This sense of invulnerability or belief that one is somehow protected from harm is referred to as "optimistic bias" (11, 14, 29, 31).

Fear

The impact of fear on behavior change has been extensively studied. Higbee (16) concluded in an extensive review on the subject that there are considerable inconsistencies among the results obtained by various fear campaigns, and he further concluded that fear is probably not a very effective means of behavior change.

Subjective Norm

The motivation to meet the expectations of others such as coworkers may have a powerful effect on the adoption of certain occupational precautionary behaviors. Within the clinical laboratory setting, several potential influences have been identified. These are patients, patients' families, attending physicians, coworkers, supervisors, infection control nurses and other health and safety personnel, the employees' own family members (especially spouses, children, and close friends), administrative policies, and finally, governmental regulations regarding safe work and infection control practices. The perceived attitudes of others, combined with the motivation to do what others want or expect, may affect compliance with infection control practices.

The influence of one's coworkers in the laboratory may be particularly important because team work and group effort are both necessary in the laboratory setting. If strict compliance with infection control practices is the group norm, then there may be a strong motivation to adhere to the same standards of practice. This may be especially true if compliance practices are assessed as part of the employee's annual personnel review. However, if the group norm leans toward poor compliance, then individuals may not only be less motivated to comply, but attempts to comply may be actively discouraged, especially if the administration does not enforce or endorse its own policies. Thus, group norms regarding preventive work behavior can be powerful influences and can serve to reinforce or subvert the recommended policies.

Sociodemographic Variables

Sociodemographic determinants have been found to be significant in the adoption of preventive behaviors (8, 18). Sociodemographic determinants that may be particularly relevant with respect to adoption of recommended work practices among laboratory workers include age, sex, education, marital status, past work history, and occupation.

Various studies have shown that there is a complex relationship between sociodemographic characteristics and preventive health behaviors. Age and gender have been found to be important determinants of certain types of preventive health behavior (26). With respect to age, it has been found that preventive health behaviors are fairly good during childhood and middle to old age, but poor during adolescence and young adulthood (30). Older workers may be adhering to old work behaviors (e.g., recapping) that are no longer recommended but difficult to change. Young unmarried men with limited education may be least likely to comply with safe work practices. However, it has been shown that older workers with fixed work habits may be unwilling to accept new safer work practices.

Lack of Organizational Factors That Promote Safe Work Practices

Although some researchers have emphasized the human element as the main cause for occupational injuries and illnesses, others have stressed the work environment (10, 28). A lack of either administrative resources or material resources may lead to a decrease in compliance with recommended work practices. Administrative resources refer to management's support of infection control and other safety programs.

Health care facilities have consistently been found deficient with respect to infection control and employee health and safety programs. In 1976, a comprehensive survey of U.S. hospital employee health and safety programs was conducted by the National Institute for Occupational Safety and Health. Only 8% of the nation's hospitals were found to have even the basic elements of employee health and safety programs. Only 65% of the large hospitals (more than 300 beds) surveyed had an employee health professional on staff, and only 10% of the small-sized hospitals (less than 99 beds) reported an employee health service provider. The criteria that the National Institute for Occupational Safety and Health used included the following basic components: (i) preplacement physical examinations, including a complete medical history; (ii) periodic health appraisal examinations; (iii) health and safety education; (iv) immunizations; (v) care for illness and injury at work; (vi) health counseling; (vii) environmental control and surveillance; (viii) health and safety record keeping system; and (ix) coordinated health and safety planning with all hospital departments.

Material Resources

The lack of availability of PPE such as disposable gloves and protective eyewear can obviously act as a barrier to its use. Poor quality, poor fit, and inaccessible PPE and other safety equipment such as sharps containers may also lead to noncompliance with their use.

SAFETY TRAINING PROGRAM DEVELOPMENT

Regulatory Requirements for Training

Safety training programs for the laboratory are recommended or required by the following:

Centers for Disease Control and Prevention (CDC)
College of American Pathologists
Environmental Protection Agency
Joint Committee on Accreditation of Healthcare Organizations
NIH/CDC (biosafety guidelines)
National Institutes of Health, Office of Recombinant DNA
National Research Council
OSHA

There are so many different requirements for safety training that it is difficult to address them all. Generally, most hospitals are attempting to meet all the training requirements by providing workers with annual 1- to 2-h training sessions. It is difficult to provide adequate training in only one session. It is best to split up the training into several sessions such as (i) fire and disaster response; (ii) bloodborne-related, and (iii) chemical hazards. More progressive hospitals are organizing a monthly safety training day, with several safety sessions held over an 8-h period. Each employee may be given 1 day a year to attend the required sessions.

Training Needs Determination

It is important to verify that all personnel have been provided with accurate information to perform their jobs safely. Ideally, safety and health training should be provided at orientation, when new standards or laboratory materials are introduced, and whenever specific problems are observed. Also, follow-up and retraining should be conducted periodically to determine that the standards and procedures are being followed. When a specific problem is found to exist, it is necessary to provide specific, corrective, and targeted training to correct the problem as soon as possible. The OSHA training model (Table 2) is a good starting point.

TABLE 2 OSHA's training model

1. Determine if training is needed
2. Identify training needs
3. Identify goals and objectives
4. Develop learning activities
5. Conduct the training
6. Evaluate effectiveness
7. Improve the program

Identifying Training Needs

Needs assessments play a pivotal role in the identification of training needs because it is often difficult to pinpoint the exact training requirements of various employee groups. Assessments are undertaken to gauge accurately the extent and nature of the lack of knowledge as exhibited by unsafe behaviors and to determine the best training strategy necessary to remedy the situation.

To design an appropriate training program or to revise an existing one, the following questions need to be addressed:

1. What is the extent of the unsafe behaviors? How many workers exhibit the same behaviors?
2. What are the job descriptions of those in need of training? What are the job-related functions that they perform and how are they related to others in the workplace?
3. What portion of the job performance results in unsafe behaviors in the workplace? Is the problem related to unsafe acts on the part of the worker, such as unsafe shortcuts, bypassing safety devices, or failure to use PPE? Could a step-by-step job analysis provide useful information?
4. Is the job hazard related to unsafe managerial practices, such as failure to provide adequate supervision
5. What information does the worker need to perform the job better and more safely?
6. What information might the supervisor need to ensure that the workers can perform the job more safely?

Some of this information can be gained from employee records of accident and injuries on the job. Others can be obtained from anecdotal reports of the workers themselves. Observations of workers by supervisory or other personnel can help to provide the answers. Finally, examination of training programs for workers in other hospitals with similar needs can provide answers. Above all, the training must be designed to meet the needs of the employee.

Identifying Goals and Objectives

The goals and objectives of the training program will be determined by the specific need for training as identified from the above procedures. The overall goal of training will be to have a better-informed work force that will be able to use safe laboratory procedures. Initially, the most immediate goals will be to increase knowledge and improve attitudes and practices with regard to safe workplace behaviors. Intermediate goals of the training program might be to create a more positive attitude toward the continued use of such practices so that compliance and safe behaviors are the norm. Finally, a long-range goal of the training might be that workers will want to ensure that others in the workplace follow the safety procedures for the benefit of themselves and others and will be able to help train newcomers to the work environment.

Training objectives should be written in clear, measurable terms. The objectives will state what is the desired skill or behavior that the employee is expected to learn and how that specific skill will be exhibited and measured. For example, "Disposable gloves will be worn whenever there is the possibility for direct contact with blood or other human body fluids."

Developing and Implementing the Training Program

All training programs should be designed with the active participation of representatives of both employee and employer groups. When this occurs, all participants tacitly agree to the terms of the training program. Ideally, training should meet the following criteria: (i) it is an ongoing regular part of an employee's schedule; (ii) it is a paid activity (the employee is not penalized for attending the training program but comes to value it as a regular part of the work program); (iii) it is mandatory (at the completion of a training module, a certificate of completion is issued that becomes part of the employee's evaluation); (iv) the materials to be used and information to be covered are agreed on, and adequate time is allocated to each topic; (v) trainers are provided, when possible, from existing workplace staff; (vi) training is provided to groups of employees with similar concerns and needs; (vii) management of information, records, and so on, is centralized; (viii) audiovisual aids are used; and (ix) there is a commitment from the administration to the program.

Several key factors that must also be considered to maximize the effectiveness of the training are listed below:

1. The trainer must be knowledgeable and experienced in speaking to audiences.
2. The message that is conveyed should be succinct; do not try to cover too many topics at once.
3. The audience must comprehend the message; aim the message at the level of their training, knowledge, and understanding (including language).
4. Environmental comfort of the audience is very important; pay attention to temperature, distractions, lighting, and so on.

EVALUATION

To ensure that the training has been effective, it is necessary that evaluation occur. Evaluation must be planned when the training curriculum or modules are designed because it will help employers determine whether the desired learning has taken place and whether employee behavior change has resulted.

There are several ways in which the results of the training program can be evaluated:

1. Pre- and post-tests are sometimes useful, and compliance officers generally view them favorably. Questionnaires or informal discussions with trainees can be conducted immediately after the training program and at specified intervals afterwards to determine retention and use of new behaviors. Questions such as "What material was already known and unnecessary?" should be asked. Was the material too complicated and thus in need of simplification? Based on their own experiences, employees should be able to let you know what was missing.
2. Supervisor observations of employee behaviors combined with self-reports by the employee can be a useful evaluation tool. Checklists can be devised with the specific behaviors addressed; other observational methods may be designed and specifically tailored to the workplace situation.
3. Analysis of workplace records of injuries and accidents.

Based on the evaluation, sometimes it is necessary to revise your program. All steps in the training process need to be reviewed to provide the level of knowledge and skill that was expected to result from the training: Was the job analysis accurate? What knowledge was missing? Were the objectives of the training clear and measurable? Was the learning activity relevant to the training objectives? Were the employees motivated and allowed to participate? Did the supervisory level personnel appropriately support the training?

Evaluation forms and a form for documenting the training should be developed. If training becomes an ongoing part of an employee's experience, it will not be viewed negatively but as a usual activity of the workplace and probably a welcome and successful one.

CONCLUSION

We have seen how various psychosocial factors may play a role in the determination of self-protective behavior in the health care workplace. The increased need for compliance with regulatory mandates in terms of safety training has changed the character of the training process itself. There is a greater need for accountability in terms of training effectiveness, and there is an increased need to make training programs more effective and less costly. This chapter reviewed some of the theoretical models that help us to understand the underlying causes and potential barriers to protective work practices. Safety training programs can be enhanced by careful planning and assessment of the training needs of the targeted audience. Although safety training programs have a very important role to play in the overall safety program of the laboratory, we have tried to stress that a multifactorial approach to safety programs is the key to developing successful safety programs. In particular, administrative support for a strong safety climate is probably the single most important factor related to safe work behaviors.

APPENDIX: TRAINING MATERIAL AND INFORMATION RESOURCES—TECHNICAL INFORMATION, SUPPORT SERVICES, AND GUIDELINES

American National Standards Institute
1430 Broadway
New York, N.Y. 10018
(212) 354-3300

American Red Cross
National Headquarters
AIDS Education Program
17th and D Streets N.W.
Washington, D.C. 20006
(202) 639-3441

American Society for Microbiology
1325 Massachusetts Avenue N.W.
Washington, D.C. 20005
(202) 737-3600

Clinical Microbiology Procedures Handbook
American Society for Microbiology, 1992

Centers for Disease Control and Prevention
Centers for Disease Control
1600 Clifton Road N.E.
Atlanta, Ga. 30333
(404) 329-3311
Laboratory Program Office
(404) 329-3232

Compressed Gas Association
1235 Jefferson Davis Highway
Arlington, Va. 22202
(703) 979-0900

Environmental Protection Agency
Office of Solid Waste Management and Emergency Response
Washington, D.C. 20460
(202) 382-4700

Joint Commission on Accreditation of Healthcare Organizations
875 North Michigan Avenue
Chicago, Ill. 60611
(312) 642-6061

Medical Device and Laboratory Product Problem Reporting Number
U.S. Food and Drug Administration
Practitioner Reporting System
United States Pharmacopoeia
12601 Twinbrook Parkway
Rockville, Md. 20852
(202) 720-2791

National Audiovisual Center
General Services Administration
Order Section CF
Washington, D.C. 20409
(301) 763-1896

National Fire Protection Association
Batterymarch Park
Quincy, Mass. 02269
(800) 344-3555

National Institutes of Health
Division of Safety
Building 31, Room 1C02
Bethesda, Md. 20892
(301) 496-2801

National Safety Council
444 North Michigan Avenue
Chicago, Ill. 60611
(312) 527-4800

OSHA
U.S. Department of Labor
Occupational Safety and Health Administration
200 Constitution Avenue
Washington, D.C. 20210
(202) 523-6091
OSHA Publications, Audiovisual Programs, and Training Guidelines
OSHA Publication No. 2019 (1991)
OSHA Publication No. 2254 (1992)

Radioisotopes
U.S. Nuclear Regulatory Commission
1717 H Street
Washington, D.C. 20555
(301) 459-7000

References

1. **Ajzen, I.** 1971. Attitudinal versus normative messages: an investigation of the differential effects of persuasive communications on behavior. *Sociometry* **34:**263–280.
2. **Ajzen, I., and M. Fishbein.** 1973. Attitudinal and normative variables as predictors of specific behaviors. *J. Pers. Soc. Psychol.* **27:**41–57.
3. **Ajzen, I., and M. Fishbein.** 1974. *Factors influencing intentions and the intention-behavior relationship. Hum. Relations* **27:**1–15.
4. **Ajzen, I., and M. Fishbein.** 1980. *Understanding attitudes and predicting social behavior.* Prentice-Hall, Inc., Englewood Cliffs, N.J.
5. **Becker, M. H.** 1974. *The health belief model and personal health behavior.* Slack, Thorofare, N.J.
6. **Becker, M. H., M. Kaback, I. Rosenstock, and M. Ruth.** 1975. Some influences of public participation in genetic screening programs. *J. Community Health* **1:**3–14.
7. **Becker, M. H., and L. A. Maiman.** 1975. Sociobehavioral determinants of compliance with health and medical care recommendations. *Med. Care* **13:**10–23.
8. **Cleary, P. D.** 1987. Why people take precautions against health risks, p. 119–149. *In* N. D. Weinstein (ed.), *Taking care: understanding and encouraging self-protective behavior.* Cambridge University Press, New York.
9. **DeJoy, D. M.** 1986. A behavioral-diagnostics model for self-protective behavior in the workplace. *Professional Safety* **12:**26–30.
10. **Douglas, B. E.** 1975. Occupation health problems in clinics and hospitals. *In* C. Zenz (ed.), *Occupational medicine: principles and practical applications.* Year Book Medical Publishers, Chicago.
11. **Earle, T. C., and G. Cvetovich.** 1983. Risk judgement and the communication of hazard information: toward a new look in the study of risk perception. *In* V. T. Covello, J. L. Mumpowere, P. J. Stallen, and V. R. R. Uppuluri (ed.), *Environmental impact assessment, technology assessment, and risk analysis.* Springer-Verlag, New York.
12. **Fishbein, A.** 1980. A theory of reasoned action: some applications and implications. *In* H. Howe and M. Page (ed.), *Nebraska Symposium on Motivation,* vol. 27. University of Nebraska Press, Lincoln.
13. **Fishbein, M., and I. Ajzen.** 1975. *Belief, attitude, intention and behavior: an introduction to theory and research.* Addison-Wesley Publishing Co., Inc., Reading, Mass.
14. **Freudenburg, W. R.** 1988. Perceived risk, real risk: social science and the art of probabilistic risk assessment. *Science* **242:**44–49.
15. **Green, L. W., M. W. Kreuter, S. G. Deeds, and K. B. Partridge.** 1980. *Health education planning: a diagnostic approach.* Mayfield Publishing, Palo Alto, Calif.
16. **Higbee, K. L.** 1969. Fifteen years of fear arousal: research on threat appeals: 1953–1968. *Psychol. Bull.* **72:**426–444.
17. **Hochbaum, G.** 1958. *Public participation in medical screening programs.* DHEW Publication No. 572, PHS. Goverment Printing Office, Washington, D.C.
18. **Kirscht, J. P.** 1983. Preventive health behavior: a review of research and issues. *Health Psychol.* **2:**277–301.
19. **Leventhal, H., G. M. Hochbaum, and I. Rosenstock.** 1960. Epidemic impact of the general population. *In The impact of Asian influenza on community life.* USHHS PHS Publication No. 706. Public Health Service, Washington, D.C.
20. **McGuire, W. J.** 1985. Attitudes and attitude change. *In* G. Lindzey and E. Aronson (ed.), *Handbook of social psychology,* 3rd ed. Random House, New York.
21. **Montgomery, S. B., J. G. Joseph, M. H. Becker, D. G. Ostrow, R. B. C. Kessler, and J. P. Kirscht.** 1989. The health belief model in understanding compliance with preventive recommendations for AIDS: how useful? *AIDS Education Prevention* **1:**303–323.
22. **O'Keefe, D. J.** 1990. *Persuasion: theory and research.* Sage Publications, Inc., Newbury Park, Calif.
23. **Occupational Safety and Health Administration.** 1989. Occupational exposure to bloodborne pathogens: proposed rule and notice of hearing. *Fed. Regist.* **54:**23041–23139.
24. **Romer, D.** 1990. Review. *AIDS Education Res. Monitor* **1:**3–4.
25. **Rosenstock, I. M., and J. Kirscht.** 1979. Why people use health services, p. 161–188. *In* G. Styone, F. Cohen, and N. Adler (ed.), *Health psychology.* Jossey-Bass, San Francisco.
26. **Schlutz, D. P., and S. E. Schultz.** 1986. *Womens work, womens health: myths and realities.* Pantheon, New York.
27. **Slovic, P.** 1987. Perception of risk. *Science* **236:**280–285.
28. **Slovic, P., B. Fischhoff, and S. Lichtenstein.** 1987. Behavioral decision theory perspectives on protective behavior, p. 14–41. *In* N. D. Weinstein (ed.), *Taking care: understanding and encouraging self-protective behavior.* Cambridge University Press, Cambridge.
29. **Slovic, P., B. Fischoff, and S. Lichtenstein.** 1988. Informing people about risk, p. 165–179. *In* L. Morris, M. Mazie, and I. Barofsky (ed.), *Product labeling and health risk.* Banbury Center, Cold Spring Harbor, N.Y.
30. **Taylor, S. E.** 1986. *Health psychology.* Random House, New York.
31. **Weinstein, N. D.** 1987. Introduction: studying self-protective behavior, p. 1–9. *In* N. D. Weinstein (ed.), *Taking care: understanding and encouraging self-protective behavior.* Cambridge University Press, Cambridge.
32. **Weinstein, N. D.** 1988. The precaution adoption process. *Health Psychol.* **4:**355–386.

Design and Implementation of Occupational Health and Safety Programs

RICHARD L. EHRENBERG AND HOWARD FRUMKIN

19

INTRODUCTION

The fundamental goal of a workplace occupational health program is to protect the health and safety of workers. To achieve this goal in the most appropriate, most cost-effective, and least disruptive manner, the occupational health program must be designed to fit workplace circumstances. This chapter describes the development and implementation of comprehensive occupational health programs appropriate to employees who work in clinical laboratories and similar settings. Factors relevant to any health and safety program are discussed as well as those that arise uniquely from the type of work performed in clinical laboratory facilities. Clinical laboratories, as defined here, include any laboratory that receives patient specimens for analysis or evaluation—whether hospital-based or free-standing, private or public sector. Other settings in which the material in this chapter might prove helpful include industry, university, or government facilities that perform toxicity testing or biotechnical research and production processes. Space does not permit an encyclopedic discussion of the field of occupational safety and health nor detailed prescriptions to address all specific workplace circumstances that may arise in the clinical laboratory setting. Rather, we highlight the issues to be addressed in the course of developing programs designed to protect the health and safety of laboratory workers, including those whose work or studies may bring them temporarily or intermittently into this environment (e.g., students and residents, maintenance staff, janitorial staff).

The overall organization of a program must reflect a logical scheme for approaching the general problem of workplace health and safety. In brief, this scheme entails (i) the identification and enumeration of all potential hazards (physical, chemical, and biological) present in the worksite, as has been described in previous chapters; (ii) the design and implementation of procedures to address the hazards contained in this hazard inventory, as has also been described previously; and (iii) the institution of a process for ongoing monitoring of the effectiveness and completeness of these activities, as described by Songer (Chapter 17) and in this chapter. All these efforts must be performed in the context of an increasingly complicated and comprehensive set of legal and regulatory requirements. Finally, as the main focus of this chapter, consideration must be given to the issue of providing certain preventive and even health promotional services, which can be beneficial to employees (and may be indirectly so to employers as well). Certain relevant terms, as used in this chapter, are defined in the glossary at the end of this chapter.

Use of Hazard Assessment in Design of Medical Programs

Once the inventory of hazards has been compiled and assessed and those with potential for meaningful degrees of exposure have been identified, the health and safety staff can assess the spectrum of adverse health effects that can reasonably be anticipated to occur among the workers as a consequence of their work. Using this information, it is possible to design an appropriate workplace medical program. This program should comprise a set of examinations and procedures (e.g., medical histories, physical examination, laboratory tests) specifically selected to be capable of identifying and responding to the occurrence of these anticipated illnesses or conditions. It is not necessary to include procedures to detect conditions not associated with the materials in the hazard inventory. A targeted program can avoid unnecessary and expensive procedures or tests without loss of overall effectiveness. If it is desired, the program can use the hazard assessment information to tailor examinations to the specific needs of each employee or job category according to the specific risks associated with that individual's or group's work responsibilities.

MEDICAL PROGRAM

The medical program includes evaluations at the time of initial hiring, periodic evaluations thereafter, and evaluations triggered by episodes of exposure, injury, illness, or absence. In each of these evaluations, certain principles of confidentiality, patient communication, record keeping, and data analysis must be observed.

Occupational Medical Evaluations

Preplacement Examinations

Medical examinations have traditionally been used before hiring as a means of selecting job applicants thought to be most fit or at least risk in a particular job. However, this approach has changed in recent years, with an increased social commitment to eliminating employment discrimination and with relevant federal legislation (the Civil Rights Act of 1964, the Rehabilitation Act of 1973, and the Americans with Disabilities Act of 1990 and current revisions). Rather than *pre-employment* examinations designed to identify those who may or may not be hired, the emphasis in now on *preplacement* examinations designed to evaluate worker capabilities and needs. The goal is to place each new employee in an appropriate job, with any reasonable accommodations that may be necessary. Also, as described in the glossary, preplacement examinations yield baseline data that can be useful in evaluating any abnormalities that may be noted later. Finally, preplacement examinations serve a less tangible function. They introduce the new employee to the occupational health providers, and they establish early that employee health is a priority.

Assessing the employee's capability of performing his or her specific responsibilities requires a thorough knowledge of the job. Any health care provider who performs these examinations should be familiar with the working conditions, the job requirements, and any hazardous exposures that may exist. Effective interactions must be established among the worker, the supervisor, the human resources staff, the safety and environmental health staff, and the occupational medical program, which can communicate job requirements to the health care provider and job restrictions to the employer.

The preplacement examination should ideally include a complete medical history, an occupational history, a physical examination, and appropriate blood and urine work based on the potential workplace hazards.

Most laboratory personnel will encounter organic solvents, which can be toxic to the liver, kidneys, bone marrow, and central nervous system; baseline blood counts and liver and kidney function test results may be useful. With modern automated testing procedures, these data are most economically collected with a chemistry panel and a complete blood count. (However, such an approach includes many tests that are not indicated, such as uric acid and serum protein levels. Positive results on these tests sometimes lead to further testing, an important disadvantage of automated test panels.) However, few laboratories pose any exposure to hazardous levels of noise, so a traditional pre-employment test such as audiometry is rarely warranted. Similarly, there is rarely a direct rationale for pre-employment electrocardiograms.

Specific hazards in a particular laboratory should lead to specific inquiries, either on the initial history or in the initial physical examination and laboratory tests. Several examples appear in Table 1.

The preplacement examination results in one of several conclusions: The employee is fit for the job, the employee is fit for the job with certain restrictions or modifications, or the employee is unfit for the job. The latter two conclusions may be temporary or permanent in nature.

In general, the response to a finding such as those in Table 1 is to implement appropriate safeguards. A G-6-PD-deficient employee should be assigned to

TABLE 1 Examples of hazardous laboratory exposures relevant during preplacement evaluation

Exposure	Employee risk factor	Preplacement test
Oxidizing chemicals	G-6-PD deficiency	G-6-PD level
Breaking glassware	Essential tremor	Physical examination
Rubella virus	Nonimmune women	History, titers
Pulmonary sensitizers	Asthma, atopy	History, pulmonary function tests
Tuberculosis bacillus	All employees	Tuberculin skin test
Hepatitis B	Nonimmune persons	History, serology
Specific allergens	Allergy	History, skin testing, serum precipitins

tasks that do not involve handling oxidizing chemicals. A tremulous worker should not be assigned to handle delicate glassware. A woman who is of childbearing age and who is not immune to rubella should be immunized appropriately, as should an individual who will have contact with body fluids and who lacks hepatitis B antibodies. If for some reason the person cannot be immunized, the exposure should be avoided (17). In all cases, the results and rationale for any intervention should be shared with the employee.

Immunizations

The most important immunization for employees in the clinical laboratory setting is the hepatitis B vaccination. The Occupational Safety and Health Administration (OSHA) blood-borne pathogen standard (7) requires that "all employees who have occupational exposure" to hepatitis B virus must be offered the hepatitis B vaccine. It must be offered at no cost, at a reasonable time and place, and within 10 days of commencing exposed work. If booster doses are recommended by the U.S. Public Health Service at some future date, the OSHA standard requires that these be made available as well. The general recommendations of the U.S. Public Health Service regarding immunizations for adults should be followed, particularly with respect to organisms likely to be encountered in the clinical laboratory setting (15–17).

Periodic Examinations

Periodic examinations encompass both screening and surveillance purposes, as defined in the glossary. Such programs have several goals (10), but the primary one remains the early detection and treatment of work-related disease. Periodic examinations of workers are also useful in evaluating whether exposure controls have functioned adequately and in detecting previously unrecognized adverse health effects or the occurrence of such effects at exposure levels previously considered safe. Finally, periodic examinations may serve a general medical function, such as breast cancer screening or hypertension detection, unrelated to specific occupational exposures.

Several general principles apply to periodic medical screening (4). These derive from the goal of detecting diseases early, before they become clinically evident, so that early intervention can be provided and the ultimate outcome improved. Accordingly, the condition to be detected through screening should be important (i.e., have potential for substantial morbidity and/or mortality) and reasonably prevalent. The natural history of the condition should include a "detectable preclinical phase," when early detection before the affected individual would normally seek medical attention is possible. The condition should be reversible (or, at least, its progression arrestable) with intervention (exposure removal and/or medical therapy), and intervention at this stage should be simpler, more effective, or less risky than that initiated at a later stage. The ideal screening *test* should be sensitive, specific, inexpensive, simple, low-risk, noninvasive, and acceptable to subjects. Finally, a screening *program* must include adequate follow-up, such as notifying participants of results and directing them promptly to appropriate treatment. Examples of screening tests in general use that meet these criteria include mammography for breast cancer, sigmoidoscopy for colon cancer, and Pap smears for cervical cancer (18). In contrast, the annual comprehensive physical examination, a time-honored tradition, is generally not justified by these principles.

In the occupational setting, these principles are often modified (10). Even if no beneficial early treatment is possible, workplace screening may still have a role, by providing evidence of unrecognized excessive exposures that require environmental controls to prevent additional exposures. Similarly, a complicated or expensive screening test that would be unsuitable in the community setting might be appropriate in the workplace because of regulatory requirements, liability considerations, employer responsibility, an increased risk of illness in a specific population, or other factors.

Several classes of examination procedures deserve specific mention. Standard *clinical tests,* such as complete blood counts and liver enzyme assays, monitor adverse health effects on specific organ systems and may be appropriate after exposures with corresponding toxicities. *Biological monitoring,* as most commonly defined, refers to assays that measure levels of foreign materials, or their metabolites, in body fluids or excreta; an example is the blood lead level. Another class of tests, sometimes considered as a form of biological monitoring and sometimes termed monitoring of nonadverse biological effects, measures effects of a toxin that are not generally considered to be direct signs of disease or impaired function—e.g., assays of red blood cell cholinesterase levels after exposure to organophosphate pesticides. Finally, *molecular biomarkers* represent a new class of laboratory tests, derived from advances in molecular biology, that evaluate evidence of exposure at the molecular level. Examples include DNA adducts associated with benzo[α]-pyrene exposure and sister chromatid exchanges and hemoglobin adducts associated with ethylene oxide exposure. Assays of these biomarkers are generally too new to be in common clinical use, but they may be expected to assume a role in workplace screening and surveillance in coming years.

Postexposure Examinations

Postexposure examinations refer to the medical evaluations that follow acute exposure incidents. In the general occupational setting, these occur most often in the context of a spill or other nonroutine release that results in short-term, potentially hazardous biological, chemical, or radiologic exposures. Postexposure medical follow-up is intended to assess exposed workers for evidence of adverse health effects, both acute and persisting. The examination should be individualized and directed toward detecting effects specifically associated with the exposure(s) in question and should take into account the relevant induction or incubation times (i.e., should not test for an effect before it is likely to have manifested or after it is likely to have resolved). In addition to such exposure incidents, workers in clinical laboratories, hospitals, and similar settings are at risk of incidents of percutaneous exposure to infectious agents, such as with needlesticks. The procedures to follow in these circumstances are based on standard medical practice and on requirements of the blood-borne pathogen standard (7). For further information on specific blood-borne pathogens, see Chapters 3 and 4.

After an incident of acute exposure to a blood-borne pathogen, the employer must document the route(s) of exposure and the circumstances under which the incident occurred. The source individual must be identified, if this is feasible and permitted under state and local law, and consent of the source individual obtained, if required, before testing for hepatitis B and human immunodeficiency virus infectivity. (If consent is required and the source individual refuses to give consent, that fact must be documented.) The results of the source individual's test must be provided to the exposed employee, who must then be offered blood testing, counseling, and postexposure prophylaxis. After exposure to other infectious agents (not covered by the OSHA blood-borne pathogen standard), other testing procedures or antibiotic treatment may be appropriate.

Return-to-Work Examinations

After a work absence for an occupational or non-occupational ailment, medical evaluation may be required. The principles that govern these evaluations are identical to those in the preplacement context. Fitness to work is assessed by reference to specific job requirements. If an employee is temporarily or permanently impaired, job modifications might be necessary, ranging from limited work hours or "light duty" to permanent changes in job responsibilities. At this point, rehabilitation or retraining should be initiated if necessary.

Termination Examinations

When an employee resigns, retires, or is discharged, a termination examination may be conducted. The purpose of this examination combines those of preplacement and periodic examinations: documentation of "baseline" health status at the time of departure from the workplace (for use in future follow-up), early detection of disease, and detection of any failures of exposure controls.

Guidelines for Managing Medical Evaluations and Information

Confidentiality and Informed Consent

A cardinal principle of occupational health practice, as with all medical practice, is that confidentiality of medical records must be maintained. The recently revised Code of Ethical Conduct of the American College of Occupational and Environmental Medicine (2), for example, states that the physician should

> keep confidential all individual medical information, releasing such information only when required by law, over-riding public health considerations, or to other physicians at the request of the individual according to accepted practice, or to others at the request of the individual.

The Code of Ethics of the American Association of Occupational Health Nurses (1) includes a similar provision, requiring that occupational health nurses should "safeguard the employee's right to privacy by protecting confidential information and releasing information only upon written consent of the employee or as required by law."

Similar respect for confidentiality is part of OSHA policy, as reflected in the *Rules of Agency Practice and Procedure Concerning OSHA Access to Employee Medical Records* (6). Although OSHA inspections (and National Institute for Occupational Safety and Health investigations) may involve a review of individual medical records, the circumstances in which this is permitted are quite restricted, and extensive privacy safeguards are built into the process.

Practically speaking, this means that medical information should be maintained by the health care facility that examines the employee, whether this facility is part of the same organization or is an outside concern. Personnel and management staff should not have access to medical records, and the personnel department should receive only a statement of an employee's fitness to work, with an additional statement of what restrictions or accommodations, if any, are recommended. The employer may have a legitimate need for medical information deemed necessary for the evaluation of the effectiveness of environmental exposure controls or the identification of problem areas (e.g., information on tuberculosis skin test conversions). In such circumstances, only the required information should be made available, and this should be done, to the maximum extent feasible, in a manner that preserves confidentiality and does not identify individual workers with their specific results (12). In some situations, the laboratory director may need specific results to identify the cause of an exposure (i.e., to tuberculosis). Conversion of an employee TB skin test to positive is a sentinel health event, alerting the laboratory director to poor work practices or ineffective safety devices. Information is needed to allow the investigation to proceed to prevent further workplace exposures.

A second central precept of medical practice is the obligation to obtain voluntary informed consent from an individual before performing a diagnostic or therapeutic action. In the context of this chapter, this concept requires that workers be fully informed, before their participation, of the purposes of all examinations, how the examination results will be used, who will have access to the information (and in what circumstances), and what actions may be taken as a consequence of the examinations (12). The provisions for confidentiality (and any exceptions to confidentiality) should be explained, as should the possible consequences of declining to participate.

Patient Communication

With respect to patient communication, occupational health evaluations are essentially no different from other medical encounters. Medical information assembled in the course of an occupational health evaluation should be shared, in an understandable manner, with the patient. This principle is included in the American College of Occupational and Environmental Medicine Code of Ethics (2), which requires that the physician should "communicate understandably to those they serve any significant observations about their health, recommending further study, counsel or treatment when indicated."

OSHA also mandates employee access to medical records maintained by employers. The standard on access to employee exposure and medical records (5) requires that "each employer shall, upon request, assure the access of each employee to employee medical records of which the employee is the subject."

Similarly, when a patient requests, medical information collected in an occupational health evaluation should be shared with his or her personal physician. The personal physician should also be made aware of the work-related hazards that could be encountered by the patient. Workers who choose to go to their personal physician after a work-related exposure may not receive timely, appropriate treatment unless the occupational health physician is involved in the process.

Record keeping

Medical record keeping in the workplace setting takes several forms. First, medical records should be maintained as in conventional medical practice and should include documentation of all patient visits, examinations and tests performed, diagnoses reached, and treatments and referrals ordered. OSHA requires that these records be maintained for the duration of employment plus 30 years, except for insurance claims, first-aid records, and medical records of employees who worked less than 1 year and were provided their records on termination (5).

OSHA also requires that work-related injuries and illnesses be recorded when they occur. Two forms are used for this purpose. The OSHA Form 200, Log and Summary of Occupational Injuries and Illnesses, is a tabular summary that includes information on type of injury or illness and severity. The OSHA Form 101, *Supplementary Record of Occupational Injuries and Illnesses,* includes more detail. These forms must be maintained for 5 years after they are completed.

State workers' compensation systems require recording and reporting of occupational injuries and illnesses. Each state has a procedure that must be followed, usually with a standard form that must be completed. When the state workers' compensation form includes all the information on the OSHA Form 101, it may replace the OSHA form.

Data Analysis

As computers and analytical software packages become readily available, epidemiologic analysis of routinely collected occupational health data becomes increasingly feasible. With a properly constructed data base and using appropriate software, occupational health personnel can, for example, review all injuries or illnesses of a particular type over a defined time interval, identify the incidence of injuries or cases for various workplace locations or specific job categories, or link employee risk factors with specific injuries or illnesses. For such analyses to be feasible and reliable, it is essential that data be collected in a standardized manner and stored in a form conducive to analysis, including standard coding schemes for department and job, exposures, and injury or diagnosis; medical results must be linked with exposure information and job histories. When properly performed, analyses of this sort can be powerful tools in identifying and abating hazards.

A discussion of available data management and analysis systems is beyond the scope of this chapter. However, the reader is referred to published sources for further information (8, 13).

ADMINISTRATIVE ISSUES

Written Protocols

To the maximum extent feasible, the organization and operation of the health and safety program should be governed by written sets of detailed procedures and protocols. Written materials document prescribed plans and procedures; standardize activities (particularly over time and from incident to incident); and increase the likelihood that predetermined policies and procedures will, in fact, be carried out as intended. This recommendation applies to all aspects of the program including the hazard identification and assessment process; the specifics of the medical program; the response, both medical and administrative, to incidents of work-related injury or illness among employees; and the emergency response procedures for worksite spills or disasters. These procedures and protocols should all be compiled and should be readily available in a central location (generally, the health and safety office). All health and safety staff, as well as those with initial emergency response responsibilities, should know how to access this information—particularly the protocols that govern procedures to be used in the event of injuries, exposure incidents, and so on. Training on the use of these procedures is essential, and the procedures should be routinely reviewed on a scheduled basis so that they can be updated as required.

Team Approach

All staff involved in the occupational safety and health program should work together using a team approach. Close and active cooperation is especially important between the medical personnel and the staff responsible for environmental and industrial hygiene assessments, particularly if these two groups operate from different jurisdictions in the facility's organizational structure. In particular, medical staff should be familiar with the principles and methodologies of industrial hygiene and environmental engineering, the hierarchy of controls of workplace hazards, and so on. Staff should collaborate in the design and implementation of worksite interventions intended to control exposures and reduce hazards, whether through the application of engineering controls, changes in worksite practices, use of personal protective equipment, or other approaches.

Resources

It is difficult to generalize regarding resource requirements (physical facilities, equipment, and staffing) necessary to support the comprehensive occupational safety and health programs outlined in this chapter. The precise requirements will depend on the specific circumstances in which each facility functions, which may range from small stand-alone laboratories with relatively small staffs to sizeable units that function as components of large institutions. It is not economically feasible for small independent facilities to provide the required services through full-time in-house resources; the use of contract or referral mechanisms is necessary. Nevertheless, it is incumbent on management of such organizations to make available (through whatever combination of arrangements) the full range of services, and it is crucial that one well-identified, qualified individual have final, recognized responsibility for coordinating the entire occupational safety and health program.

Occupational Medicine Staffing and Facility Size
Various publications provide guidelines regarding provision of occupational health resources (as a function of facility size) as well as suggestions for health unit floor plans, access and privacy requirements, and equipment lists (8, 9, 14). For example, suggested minimum health unit sizes often start at about 200 ft^2 for facilities with approximately 200 employees, with increments of an additional 1 to 1.5 ft^2 per employee beyond 200 (8, 9, 14). Similar recommendations are available regarding staffing of health units. For facilities with fewer than 200 employees, full-time, on-site staffing is probably impractical. At about 300 employees, hiring a full-time occupational nurse is suggested, with an additional nurse for each 750 additional employees. A full-time industrial hygienist or environmental engineer is recommended at about the 500-employee level. A part-time occupational physician should be hired for facilities with 1,000 or more employees; more than 2,000 employees require a full-time physician (9). Consideration such as the anticipated number of employee visits, likely types of injuries and illnesses, and administrative factors may modify these recommendations. Finally, these estimates were originally devised in an industrial context and may require some adjustment, particularly because applicability of the recommendations for even the minimum number of employees cited is likely to occur only in the circumstance in which the laboratory facility functions as part of a larger unit.

Occupational Medicine Skills
Irrespective of facility size and the applicability of the above guidelines, reliable availability of a defined set of skills is necessary. It is mandatory to have convenient and timely access to a well-qualified physician who is both knowledgeable in the fields of occupational medicine and laboratory safety and familiar with the specific facility and its operations—whether this individual is on-staff or available in the community on a contract basis; this physician should also have clear responsibility for general supervision of the medical program and any medical staff employed at the facility (e.g., nurses, physician's assistants, medical technicians involved in diagnostic work on employee specimens). Similarly, arrangements must be in place for timely access when necessary to all appropriate medical consultants (e.g., orthopedists, neurologists, infectious disease specialists).

Education and Training

Worker education and training represent important components of any occupational health and safety program, and adequate provision should be made in the design of the program to incorporate this aspect. Workers should receive information (at an appropriate educational level—and in the appropriate language, in circumstances in which this factor is a consideration) regarding the hazards they may encounter through their jobs and the actions they need to take to protect themselves. This information should include purpose and proper use of engineering controls and safety devices, the importance of the use of proper work practices, the proper use and maintenance of personal protective equipment, and the function of the medical program. This information should be provided, in a systematic manner, to all workers at the inception of the occupational safety and health program and to all new workers at the time of hire. Refresher training and education should be provided to all workers on a periodic basis and as other circumstances dictate (e.g., adoption of new processes or equipment, use of new materials).

Nonoccupational Services

Health and safety staff who have responsibility for designing and implementing occupational medical programs should recognize that these programs also have an opportunity to provide additional services for the workers covered by the program. These services extend beyond those that are dictated by the results of the workplace hazard assessment and consist of a variety of nonoccupational preventive and health-promoting activities. Possible components include performance of screening examinations for nonoccupational risk factors such as hypertension and elevated cholesterol and provision (whether directly or through a referral system) of counseling or assistance programs to address lifestyle problems (e.g., smoking cessation or substance abuse programs). Such activities are by no means necessary elements of an occupational health program, but they may be desirable. Although they increase the cost of the program and add complexity, the indirect benefits (e.g., improved employee relations, a healthier, more productive work force) may outweigh these costs. Furthermore, this approach may ultimately produce more measurable benefits, such as lower costs associated with absenteeism and reductions in health insurance premiums.

LEGAL AND ETHICAL ISSUES

Medical personnel and staff with responsibility for the design, implementation, and administration of occupational safety and health programs should be aware of many legal obligations and ethical considerations that are inherent in these activities. The legal obligations result from many federal laws and regulations as well as various state and local ordinances, which vary from locality to locality. The ethical issues grow out of the requirements of the physician-patient relationship when it is situated in the unique context of the workplace.

Although discussion of the full spectrum of the legal and ethical obligations lies beyond the scope of this chapter, staff involved with such programs should be aware of their existence. Some have previously been discussed in some detail; additional information can be found in a number of texts (8, 9, 11) as well as in the regulations themselves.

As alluded to earlier, the physician-patient relationship imposes certain ethical responsibilities on the physician (as well as the medical and administrative staff who support the medical operation). Among these are obligations to obtain informed consent from the patient regarding the examinations and procedures to be performed, their purposes and risks (if any), how the results will be used, and who will have access to them; to notify the patient of the results and interpretations of any examinations or laboratory procedures; and to maintain the confidentiality of all information acquired through this relationship (and the records thus generated), releasing this only with the explicit consent of the affected individual.

In the occupational context, however, these issues acquire additional complexity, in part because a closely involved third party, with a separate set of legitimate interests, is introduced into the usual two-way relationship. Furthermore, the physician (or his or her surrogate) is generally employed by (or at least paid by) this third party and has a separate set of obligations to this third party. Thus, for example, these circumstances can introduce pressures to attenuate the usual absolute confidentiality of medical records. Employers may need access to some information contained in medical records—where these data convey information about possible worksite overexposures or the occurrence of work-related health effects—to enable them better to protect the health of the work force; this is particularly true for biological monitoring results (e.g., blood lead levels), which can represent a means of measuring exposure and medical information about the individual. Medical personnel must be aware of the issues and carefully consider their implications in advance. In general, it is recommended that the medical staff, with the prior understanding and consent of the affected individuals, release only the minimal amount of information (without individual identifiers) necessary to direct workplace intervention activities or determine fitness to perform work duties; in no instance should details of medical conditions be released. Finally, medical records should be kept in secure facilities, separate from other personnel records and not accessible by supervisors or administrative staff.

Finally, brief mention should be made of the workers' compensation system. This system was developed to provide a mechanism for rapidly compensating workers who are injured on the job or develop work-related illnesses. The system is intended to compensate for medical care expenses, rehabilitation expenses, and some portion of lost wages. It is designed to do this through an automatic, no-fault administrative process that eliminates litigation. The regulations governing the system and the specific procedures differ from state to state, and health and safety staff need to be conversant with the system's operations and their obligations (e.g., the filing of reports in conjunction with a worker's filing of a claim) in their specific locality (8).

Workers with Disabilities

The Americans with Disabilities Act (ADA), P.L. 101–336, was passed in July 1990 and took effect in mid-1992. This law is intended to prevent discrimination and to safeguard civil rights in employment, public services and accommodations, transportation, and telecommunications for those defined as mentally or physically disabled. A disabled person is defined in the ADA as an individual who has an impairment that substantially limits one or more major life activities, has a record of such an impairment, or is regarded by others as having such an impairment. For the purposes of this definition, major life activities include caring for oneself, performing manual tasks, walking, seeing, speaking, hearing, learning, and working; homosexuality, bisexuality, sexual behavior disorders, and drug use are specifically excluded from the definition of disability.

The ADA prohibits employers from discriminating against a "qualified individual with a disability" in hiring, promotion, discharge, pay, training, conditions, and privileges of employment. Moreover, "reasonable accommodations" must be made to enable a disabled individual to perform his or her job. Such accommodation might include adjusting work schedules, physically modifying the work environment, or redefining an individual's responsibilities.

Several aspects of ADA will be relevant to occupational health practice. Pre-employment physical examinations will be permitted only if they are job-related, consistent with business necessity, and administered after a job has been offered. However, these examinations may be performed before the new hire begins work, and the job offer may be conditional on the results of the examination if all employees are similarly examined and if confidentiality is maintained (3).

The overall thrust of these provisions is that individuals who are able to perform a job's requirements may not be excluded from the job because of their disabilities. Drug testing is still permitted under ADA, and those who test positive are not protected from discharge. In general, medical evaluations will therefore focus on job *fitness* rather than *risk*. The concept of a *direct threat* to the employee from the hazard is also important in interpreting these provisions.

GLOSSARY

Biological monitoring: Measurement of the level of a chemical or its metabolites in a biological specimen (e.g., urine or blood). Biological monitoring provides an estimate of the total absorbed or internal dose. Thus, it accounts for absorption via all routes of entry (inhalation, ingestion, and percutaneous) and other factors that may affect absorption (e.g., activity levels, use of protective equipment).

Induction time or latency: The time period between initial exposure to an agent (physical, chemical, or biological) and the development of detectable signs of the illness or injury produced by that agent (analogous to the incubation period of an infectious disease).

Medical screening: The performance of one or more medical examinations or procedures to identify unrecognized diseases or medical conditions at an early (often asymptomatic) stage when they are more likely to be reversible or more easily treatable. Screening is distinguished from surveillance in that screening activities are directed at the individual for the specific benefit of that individual, whereas surveillance activities are directed at a group for the overall benefit of the group (and similar groups elsewhere), without necessarily benefiting all individual members of the group.

Natural history (of a disease or condition): The time course and outcome an illness or injury typically follows in the absence of any active preventive or therapeutic intervention.

Periodic or interval examinations: Medical examinations performed at regular predetermined intervals during the course of employment, intended to confirm continued ability to perform given job duties as well as detect early signs of the development of new medical conditions, particularly those that may be related to (or aggravated by) workplace exposures or activities.

Preplacement examinations: Medical examinations performed after initial hiring, intended to determine fitness for employment and capacity to perform specific job duties. It also serves to obtain baseline medical data, against which to compare the results of subsequent medical examinations for evidence of changes over time, development of new medical conditions, and so on.

Surveillance: The epidemiologic collection and analysis of medical data about a group, with the timely dissemination of the results to those who need to know. The purpose of surveillance is to measure the occurrence of illness or injury, to identify changes in trends or distribution of cases to direct investigative and control (preventive) measures. In contrast to screening activities, surveillance efforts may not necessarily benefit individuals identified by the surveillance activity but may benefit the group by leading to prevention of future cases of disease or injury.

Workers' compensation: State-based, no-fault insurance systems intended to compensate workers who are injured on the job or develop a job-related illness. Compensation is provided for medical expenses and lost wages. Participation in a workers' compensation system replaces tort or liability litigation.

References

1. **American Association of Occupational Health Nurses.** Code of Ethics, adopted June 1986.
2. **American College of Occupational and Environmental Medicine, Committee on Ethical Practice in Occupational Medicine.** 1994. The new ACOEM code of ethical conduct. *J. Occup. Med.* **36:**27–30.
3. **Bureau of National Affairs.** 1990. *The Americans with Disabilities Act: a practical and legal guide to impact, en-*

forcement, and compliance. Bureau of National Affairs, Washington, D.C.

4. **Cole, P., and A. Morrison.** 1980. Basic issues in population screening for cancer. *JNCI* **64:**1263–1272.
5. **Federal Register.** 1991. *Code of federal regulations: 29 CFR 1910.20(e)(2)(i). Subpart C: General safety and health provisions—access to employee exposure and medical records.* Office of the Federal Register, National Archives and Records Administration, Washington, D.C.
6. **Federal Register.** 1991. *Code of federal regulations: 29 CFR 1913.10. Rules of agency practice and procedure concerning OSHA access to employee medical records.* Office of the Federal Register, National Archives and Records Administration, Washington, D.C.
7. **Federal Register.** 1992. *Code of federal regulations: 29 CFR 1910.1030. Occupational exposure to bloodborne pathogens.* Office of the Federal Register, National Archives and Records Administration, Washington, D.C.
8. **Felton, J. S.** 1990. Medical records and documentation, p. 271–332. *In* J. S. Felton (ed.), *Occupational medical management: a guide to the organization and operation of in-plant occupational health services.* Little, Brown & Co., Boston.
9. **Guidotti, T. L., J. W. F. Cowell, and G. G. Jamieson.** 1989. *Occupational health services: a practical approach.* American Medical Association, Chicago.
10. **Halperin, W. E., J. Ratcliffe, T. M. Frazier, L. Wilson, S. P. Becker, and P. A. Schulte.** 1986. Medical screening in the workplace: proposed principles. *J. Occup. Med.* **28:**547–553.
11. **Levy, B. S., and D. H. Wegman.** 1988. *Occupational health: recognizing and preventing work-related disease,* 2nd ed. Little, Brown & Co., Boston.
12. **Matte, T. D., L. Fine, T. J. Meinhardt, and E. L. Baker.** 1990. Guidelines for medical screening in the workplace. *Occup. Med.* **5**(3)**:**439–456.
13. **Peterson, K. W., and L. F. David (ed.).** 1991. *Directory of occupational health and safety software,* version 4. American College of Occupational and Environmental Medicine, Chicago.
14. **Travers, P. H.** 1987. *A comprehensive guide for establishing an occupational health service.* American Association of Occupational Health Nurses, Atlanta.
15. **U.S. Department of Health and Human Services, Public Health Service, Centers for Disease Control, and National Institutes of Health.** 1993. *Biosafety in microbiological and biomedical laboratories,* 3rd ed. HHS Publication No. (CDC)93–8395. U.S. Department of Health and Human Services, Washington, D.C.
16. **U.S. Department of Health and Human Services, Public Health Service, Centers for Disease Control and Prevention.** 1993. Recommendations of the Advisory Committee on Immunization Practices (ACIP): Use of Vaccine and Immune Globulins in Persons with Altered Immunocompetence. *Morbid. Mortal. Weekly Rep.* **42**(RR-4)**:**1–18.
17. **U.S. Department of Health and Human Services, Public Health Service, National Center for Prevention Services, National Center for Infectious Diseases.** 1991. Update on adult immunization: recommendations of the Immunization Practices Advisory Committee (ACIP). *Morbid. Mortal. Weekly Rep.* **40**(RR-12)**:**1–94.
18. **U.S. Preventive Services Task Force.** 1989. *Guide to clinical preventive services.* Williams & Wilkins, Baltimore.

EPILOG
Current and Emerging Trends

The potential for laboratory-acquired infection has been recognized for many years as an occupational hazard. In recent years, we have come to appreciate the magnitude of the problem and the myriad of factors, human, environmental, and procedural, that may be involved. Considerable information has been accumulated on the main routes of worker exposure. The earliest reviews of Sulkin and Pike, as mentioned throughout this book, indicate that accidents with needles and syringes gave rise to laboratory-acquired infections. This is as true today as it was 40 years ago. The illnesses that we see now are not just brucellosis, Q fever, and salmonellosis. The diseases we are now addressing are AIDS, hepatitis B, new strains of hantavirus, and a re-emergence of tuberculosis, including strains that are multiple drug-resistant. The Centers for Disease Control and Prevention and the National Institutes of Health in 1993 advocated the development and implementation of procedures appropriate for the level of risk on a case-by-case basis (i.e., each experiment) by the worker.

Physical controls alone cannot create or generate a laboratory facility that is safe. Safety, achieved in part through a combination of equipment, facilities, engineering features, and the strict adherence to proper procedures and practices, also demands an appropriate attitude and behavior pattern on the part of the worker as discussed in Chapter 18. The history of laboratory-acquired infection teaches us that simply providing facilities, equipment, procedures, and practices is insufficient. Laboratory illnesses were, are, and always will be an occupational problem as long as the workers who perform the various tasks and duties remain complacent about how they perform their assignments. Do workers fail to wear gloves? Do they disregard material spilled on benches? Do they disregard the use of eye and face protection? Do they mouth-pipette? Do they eat and drink in the laboratory? Do they fail to use syringes and needles properly? Although not all laboratory workers demonstrate such unsafe behavior, a sufficient number ignore prudent biosafety practices for laboratory-acquired infections to remain a concern in the future.

Appendixes

APPENDIX

Biosafety in Microbiological and Biomedical Laboratories†

U.S. DEPARTMENT OF HEALTH AND HUMAN SERVICES, PUBLIC HEALTH SERVICE, CENTERS FOR DISEASE CONTROL AND PREVENTION, AND NATIONAL INSTITUTES OF HEALTH

I

SECTION I: INTRODUCTION

Microbiology laboratories are special, often unique, work environments that may pose identifiable infectious disease risks to persons in or near them. Infections have been contracted in the laboratory throughout the history of microbiology. Published reports around the turn of the century described laboratory-associated cases of typhoid, cholera, glanders, brucellosis, and tetanus.[192] In 1941 Meyer and Eddie[125] published a survey of 74 laboratory-associated brucellosis infections that had occurred in the United States, and concluded that the "handling of cultures or specimens or the inhalation of dust containing *Brucella* organisms is eminently dangerous to laboratory workers." A number of cases were attributed to carelessness or poor technique in the handling of infectious materials.

In 1949, Sulkin and Pike[179] published the first in a series of surveys of laboratory-associated infections summarizing 222 viral infections—21 of which were fatal. In at least a third of the cases the probable source of infection was considered to be associated with the handling of infected animals and tissues. Known accidents were recorded in 27 (12%) of the reported cases.

†This is HHS Publication No. (CDC) 93-8395, Third Edition (May 1993). Numbers in the margins correspond to page numbers in the original publication.

In 1951, Sulkin and Pike[180] published the second of a series of summaries of laboratory-associated infections based on a questionnaire sent to 5,000 laboratories. Only one-third of the 1,342 cases cited had been reported in the literature. Brucellosis outnumbered all other reported laboratory-acquired infections and, together with tuberculosis, tularemia, typhoid, and streptococcal infection, accounted for 72% of all bacterial infections and for 31% of infections caused by all agents. The overall case fatality rate was 3%. Only 16% of all infections reported were associated with a documented accident. The majority of these were related to mouth pipetting and the use of needle and syringe.

This survey was updated in 1965,[154] adding 641 new or previously unreported cases, and again in 1976,[151] summarizing a cumulative total of 3,921 cases. Brucellosis, typhoid, tularemia, tuberculosis, hepatitis, and Venezuelan equine encephalitis were the most commonly reported. Fewer than 20% of all [2] cases were associated with a known accident. Exposure to infectious aerosols was considered to be a plausible but unconfirmed source of infection for the more than 80% of the reported cases in which the infected person had "worked with the agent."

In 1967 Hanson et al.[86] reported 428 overt laboratory-associated infections with arboviruses. In some instances the ability of a given arbovirus to produce human disease was first confirmed as the result of unintentional infection of laboratory personnel. Exposure to infectious aerosols was considered the most common source of infection.

In 1974 Skinhoj[170] published the results of a survey which showed that personnel in Danish clinical chemistry laboratories had a reported incidence of hepatitis (2.3 cases per year per 1,000 employees), seven times higher than that of the general population. Similarly, a 1976 survey by Harrington and Shannon[88] indicated that medical laboratory workers in England had "a five times increased risk of acquiring tuberculosis compared with the general population." Hepatitis B and shigellosis were also shown to be continuing occupational risks and, along with tuberculosis, were the three most commonly reported occupation-associated infections in Britain.

Although these reports suggest that laboratory personnel were at increased risk of being infected by the agents they handle, actual rates of infection are typically not available. However, the studies of Harrington and Shannon[88] and of Skinhoj[170] indicate that laboratory personnel had higher rates of tuberculosis, shigellosis, and hepatitis B than does the general population.

In contrast to the documented occurrence of laboratory-acquired infections in laboratory personnel, laboratories working with infectious agents have not been shown to represent a threat to the community. For example, although 109 laboratory-associated infections were recorded at the Centers for Disease Control and Prevention from 1947–1973,[159] no secondary cases were reported in family members or community contacts. The National Animal Disease Center reported a similar experience,[181] with no secondary cases occurring in laboratory and nonlaboratory contacts of 18 laboratory-associated cases occurring from 1960–1975. A secondary case of Marburg disease in the wife of a primary case was presumed to have been transmitted sexually two months after dismissal from the hospital.[117] Three secondary cases of smallpox were reported in two laboratory-associated outbreaks in England in 1973[157] and 1978.[202] There were earlier reports of six cases of Q fever in a commercial laundry cleaning linens and uniforms from a laboratory working with the agent,[140] one case of Q fever in a visitor to a laboratory,[140] and two cases of Q fever in household contacts of a rickettsiologist.[10] One case of Monkey B virus transmission from an infected animal care giver to his wife has been reported, apparently due to contact of the virus with broken skin.[92] These cases are representative of the sporadic nature and infrequency of community infections in laboratory personnel working with infectious agents.

In his 1979 review,[153] Pike concluded "the knowledge, the techniques, and the equipment to prevent most laboratory infections are available." In the United States, however, no single code of practice, standards, guidelines, or other publication provided detailed descriptions of techniques, equipment, and other considerations or recommendations for the broad scope of laboratory activities conducted with a variety of indigenous and exotic infectious agents. The booklet, *Classification of Etiologic Agents on the Basis of Hazard,*[23] served as a general reference for some laboratory activities utilizing infectious agents. This booklet, and the concept of categorizing infectious agents and laboratory activities into four classes or levels, served as a basic format for earlier editions of *Biosafety in Microbiological and Biomedical Laboratories* (BMBL). This third edition of the BMBL continues to specifically describe combinations of microbiological practices, laboratory facilities, and safety equipment, and recommend their use in four categories or biosafety levels of laboratory operation with selected agents infectious to humans.

The descriptions of Biosafety Levels 1–4 parallel those in the *NIH Guidelines for Research Involving Recombinant DNA,*[71,72,139] and are consistent with the general criteria originally used in assigning agents to Classes 1–4 in *Classification of Etiologic Agents on the Basis of Hazards.*[23] Four biosafety levels are also described for infectious disease activities utilizing small laboratory animals. Recommendations for biosafety levels for specific agents are made on the basis of the potential hazard of the agent and of the laboratory function or activity.

Since the early 1980's, laboratories have applied these fundamental guidelines in activities associated with manipulations involving the human immunodeficiency virus (HIV). Even before HIV was identified as the causative agent of acquired immunodeficiency syndrome (AIDS), the principles for manipulating a bloodborne pathogen were suitable for safe laboratory work. Guidelines were also promulgated for health care workers under the rubric of Universal Precautions.[43] Indeed, Universal Precautions and this publication have become the basis for safe handling of blood and body fluids, as described in the recent OSHA publication *Bloodborne Pathogen Standard.*[187]

In the late 1980's, considerable public concern was expressed about medical wastes, which led to the promulgation of the Medical Waste Tracking Act of 1988.[186] The principles established in the earlier volumes of the BMBL for handling potentially infectious wastes as an occupational hazard were reinforced by the National Research Council's *Biosafety in the Laboratory: Prudent Practices for the Handling and Disposal of Infectious Materials.*[12]

As this edition goes to press, there is growing concern about safe practices, procedures and facilities integral to the issues associated with the reemergence of tuberculosis and worker safety in laboratory and health care settings. The underlying principles of the BMBL are applicable in the control of this airborne pathogen, including multi-drug resistant strains of *M. tuberculosis*.[47,54]

Experience has demonstrated the prudence of the Biosafety Level 1–4 practices, procedures and facilities described for manipulations of etiologic agents in laboratory settings and animal facilities. Although no national reporting system exists for reporting laboratory-associated infections, anecdotal information suggests that strict adherence to these guidelines does contribute to a healthier and safer work environment for laboratorians, their co-workers and the surrounding community. The guidelines presented here can be customized for each individual laboratory, and can be used in conjunction with other available scientific information on risk assessment, to further minimize the potential for laboratory-associated infections.

SECTION II: PRINCIPLES OF BIOSAFETY

The term "containment" is used in describing safe methods for managing infectious agents in the laboratory environment where they are being handled or maintained. The purpose of containment is to reduce or eliminate exposure of laboratory workers, other persons, and the outside environment to potentially hazardous agents.

Primary containment, the protection of personnel and the immediate laboratory environment from exposure to infectious agents, is provided by both good microbiological technique and the use of appropriate safety equipment. The use of vaccines may provide an increased level of personal protection. Secondary containment, the protection of the environment external to the laboratory from exposure to infectious materials, is provided by a combination of facility design and operational practices. Therefore, the three elements of containment include laboratory practice and technique, safety equipment, and facility design. The risk assessment of the work to be done with a specific agent will determine the appropriate combination of these elements.

Laboratory Practice and Technique

The most important element of containment is strict adherence to standard microbiological practices and techniques. Persons working with infectious agents or potentially infected materials must be aware of potential hazards, and must be trained and proficient in the practices and techniques required for handling such material safely. The director or person in charge of the laboratory is responsible for providing or arranging for appropriate training of personnel.

Each laboratory should develop or adopt a biosafety or operations manual which identifies the hazards that will or may be encountered, and which specifies practices and procedures designed to minimize or eliminate risks. Personnel should be advised of special hazards and should be required to read and to follow the required practices and procedures. A scientist trained and knowledgeable in appropriate laboratory techniques, safety procedures, and hazards associated with handling infectious agents must direct laboratory activities.

When standard laboratory practices are not sufficient to control the hazard associated with a particular agent or laboratory procedure, additional measures may be needed. The laboratory director is responsible for selecting additional safety practices, which must be in keeping with the hazard associated with the agent or procedure.

Laboratory personnel, safety practices, and techniques must be supplemented by appropriate facility design and engineering features, safety equipment, and management practices.

Safety Equipment (Primary Barriers)

Safety equipment includes biological safety cabinets (BSCs), enclosed containers, and other engineering controls designed to remove or minimize exposures to hazardous biological materials. The biological safety cabinet (BSC) is the principal device used to provide containment of infectious splashes or aerosols generated by many microbiological procedures. Three types of biological safety cabinets (Class I, II, III) used in microbiological laboratories are described and illustrated in Appendix A. Open-fronted Class I and Class II biological safety cabinets are primary barriers which offer significant levels of protection to laboratory personnel and to the environment when used with good microbiological techniques. The Class II biological safety cabinet also provides protection from external contamination of the materials (e.g., cell cultures, microbiological stocks) being manipulated inside the cabinet. The gas-tight Class III biological safety cabinet provides the highest attainable level of protection to personnel and the environment.

An example of another primary barrier is the safety centrifuge cup, an enclosed container designed to prevent aerosols from being released during centrifugation. To minimize this hazard, containment controls such as BSCs or centrifuge cups must be used for handling infectious agents that can be transmitted through the aerosol route of exposure.

Safety equipment also may include items for personal protection such as gloves, coats, gowns, shoe covers, boots, respirators, face shields, safety glasses, or goggles. Personal protective equipment is often used in combination with biological safety cabinets and other devices which contain the agents, animals, or materials being worked with. In some situations in which it is impractical to work in biological safety cabinets, personal protective equipment may form the primary barrier between personnel and the infectious materials. Examples include certain animal studies, animal necropsy, agent production activities, and activities relating to maintenance, service, or support of the laboratory facility.

Facility Design (Secondary Barriers)

The design of the facility is important in providing a barrier to protect persons working inside and outside of the laboratory within the facility, and to protect persons or animals in the community from infectious agents which may be accidentally released from the laboratory. Laboratory management is responsible for providing facilities commensurate with the laboratory's function and the recommended biosafety level for the agents being manipulated.

The recommended secondary barrier(s) will depend on the risk of transmission of specific agents. For example, the exposure risks for most laboratory work in Biosafety Level 1 and 2 facilities will be direct contact with the agents, or inadvertent contact exposures through contaminated work environments. Secondary barriers in these laboratories may include separation of the laboratory work area from public access, availability of a decontamination facility (e.g., autoclave), and handwashing facilities.

As the risk for aerosol transmission increases, higher levels of primary containment and multiple secondary barriers may become necessary to prevent infectious agents from escaping into the environment. Such design features could include specialized ventilation systems to assure directional air flow, air treatment systems to decontaminate or remove agents from exhaust air, controlled access zones, airlocks as laboratory entrances, or separate buildings or modules for isolation of the laboratory. Design engineers for laboratories may refer to specific ventilation recommendations as found in the *Applications Handbook for Heating, Ventilation, and Air-Conditioning (HVAC)* published by the American Society of Heating, Refrigerating, and Air-Conditioning Engineers (ASHRAE).[2]

Biosafety Levels

Four biosafety levels (BSLs) are described which consist of combinations of laboratory practices and techniques, safety equipment, and laboratory facilities. Each combination is specifically appropriate for the operations performed, the documented or suspected routes of transmission of the infectious agents, and for the laboratory function or activity.

The recommended biosafety level(s) for the organisms in Section VII (Agent Summary Statements) represent those conditions under which the agent can ordinarily be safely handled. The laboratory director is specifically and primarily responsible for assessing risks and for appropriately applying the recommended biosafety levels. Generally, work with known agents should be conducted at the biosafety level recommended in Section VII. When specific information is available to suggest that virulence, pathogenicity, antibiotic resistance patterns, vaccine and treatment availability, or other factors are significantly altered, more (or less) stringent practices may be specified.

Biosafety Level 1 practices, safety equipment, and facilities are appropriate for undergraduate and secondary educational training and teaching laboratories, and for other facilities in which work is done with defined and characterized strains of viable microorganisms not known to cause disease in healthy adult humans. *Bacillus subtilis, Naegleria gruberi,* and infectious canine hepatitis virus are representative of those microorganisms meeting these criteria. Many agents not ordinarily associated with disease processes in humans are, however, opportunistic pathogens and may cause infection in the young, the aged, and immunodeficient or immunosuppressed individuals. Vaccine strains which have undergone multiple *in vivo* passages should not be considered avirulent simply because they are vaccine strains.

Biosafety Level 1 represents a basic level of containment that relies on standard microbiological practices with no special primary or secondary barriers recommended, other than a sink for handwashing.

Biosafety Level 2 practices, equipment, and facilities are applicable to clinical, diagnostic, teaching and other facilities in which work is done with the broad spectrum of indigenous moderate-risk agents present in the community and associated with human disease of varying severity. With good microbiological techniques, these agents can be used safely in activities conducted on the open bench, provided the potential for producing splashes or aerosols is low. Hepatitis B virus, the salmonellae, and *Toxoplasma* spp. are representative of microorganisms assigned to this containment level. Biosafety Level 2 is appropriate when work is done with any human-derived blood, body fluids, or tissues where the presence of an infectious agent may be unknown. (Laboratory personnel working with human-derived materials should refer to the *Bloodborne Pathogen Standard*[187] for specific, required precautions).

Primary hazards to personnel working with these agents relate to accidental percutaneous or mucous membrane exposures, or ingestion of infectious materials. Extreme precaution with contaminated needles or sharp instruments must be emphasized. Even though organisms routinely manipulated at BSL2 are not known to be transmissible by the aerosol route, procedures with aerosol or high splash potential that may increase the risk of such personnel exposure must be conducted in primary containment equipment, or devices such as a BSC or safety centrifuge cups. Other primary barriers should be used as appropriate, such as splash shields, face protection, gowns, and gloves.

Secondary barriers such as handwashing and waste decontamination facilities must be available to reduce potential environmental contamination.

Biosafety Level 3 practices, safety equipment, and facilities are applicable to clinical, diagnostic, teaching, research, or production facilities in which work is done with indigenous or exotic agents with a potential for respiratory transmission, and which may cause serious and potentially lethal infection. *Mycobacterium tuberculosis,* St. Louis encephalitis virus, and *Coxiella burnetii* are representative of microorganisms assigned to this level. Primary hazards to personnel working with these agents relate to autoinoculation, ingestion, and exposure to infectious aerosols.

At Biosafety Level 3, more emphasis is placed on primary and secondary barriers to protect personnel in contiguous areas, the community, and the environment from exposure to potentially infectious aerosols. For example, all laboratory manipulations should be performed in a BSC or other enclosed equipment, such as a gas-tight aerosol generation chamber. Secondary barriers for this level include controlled access to the laboratory and a specialized ventilation system that minimizes the release of infectious aerosols from the laboratory.

Biosafety Level 4 practices, safety equipment, and facilities are applicable for work with dangerous and exotic agents which pose a high individual risk of life-threatening disease, which may be transmitted via the aerosol route, and for which there is no available vaccine or therapy. Additionally, agents with a close or identical antigenic relationship to Biosafety Level 4 agents should also be handled at this level. When sufficient data are obtained, work with these agents may continue at this level or at a lower level. Viruses such as Marburg or Congo-Crimean hemorrhagic fever are manipulated at Biosafety Level 4.

The primary hazards to personnel working with Biosafety Level 4 agents are respiratory exposure to infectious aerosols, mucous membrane exposure to infectious droplets, and autoinoculation. All manipulations of potentially infectious diagnostic materials, isolates, and naturally or experimentally infected animals pose a high risk of exposure and infection to laboratory personnel, the community, and the environment.

The laboratory worker's complete isolation of aerosolized infectious materials is accomplished primarily by working in a Class III BSC or a full-body, air-supplied positive-pressure personnel suit. The Biosafety Level 4 facility itself is generally a separate building or completely isolated zone with complex, specialized ventilation and waste management systems to prevent release of viable agents to the environment.

The laboratory director is specifically and primarily responsible for the safe operation of the laboratory. His/her knowledge and judgment are critical in assessing risks and appropriately applying these recommendations. The recommended biosafety level represents those conditions under which the agent can ordinarily be safely handled. Special characteristics of the agents used, the training and experience of personnel, and the nature or function of the laboratory may further influence the director in applying these recommendations.

Animal Facilities

Four biosafety levels are also described for activities involving infectious disease work with experimental mammals. These four combinations of practices, safety equipment, and facilities are designated *Animal Biosafety Levels 1, 2, 3, and 4,* and provide increasing levels of protection to personnel and the environment.

Clinical Laboratories

Clinical laboratories, especially those in health care facilities, receive clinical specimens with requests for a variety of diagnostic and clinical support services. Typically, the infectious nature of clinical material is unknown, and specimens are often submitted with a broad request for microbiological examination for multiple agents (e.g., sputa submitted for "routine," acid-fast, and fungal cultures). It is the responsibility of the laboratory director to establish standard procedures in the laboratory which realistically address the issue of the infective hazard of clinical specimens.

Except in extraordinary circumstances (e.g., suspected hemorrhagic fever), the initial processing of clinical specimens and identification of isolates can be done safely at Biosafety Level 2, the recommended level for work with bloodborne pathogens such as hepatitis B virus and HIV. The containment elements described in Biosafety Level 2 are consistent with the *Occupational Exposure to Bloodborne Pathogens Standard*[187] from the Occupational Safety and Health Administration (OSHA), that requires the use of specific precautions with *all* clinical specimens of blood or other potentially infectious material (Universal Precautions).[43] Additionally, other recommendations specific for clinical laboratories may be obtained from the National Committee for Clinical Laboratory Standards.[134]

Biosafety Level 2 recommendations and OSHA requirements focus on the prevention of percutaneous and mucous membrane exposures to clinical material. Primary barriers such as biological safety cabinets (Class I or II) should be used when performing procedures that might cause splashing, spraying, or splattering of droplets. Biological safety cabinets should also be used for the initial processing of clinical specimens when the nature of the test requested or other information is suggestive that an agent readily transmissible by infectious aerosols is likely to be present (e.g., *M. tuberculosis*), or when the use of a biological safety cabinet (Class II) is indicated to protect the integrity of the specimen.

The segregation of clinical laboratory functions and limiting or restricting access to such areas is the responsibility of the laboratory director. It is also the director's responsibility to establish standard, written procedures that address the potential hazards and the required precautions to be implemented.

Importation and Interstate Shipment of Certain Biomedical Materials

The importation of etiologic agents and vectors of human diseases is subject to the requirements of the Public Health Service Foreign Quarantine regulations. Companion regulations of the Public Health Service and the Department of Transportation specify packaging, labeling, and shipping requirements for etiologic agents and diagnostic specimens shipped in interstate commerce (see Appendix D).

The U. S. Department of Agriculture regulates the importation and interstate shipment of animal pathogens and prohibits the importation, possession, or use of certain exotic animal disease agents which pose a serious disease threat to domestic livestock and poultry (see Appendix E).

SECTION III: LABORATORY BIOSAFETY LEVEL CRITERIA

The essential elements of the four biosafety levels for activities involving infectious microorganisms and laboratory animals are summarized in Tables 1 and 2. The levels are designated in ascending order, by degree of protection provided to personnel, the environment, and the community.

Biosafety Level 1

Biosafety Level 1 is suitable for work involving well-characterized agents not known to cause disease in healthy adult humans, and of minimal potential hazard to laboratory personnel and the environment. The laboratory is not necessarily separated from the general traffic patterns in the building. Work is generally conducted on open bench tops using standard microbiological practices. Special containment equipment or facility design is not required nor generally used. Laboratory personnel have specific training in the procedures conducted in the laboratory and are supervised by a scientist with general training in microbiology or a related science.

The following standard and special practices, safety equipment and facilities apply to agents assigned to Biosafety Level 1:

A. Standard Microbiological Practices

1. Access to the laboratory is limited or restricted at the discretion of the laboratory director when experiments or work with cultures and specimens are in progress.
2. Persons wash their hands after they handle viable materials and animals, after removing gloves, and before leaving the laboratory.
3. Eating, drinking, smoking, handling contact lenses, and applying cosmetics are not permitted in the work areas where there is reasonable likelihood of exposure to potentially

TABLE 1 Summary of recommended biosafety levels for infectious agents

Biosafety Level	Agents	Practices	Safety Equipment (Primary Barriers)	Facilities (Secondary Barriers)
1	Not known to cause disease in healthy adults.	Standard Microbiological Practices	None required	Open bench top sink required
2	Associated with human disease, hazard = autoinoculation, ingestion, mucous membrane exposure	BSL-1 practice plus: • Limited access • Biohazard warning signs • "Sharps" precautions • Biosafety manual defining any needed waste decontamination or medical surveillance policies	Primary barriers = Class I or II BSCs or other physical containment devices used for all manipulations of agents that cause splashes or aerosols of infectious materials; PPEs: laboratory coats; gloves; face protection as needed	BSL-1 plus: Autoclave available
3	Indigenous or exotic agents with potential for aerosol transmission; disease may have serious or lethal consequences	BSL-2 practice plus: • Controlled access • Decontamination of all waste • Decontamination of lab clothing before laundering • Baseline serum	Primary barriers = Class I or II BCSs or other physical containment devices used for all manipulations of agents; PPEs: protective lab clothing; gloves; respiratory protection as needed	BSL-2 plus: • Physical separation from access corridors • Self-closing, double door access • Exhausted air not recirculated • Negative airflow into laboratory
4	Dangerous/exotic agents which pose high risk of life-threatening disease, aerosol-transmitted lab infections; or related agents with unknown risk of transmission	BSL-3 practices plus: • Clothing change before entering • Shower on exit • All material decontaminated on exit from facility	Primary barriers = All procedures conducted in Class III BSCs or Class I or II BSCs <u>in combination with</u> full-body, air-supplied, positive pressure personnel suit	BSL-3 plus: • Separate building or isolated zone • Dedicated supply/exhaust, vacuum, and decon systems • Other requirements outlined in the text

infectious materials. Persons who wear contact lenses in laboratories should also wear goggles or a face shield. Food is stored outside the work area in cabinets or refrigerators designated and used for this purpose only.

4. Mouth pipetting is prohibited; mechanical pipetting devices are used.
5. All procedures are performed carefully to minimize the creation of splashes or aerosols.
6. Work surfaces are decontaminated at least once a day and after any spill of viable material.
7. All cultures, stocks, and other regulated wastes are decontaminated before disposal by an approved decontamination method, such as autoclaving. Materials to be decontaminated outside of the immediate laboratory are to be placed in a durable, leakproof container and closed for transport from the laboratory. Materials to be decontaminated at off-site from the laboratory are packaged in accordance with applicable local, state, and federal regulations, before removal from the facility.
8. An insect and rodent control program is in effect.

B. Special Practices: None

C. Safety Equipment (Primary Barriers)

1. Special containment devices or equipment such as a biological safety cabinet are generally not required for manipulations of agents assigned to Biosafety Level 1.
2. It is recommended that laboratory coats, gowns, or uniforms be worn to prevent contamination or soiling of street clothes.
3. Gloves should be worn if the skin on the hands is broken or if a rash exists.
4. Protective eyewear should be worn for anticipated splashes of microorganisms or other hazardous materials to the face.

D. Laboratory Facilities (Secondary Barriers)

1. Each laboratory contains a sink for handwashing.
2. The laboratory is designed so that it can be easily cleaned. Rugs in laboratories are not appropriate, and should not be used because proper decontamination following a spill is extremely difficult to achieve.
3. Bench tops are impervious to water and resistant to acids, alkalis, organic solvents, and moderate heat.
4. Laboratory furniture is sturdy. Spaces between benches, cabinets, and equipment are accessible for cleaning.

TABLE 2 Summary of recommended biosafety levels for activities in which experimentally or naturally infected vertebrate animals are used

15

Biosafety Level	Agents	Practices	Safety Equipment (Primary Barriers)	Facilities (Secondary Barriers)
1	Not known to cause disease in healthy human adults.	Standard animal care and management practices, including appropriate medical surveillance programs	As required for normal care of each species.	Standard animal facility • non recirculation of exhaust air • directional air flow recommended
2	Associated with human disease. Hazard: percutaneous exposure, ingestion, mucous membrane exposure.	ABSL-1 practices plus: • limited access • biohazard warning signs • sharps precautions • biosafety manual • decontamination of all infectious wastes and of animal cages prior to washing	ABSL-1 equipment plus primary barriers: containment equipment appropriate for animal species; PPES: laboratory coats, gloves, face and respiratory protection as needed.	ABSL-1 facility plus: • autoclave available • handwashing sink available in the animal room.
3	Indigenous or exotic agents with potential for aerosol transmission; disease may have serious health effects.	ABSL-2 practices plus: • controlled access • decontamination of clothing before laundering • cages decontaminated before bedding removed • disinfectant foot bath as needed	ABSL-2, equipment plus: • containment equipment for housing animals and cage dumping activities • Class I or II BSCs available for manipulative procedures (inoculation, necropsy) that may create infectious aerosols. PPEs: appropriate respiratory protection	ABSL-2, facility plus: • physical separation from access corridors • self-closing, double door access. • sealed penetrations • sealed windows • autoclave available in facility
4	Dangerous/exotic agents which pose high risk of life threatening disease; aerosol transmission, or related agents with unknown risk of transmission.	ABSL-3 practices plus: • entrance through change room where personal clothing is removed and laboratory clothing is put on; shower on exiting; • all wastes are decontaminated before removal from the facility	ABSL-3 equipment plus: • Maximum containment equipment (i.e., Class III BSC or partial containment equipment in combination withfull body, air-supplied positive-pressure personnel suit) used for all procedures andactivities	ABSL-3 facility plus: • separate building or isolated zone. • dedicated supply/exhaust, vacuum and decontamination systems. • other requirements outlined in the text.

5. If the laboratory has windows that open, they are fitted with fly screens.

Biosafety Level 2

Biosafety Level 2 is similar to Level 1 and is suitable for work involving agents of moderate potential hazard to personnel and the environment. It differs in that (1) laboratory personnel have specific training in handling pathogenic agents and are directed by competent scientists, (2) access to the laboratory is limited when work is being conducted, (3) extreme precautions are taken with contaminated sharp items, and (4) certain procedures in which infectious aerosols or splashes may be created are conducted in biological safety cabinets or other physical containment equipment.

The following standard and special practices, safety equipment, and facilities apply to agents assigned to Biosafety Level 2:

19

A. Standard Microbiological Practices

1. Access to the laboratory is limited or restricted at the discretion of the laboratory director when experiments are in progress.
2. Persons wash their hands after they handle viable materials and animals, after removing gloves, and before leaving the laboratory.
3. Eating, drinking, smoking, handling contact lenses, and applying cosmetics are not permitted in the work areas. Persons who wear contact lenses in laboratories should also wear goggles or a face shield. Food is stored outside the work area in cabinets or refrigerators designated for this purpose only.
4. Mouth pipetting is prohibited; mechanical pipetting devices are used.
5. All procedures are performed carefully to minimize the creation of splashes or aerosols.
6. Work surfaces are decontaminated at least once a day and after any spill of viable material.
7. All cultures, stocks, and other regulated wastes are decontaminated before disposal by an approved decontamination method, such as autoclaving. Materials to be decontaminated outside of the immediate laboratory are to be placed in a durable, leakproof container and closed for transport from the laboratory. Materials to be decontaminated at off-site from the laboratory are packaged in accordance with applicable local, state, and federal regulations, before removal from the facility.
8. An insect and rodent control program is in effect.

B. Special Practices

1. Access to the laboratory is limited or restricted by the laboratory director when work with infectious agents is in progress. In general, persons who are at increased risk of acquiring infection or for whom infection may be unusually hazardous are not allowed in the laboratory or animal rooms. For example, persons who are immunocompromised or immunosuppressed may be at risk of acquiring infections. The laboratory director has the final responsibility for assessing each circumstance and determining who may enter or work in the laboratory.
2. The laboratory director establishes policies and procedures whereby only persons who have been advised of the potential hazard and meet specific entry requirements (e.g., immunization) enter the laboratory or animal rooms.
3. When the infectious agent(s) in use in the laboratory require special provisions for entry (e.g., immunization), a hazard warning sign incorporating the universal biohazard symbol is posted on the access door to the laboratory work area. The hazard warning sign identifies the infectious agent, lists the name and telephone number of the laboratory director or other responsible person(s), and indicates the special requirement(s) for entering the laboratory.
4. Laboratory personnel receive appropriate immunizations or tests for the agents handled or potentially present in the laboratory (e.g., hepatitis B vaccine or TB skin testing).
5. When appropriate, considering the agent(s) handled, baseline serum samples for laboratory and other at-risk personnel are collected and stored. Additional serum specimens may be collected periodically, depending on the agents handled or the function of the facility.
6. A biosafety manual is prepared or adopted. Personnel are advised of special hazards and are required to read and to follow instructions on practices and procedures.
7. Laboratory personnel receive appropriate training on the potential hazards associated with the work involved, the necessary precautions to prevent exposures, and the exposure evaluation procedures. Personnel receive annual updates, or additional training as necessary for procedural or policy changes.
8. A high degree of precaution must always be taken with any contaminated sharp items, including needles and syringes, slides, pipettes, capillary tubes, and scalpels. Needles and syringes or other sharp instruments should be restricted in the laboratory for use only when there is no alternative, such as parenteral injection, phlebotomy, or aspiration of fluids from laboratory animals and diaphragm bottles. Plasticware should be substituted for glassware whenever possible.
   - **a.** Only needle-locking syringes or disposable syringe-needle units (i.e., needle is integral to the syringe) are used for injection or aspiration of infectious materials. Used disposable needles must not be bent, sheared, broken, recapped, removed from disposable syringes, or otherwise manipulated by hand before disposal; rather, they must be carefully placed in conveniently located puncture-resistant containers used for sharps disposal. Nondisposable sharps must be placed in a hard-walled container for transport to a processing area for decontamination, preferably by autoclaving.
   - **b.** Syringes which re-sheathe the needle, needle-less systems, and other safe devices should be used when appropriate.
   - **c.** Broken glassware must not be handled directly by hand, but must be removed by mechanical means such as a brush and dustpan, tongs, or forceps. Containers of contaminated needles, sharp equipment, and broken glass are decontaminated before disposal, according to any local, state, or federal regulations.
9. Cultures, tissues, or specimens of body fluids are placed in a container that prevents leakage during collection, handling, processing, storage, transport, or shipping.
10. Laboratory equipment and work surfaces should be decontaminated with an appropriate disinfectant on a routine basis, after work with infectious materials is finished, and especially after overt spills, splashes, or other contamination by infectious materials. Contaminated equipment must be decontaminated according to any local, state, or federal regulations before it is sent for repair or maintenance or packaged for transport in accordance with applicable local, state, or federal regulations, before removal from the facility.

11. Spills and accidents which result in overt exposures to infectious materials are immediately reported to the laboratory director. Medical evaluation, surveillance, and treatment are provided as appropriate and written records are maintained.
12. Animals not involved in the work being performed are not permitted in the lab.

C. Safety Equipment (Primary Barriers)

1. Properly maintained biological safety cabinets, preferably Class II, or other appropriate personal protective equipment or physical containment devices are used whenever:
   a. Procedures with a potential for creating infectious aerosols or splashes are conducted. These may include centrifuging, grinding, blending, vigorous shaking or mixing, sonic disruption, opening containers of infectious materials whose internal pressures may be different from ambient pressures, inoculating animals intranasally, and harvesting infected tissues from animals or eggs.
   b. High concentrations or large volumes of infectious agents are used. Such materials may be centrifuged in the open laboratory if sealed rotor heads or centrifuge safety cups are used, and if these rotors or safety cups are opened only in a biological safety cabinet.
2. Face protection (goggles, mask, faceshield or other splatter guards) is used for anticipated splashes or sprays of infectious or other hazardous materials to the face, when the microorganisms must be manipulated outside the BSC.
3. Protective laboratory coats, gowns, smocks, or uniforms designated for lab use are worn while in the laboratory. This protective clothing is removed and left in the laboratory before leaving for non-laboratory areas (e.g., cafeteria, library, administrative offices). All protective clothing is either disposed of in the laboratory or laundered by the institution; it should never be taken home by personnel.
4. Gloves are worn when handling infected animals and when hands may contact infectious materials, contaminated surfaces or equipment. Wearing two pairs of gloves may be appropriate; if a spill or splatter occurs, the hand will be protected after the contaminated glove is removed. Gloves are disposed of when contaminated, removed when work with infectious materials is completed, and are not worn outside the laboratory. Disposable gloves are not washed or reused.

D. Laboratory Facilities (Secondary Barriers)

1. Each laboratory contains a sink for handwashing.
2. The laboratory is designed so that it can be easily cleaned. Rugs in laboratories are not appropriate, and should not be used because proper decontamination following a spill is extremely difficult to achieve.
3. Bench tops are impervious to water and resistant to acids, alkalis, organic solvents, and moderate heat.
4. Laboratory furniture is sturdy, and spaces between benches, cabinets, and equipment are accessible for cleaning.
5. If the laboratory has windows that open, they are fitted with fly screens.
6. A method for decontamination of infectious or regulated laboratory wastes is available (e.g., autoclave, chemical disinfection, incinerator, or other approved decontamination system).
7. An eyewash facility is readily available.

Biosafety Level 3

Biosafety Level 3 is applicable to clinical, diagnostic, teaching, research, or production facilities in which work is done with indigenous or exotic agents which may cause serious or potentially lethal disease as a result of exposure by the inhalation route. Laboratory personnel have specific training in handling pathogenic and potentially lethal agents, and are supervised by competent scientists who are experienced in working with these agents.

All procedures involving the manipulation of infectious materials are conducted within biological safety cabinets or other physical containment devices, or by personnel wearing appropriate personal protective clothing and equipment. The laboratory has special engineering and design features.

It is recognized, however, that many existing facilities may not have all the facility safeguards recommended for Biosafety Level 3 (e.g. access zone, sealed penetrations, and directional airflow, etc.). In these circumstances, acceptable safety may be achieved for routine or repetitive operations (e.g. diagnostic procedures involving the propagation of an agent for identification, typing, and susceptibility testing) in Biosafety Level 2 facilities. However, the recommended Standard Microbiological Practices, Special Practices, and Safety Equipment for Biosafety Level 3 must be rigorously followed. The decision to implement this modification of Biosafety Level 3 recommendations should be made only by the laboratory director.

The following standard and special safety practices, equipment and facilities apply to agents assigned to Biosafety Level 3:

A. Standard Microbiological Practices

1. Access to the laboratory is limited or restricted at the discretion of the laboratory director when experiments are in progress.
2. Persons wash their hands after handling infectious materials and animals, after removing gloves, and when they leave the laboratory.
3. Eating, drinking, smoking, handling contact lenses, and applying cosmetics are not permitted in the laboratory. Persons who wear contact lenses in laboratories should also wear goggles or a face shield. Food is stored outside the work area in cabinets or refrigerators designated for this purpose only.
4. Mouth pipetting is prohibited; mechanical pipetting devices are used.
5. All procedures are performed carefully to minimize the creation of aerosols.
6. Work surfaces are decontaminated at least once a day and after any spill of viable material.
7. All cultures, stocks, and other regulated wastes are decontaminated before disposal by an approved decontamination method, such as autoclaving. Materials to be decontaminated outside of the immediate laboratory are to be placed in a durable, leakproof container and closed for transport from the laboratory. Materials to be decontaminated at off-site from the laboratory are packaged in accordance with applicable local, state, and federal regulations, before removal from the facility.
8. An insect and rodent control program is in effect.

B. Special Practices

1. Laboratory doors are kept closed when experiments are in progress.
2. The laboratory director controls access to the laboratory and restricts access to persons whose presence is required for program or support purposes. For example, persons who are immunocompromised or immunosuppressed may be at risk of acquiring infections. Persons who are at increased risk of acquiring infection or for whom infection may be unusually hazardous are not allowed in the laboratory or animal rooms. The director has the final responsibility for assessing each circumstance and determining who may enter or work in the laboratory.
3. The laboratory director establishes policies and procedures whereby only persons who have been advised of the potential biohazard, who meet any specific entry requirements (e.g., immunization), and who comply with all entry and exit procedures, enter the laboratory or animal rooms.
4. When infectious materials or infected animals are present in the laboratory or containment module, a hazard warning sign, incorporating the universal biohazard symbol, is posted on all laboratory and animal room access doors. The hazard warning sign identifies the agent, lists the name and telephone number of the laboratory director or other responsible person(s), and indicates any special requirements for entering the laboratory, such as the need for immunizations, respirators, or other personal protective measures.
5. Laboratory personnel receive the appropriate immunizations or tests for the agents handled or potentially present in the laboratory (e.g., hepatitis B vaccine or TB skin testing).
6. Baseline serum samples are collected and stored for all laboratory and other at-risk personnel. Additional serum specimens may be collected periodically, depending on the agents handled or the function of the laboratory.
7. A biosafety manual is prepared or adopted. Personnel are advised of special hazards and are required to read and to follow instructions on practices and procedures.
8. Laboratory personnel receive appropriate training on the potential hazards associated with the work involved, the necessary precautions to prevent exposures, and the exposure evaluation procedures. Personnel receive annual updates, or additional training as necessary for procedural changes.
9. The laboratory director is responsible for insuring that, before working with organisms at Biosafety Level 3, all personnel demonstrate proficiency in standard microbiological practices and techniques, and in the practices and operations specific to the laboratory facility. This might include prior experience in handling human pathogens or cell cultures, or a specific training program provided by the laboratory director or other

competent scientist proficient in safe microbiological practices and techniques.

10. A high degree of precaution must always be taken with any contaminated sharp items, including needles and syringes, slides, pipettes, capillary tubes, and scalpels. Needles and syringes or other sharp instruments should be restricted in the laboratory for use only when there is no alternative, such as parenteral injection, phlebotomy, or aspiration of fluids from laboratory animals and diaphragm bottles. Plasticware should be substituted for glassware whenever possible.
    a. Only needle-locking syringes or disposable syringe-needle units (i.e., needle is integral to the syringe) are used for injection or aspiration of infectious materials. Used disposable needles must not be bent, sheared, broken, recapped, removed from disposable syringes, or otherwise manipulated by hand before disposal; rather, they must be carefully placed in conveniently located puncture-resistant containers used for sharps disposal. Non-disposable sharps must be placed in a hard-walled container for transport to a processing area for decontamination, preferably by autoclaving.
    b. Syringes which re-sheathe the needle, needle-less systems, and other safe devices should be used when appropriate.
    c. Broken glassware must not be handled directly by hand, but must be removed by mechanical means such as a brush and dustpan, tongs, or forceps. Containers of contaminated needles, sharp equipment, and broken glass should be decontaminated before disposal, according to any local, state, or federal regulations.

29

11. All manipulations involving infectious materials are conducted in biological safety cabinets or other physical containment devices within the containment module. No work in open vessels is conducted on the open bench.
12. Laboratory equipment and work surfaces should be decontaminated with an appropriate disinfectant on a routine basis, after work with infectious materials is finished, and especially after overt spills, splashes, or other contamination with infectious materials. Contaminated equipment should also be decontaminated before it is sent for repair or maintenance or package for transport in accordance with applicable local, state, or federal regulations, before removal from the facility. Plastic-backed paper toweling used on non-perforated work surfaces within biological safety cabinets facilitates clean-up.
13. Cultures, tissues, or specimens of body fluids are placed in a container that prevents leakage during collection, handling, processing, storage, transport, or shipping.
14. All potentially contaminated waste materials (e.g., gloves, lab coats, etc.) from laboratories or animal rooms are decontaminated before disposal or reuse.
15. Spills of infectious materials are decontaminated, contained and cleaned up by appropriate professional staff, or others properly trained and equipped to work with concentrated infectious material.
16. Spills and accidents which result in overt or potential exposures to infectious materials are immediately reported to the laboratory director. Appropriate medical evaluation, surveillance, and treatment are provided and written records are maintained.
17. Animals and plants not related to the work being conducted are not permitted in the laboratory.

C. Safety Equipment (Primary Barriers)

30

1. Properly maintained biological safety cabinets are used (Class II or III—see Appendix A) for all manipulation of infectious materials.
2. Outside of a BSC, appropriate combinations of personal protective equipment are used (e.g., special protective clothing, masks, gloves, face protection, or respirators), in combination with physical containment devices (e.g., centrifuge safety cups, sealed centrifuge rotors, or containment caging for animals).
3. This equipment must be used for manipulations of cultures and of those clinical or environmental materials which may be a source of infectious aerosols; the aerosol challenge of experimental animals; harvesting of tissues or fluids from infected animals and embryonated eggs, and necropsy of infected animals.
4. Face protection (goggles and mask, or faceshield) is worn for manipulations of infectious materials outside of a biological safety cabinet.
5. Respiratory protection is worn when aerosols cannot be safely contained (i.e, outside of a biological safety cabinet), and in rooms containing infected animals.

6. Protective laboratory clothing such as solid-front or wrap-around gowns, scrub suits, or coveralls must be worn in, and not worn outside, the laboratory. Reusable laboratory clothing is to be decontaminated before being laundered.
7. Gloves must be worn when handling infected animals and when hands may contact infectious materials and contaminated surfaces or equipment. Disposable gloves should be discarded when contaminated, and never washed for reuse.

D. Laboratory Facilities (Secondary Barriers)

1. The laboratory is separated from areas which are open to unrestricted traffic flow within the building. Passage through two sets of self-closing doors is the basic requirement for entry into the laboratory from access corridors or other contiguous areas. A clothes change room (shower optional) may be included in the passage way.
2. Each laboratory contains a sink for handwashing. The sink is foot, elbow, or automatically operated and is located near the laboratory exit door.
3. The interior surfaces of walls, floors, and ceilings are water resistant so that they can be easily cleaned. Penetrations in these surfaces are sealed or capable of being sealed to facilitate decontamination.
4. Bench tops are impervious to water and resistant to acids, alkalis, organic solvents, and moderate heat.
5. Laboratory furniture is sturdy, and spaces between benches, cabinets, and equipment are accessible for cleaning.
6. Windows in the laboratory are closed and sealed.
7. A method for decontaminating all laboratory wastes is available, preferably within the laboratory (i.e, autoclave, chemical disinfection, incineration, or other approved decontamination method).
8. A ducted exhaust air ventilation system is provided. This system creates directional airflow that draws air from "clean" areas into the laboratory toward "contaminated" areas. The exhaust air is not recirculated to any other area of the building, and is discharged to the outside with filtration and other treatment optional. The outside exhaust must be dispersed away from occupied areas and air intakes. Laboratory personnel must verify that the direction of the airflow (into the laboratory) is proper.
9. The High Efficiency Particulate Air (HEPA)-filtered exhaust air from Class II or Class III biological safety cabinets is discharged directly to the outside or through the building exhaust system. If the HEPA-filtered exhaust air from Class II or III biological safety cabinets is to be discharged to the outside through the building exhaust air system, it is connected to this system in a manner (e.g., thimble unit connection)[136] that avoids any interference with the air balance of the cabinets or building exhaust system. Exhaust air from Class II biological safety cabinets may be recirculated within the laboratory if the cabinet is tested and certified at least every twelve months.
10. Continuous flow centrifuges or other equipment that may produce aerosols are contained in devices that exhaust air through HEPA filters before discharge into the laboratory.
11. Vacuum lines are protected with liquid disinfectant traps and HEPA filters, or their equivalent, which are routinely maintained and replaced as needed.
12. An eyewash facility is readily available.

Biosafety Level 4

Biosafety Level 4 is required for work with dangerous and exotic agents which pose a high individual risk of aerosol-transmitted laboratory infections and life-threatening disease. Agents with a close or identical antigenic relationship to Biosafety Level 4 agents are handled at this level until sufficient data are obtained either to confirm continued work at this level, or to work with them at a lower level. Members of the laboratory staff have specific and thorough training in handling extremely hazardous infectious agents; and they understand the primary and secondary containment functions of the standard and special practices, the containment equipment, and the laboratory design characteristics. They are supervised by competent scientists who are trained and experienced in working with these agents. Access to the laboratory is strictly controlled by the laboratory director. The facility is either in a separate building or in a controlled area within a building, which is completely isolated from all other areas of the building. A specific facility operations manual is prepared or adopted.

Within work areas of the facility, all activities are confined to Class III biological safety cabinets, or Class II biological safety cabinets used with one-piece positive pressure personnel suits ventilated by a life support system. The Biosafety Level 4 laboratory has special engineering and design features to prevent microorganisms from being disseminated into the environment.

The following standard and special safety practices equipment, and facilities apply to agents assigned to Biosafety Level 4:

A. Standard Microbiological Practices

1. Access to the laboratory is limited or restricted at the discretion of the laboratory director when experiments are in progress.
2. Persons wash their hands after handling infectious materials and animals; they take a decontaminating shower when they leave the laboratory.
3. Eating, drinking, smoking, handling contact lenses, and applying cosmetics are not permitted in the laboratory. Persons who wear contact lenses in laboratories should also wear goggles or a face shield. Food is stored outside the work area in cabinets or refrigerators designated for this purpose only.
4. Mouth pipetting is prohibited; only mechanical pipetting devices are used.
5. All procedures are performed carefully to minimize the creation of aerosols.
6. Work surfaces are decontaminated at least once a day and after any spill of viable material.
7. An insect and rodent control program is in effect.

B. Special Practices

1. Only persons whose presence in the facility or individual laboratory rooms is required for program or support purposes are authorized to enter. Persons who are immunocompromised or immunosuppressed may be at risk of acquiring infections. Therefore, persons who may be at increased risk of acquiring infection or for whom infection may be unusually hazardous, such as children or pregnant women, are not allowed in the laboratory or animal rooms.

   The supervisor has the final responsibility for assessing each circumstance and determining who may enter or work in the laboratory. Access to the facility is limited by means of secure, locked doors; accessibility is managed by the laboratory director, biohazards control officer, or other person responsible for the physical security of the facility. Before entering, persons are advised of the potential biohazards and instructed as to appropriate safeguards for insuring their safety. Authorized persons comply with the instructions and all other applicable entry and exit procedures. A logbook, signed by all personnel, indicates the date and time of each entry and exit. Practical and effective protocols for emergency situations are established.
2. When infectious materials or infected animals are present in the laboratory or animal rooms, hazard warning signs, incorporating the universal biohazard symbol, are posted on all access doors. The sign identifies the agent, lists the name of the laboratory director or other responsible person(s), and indicates any special requirements for entering the area (e.g., the need for immunizations or respirators).
3. The laboratory director is responsible for insuring that, before working with organisms at Biosafety Level 4, all personnel demonstrate a high proficiency in standard microbiological practices and techniques, and in the special practices and operations specific to the laboratory facility. This might include prior experience in handling human pathogens or cell cultures, or a specific training program provided by the laboratory director or other competent scientist proficient in these unique safe microbiological practices and techniques.
4. Laboratory personnel receive available immunizations for the agents handled or potentially present in the laboratory.
5. Baseline serum samples for all laboratory and other at-risk personnel are collected and stored. Additional serum specimens may be collected periodically, depending on the agents handled or the function of the laboratory. The decision to establish a serologic surveillance program takes into account the availability of methods for the assessment of antibody to the agent(s) of concern. The program provides for the testing of serum samples at each collection interval and the communication of results to the participants.
6. A biosafety manual is prepared or adopted. Personnel are advised of special hazards and are required to read and to follow instructions on practices and procedures.
7. Laboratory personnel receive appropriate training on the potential hazards associated with the work involved, the necessary precautions to prevent exposures, and the exposure evaluation procedures. Personnel receive annual updates, or additional training as necessary for procedural changes.

8. Personnel enter and leave the facility only through the clothing change and shower rooms, and shower each time they leave the facility. Personnel use the airlocks to enter or leave the laboratory only in an emergency.
9. Personal clothing is removed in the outer clothing change room and kept there. Complete laboratory clothing, including undergarments, pants and shirts or jumpsuits, shoes, and gloves, is provided and used by all personnel entering the facility. When leaving the laboratory and before proceeding into the shower area, personnel remove their laboratory clothing in the inner change room. Soiled clothing is autoclaved before laundering.
10. Supplies and materials needed in the facility are brought in by way of the double-doored autoclave, fumigation chamber, or airlock, which is appropriately decontaminated between each use. After securing the outer doors, personnel within the facility retrieve the materials by opening the interior doors of the autoclave, fumigation chamber, or airlock. These doors are secured after materials are brought into the facility.
11. A high degree of precaution must always be taken with any contaminated sharp items, including needles and syringes, slides, pipettes, capillary tubes, and scalpels. Needles and syringes or other sharp instruments are restricted in the laboratory for use only when there is no alternative, such as for parenteral injection, phlebotomy, or aspiration of fluids from laboratory animals and diaphragm bottles. Plasticware should be substituted for glassware whenever possible.
    a. Only needle-locking syringes or disposable syringe-needle units (i.e., needle is integral to the syringe) are used for injection or aspiration of infectious materials. Used disposable needles must not be bent, sheared, broken, recapped, removed from disposable syringes, or otherwise manipulated by hand before disposal; rather, they must be carefully placed in conveniently located puncture-resistant containers used for sharps disposal. Non-disposable sharps must be placed in a hard-walled container for transport to a processing area for decontamination, preferably by autoclaving.
    b. Syringes which re-sheath the needle, needle-less systems, and other safe devices should be used when appropriate.
    c. Broken glassware must not be handled directly by hand, but must be removed by mechanical means such as a brush and dustpan, tongs, or forceps. Containers of contaminated needles, sharp equipment, and broken glass should be decontaminated before disposal, according to any local, state, or federal regulations.
12. Biological materials to be removed from the Class III cabinet or from the Biosafety Level 4 laboratory in a viable or intact state are transferred to a nonbreakable, sealed primary container and then enclosed in a nonbreakable, sealed secondary container. This is removed from the facility through a disinfectant dunk tank, fumigation chamber, or an airlock designed for this purpose.
13. No materials, except for biological materials that are to remain in a viable or intact state, are removed from the Biosafety Level 4 laboratory unless they have been autoclaved or decontaminated before they leave the facility. Equipment or material which might be damaged by high temperatures or steam may be decontaminated by gaseous or vapor methods in an airlock or chamber designed for this purpose.
14. Laboratory equipment is decontaminated routinely after work with infectious materials is finished, and especially after overt spills, splashes, or other contamination with infectious materials. Contaminated equipment is also decontaminated before it is sent for repair or maintenance.
15. Spills of infectious materials are contained and cleaned up by appropriate professional staff or others properly trained and equipped to work with concentrated infectious material.
16. A system is set up for reporting laboratory accidents and exposures and employee absenteeism, and for the medical surveillance of potential laboratory-associated illnesses. Written records are prepared and maintained. An essential adjunct to such a reporting-surveillance system is the availability of a facility for the quarantine, isolation, and medical care of personnel with potential or known laboratory-associated illnesses.
17. Materials (e.g., plants, animals, and clothing) not related to the experiment being conducted are not permitted in the facility.

C. Safety Equipment (Primary Barriers)
   1. All procedures within the facility with agents assigned to Biosafety Level 4 are conducted in the Class III biological safety cabi-

net or in Class II biological safety cabinets used in conjunction with one-piece positive pressure personnel suits ventilated by a life support system.

Activities with viral agents that require Biosafety Level 4 secondary containment capabilities can be conducted within Class II biological safety cabinets within the facility, without the one-piece positive pressure personnel suit being used if (a) the facility has been decontaminated, (b) no work is being conducted in the facility with other agents assigned to Biosafety Level 4, (c) all personnel are immunized against the specific agent being manipulated and demonstrate protective antibody levels, and (d) all other standard and special practices are followed.

2. All personnel entering the facility will don complete laboratory clothing, including undergarments, pants, and shirts or jumpsuits, shoes, and gloves. All such personal protective equipment is removed in the change room before showering and leaving the laboratory.

D. Laboratory Facility (Secondary Barriers)

1. The Biosafety Level 4 facility consists of either a separate building or a clearly demarcated and isolated zone within a building. Outer and inner change rooms separated by a shower are provided for personnel entering and leaving the facility. A double-doored autoclave, fumigation chamber, or ventilated airlock is provided for passage of those materials, supplies, or equipment which are not brought into the facility through the change room.
2. Walls, floors, and ceilings of the facility are constructed to form a sealed internal shell which facilitates fumigation and is animal and insect proof. The internal surfaces of this shell are resistant to liquids and chemicals, thus facilitating cleaning and decontamination of the area. All penetrations in these structures and surfaces are sealed. Any drains in the floors contain traps filled with a chemical disinfectant of demonstrated efficacy against the target agent, and they are connected directly to the liquid waste decontamination system. Sewer vents and other ventilation lines contain HEPA filters.
3. Internal facility appurtenances, such as light fixtures, air ducts, and utility pipes, are arranged to minimize the horizontal surface area on which dust can settle.
4. Bench tops have seamless surfaces which are impervious to water and resistant to acids, alkalis, organic solvents, and moderate heat.
5. Laboratory furniture is of simple and sturdy construction, and spaces between benches, cabinets, and equipment are accessible for cleaning.
6. A foot, elbow, or automatically operated handwashing sink is provided near the door of each laboratory room in the facility.
7. If there is a central vacuum system, it does not serve areas outside the facility. In-line HEPA filters are placed as near as practicable to each use point or service cock. Filters are installed to permit in-place decontamination and replacement. Other liquid and gas services to the facility are protected by devices that prevent backflow.
8. If water fountains are provided, they are foot operated and are located in the facility corridors outside the laboratory. The water service to the fountain is not connected to the backflow-protected distribution system supplying water to the laboratory areas.
9. Access doors to the laboratory are self-closing and lockable.
10. Any windows are breakage resistant.
11. A double-doored autoclave is provided for decontaminating materials passing out of the facility. The autoclave door which opens to the area external to the facility is sealed to the outer wall, and automatically controlled so that the outside door can only be opened after the autoclave "sterilization" cycle has been completed.
12. A pass-through dunk tank, fumigation chamber, or an equivalent decontamination method is provided so that materials and equipment that cannot be decontaminated in the autoclave can be safely removed from the facility.
13. Liquid effluents from laboratory sinks, biological safety cabinets, floor drains (if used), and autoclave chambers are decontaminated by heat treatment before being discharged to the sanitary sewer. Effluents from showers and toilets may be discharged to the sanitary sewer without treatment. The process used for decontamination of liquid wastes must be validated physically and biologically by use of a constant recording temperature sensor in conjunction with an indicator microorganism having a defined heat susceptibility profile.

TABLE 3 Comparison of biological safety cabinets

43

Cabinets			Applications		
Type	**Face velocity (lfpm)**	**Airflow Pattern**	**Radionuclides/ Toxic Chemicals**	**Biosafety Level (s)**	**Product Protection**
Class I*, open front	75	In at front; out rear and top through HEPA filter	NO	2,3	NO
Class II: Type A	75	70% recirculated through HEPA; exhaust through HEPA	NO	2,3	YES
Type B1	100	30% recirculated through HEPA; exhaust via HEPA and hard-ducted	YES (Low levels/ volatility)	2,3	YES
Type B2	100	No recirculation; total exhaust via HEPA and hard-ducted	YES	2,3	YES
Type B3	100	Same as IIA, but plena under negative pressure to room and exhaust air is ducted	YES	2,3	YES
Class III	NA	Supply air inlets and exhaust through 2 HEPA filters	YES	3,4	YES

*Glove panels may be added and will increase face velocity to 150 lfpm; gloves may be added with an inlet air pressure release that will allow work with chemicals/radionuclides.

14. A dedicated non-recirculating ventilation system is provided. The supply and exhaust components of the system are balanced to assure directional airflow from the area of least hazard to the area(s) of greatest potential hazard. The differential pressure/directional airflow between adjacent areas is monitored and alarmed to indicate malfunction of the system. The airflow in the supply and exhaust components is monitored and the components interlocked to assure inward (or zero) airflow is maintained.
15. The general room exhaust air from a facility in which the work is conducted in a Class III cabinet system is treated by a passage through a HEPA filter(s) prior to discharge to the outside. The air is discharged away from occupied spaces and air intakes. The HEPA filter(s) are located as near as practicable to the source in order to minimize the length of potentially contaminated ductwork. The HEPA filter housings are designed to allow for *in situ* decontamination of the filter prior to removal, or removal of the filter in a sealed gas-tight primary container for subsequent decontamination and/or destruction by incineration. The design of the HEPA filter housing should facilitate validation of the filter installation. The use of pre-certified HEPA filters can be an advantage. The service-life of the exhaust HEPA filters can be extended through adequate filtration of the supply air.
16. A specially designed suit area may be provided in the facility to provide personnel protection equivalent to that provided by Class III cabinets. Personnel who enter this area wear a one-piece positive pressure suit that is ventilated by a life support system. The life support system includes alarms and emergency backup breathing air tanks. Entry to this area is through an airlock fitted with airtight doors. A chemical shower is provided to decontaminate the surface of the suit before the worker leaves the area. The exhaust air from the suit area is filtered by two sets of HEPA filters installed in series. A duplicate filtration unit, exhaust fan, and an automatically starting emergency power source are provided. The air pressure within the suit area is lower than that of any adjacent area. Emergency lighting and communication systems are provided. All penetrations into the internal shell of the suit area

42

are sealed. A double-doored autoclave is provided for decontaminating waste materials to be removed from the suit area.

17. The treated exhaust air from Class II biological safety cabinets, located in a facility in which workers wear a positive pressure suit, may be discharged into the animal room environment or to the outside through the facility air exhaust system. The biological safety cabinets are tested and certified at 12-month intervals. The air exhausted from Class III biological safety cabinets is passaged through two HEPA filter systems (in series) prior to discharge to the outside. If the treated exhaust is discharged to the outside through the facility exhaust system, it is connected to this system in a manner that avoids any interference with the air balance of the cabinets or the facility exhaust system.

SECTION IV: VERTEBRATE ANIMAL BIOSAFETY LEVEL CRITERIA

If experimental animals are used, institutional management must provide facilities and staff and establish practices which reasonably assure appropriate levels of environmental quality, safety, and care. Laboratory animal facilities in many ways are extensions of the laboratory. As a general principle, the biosafety level (facilities, practices, and operational requirements) recommended for working with infectious agents *in vivo* and *in vitro* are comparable. It is well to remember, however, that the animal room is not the laboratory, and can present some unique problems. In the laboratory, hazardous conditions are caused by personnel or the equipment that is being used. In the animal room the activities of the animals themselves can introduce new hazards. Animals may produce aerosols, and they may also infect and traumatize animal handlers by biting and scratching.

These recommendations presuppose that laboratory animal facilities, operational practices, and quality of animal care meet applicable standards and regulations and that appropriate species have been selected for animal experiments (e.g., *Guide for the Care and Use of Laboratory Animals*, HEW Publication No. (NIH) 86-23, Rev. 1985, and *Laboratory Animal Welfare Regulations*—9 CFR, Subchapter A, Parts 1, 2 and 3).

Ideally, facilities for laboratory animals used for studies of infectious or noninfectious disease should be physically separate from other activities such as animal production and quarantine, clinical laboratories, and especially from facilities that provide patient care. Animal facilities should be designed and constructed to facilitate cleaning and housekeeping. Traffic flow that will minimize the risk of cross contamination should be considered in the plans. A "clean/dirty hall" layout is useful in achieving this. Floor drains should be installed in animal facilities only on the basis of clearly defined needs. If floor drains are installed, the drain trap should always contain water or a suitable disinfectant.

These recommendations describe four combinations of practices, safety equipment, and facilities for experiments on animals infected with agents which produce, or may produce, human infection. These four combinations provide increasing levels of protection to personnel and to the environment, and are recommended as minimal standards for activities involving infected laboratory animals. These four combinations, designated Animal Biosafety Levels (ABSL) 1–4, describe animal facilities and practices applicable to work on animals infected with agents assigned to corresponding Biosafety Levels 1–4.

Facility standards and practices for invertebrate vectors and hosts are not specifically addressed in standards written for commonly used laboratory animals. "Laboratory Safety for Arboviruses and Certain other Viruses of Vertebrates,"[178] prepared by the Subcommittee on Arbovirus Laboratory Safety of the American Committee on Arthropod-Borne Viruses, serves as a useful reference in the design and operation of facilities using arthropods.

Animal Biosafety Level 1

A. Standard Practices

1. Access to the animal facility is limited or restricted at the discretion of the laboratory or animal facility director.
2. Personnel wash their hands after handling cultures and animals, after removing gloves, and before leaving the animal facility.
3. Eating, drinking, smoking, handling contact lenses, applying cosmetics, and storing food for human use are not permitted in animal rooms. Persons who wear contact lenses in animal rooms should also wear goggles or a face shield.
4. All procedures are carefully performed to minimize the creation of aerosols.
5. Work surfaces are decontaminated after use or after any spill of viable materials.
6. Doors to animal rooms open inward, are self-closing and are kept closed when experimental animals are present.
7. All wastes from the animal room are appropriately decontaminated, preferably by au-

toclaving, before disposal. Infected animal carcasses are incinerated after being transported from the animal room in leakproof, covered containers.

8. An insect and rodent control program is in effect.

B. Special Practices

1. The laboratory or animal facility director limits access to the animal room to personnel who have been advised of the potential hazard and who need to enter the room for program or service purposes when work is in progress. In general, persons who may be at increased risk of acquiring infection, or for whom infection might be unusually hazardous, are not allowed in the animal room.
2. The laboratory or animal facility director establishes policies and procedures whereby only persons who have been advised of the potential hazard and meet any specific requirements (e.g., immunization) may enter the animal room.
3. Bedding materials from animal cages are removed in such a manner as to minimize the creation of aerosols, and are disposed of in compliance with applicable institutional or local requirements.
4. Cages are washed manually or in a cage washer. Temperature of final rinse water in a mechanical washer should be 180°F.
5. The wearing of laboratory coats, gowns, or uniforms in the animal facility is recommended. It is further recommended that laboratory coats worn in the animal facility not be worn in other areas.
6. A biosafety manual is prepared or adopted. Personnel are advised of special hazards, are required to read and to follow instructions on practices and procedures.

C. Safety Equipment (Primary Barriers)

Special containment equipment is not required for animals infected with agents assigned to Biosafety Level 1.

D. Animal Facilities (Secondary Barriers)

1. The animal facility is designed and constructed to facilitate cleaning and housekeeping.
2. A handwashing sink is available in the animal facility.
3. If the animal facility has windows that open, they are fitted with fly screens.
4. Exhaust air is discharged to the outside without being recirculated to other rooms, and it is recommended, but not required, that the direction of airflow in the animal facility is inward.

Animal Biosafety Level 2

A. Standard Practices

1. Access to the animal facility is limited or restricted at the discretion of the laboratory or animal facility director.
2. Personnel wash their hands after handling cultures and animals, after removing gloves, and before leaving the animal facility.
3. Eating, drinking, smoking, handling contact lenses, applying cosmetics, and storing food for human use are not permitted in animal rooms. Persons who wear contact lenses in animal rooms should also wear goggles or a face shield.
4. All procedures are carefully performed to minimize the creation of aerosols.
5. Work surfaces are decontaminated after use or after any spill of viable materials.
6. Doors to animal rooms open inward, are self-closing and are kept closed when experimental animals are present.
7. All wastes from the animal room are appropriately decontaminated, preferably by autoclaving, before disposal. Infected animal carcasses are incinerated after being transported from the animal room in leakproof, covered containers.
8. An insect and rodent control program is in effect.

B. Special Practices

1. The laboratory or animal facility director limits access to the animal room to personnel who have been advised of the potential hazard and who need to enter the room for program or service purposes when work is in progress. In general, persons who may be at increased risk of acquiring infection, or for whom infection might be unusually hazardous, are not allowed in the animal room.
2. The laboratory or animal facility director establishes policies and procedures whereby only persons who have been advised of the potential hazard and meet any specific requirements (e.g., immunization) may enter the animal room.
3. When the infectious agent(s) in use in the animal room requires special entry provisions (e.g., the need for immunizations and respirators) a hazard warning sign, incorporating the universal biohazard symbol, is posted on the access door to the animal

room. The hazard warning sign identifies the infectious agent(s) in use, lists the name and telephone number of the animal facility supervisor or other responsible person(s), and indicates the special requirement(s) for entering the animal room.

4. Laboratory personnel receive appropriate immunizations or tests for the agents handled or potentially present in the laboratory (e.g., hepatitis B vaccine or TB skin testing).
5. When appropriate, considering the agents handled, baseline serum samples from animal care and other at-risk personnel are collected and stored. Additional serum samples may be collected periodically depending on the agents handled or the function of the facility. The decision to establish a serologic surveillance program must take into account the availability of methods for the assessment of antibody to the agent(s) of concern. The program should provide for the testing of serum samples at each collection interval and the communication of results to the participants.
6. A biosafety manual is prepared or adopted. Personnel are advised of special hazards, and are required to read and to follow instructions on practices and procedures.
7. Laboratory personnel receive appropriate training on the potential hazards associated with the work involved, the necessary precautions to prevent exposures, and the exposure evaluation procedures. Personnel receive annual updates, or additional training as necessary for procedural or policy changes.
8. A high degree of precaution must always be taken with any contaminated sharp items, including needles and syringes, slides, pipettes, capillary tubes, and scalpels. Needles and syringes or other sharp instruments are restricted in the animal facility for use only when there is no alternative, such as for parenteral injection, blood collection, or aspiration of fluids from laboratory animals and diaphragm bottles. Plasticware should be substituted for glassware whenever possible.
   a. Only needle-locking syringes or disposable syringe-needle units (i.e., needle is integral to the syringe) are used for injection or aspiration of infectious materials. Used disposable needles must not be bent, sheared, broken, recapped, removed from disposable syringes, or otherwise manipulated by hand before disposal; rather, they must be carefully placed in conveniently located puncture-resistant containers used for sharps disposal. Non-disposable sharps must be placed in a hard-walled container for transport to a processing area for decontamination, preferably by autoclaving.
   b. Syringes which re-sheathe the needle, needle-less systems, and other safe devices should be used when appropriate.
   c. Broken glassware must not be handled directly by hand, but must be removed by mechanical means such as a brush and dustpan, tongs, or forceps. Containers of contaminated needles, sharp equipment, and broken glass should be decontaminated before disposal, according to any local, state, or federal regulations.
9. Cultures, tissues, or specimens of body fluids are placed in a container that prevents leakage during collection, handling, processing, storage, transport, or shipping.
10. Cages are appropriately decontaminated, preferably by autoclaving, before they are cleaned and washed. Equipment and work surfaces should be decontaminated with an appropriate disinfectant on a routine basis, after work with infectious materials is finished, and especially after overt spills, splashes, or other contamination by infectious materials. Contaminated equipment must be decontaminated according to any local, state, or federal regulations before it is sent for repair or maintenance or packaged for transport in accordance with applicable local, state, or federal regulations, before removal from the facility.
11. Spills and accidents which result in overt exposures to infectious materials are immediately reported to the laboratory director. Medical evaluation, surveillance, and treatment are provided as appropriate and written records are maintained.
12. Animals not involved in the work being performed are not permitted in the lab.

C. Safety Equipment (Primary Barriers)

1. Biological safety cabinets, other physical containment devices, and/or personal protective equipment (e.g., respirators, face shields) are used whenever procedures with a high potential for creating aerosols are conducted.[139] These include necropsy of infected animals, harvesting of tissues or fluids from infected animals or eggs, intranasal inoculation of animals, and manipulations

of high concentrations or large volumes of infectious materials.

2. Appropriate face/eye and respiratory protection is worn by all personnel entering animal rooms housing nonhuman primates.
3. Laboratory coats, gowns, or uniforms are worn while in the animal room. This protective clothing is removed before leaving the animal facility.
4. Special care is taken to avoid skin contamination with infectious materials; gloves are worn when handling infected animals and when skin contact with infectious materials is unavoidable.

D. Animal Facilities (Secondary Barriers)

1. The animal facility is designed and constructed to facilitate cleaning and housekeeping.
2. A handwashing sink is available in the room where infected animals are housed.
3. If the animal facility has windows that open, they are fitted with fly screens.
4. If floor drains are provided, the drain traps are always filled with water or a suitable disinfectant.
5. Exhaust air is discharged to the outside without being recirculated to other rooms, and it is recommended, but not required, that the direction of airflow in the animal facility is inward.
6. An autoclave which can be used for decontaminating infectious laboratory waste is available in the building with the animal facility.

Animal Biosafety Level 3

A. Standard Practices

1. Access to the animal facility is limited or restricted at the discretion of the laboratory or animal facility director.
2. Personnel wash their hands after handling cultures and animals, after removing gloves, and before leaving the animal facility.
3. Eating, drinking, smoking, handling contact lenses, applying cosmetics, and storing food for human use are not permitted in animal rooms. Persons who wear contact lenses in animal rooms should also wear goggles or a face shield.
4. All procedures are carefully performed to minimize the creation of aerosols.
5. Work surfaces are decontaminated after use or after any spill of viable materials.
6. Doors to animal rooms open inward, are self-closing and are kept closed when experimental animals are present.
7. All wastes from the animal room are appropriately decontaminated, preferably by autoclaving, before disposal. Infected animal carcasses are incinerated after being transported from the animal room in leakproof, covered containers.
8. An insect and rodent control program is in effect.

B. Special Practices

1. The laboratory director or other responsible person restricts access to the animal room to personnel who have been advised of the potential hazard and who need to enter the room for program or service purposes when infected animals are present. Persons who are at increased risk of acquiring infection, or for whom infection might be unusually hazardous, are not allowed in the animal room. Persons at increased risk may include children, pregnant women, and persons who are immunodeficient or immunosuppressed. The supervisor has the final responsibility for assessing each circumstance and determining who may enter or work in the facility.
2. The laboratory director or other responsible person establishes policies and procedures whereby only persons who have been advised of the potential hazard and meet any specific requirements (e.g., for immunization) may enter the animal room.
3. When the infectious agent(s) in use in the animal room requires special entry provisions (e.g., the need for immunizations and respirators) a hazard warning sign, incorporating the universal biohazard symbol, is posted on the access door to the animal room. The hazard warning sign identifies the infectious agent(s) in use, lists the name and telephone number of the animal facility supervisor or other responsible person(s), and indicates the special requirement(s) for entering the animal room.
4. Laboratory personnel receive appropriate immunizations or tests for the agents handled or potentially present in the laboratory (e.g., hepatitis B vaccine or TB skin testing).
5. Baseline serum samples from all personnel working in the facility and other at-risk personnel should be collected and stored. Additional serum samples may be collected periodically and stored. The serum surveil-

lance program must take into account the availability of methods for the assessment of antibody to the agent(s) of concern. The program should provide for the testing of serum samples at each collection interval and the communication of results to the participants.

6. A biosafety manual is prepared or adopted. Personnel are advised of special hazards, and are required to read and to follow instructions on practices and procedures.
7. Laboratory personnel receive appropriate training on the potential hazards associated with the work involved, the necessary precautions to prevent exposures, and the exposure evaluation procedures. Personnel receive annual updates, or additional training as necessary for procedural or policy changes.
8. A high degree of precaution must always be taken with any contaminated sharp items, including needles and syringes, slides, pipettes, capillary tubes, and scalpels. Needles and syringes or other sharp instruments are restricted in the laboratory for use only when there is no alternative, such as for parenteral injection, blood collection, or aspiration of fluids from laboratory animals and diaphragm bottles. Plasticware should be substituted for glassware whenever possible.
   a. Only needle-locking syringes or disposable syringe-needle units (i.e., needle is integral to the syringe) are used for injection or aspiration of infectious materials. Used disposable needles must not be bent, sheared, broken, recapped, removed from disposable syringes, or otherwise manipulated by hand before disposal; rather, they must be carefully placed in conveniently located puncture-resistant containers used for sharps disposal. Nondisposable sharps must be placed in a hard-walled container, preferably containing a suitable disinfectant, for transport to a processing area for decontamination, preferably by autoclaving.
   b. Syringes which re-sheathe the needle, needle-less systems, and other safe devices should be used when appropriate.
   c. Broken glassware must not be handled directly by hand, but must be removed by mechanical means such as a brush and dustpan, tongs, or forceps. Containers of contaminated needles, sharp equipment, and broken glass should be decontaminated before disposal, according to any local, state, or federal regulations.
9. Cultures, tissues, or specimens of body fluids are placed in a container that prevents leakage during collection, handling, processing, storage, transport, or shipping.
10. Cages are autoclaved or thoroughly decontaminated before bedding is removed or before they are cleaned and washed. Equipment and work surfaces should be decontaminated with an appropriate disinfectant on a routine basis, after work with infectious materials is finished, and especially after overt spills, splashes, or other contamination by infectious materials. Contaminated equipment must be decontaminated according to any local, state, or federal regulations before it is sent for repair or maintenance or packaged for transport in accordance with applicable local, state, or federal regulations, before removal from the facility.
11. Spills and accidents which result in overt exposures to infectious materials are immediately reported to the laboratory director. Medical evaluation, surveillance, and treatment are provided as appropriate and written records are maintained.
12. All wastes from the animal room are autoclaved before disposal. All animal carcasses are incinerated. Dead animals are transported from the animal room to the incinerator in leakproof covered containers.
13. Animals not involved in the work being performed are not permitted in the lab.

C. Safety Equipment (Primary Barriers)

1. Personal protective equipment is used for all activities involving manipulations of infectious materials or infected animals.
   a. Wrap-around or solid-front gowns or uniforms are worn by personnel entering the animal room. Front-button laboratory coats are unsuitable. Protective gowns should be appropriately contained until decontamination or disposal.
   b. Personnel wear gloves when handling infected animals. Gloves are removed aseptically and autoclaved with other animal room wastes before disposal.
   c. Appropriate face/eye and respiratory protection is worn by all personnel entering animal rooms housing nonhuman primates.
   d. Boots, shoe covers, or other protective footwear, and disinfectant footbaths are available and used when indicated.

2. Physical containment devices and equipment appropriate for the animal species are used for all procedures and manipulations of infectious materials or infected animals.
3. The risk of infectious aerosols from infected animals or their bedding also can be reduced if animals are housed in partial containment caging systems, such as open cages placed in ventilated enclosures (e.g., laminar flow cabinets), solid wall and bottom cages covered with filter bonnets, or other equivalent primary containment systems.

D. Animal Facilities (Secondary Barriers)

1. The animal facility is designed and constructed to facilitate cleaning and housekeeping, and is separated from areas which are open to unrestricted personnel traffic within the building. Passage through two sets of doors is the basic requirement for entry into the animal room from access corridors or other contiguous areas. Physical separation of the animal room from access corridors or other activities may also be provided by a double-doored clothes change room (showers may be included), airlock, or other access facility which requires passage through two sets of doors before entering the animal room.
2. The interior surfaces of walls, floors, and ceilings are water resistant so that they may be easily cleaned. Penetrations in these surfaces are sealed or capable of being sealed to facilitate fumigation or space decontamination.
3. A foot, elbow, or automatically operated handwashing sink is provided in each animal room near the exit door.
4. If vacuum service (i.e., central or local) is provided, each service connection should be fitted with liquid disinfectant traps and a HEPA filter.
5. If floor drains are provided, they are protected with liquid traps that are always filled with water or disinfectant.
6. Windows in the animal room are non-operating and sealed.
7. Animal room doors are self-closing and are kept closed when infected animals are present.
8. An autoclave for decontaminating wastes is available, preferably within the animal facility. Materials are transferred to the autoclave in a covered leakproof container whose outer surface has been decontaminated.
9. A non-recirculating ventilation system is provided. The supply and exhaust components of the system are balanced to provide for directional flow of air into the animal room. The exhaust air is discharged directly to the outside and clear of occupied areas and air intakes. Exhaust air from the room can be discharged without filtration or other treatment. Personnel must periodically validate that proper directional airflow is maintained.
10. The HEPA filtered exhaust air from Class I or Class II biological safety cabinets or other primary containment devices is discharged directly to the outside or through the building exhaust system. Exhaust air from these primary containment devices may be recirculated within the animal room if the device is tested and certified at least every 12 months. If the HEPA filtered exhaust air from Class I or Class II biological safety cabinets is discharged to the outside through the building exhaust system, it is connected to this system in a manner (e.g., thimble unit connection)[134] that avoids any interference with the performance of either the cabinet or building exhaust system

Animal Biosafety Level 4

A. Standard Practices

1. Access to the animal facility is limited or restricted at the discretion of the laboratory or animal facility director.
2. Personnel wash their hands after handling cultures and animals, after removing gloves, and before leaving the animal facility.
3. Eating, drinking, smoking, handling contact lenses, applying cosmetics, and storing food for human use are not permitted in animal rooms. Persons who wear contact lenses in animal rooms should also wear goggles or a face shield.
4. All procedures are carefully performed to minimize the creation of aerosols.
5. Work surfaces are decontaminated after use or after any spill of viable materials.
6. Doors to animal rooms open inward, are self-closing and are kept closed when experimental animals are present.
7. All wastes from the animal room are appropriately decontaminated, preferably by autoclaving, before disposal. Infected animal carcasses are incinerated after being transported from the animal room in leakproof, covered containers.
8. Cages are autoclaved before bedding is removed and before they are cleaned and washed. When feasible, disposable cages

that do not require cleaning are recommended; however, these cages are also autoclaved before disposal. Equipment and work surfaces should be decontaminated with an appropriate disinfectant on a routine basis, after work with infectious materials is finished, and especially after overt spills, splashes, or other contamination by infectious materials. Contaminated equipment must be decontaminated according to any local, state, or federal regulations before it is sent for repair or maintenance or packaged for transport in accordance with applicable local, state, or federal regulations, before removal from the facility.

9. An insect and rodent control program is in effect.

B. Special Practices

1. Only persons whose entry into the facility or individual animal room is required for program or support purposes are authorized to enter. Persons who may be at increased risk of acquiring infection or for whom infection might be unusually hazardous are not allowed in the animal facility. Persons at increased risk may include children, pregnant women, and persons who are immunodeficient or immunosuppressed. The supervisor has the final responsibility for assessing each circumstance and determining who may enter or work in the facility. Access to the facility is limited by secure, locked doors; accessibility is controlled by the animal facility supervisor, biohazards control officer, or other person responsible for the physical security of the facility. Before entering, persons are advised of the potential biohazards and instructed as to appropriate safeguards. Personnel comply with the instructions and all other applicable entry and exit procedures. Practical and effective protocols for emergency situations are established.
2. Laboratory personnel receive appropriate immunizations or tests for the agents handled or potentially present in the laboratory (e.g., hepatitis B vaccine or TB skin testing).
3. Baseline serum samples are collected and stored for all laboratory and other at-risk personnel. Additional serum specimens may be collected periodically, depending on the agents handled or the function of the laboratory. The decision to establish a serologic surveillance program takes into account the availability of methods for the assessment of antibody to the agent(s) of concern. The program provides for the testing of serum samples at each collection interval and the communication of results to the participants.
4. A biosafety manual is prepared or adopted. Personnel are advised of special hazards, and are required to read and to follow instructions on practices and procedures.
5. When the infectious agent(s) in use in the animal room requires special entry provisions (e.g., the need for immunizations and respirators) a hazard warning sign, incorporating the universal biohazard symbol, is posted on the access door to the animal room. The hazard warning sign identifies the infectious agent(s) in use, lists the name and telephone number of the animal facility supervisor or other responsible person(s), and indicates the special requirement(s) for entering the animal room.
6. Laboratory personnel receive appropriate training on the potential hazards associated with the work involved, the necessary precautions to prevent exposures, and the exposure evaluation procedures. Personnel receive annual updates, or additional training as necessary for procedural or policy changes.
7. Hypodermic needles and syringes are used only for gavage, for parenteral injection, and aspiration of fluids from diaphragm bottles or well-restrained laboratory animals.
   a. A high degree of precaution must always be taken with any contaminated sharp items, including needles and syringes, slides, pipettes, capillary tubes, and scalpels. Needles and syringes or other sharp instruments are restricted in the laboratory for use only when there is no alternative, such as for parenteral injection, blood collection, or aspiration of fluids from laboratory animals and diaphragm bottles. Plasticware is substituted for glassware whenever possible.
   b. Only needle-locking syringes or disposable syringe-needle units (i.e., needle is integral to the syringe) are used for injection or aspiration of infectious materials. Used disposable needles are not bent, sheared, broken, recapped, removed from disposable syringes, or otherwise manipulated by hand before disposal; rather, they are carefully placed in conveniently located puncture-resistant containers used for sharps disposal. Non-disposable

sharps are placed in a hard-walled container, preferably containing a suitable disinfectant, for transport to a processing area for decontamination, preferably by autoclaving.

c. Syringes which re-sheathe the needle, needle-less systems, and other safe devices should be used when appropriate.

d. Broken glassware is not handled directly by hand, but is removed by mechanical means such as a brush and dustpan, tongs, or forceps. Containers of contaminated needles, sharp equipment, and broken glass are decontaminated before disposal, according to any local, state, or federal regulations.

8. Cultures, tissues, or specimens of body fluids are placed in a container that prevents leakage during collection, handling, processing, storage, transport, or shipping.

9. Spills and accidents which result in overt exposures to infectious materials are immediately reported to the laboratory director. Medical evaluation, surveillance, and treatment are provided as appropriate, and written records are maintained.

10. Personnel enter and leave the facility only through the clothing change and shower rooms. Personnel shower each time they leave the facility. Head covers are provided to personnel who do not wash their hair during the exit shower. Except in an emergency, personnel do not enter or leave the facility through the airlocks.

11. Personal clothing is removed in the outer clothing change room and kept there. Complete laboratory clothing, including undergarments, pants and shirts or jumpsuits, shoes, and gloves, are provided and used by all personnel entering the facility. When exiting, personnel remove laboratory clothing in the inner change room before entering the shower area. Soiled clothing is autoclaved before laundering.

12. Supplies and materials are brought into the facility by way of a double-door autoclave, fumigation chamber, or airlock. After securing the outer doors, personnel inside the facility retrieve the materials by opening the interior door of the autoclave, fumigation chamber, or airlock. This inner door is secured after materials are brought into the facility. The autoclave fumigation chamber or airlock is decontaminated before the outer door is opened.

13. A system is established for the reporting of animal facility accidents and exposures, employee absenteeism, and for the medical surveillance of potential laboratory-associated illnesses. An essential adjunct to such a reporting-surveillance system is the availability of a facility for the quarantine, isolation, and medical care of persons with potential or known laboratory-associated illnesses.

14. Materials (e.g., plants, animals, clothing) not related to the experiment are not permitted in the facility.

C. Safety Equipment (Primary Barriers)

Laboratory animals, infected with agents assigned to Biosafety Level 4, are housed in a Class III biological safety cabinet or in a partial containment caging system (such as open cages placed in ventilated enclosures, solid wall and bottom cages covered with filter bonnets, or other equivalent primary containment systems), in specially designed areas in which all personnel are required to wear one-piece positive pressure suits ventilated with a life support system.

Animal work with viral agents that require Biosafety Level 4 secondary containment, and for which highly effective vaccines are available and used, may be conducted with partial containment cages and without the one-piece positive pressure personnel suit if: the facility has been decontaminated, no concurrent experiments are being done in the facility which require Biosafety Level 4 primary and secondary containment, and all other standard and special practices are followed.

D. Animal Facility (Secondary Barriers)

1. The animal rooms are located in a separate building or in a clearly demarcated and isolated zone within a building. Outer and inner change rooms separated by a shower are provided for personnel entering and leaving the facility. A double-doored autoclave, fumigation chamber, or ventilated airlock is provided for passage of materials, supplies, or equipment which are not brought into the facility through the change room.

2. Walls, floors, and ceilings of the facility are constructed to form a sealed internal shell which facilitates decontamination and is animal and insect proof. The internal surfaces of this shell are resistant to liquids and chemicals, thus facilitating cleaning and de-

contamination of the area. All penetrations in these structures and surfaces are sealed.

65

3. Internal facility appurtenances, such as light fixtures, air ducts, and utility pipes, are arranged to minimize horizontal surface areas on which dust can settle.
4. A foot, elbow, or automatically operated handwashing sink is provided in each animal room near the exit door.
5. If there is a central vacuum system, it does not serve areas outside of the facility. The vacuum system has in-line HEPA filters placed as near as practicable to each use point or service cock. Filters are installed to permit in-place decontamination and replacement. Other liquid and gas services for the facility are protected by devices that prevent backflow.
6. External animal facility doors are self-closing and self-locking.
7. Any windows must be resistant to breakage and sealed.
8. A double-doored autoclave is provided for decontaminating materials that leave the facility. The autoclave door which opens to the area external to the facility is automatically controlled so that it can only be opened after the autoclave "sterilization" cycle is completed.
9. A pass-through dunk tank, fumigation chamber, or an equivalent decontamination method is provided so that materials and equipment that cannot be decontaminated in the autoclave can be safely removed from the facility.
10. Liquid effluents from laboratory sinks, biological safety cabinets, floor drains (if used), and autoclave chambers are decontaminated by heat treatment before being discharged to the sanitary sewer. Effluents from showers and toilets may be discharged to the sanitary sewer without treatment. The process used for decontamination of liquid wastes must be validated physically and biologically by use of a constant recording temperature sensor in conjunction with an indicator microorganism having a defined heat susceptibility profile.

66

11. A dedicated non-recirculating ventilation system is provided. The supply and exhaust components of the system are balanced to assure directional airflow from the area of least hazard to the area(s) of greatest potential hazard. The differential pressure/directional airflow between adjacent areas is monitored and alarmed to indicate malfunction of the system. The airflow in the supply and exhaust components is monitored and the components interlocked to assure inward (or zero) airflow is maintained.
12. The general room exhaust air from a facility in which the work is conducted in a Class III cabinet system is treated by a passage through a HEPA filter(s) prior to discharge to the outside. The air is discharged away from occupied spaces and air intakes. The HEPA filter(s) are located as near as practicable to the source in order to minimize the length of potentially contaminated ductwork. The HEPA filter housings are designed to allow for *in situ* decontamination of the filter prior to removal, or removal of the filter in a sealed gas-tight primary container for subsequent decontamination and/or destruction by incineration. The design of the HEPA filter housing should facilitate validation of the filter installation. The use of pre-certified HEPA filters can be an advantage. The service-life of the exhaust HEPA filters can be extended through adequate filtration of the supply air.
13. The treated exhaust air from Class II biological safety cabinets located in a facility in which workers wear a positive pressure suit may be discharged into the animal room environment or to the outside through the facility air exhaust system. The biological safety cabinets are tested and certified at 9-month intervals. The air exhausted from Class III biological safety cabinets is passaged through two HEPA filter systems (in series) prior to discharge to the outside. If the treated exhaust is discharged to the outside through the facility exhaust system, it is connected to this system in a manner that avoids any interference with the air balance of the cabinets or the facility exhaust system.

67

14. A specially designed suit area may be provided in the facility. Personnel who enter this area wear a one-piece positive pressure suit that is ventilated by a life support system. The life support system is provided with alarms and emergency backup breathing air tanks. Entry to this area is through an airlock fitted with airtight doors. A chemical shower is provided to decontaminate the surface of the suit before the worker leaves the area. The exhaust air from the area in which the suit is used is filtered by two sets of HEPA filters installed in series. Duplicate filtration units and exhaust fans are pro-

vided. An automatically starting emergency power source is provided. The air pressure within the suit area is lower than that of any adjacent area. Emergency lighting and communication systems are provided. All penetrations into the inner shell of the suit area are sealed. A double-doored autoclave is provided for decontaminating waste materials to be removed from the suit area.

68

SECTION V: RECOMMENDED BIOSAFETY LEVELS FOR INFECTIOUS AGENTS AND INFECTED ANIMALS

Selection of an appropriate biosafety level for work with a particular agent or animal study depends upon a number of factors. Some of the most important are: the virulence, pathogenicity, biological stability, route of spread, and communicability of the agent; the nature or function of the laboratory; the procedures and manipulations involving the agent; the endemicity of the agent; and the availability of effective vaccines or therapeutic measures.

Agent summary statements in this section provide guidance for the selection of appropriate biosafety levels. Specific information on laboratory hazards associated with a particular agent, and recommendations regarding practical safeguards that can significantly reduce the risk of laboratory-associated diseases, are included. Agent summary statements are presented for agents which meet one or more of the following criteria: the agent is a proven hazard to laboratory personnel working with infectious materials (e.g., hepatitis B virus, *M. tuberculosis*); the potential for laboratory associated infections is high, even in the absence of previously documented laboratory-associated infections (e.g., exotic arboviruses); or, the consequences of infection are grave.

Recommendations for the use of vaccines and toxoids are included in agent summary statements when such products are available, either as licensed or Investigational New Drug (IND) products. When applicable, recommendations for the use of these products are based on current recommendations of the Public Health Service Advisory Committee on Immunization Practice, and are specifically targeted to at-risk laboratory personnel and others who must work in or enter laboratory areas. These specific recommendations should in no way preclude the routine use of such products as diphtheria-tetanus toxoids, poliovirus vaccine, influenza vaccine and others, because of the potential risk of community exposures irrespective of any laboratory risks. Appropriate precautions should be taken in the administration of live attenuated virus vaccines in individuals with altered immunocompetence, or other medical condition (e.g., pregnancy), in which a viral infection could result in adverse consequences.

69

Risk assessments and biosafety levels recommended in the agent summary statements presuppose a population of immunocompetent individuals. Persons with altered immunocompetence may be at an increased risk when exposed to infectious agents. Immunodeficiency may be hereditary, congenital, or induced by a number of neoplastic or infectious diseases, by therapy, or by radiation. The risk of becoming infected or the consequence of infection may also be influenced by such factors as age, sex, race, pregnancy, surgery (e.g., splenectomy, gastrectomy), predisposing diseases (e.g., diabetes, lupus erythematosus) or altered physiological function. These and other variables must be considered in applying the generic risk assessments of the agent summary statements to specific activities of selected individuals.

The biosafety level assigned to an agent is based on the activities typically associated with the growth and manipulation of the quantities and concentrations of infectious agents required to accomplish identification or typing. If activities with clinical materials pose a lesser risk to personnel than those activities associated with manipulation of cultures, a lower biosafety level is recommended. On the other hand, if the activities involve large volumes and/or concentrated preparations ("production quantities"), or manipulations which are likely to produce aerosols or which are otherwise intrinsically hazardous, additional personnel precautions and increased levels of primary and secondary containment may be indicated.

"Production quantities" refers to large volumes or concentrations of infectious agents considerably in excess of those typically used for identification and typing activities. Propagation and concentration of infectious agents as occurs in large-scale fermentations, antigen and vaccine production, and a variety of other commercial and research activities, clearly deal with significant masses of infectious agents that are reasonably considered "production quantities". However, in terms of potentially increased risk as a function of the mass of infectious agents, it is not possible to define "production quantities" in finite volumes or concentrations for any given agent. Therefore, the laboratory director must make an assessment of the activities conducted and select practices, containment equipment, and facilities appropriate to the risk, irrespective of the volume or concentration of agent involved.

70

Occasions will arise when the laboratory director should select a biosafety level higher than that recommended. For example, a higher biosafety level may be indicated by the unique nature of the proposed activity (e.g., the need for special containment for experimentally generated aerosols for inhalation studies) or by the proximity of the laboratory to areas of special concern (e.g., a diagnostic laboratory located near patient care areas). Similarly, a recommended biosafety level may be adapted to compensate for the absence of certain recommended safeguards. For example, in those situations where Biosafety Level 3 is recommended, acceptable safety may be achieved for routine or repetitive operations (e.g., diagnostic procedures involving the propagation of an agent for identification, typing and susceptibility testing) in laboratories where facility features satisfy Biosafety Level 2 recommendations, provided the recommended Standard Microbiological Practices, Special Practices, and Safety Equipment for Biosafety Level 3 are rigorously followed.

One example involves work with the Human Immunodeficiency Viruses (HIVs). Routine diagnostic work with clinical specimens can be done safely at Biosafety Level 2, using Biosafety Level 2 practices and procedures. Research work (including co-cultivation, virus replication studies, or manipulations involving concentrated virus) can be done in a BSL-2 facility, using BSL-3 practices and procedures. Virus production activities, including virus concentrations, require a BSL-3 facility and use of BSL-3 practices and procedures (see Agent Summary Statement).

The decision to adapt Biosafety Level 3 recommendations in this manner should be made only by the laboratory director. This adaptation, however, is not suggested for agent production operations or activities where procedures are frequently changing. The laboratory director should also give special consideration to selecting appropriate safeguards for materials that may contain a suspected agent. For example, sera of human origin may contain hepatitis B virus, and therefore, all blood or blood-derived fluids should be handled under conditions which

71

reasonably preclude cutaneous, mucous membrane or parenteral exposure of personnel. Sputa submitted to the laboratory for tubercle bacilli assay should be handled under conditions which reasonably preclude the generation of aerosols during the manipulation of clinical materials or cultures.

The infectious agents which meet the previously stated criteria are listed by category of agent in Section VII. To use these summaries, first locate the agent in the listing under the appropriate category of agent. Second, utilize the practices, safety equipment, and type of facilities recommended in the agent summary statement as described in Section VII for working with clinical materials, cultures or infectious agents, or infected animals.

The laboratory director is also responsible for appropriate risk assessment and for utilization of appropriate practices, containment equipment, and facilities for agents not included in the agent summary statements.

SECTION VI: RISK ASSESSMENT

72

The assessment of risks associated with laboratory activities involving the use of infectious microorganisms is ultimately a subjective process. The risks associated with the agent, as well as with the activity to be conducted, must be considered in the assessment. The characteristics of infectious agents and the primary laboratory hazards of working with the agents are described generically for agents in Biosafety Levels 1–4 and specifically for individual agents or groups of agents in Section VII of this publication.

Hepatitis B (HBv) is an appropriate model for illustrating the risk assessment process. HBv is among the most ubiquitous of human pathogens and most prevalent of laboratory-associated infections. The agent has been demonstrated in a variety of body secretions and excretions. Blood, saliva, and semen have been shown to contain the virus. Natural transmission is associated with parenteral inoculation or with contamination of broken skin or mucous membranes with infectious body fluids. There is no evidence of airborne or interpersonal spread through casual contact. Prophylactic measures include the use of a licensed vaccine in high-risk groups and the use of hepatitis B immune globulin following overt exposures.

The primary risks of HBv infection for laboratory personnel are accidental parenteral inoculation, exposure of broken skin or the mucous membranes of eyes, nose, or mouth. These risks are typical of those described for Biosafety Level 2 agents, and are addressed by using the recommended standard and special microbiological practices to minimize or eliminate these overt exposures.

The routes of infection with hepatitis C virus (non A-non B) and human immunodeficiency virus are similar for laboratory personnel. The prudent practices recommended for HBv are applicable to these two pathogens, as well as to the routine laboratory manipulation of clinical materials of domestic origin.

73

The described risk assessment process is also applicable to laboratory operations other than those involving the use of primary agents of human disease. Microbiological studies of animal host-specific pathogens, soil, water, food, feeds, and other natural or manufactured materials, by comparison, pose substantially lower risks for the laboratory worker. Microbiologists and other scientists working with such materials may, nevertheless, find the practices, containment equipment, and facility recommendations described in this publication of value in developing operational standards to meet their own assessed needs.

SECTION VII: AGENT SUMMARY STATEMENTS

Parasitic Agents

Agent: Nematode Parasites of Humans

Laboratory-associated infections with *Ascaris* spp.; *Strongyloides* spp.; hookworms; and *Enterobius* spp. have been reported.[91,110,151] Allergic reactions to various antigenic components of nematodes (e.g., aerosolized *Ascaris* antigens) may represent an individual risk to sensitized persons. Laboratory animal-associated infections (including arthropods) have not been reported, but infective larvae in the feces of nonhuman primates infected with *Strongyloides* spp. are a potential infection hazard for laboratory and animal care personnel.

Laboratory Hazards: Eggs and larvae in freshly passed feces of infected hosts are usually not infective; development to the infective stages may take periods of one day to several weeks. *Trichinella* is of concern since fresh or digested tissue may contain larvae and would be infective if ingested. Ingestion of the infective eggs or skin penetration of infective larvae are the primary hazards to laboratory and animal care personnel. Arthropods infected with filarial parasites pose a potential hazard to laboratory personnel. In laboratory personnel with frequent exposure to aerosolized antigens of *Ascaris* spp. development of hypersensitivity is common.

Recommended Precautions: Biosafety Level 2 practices and facilities are recommended for activities with infective stages listed. Exposure to aerosolized sensitizing antigens of *Ascaris* spp. should be avoided. Primary containment (e.g., biological safety cabinet) may be required for work with these materials by hypersensitive individuals. Appropriate treatment for most nematode infections exists, and information on dosage, source of drugs, etc. is available.[5]

Agent: Protozoal Parasites of Humans

Laboratory-associated infections with *Toxoplasma* spp.; *Plasmodium* spp. (including *P. cynomologi*); *Trypanosoma* spp.; *Entamoeba* spp.; *Coccidia* spp.; *Giardia* spp.; *Leishmania* spp.; *Sarcocystis* spp.; and *Cryptosporidia* spp. have been reported.[29,68,91,110,151,162] In addition, no laboratory infections with *Babesia* spp. or *Microsporidia* spp. have been reported but could result from accidental needlestick or ingestion of cysts, oocysts, or spores in feces.

Although laboratory animal-associated infections have not been reported, a direct source of infection for laboratory personnel may be contact with lesion material from rodents with cutaneous leishmaniasis and with feces or blood of experimentally or naturally infected animals.

Laboratory-related infections with *Cryptosporidia* have occurred with regularity in almost every laboratory working with this agent, especially those in which calves are utilized as the source of oocysts. Other experimentally-infected animals pose potential risks as well. There is circumstantial evidence that airborne transmission of oocysts of this small organism may occur. Rigid adherence to protocol should reduce the occurrence in laboratory and animal care personnel.

Laboratory Hazards: Infective stages may be present in blood, feces, lesion exudates, and infected arthropods. Depending on the parasite, ingestion, skin penetration through wounds or microabrasions, accidental parenteral inoculation, and transmission by arthropod vectors are the primary laboratory hazards. Aerosol or droplet exposure of the mucous membranes of the eyes, nose, or mouth with trophozoites are potential hazards when working with cultures of *Naegleria fowleri*, *Leishmania* spp., *T. cruzi*, or with tissue homogenates or blood containing hemoflagellates. Immunocompromised individuals should avoid working with live organisms. Because of the grave consequences of toxoplasmosis in the developing fetus, serologically negative women of childbearing age who might become pregnant should not work with *Toxoplasma* in the same laboratory room where these materials are handled.

Recommended Precautions: Biosafety Level 2 practices and facilities are recommended for activities with infective stages of the parasites listed. Infected arthropods should be maintained in facilities which reasonably preclude the exposure of personnel or their escape to the outside. Primary containment (e.g., biological safety cabinet) or personal protection (e.g., face shield) may be indicated when working with cultures of *Naegleria fowleri*, *Leishmania* spp., *T. cruzi* or with tissue homogenates or blood containing hemoflagellates.[81] Gloves are recommended for activities where there is the likelihood of direct skin contact with infective stages of the parasites listed. Appropriate treatment for most protozoal infections exists, and information on dosage, source of drugs, etc., is available.[5]

Agent: Trematode Parasites of Humans (*Schistosoma* spp. and *Fasciola* spp.)

Laboratory-associated infections with *Schistosoma* spp. and *Fasciola* spp. have been reported, none associated directly with laboratory animals.[91,110,151]

Laboratory Hazards: Infective stages of *Schistosoma* spp. (cercariae) and *Fasciola* spp. (metacercaria) may be found, respectively, in the water or encysted on aquatic plants in laboratory aquaria used to maintain snail intermediate hosts. Skin penetration by schistosome cercariae and ingestion of fluke metacercaria are the primary laboratory hazards. Dissection or crushing of schistosome-infected snails may also result in exposure of skin or mucous membrane to cercariae-containing droplets. Additionally, metacercaria may be inadvertently transferred from hand to mouth by fingers or gloves following contact with contaminated aquatic vegetation or surfaces of aquaria. Most laboratory exposures to *Schistosoma* spp. would predictably result in low worm burdens with minimal disease potential. Safe and effective drugs are available for the treatment of schistosomiasis.

Recommended Precautions: Biosafety Level 2 practices and facilities are recommended for activities with infective stages of the parasites listed. Gloves should be worn when there may be direct contact with a water containing cercariae, or vegetation containing metacercaria from naturally or experimentally infected snail intermediate hosts. Long-sleeved laboratory coats or other protective garb should be worn when working arouquaria or other water sources that may contain schistosome cercariae. Snails and cercariae in the water of laboratory aquaria should be killed by chemicals (e.g., hypochlorites, iodine) or heat before discharge to sewers. Appropriate treatment for most trematode infections exists, and information on source of drugs, dosage, etc. is available.[5]

Agent: Cestode Parasites of Humans—*Echinococcus granulosus, Taenia solium* (*Cysticercus cellulosae*) and *Hymenolepis nana*

Although no laboratory-associated infections with either *E. granulosus* or *T. solium* have been reported, the consequences of such infections following the ingestion of infective eggs of *E. granulosus* or *T. solium* are potentially grave. *H. nana* is a very cosmopolitan parasite, does not require an intermediate host, and is directly transmissible by ingestion of feces of infected humans or rodents.

Laboratory Hazards: Infective eggs may be present in the feces of dogs or other canids (the definitive hosts of *E. granulosus*), or in the feces of humans (the definitive host of *T. solium*). Ingestion of infective eggs from these sources are the primary laboratory hazard. Cysts and cyst fluids of *E. granulosus* are not infectious for humans. Ingestion of cysts containing the larval stage of *T. solium (Cysticercus cellulosae)* readily produces human infection with the adult tapeworm. With either parasite, the ingestion of a single infective egg from the feces of the definitive host could potentially result in serious disease. Ingestion of the eggs of *H. nana* in the feces of the definitive host could result in intestinal infection.

Recommended Precautions: Biosafety Level 2 practices and facilities are recommended for work with infective stages of these parasites. Special attention should be given to personal hygiene practices (e.g., handwashing) and avoidance of ingestion of infective eggs. Gloves are recommended when there may be direct contact with feces or surfaces contaminated with fresh feces of dogs infected with *E. granulosus,* humans infected with *T. solium* adults, or humans or rodents infected with *H. nana*. Appropriate treatment for many cestode infections exists, and information concerning source of drugs, dosage, etc., is available.[5]

Fungal Agents

Agent: *Blastomyces dermatitidis*

Laboratory-associated local infections following accidental parenteral inoculation with infected tissues or cultures containing yeast forms of *B. dermatitidis*[67,87,107,108,168,199] have been reported. Pulmonary infections have occurred following the presumed inhalation of conidia; two developed pneumonia and one had an osteolytic lesion from which *B. dermatitidis* was cultured.[9,58] Presumably, pulmonary infections are associated only with sporulating mold forms (conidia).

Laboratory Hazards: Yeast forms may be present in the tissues of infected animals and in clinical specimens. Parenteral (subcutaneous) inoculation of these materials may cause local granulomas. Mold form cultures of *B. dermatitidis* containing infectious conidia may pose a hazard of aerosol exposure.

Recommended Precautions: Biosafety Level 2 and Animal Biosafety Level 2 practices and facilities are recommended for activities with clinical materials, animal tissues, cultures, and infected animals.

Agent: *Coccidioides immitis*

Laboratory-associated coccidioidomycosis is a documented hazard.[18,56,59,60,61,105,113,133,171,172,173] Smith reported that 28 of 31 (90%) laboratory-associated infections in his institution resulted in clinical disease, whereas more than half of infections acquired in nature were asymptomatic.[200]

Laboratory Hazards: Because of the size (2–5 millimicrons), the arthroconidia are conducive to ready dispersal in air and retention in the deep pulmonary spaces. The much larger size of the spherule (30–60 millimicrons) considerably reduces the effectiveness of this form of the fungus as an airborne pathogen.

Spherules of the fungus may be present in clinical specimens and animal tissues, and infectious arthroconidia in mold cultures and soil samples. Inhalation of arthroconidia from soil samples, mold cultures, or following transformation from the spherule form in clinical materials, is the primary laboratory hazard. Accidental percutaneous inoculation of the spherule form may result in local granuloma formation.[184] Disseminated disease occurs at a much greater frequency in blacks and Filipinos than whites.

Recommended Precautions: Biosafety Level 2 practices and facilities are recommended for handling and processing clinical specimens, identifying isolates, and processing animal tissues. Animal Biosafety Level 2 practices and facilities are recommended for experimental animal studies when the route of challenge is parenteral.

Biosafety Level 3 practices and facilities are recommended for propagating and manipulating sporulating cultures already identified as *C. immitis* and for processing soil or other environmental materials known or likely to contain infectious arthroconidia.

Agent: ***Cryptococcus neoformans***

A single account is reported of a laboratory exposure to *Cryptococcus neoformans* as a result of a laceration by a scalpel blade heavily contaminated with encapsulated cells.[83] This vigorous exposure, which did not result in local or systemic evidence of infection, suggests that the level of pathogenicity for normal immunocompetent adults is low. Respiratory infections as a consequence of laboratory exposure have not been recorded.

Laboratory Hazards: Accidental parenteral inoculation of cultures or other infectious materials represents a potential hazard to laboratory personnel—particularly to those that may be immunocompromised. Bites by experimentally infected mice and manipulations of infectious environmental materials (e.g., pigeon droppings) may also represent a potential hazard to laboratory personnel.

Recommended Precautions: Biosafety Level 2 and Animal Biosafety Level 2 practices and facilities are recommended, respectively, for activities with known or potentially infectious clinical, environmental, or culture materials and with experimentally infected animals.

The processing of soil or other environmental materials known or likely to contain infectious yeast cells should be conducted in a Class I or Class II biological safety cabinet. This precaution is also indicated for culture of the perfect or sexual state of the agent.

Agent: ***Histoplasma capsulatum***

Laboratory-associated histoplasmosis is a documented hazard in facilities conducting diagnostic or investigative work.[151,152] Pulmonary infections have resulted from handling mold form cultures.[132] Local infection has resulted from skin puncture during autopsy of an infected human[185] and from accidental needle inoculation of a viable culture.[182] Collecting and processing soil samples from endemic areas has caused pulmonary infections in laboratory workers. Encapsulated spores are resistant to drying and may remain viable for long periods of time. The small size of the infective conidia (less than 5 microns) is conducive to airborne dispersal and intrapulmonary retention. Furcolow reported that 10 spores were almost as effective as a lethal inoculum in mice as 10,000 to 100,000 spores.[75]

Laboratory Hazards: The infective stage of this dimorphic fungus (conidia) is present in sporulating mold form cultures and in soil from endemic areas. The yeast form in tissues or fluids from infected animals may produce local infection following parenteral inoculation.

Recommended Precautions: Biosafety Level 2 and Animal Biosafety Level 2 practices and facilities are recommended for handling and processing clinical specimens, identifying isolates, animal tissues and mold cultures, identifying cultures in routine diagnostic laboratories, and for experimental animal studies when the route of challenge is parenteral.

Biosafety Level 3 practices and facilities are recommended for propagating and manipulating cultures already identified as *H. capsulatum*, as well as processing soil or other environmental materials known or likely to contain infectious conidia.

Agent: ***Sporothrix schenckii***

S. schenckii has caused a substantial number of local skin or eye infections in laboratory personnel. Most cases have been associated with accidents and have involved splashing culture material into the eye,[69,197] scratching[21] or injecting[183] infected material into the skin or being bitten by an experimentally infected animal.[100,101] Skin infections have resulted also from handling cultures[57,124,137] or necropsy of animals[73] without a known break in technique. No pulmonary infections have been reported to result from laboratory exposure, although naturally occurring lung disease is thought to result from inhalation.

Recommended Precautions: Biosafety Level 2 and Animal Biosafety Level 2 practices and facilities are recommended for all laboratory and experimental animal activities with *S. schenckii*. Gloves should be worn when handling experimentally infected animals, and during operations with broth cultures that might result in hand contamination.

Agents: Pathogenic Members of the Genera *Epidermophyton, Microsporum* and *Trichophyton*

Although skin, hair and nail infections by these dermatophytid molds are among the most prevalent of human infections, the processing of clinical material has not been associated with laboratory infections. Infections have been acquired through contacts with naturally or experimentally infected laboratory animals (mice, rabbits, guinea pigs, etc.) and, rarely, with handling cultures.[84,119,151]

Laboratory Hazards: Agents are present in the skin, hair and nails of human and animal hosts. Contact with infected laboratory animals with inapparent or apparent infections is the primary hazard to laboratory personnel. Cultures and clinical materials are not an important source of human infection.

Recommended Precautions: Biosafety Level 2 and Animal Biosafety Level 2 practices and facilities are recommended for all laboratory and experimental animal activities with dermatophytes. Experimentally infected animals should be handled with disposable gloves.

Agent: Miscellaneous Molds

Several molds have caused serious infection in immunocompetent hosts following presumed inhalation or accidental subcutaneous inoculation from environmental sources. These agents are *Cladosporium (Xylohypha) trichoides, Cladosporium bantianum, Penicillium marnefii, Exophiala (Wangiella) dermatitidis, Fonsecaea pedrosoi* and *Dactylaria gallopava (Ochroconis gallopavum)*. Even though no laboratory acquired infections appear to have been reported with most of these agents, the gravity of naturally acquired illness is sufficient to merit special precautions in the laboratory. *Penicillium marnefii* has caused a local inoculation infection in a laboratory worker.[169]

Laboratory Hazards: Inhalation of conidia from sporulating mold cultures or accidental injection into the skin during infection or experimental animals is a theoretical risk to laboratory personnel.

Recommended Precautions: Biosafety Level 2 practices and facilities are recommended for propagating and manipulating cultures known to contain these agents.

Bacterial Agents

Agent: *Bacillus anthracis*

Forty (40) cases of laboratory-associated anthrax, occurring primarily at facilities conducting anthrax research, have been reported.[66,151] No laboratory-associated cases of anthrax have been reported in the United States since the late 1950's when human anthrax vaccine was introduced.

Naturally and experimentally infected animals pose a potential risk to laboratory and animal care personnel.

Laboratory Hazards: The agent may be present in blood, skin lesion exudates, cerebrospinal fluid, pleural fluid, sputum, and rarely, in urine and feces. Direct and indirect contact of the intact and broken skin with cultures and contaminated laboratory surfaces, accidental parenteral inoculation, and rarely, exposure to infectious aerosols are the primary hazards to laboratory personnel.

Recommended Precautions: Biosafety Level 2 practices, containment equipment and facilities are recommended for activities using clinical materials and diagnostic quantities of infectious cultures. Animal Biosafety Level 2 practices, containment equipment and facilities are recommended for studies utilizing experimentally infected laboratory rodents. A licensed vaccine is available through the Centers for Disease Control and Prevention; however, immunization of laboratory personnel is not recommended unless frequent work with clinical specimens or diagnostic cultures is anticipated (e.g., animal disease diagnostic laboratory). Biosafety Level 3 practices, containment equipment and facilities are recommended for work involving production volumes or concentrations of cultures, and for activities which have a high potential for aerosol production. In these facilities immunization is recommended for all persons working with the agent, all persons working in the same laboratory room where the cultures are handled, and persons working with infected animals.

Agent: *Bordetella pertussis*

Bordetella pertussis, a human respiratory pathogen of worldwide distribution, is the causative agent of whooping cough. The disease is typically a childhood illness; however, the agent has been associated, with increased frequency, in adult illness.[106,112,130] Several outbreaks in health-care workers have been reported in the literature.[106,112] Adolescents and adults with atypical or undiagnosed disease can serve as reservoirs of infection and transmit the organism to infants and children.[135] Eight cases of infection with *B. pertussis* in adults have been documented at a large research institution. The individuals involved did not work directly with the organism, but had access to common laboratory spaces where the organism was manipulated. One case of secondary transmission to a family member was documented.[122] A similar incident occurred at a large midwestern university resulting in two documented cases of laboratory-acquired infection and one documented case of secondary transmission.[146] Other laboratory-acquired infections with *B. pertussis* have been reported, as well as adult-to-adult transmission in the workplace.[19,35] Laboratory-acquired infections resulting from the manipulation of clinical specimens or isolates have not been reported. The attack rate of this airborne infection is influenced by intimacy and frequency of exposure of susceptible individuals.

Laboratory Hazards: The agent may be present in respiratory secretions, but is not found in blood or tissues. Since the natural mode of transmission is by the respiratory route, the greatest potential hazard is aerosol generation during the manipulation of cultures or concentrated suspensions of the organism.

Recommended Precautions: Biosafety Level 2 practices, containment equipment, and facilities are recommended for all activities involving the use or manipulation of known or potentially infectious clinical materials or cultures. Animal Biosafety Level 2 should be used for the housing of infected animals. Primary containment devices and equipment (e.g., biological safety cabinets, centrifuge safety cups, or specially designed safety centrifuges) should be used for activities likely to generate potentially infectious aerosols. Biosafety Level 3 practices, procedures, and facilities are appropriate when engaged in large scale production operations. The current pertussis vaccine may not provide complete and permanent immunity; however, a booster dose of pertussis vaccine is not recommended for use in persons who have passed their seventh birthday.[50]

Agent: *Brucella (B. abortus, B. canis, B. melitensis, B. suis)*

B. abortus, B. canis, B. melitensis, and *B. suis* have all caused illness in laboratory personnel.[129,151,176] Brucellosis is the most commonly reported laboratory-associated bacterial infection.[127,143,151] Hypersensitivity to *Brucella* antigens is also a hazard to laboratory personnel. Occasional cases have been attributed to exposure to experimentally and naturally infected animals or their tissues.

Laboratory Hazards: The agent may be present in blood, cerebrospinal fluid, semen, and occasionally urine. Most laboratory-associated cases have occurred in research facilities and have involved exposure to *Brucella* organisms being grown in large quantities. Cases have also occurred in a clinical laboratory setting: direct skin contact with cultures or with infectious clinical specimens from animals (e.g., blood, uterine discharges) are commonly implicated in these cases. Aerosols generated during laboratory procedures have caused large outbreaks.[95] Mouth pipetting, accidental parenteral inoculations, and sprays into eyes, nose and mouth have also resulted in infection.

Recommended Precautions: Biosafety Level 2 practices are recommended for activities with clinical specimens of human or animal origin containing or potentially containing pathogenic *Brucella* spp. Biosafety Level 3 and Animal Biosafety Level 3 practices, containment equipment and facilities are recommended, respectively, for all manipulations of cultures of the pathogenic *Brucella* spp. listed in this summary, and for experimental animal studies. Vaccines are not available for use in humans.

Agent: *Campylobacter (C. jejuni/C. coli, C. fetus* subsp. *fetus)*

C. jejuni/C. coli gastroenteritis is rarely a cause of laboratory associated illness. Three laboratory-acquired cases have been documented.[138,149,155] Numerous domestic and wild animals, including poultry, pets, farm animals, laboratory animals, and wild birds are known reservoirs and are a potential source of infection for laboratory and animal care personnel. Experimentally infected animals are also a potential source of infection.[155]

Laboratory Hazards: Pathogenic campylobacters may occur in fecal specimens in large numbers. *C. fetus* subsp. *fetus* may also be present in blood, exudates from abscesses, tissues, and sputa. Ingestion or parenteral inoculation of *C. jejuni* constitute the primary laboratory hazards. The oral ingestion of 500 organisms caused infection in one individual.[163] The importance of aerosol exposure is not known.

Recommended Precautions: Biosafety LUevel 2 practices, containment equipment and facilities are recommended for activities with cultures or potentially infectious clinical materials. Animal Biosafety Level 2 practices, containment equipment and facilities are recommended for activities with naturally or experimentally infected animals. Vaccines are not available for use in humans.

Agent: ***Chlamydia psittaci, C. pneumoniae, C. trachomatis***

Infections with psittacosis, lymphogranuloma venereum (LGV), and trachoma are documented hazards and among the most commonly reported laboratory-associated bacterial infection. In one report,[151] the majority of cases were psittacosis, occurred before 1955, and had the highest case fatality rate of all groups of infectious agents. Additional cases of laboratory-acquired psittacosis have been documented more recently.[127] Contact with and exposure to infectious aerosols in the handling, care, or necropsy of naturally or experimentally infected birds are the major sources of laboratory-associated psittacosis. Infected mice and eggs are less important sources of *C. psittaci*. Laboratory animals are not a reported source of human infection with *C. trachomatis*.

Laboratory Hazards: *C. psittaci* may be present in the tissues, feces, nasal secretions and blood, of infected birds and in blood, sputum, and tissues of infected humans. *C. trachomatis* may be present in genital, bubo, and conjunctival fluids of infected humans. Exposure to infectious aerosols and droplets, created during the handling of infected birds and tissues, are the primary hazards to laboratory personnel working with psittacosis. The primary laboratory hazards of *C. trachomatis* are accidental parenteral inoculation and direct and indirect exposure of mucous membranes of the eyes, nose, and mouth to genital, bubo, or conjunctival fluids, cell culture materials, and fluids from infected eggs. Infectious aerosols may also pose a potential source of infection.

Recommended Precautions: Biosafety Level 2 practices, containment equipment and facilities are recommended for activities involving the necropsy of infected birds and the diagnostic examination of tissues or cultures known or potentially infected with *C. psittaci* or *C. trachomatis*. Wetting the feathers of infected birds with a detergent-disinfectant prior to necropsy can appreciably reduce the risk of aerosols of infected feces and nasal secretions on the feathers and external surfaces of the bird. Animal Biosafety Level 2 practices, containment equipment and facilities and respiratory protection are recommended for personnel working with naturally or experimentally infected caged birds. Gloves are recommended for the necropsy of birds and mice, the opening of inoculated eggs, and when there is the likelihood of direct skin contact with infected tissues, bubo fluids, and other clinical materials. Additional primary containment and personnel precautions, such as those recommended for Biosafety Level 3, may be indicated for activities with high potential for droplet or aerosol production and for activities involving production quantities or concentrations of infectious materials. Vaccines are not available for use in humans.

Agent: ***Clostridium botulinum***

While there is only one report[177] of botulism associated with the handling of the agent or toxin in the laboratory or working with naturally or experimentally infected animals, the consequences of such intoxications must still be considered quite grave.

Laboratory Hazards: *C. botulinum* or its toxin may be present in a variety of food products, clinical materials (serum, feces) and environmental samples (soil, surface water). Exposure to the toxin of *C. botulinum* is the primary laboratory hazard. The toxin may be absorbed after ingestion or following contact with the skin, eyes, or mucous membranes, including the respiratory tract.[93] Accidental parenteral inoculation may also represent a significant exposure to toxin. Broth cultures grown under conditions of optimal toxin production may contain 2×10^6 mouse LD_{50} per mL.[177]

Recommended Precautions: Biosafety Level 2 practices, containment equipment and facilities are recommended for all activities with materials known or potentially containing the toxin. A pentavalent (ABCDE) botulism toxoid is available through the Centers for Disease Control and Prevention, as an investigational new drug (IND). This toxoid is recommended for personnel working with cultures of *C. botulinum* or its toxins. Solutions of sodium hypochlorite (0.1%) or sodium hydroxide (0.1N) readily inactivate the toxin and are recommended for decontaminating work surfaces and spills of cultures or toxin. Additional primary containment and personnel precautions, such as those recommended for Biosafety Level 3, are indicated for activities with a high potential for aerosol or droplet production, and those involving production quantities of toxin. Animal Biosafety Level 2 practices, containment equipment and facilities are recommended for diagnostic studies and titration of toxin.

Agent: ***Clostridium tetani***

Although the risk of infection to laboratory personnel is negligible, Pike[151] has recorded 5 incidents related to exposure of personnel during manipulation of the toxin.

Laboratory Hazards: Accidental parenteral inoculation and ingestion of the toxin are the primary hazards to laboratory personnel. It is uncertain if tetanus toxin can be absorbed through mucous membranes; consequently, the hazards associated with aerosols and droplets remain unclear.

Recommended Precautions: Biosafety Level 2 practices, containment equipment and facilities are recommended for activities involving the manipulation of cultures or toxin. While the risk of laboratory-associated tetanus is low, the administration of an adult diphtheria-tetanus toxoid at 10-year intervals further reduces the risk to laboratory and animal care personnel of toxin exposures and wound contamination, and is therefore highly recommended.[32]

Agent: *Corynebacterium diphtheriae*

Laboratory-associated infections with *C. diphtheriae* are documented. Pike[151] lists 33 cases reported in the world literature. Laboratory animal-associated infections have not been reported.

Laboratory Hazards: The agent may be present in exudates or secretions of the nose, throat (tonsil), pharynx, larynx, wounds, in blood, and on the skin. Inhalation, accidental parenteral inoculation, and ingestion are the primary laboratory hazards.

Recommended Precautions: Biosafety Level 2 practices, containment equipment and facilities are recommended for all activities utilizing known or potentially infected clinical materials or cultures. Animal Biosafety Level 2 facilities are recommended for studies utilizing infected laboratory animals. While the risk of laboratory-associated diphtheria is low, the administration of an adult diphtheria-tetanus toxoid at 10-year intervals may further reduce the risk to laboratory and animal care personnel of toxin exposures and work with infectious materials.[32]

Agent: *Francisella tularensis*

Tularemia is the third most commonly reported laboratory-associated bacterial infection.[151] Almost all cases occurred at facilities involved in tularemia research. Occasional cases have been related to work with naturally or experimentally infected animals or their ectoparasites. Although not reported, cases have occurred in clinical laboratories.

Laboratory Hazards: The agent may be present in lesion exudate, respiratory secretions, cerebrospinal fluid, blood, urine, tissues from infected animals, and fluids from infected arthropods. Direct contact of skin or mucous membranes with infectious materials, accidental parenteral inoculation, ingestion, and exposure to aerosols and infectious droplets have resulted in infection. Cultures have been more commonly associated with infection than clinical materials and infected animals. The human 25% to 50% infectious dose is approximately 10 organisms by the respiratory route.[16]

Recommended Precautions: Biosafety Level 2 practices, containment equipment and facilities are recommended for activities with clinical materials of human or animal origin containing or potentially containing *Francisella tularensis.* Biosafety Level 3 and Animal Biosafety Level 3 practices, containment equipment and facilities are recommended, respectively, for all manipulations of cultures and for experimental animal studies. An investigational live attenuated vaccine[16] is available. It is recommended for persons working with the agent or infected animals, and for persons working in or entering the laboratory or animal room where cultures or infected animals are maintained.

Agent: *Leptospira interrogans*—all serovars

Leptospirosis is a well-documented laboratory hazard. Pike[151] reported 67 laboratory-associated infections and 10 deaths, and three additional cases have been reported elsewhere.[127]

An experimentally infected rabbit was identified as the source of an infection with *L. interrogans* serovar *icterohemorrhagiae.*[159] Direct and indirect contact with fluids and tissues of experimentally or naturally infected mammals during handling, care, or necropsy is a potential source of infection. In animals with chronic kidney infections, the agent is shed in the urine in enormous numbers for long periods of time.

Laboratory Hazards: The agent may be present in urine, blood, and tissues of infected animals and humans. Ingestion, accidental parenteral inoculation, and direct and indirect contact of skin or mucous membranes with cultures or infected tissues or body fluids—especially urine—are the primary laboratory hazards. The importance of aerosol exposure is not known.

Recommended Precautions: Biosafety Level 2 practices, containment equipment and facilities are recommended for all activities involving the use or manipulation of known or potentially infectious tissues, body fluids, and cultures, and for the housing of infected animals. Gloves are recommended for the handling and necropsy of infected animals, and when there is the likelihood of direct skin contact with infectious materials. Vaccines are not available for use in humans.

Agent: *Legionella pneumophila;* other Legionella-like agents

A single documented nonfatal laboratory-associated case of legionellosis due to presumed aerosol or droplet exposure during animal challenge studies with Pontiac Fever agent (*L. pneumophila*) has been recorded.[24] Human-to-human spread has not been documented.

Experimental infections are readily produced in guinea pigs and embryonated chicken eggs.[121] Challenged rabbits develop antibodies but not clinical disease. Mice are refractory to parenteral exposure. Unpublished studies by Kaufmann, Feeley and others at the Centers for Disease Control and Prevention have shown that animal-to-animal transmission did not occur in a variety of experimentally infected mammalian and avian species.

Laboratory Hazards: The agent may be present in pleural fluid, tissue, sputum, and environmental sources (e.g., cooling tower water). Because the natural mode of transmission appears to be airborne, the greatest potential hazard is the generation of aerosols during the manipulation of cultures or of other materials containing high concentrations of infectious microorganisms (e.g., infected yolk sacs and tissues).

Recommended Precautions: Biosafety Level 2 practices, containment equipment and facilities are recommended for all activities involving the use or manipulation of known or potentially infectious clinical materials or cultures, and for the housing of infected animals. Biosafety Level 3 practices with primary containment devices and equipment (e.g., biological safety cabinets, centrifuge safety cups) are used for activities likely to generate potentially infectious aerosols and for activities involving production quantities of microorganisms. Vaccines are not available for use in humans.

Agent: ***Mycobacterium leprae***

Inadvertent parenteral human to human transmission of leprosy following an accidental needle stick in a surgeon[115] and the use of a presumably contaminated tattoo needle[147] have been reported. There are no cases reported as a result of working in a laboratory with biopsy or other clinical materials of human or animal origin. While naturally occurring leprosy or leprosy-like diseases have been reported in armadillos[189] and in nonhuman primates,[63,126] humans are the only known important reservoir of this disease.

Laboratory Hazards: The infectious agent may be present in tissues and exudates from lesions of infected humans and experimentally or naturally infected animals. Direct contact of the skin and mucous membranes with infectious materials, and accidental parenteral inoculation, are the primary laboratory hazards associated with handling infectious clinical materials.

Recommended Precautions: Biosafety Level 2 practices, containment equipment and facilities are recommended for all activities with known or potentially infectious clinical materials from infected humans and animals. Extraordinary care should be taken to avoid accidental parenteral inoculation with contaminated sharp instruments. Animal Biosafety Level 2 practices, containment equipment and facilities are recommended for animal studies utilizing rodents, armadillos, and nonhuman primates.

Agent: ***Mycobacterium*** **spp. other than** ***M. tuberculosis, M. bovis*** **or** ***M. leprae***

Pike reported 40 cases of nonpulmonary "tuberculosis" thought to be related to accidents or incidents in the laboratory or autopsy room.[151] Presumably these infections were due to mycobacteria other than *M. tuberculosis* or *M. bovis.* A number of mycobacteria which are ubiquitous in nature are associated with diseases, other than tuberculosis or leprosy, in humans, domestic animals, and wildlife. Characteristically, these organisms are infectious but not contagious. Clinically, the diseases associated with infections by these "atypical" mycobacteria can be divided into three general categories:

1. *Pulmonary diseases resembling tuberculosis* which may be associated with infection by *M. kansasii, M. avium* complex, and rarely, by *M. xenopi, M. malmoense, M. asiaticum, M. simiae* and *M. szulgai.*
2. *Lymphadenitis* which may be associated with infection by *M. scrofulaceum, M. avium* complex, and rarely, by *M. fortuitum* and *M. kansasii.*
3. *Skin ulcers and soft tissue wound infections* which may be associated with infection by *M. ulcerans, M. marinum, M. fortuitum,* and *M. chelonei.*

Laboratory Hazards: The agents may be present in sputa, exudates from lesions, tissues, and in environmental samples (e.g., soil and water). Direct contact of skin or mucous membranes with infectious materials, ingestion, and accidental parenteral inoculation are the primary laboratory hazards associated with clinical materials and cultures. Infectious aerosols, created during the manipulation of broth cultures or tissue homogenates of these organisms associated with pulmonary disease, also pose a potential infection hazard to laboratory personnel.

Recommended Precautions: Biosafety Level 2 practices, containment equipment and facilities are recommended for activities with clinical materials and cultures of *Mycobacterium* spp. other than *M. tuberculosis* or *M. bovis.* Animal Biosafety Level 2 practices, containment equipment and facilities are recommended for animal studies with mycobacteria other than *M. tuberculosis, M. bovis,* or *M. leprae.*

Agent: ***Mycobacterium tuberculosis, M. bovis***

Mycobacterium tuberculosis and *M. bovis* infections are a proven hazard to laboratory personnel as well as others who may be exposed to infectious aerosols in the laboratory.[79,127,131,151,154] The incidence of tuberculosis in laboratory personnel working with *M. tuberculosis* has been reported to be three times higher than those not working with the agent.[156] Naturally or experimentally infected nonhuman primates are a proven source of human infection (e.g., the annual tuberculin conversion rate in personnel working with infected nonhuman primates is about 70/10,000 compared with less than 3/10,000 in the general population).[102] Experimentally infected guinea pigs or mice do not pose the same problem since droplet nuclei are not produced by coughing in these species; however, litter from infected animals may become contaminated and serve as a source of infectious aerosols.

Laboratory Hazards: Tubercle bacilli may be present in sputum, gastric lavage fluids, cerebrospinal fluid, urine, and in lesions from a variety of tissues.[6] Exposure to laboratory-generated aerosols is the most important hazard encountered. Tubercle bacilli may survive in heat-fixed smears[1], and may be aerosolized in the preparation of frozen sections and during manipulation of liquid cultures. Because of the low infective dose of *M. tuberculosis* for humans (i.e., ID_{50} <10 bacilli)[160,161] and in some laboratories a high rate of isolation of acid-fast organisms from clinical specimens (>10%),[77] sputa and other clinical specimens from suspected or known cases of tuberculosis must be considered potentially infectious and handled with appropriate precautions.

Recommended Precautions: Biosafety Level 2 practices, containment equipment and facilities are required for activities at American Thoracic Society (ATS) laboratory level I,[3,4] preparation of acid-fast smears, and culturing of sputa or other clinical specimens, provided that aerosol generating manipulations of such specimens are conducted in a Class I or II biological safety cabinet. Liquification and concentration of sputa for acid-fast staining may also be conducted safely on the open bench by first treating the specimen (in a Class I or II safety cabinet) with an equal volume of 5% sodium hypochlorite solution (undiluted household bleach) and waiting 15 minutes before centrifugation.[142,174]

Biosafety Level 3 practices, containment equipment and facilities are required for laboratory activities of ATS levels II and III)[3,4] in the propagation and manipulation of cultures of *M. tuberculosis* or *M. bovis*, and for animal studies utilizing nonhuman primates experimentally or naturally infected with *M. tuberculosis* or *M. bovis*. Animal studies utilizing guinea pigs or mice can be conducted at Animal Biosafety Level 2. Skin testing with purified protein devivatie (PPD) of previously skin-tested-negative laboratory personnel can be used as a surveillance procedure. A licensed attenuated live vaccine (BCG) is available but is not routinely used in the United States for laboratory personnel.

Agent: ***Neisseria gonorrhoeae***

Four cases of laboratory-associated gonorrhoea have been reported in the United States.[62,151]

Laboratory Hazards: The agent may be present in conjunctival, urethral and cervical exudates, synovial fluid, urine, feces, and cerebrospinal fluid. Accidental parenteral inoculation and direct or indirect contact of mucous membranes with infectious clinical materials are the known primary laboratory hazards. The importance of aerosols is not determined.

Recommended Precautions: Biosafety Level 2 practices, containment equipment and facilities are recommended for all activities involving the use or manipulation of clinical materials or cultures. Gloves should be worn when handling infected laboratory animals and when there is the likelihood of direct skin contact with infectious materials. Additional primary containment and personnel precautions, such as those described for Biosafety Level 3, may be indicated for aerosol or droplet production, and for activities involving production quantities or concentrations of infectious materials. Vaccines are not available for use in humans.

Agent: ***Neisseria meningitidis***

Meningococcal meningitis is a demonstrated but rare hazard to laboratory workers.[8,49,153]

Laboratory Hazards: The agent may be present in pharyngeal exudates, cerebrospinal fluid, blood, and saliva. Parenteral inoculation, droplet exposure of mucous membranes, infectious aerosol and ingestion are the primary hazards to laboratory personnel.

Recommended Precautions: Biosafety Level 2 practices, containment equipment and facilities are recommended for all activities utilizing known or potentially infectious body fluids, tissues, and cultures. Additional primary containment and personnel precautions such as those described for Biosafety Level 3, may be indicated for activities with high potential for droplet or aerosol production, and for activities involving production quantities or concentrations of infectious materials. The use of licensed polysaccharide vaccines[27] should be considered for personnel regularly working with large volumes or high concentrations of infectious materials.

Agent: ***Pseudomonas pseudomallei***

Two laboratory-associated cases of melioidosis are reported: one associated with a massive aerosol and skin exposure;[78] the second resulting from an aerosol created during the open-flask sonication of a culture presumed to be *Ps. cepacia.*[166a]

Laboratory Hazards: The agent may be present in sputum, blood, wound exudates and various tissues depending on the infection's site of localization. Direct contact with cultures and infectious materials from humans, animals, or the environment, ingestion, autoinoculation, and exposure to infectious aerosols and droplets are the primary laboratory hazards. The agent has been demonstrated in blood, sputum, and abscess materials and may be present in soil and water samples from endemic areas.

98

Recommended Precautions: Biosafety Level 2 practices, containment equipment and facilities are recommended for all activities utilizing known or potentially infectious body fluids, tissues, and cultures. Gloves should be worn when handling infected animals, during their necropsy, and when there is the likelihood of direct skin contact with infectious materials. Additional primary containment and personnel precautions, such as those described for Biosafety Level 3, may be indicated for activities with a high potential for aerosol or droplet production, and for activities involving production quantities or concentrations of infectious materials.

Agent: *Salmonella*—all serotypes except *typhi*

Salmonellosis is a documented hazard to laboratory personnel.[80,127,151] Primary reservoir hosts include a broad spectrum of domestic and wild animals, including birds, mammals, and reptiles, all of which may serve as a source of infection to laboratory personnel.

Laboratory Hazards: The agent may be present in feces, blood, urine, and in food, feed, and environmental materials. Ingestion or parenteral inoculation are the primary laboratory hazards. The importance of aerosol exposure is not known. Naturally or experimentally infected animals are a potential source of infection for laboratory and animal care personnel, and for other animals.

Recommended Precautions: Biosafety Level 2 practices, containment equipment and facilities are recommended for activities with clinical materials and cultures known or potentially containing the agents. Animal Biosafety Level 2 practices, containment equipment and facilities are recommended for activities with experimentally or naturally infected animals.

99

Agent: *Salmonella typhi*

Typhoid fever is a demonstrated hazard to laboratory personnel.[13,80,153]

Laboratory Hazards: The agent may be present in feces, blood, gallbladder (bile) and urine. Humans are the only known reservoir of infection. Ingestion or parenteral inoculation of the organism represent the primary laboratory hazards. The importance of aerosol exposure is not known.

Recommended Precautions: Biosafety Level 2 practices, containment equipment and facilities are recommended for all activities utilizing known or potentially infectious clinical materials and cultures. Biosafety Level 3 practices and procedures are recommended for activities likely to generate aerosols or for activities involving production quantities of organisms.

Licensed vaccines, which have been shown to protect 70–90% of recipients, may be a valuable adjunct to good safety practices in personnel regularly working with cultures or clinical materials which may contain *S. typhi.*[13]

Agent: *Shigella* spp.

Shigellosis is a demonstrated hazard to laboratory personnel with dozens of cases reported in the United States and Great Britain alone.[79,80,97,151] While outbreaks have occurred in captive nonhuman primates, humans are the only significant reservoir of infection. However, experimentally infected guinea pigs, other rodents, and nonhuman primates are also proven sources of infection.

Laboratory Hazards: The agent may be present in feces, and rarely, in blood of infected humans or animals. Ingestion or parenteral inoculation of the agent are the primary laboratory hazards. The oral 25%–50% infectious dose of *S. flexneri* for humans is approximately 200 organisms.[191] The importance of aerosol exposure is not known.

100

Recommended Precautions: Biosafety Level 2 practices, containment equipment and facilities are recommended for all activities utilizing known or potentially infectious clinical materials or cultures. Animal Biosafety Level 2 facilities and practices are recommended for activities with experimentally or naturally infected animals. Vaccines are not available for use in humans.

Agent: *Treponema pallidum*

Syphilis is a documented hazard to laboratory personnel who handle or collect clinical material from cutaneous lesions. Pike lists 20 cases of laboratory-associated infection.[151] Humans are the only known natural reservoir of the agent. Syphilis has been transmitted to laboratory personnel working with a concentrated suspension of *T. pallidum* obtained from an experimental rabbit orchitis.[74] Hematogenous transfer of syphilis has occurred from the transfusion of a unit of fresh blood obtained from a patient with secondary syphilis. *T. pallidum* is present in the circulation during primary and secondary syphilis. The minimum number (LD_{50}) of *T. pallidum* organisms needed to infect by subcutaneous injection is 23.[114] The concentration of *T. pallidum* in patients' blood during early syphilis, however, has not been determined.

No cases of laboratory animal-associated infections are reported; however, rabbit-adapted strains of *T. pallidum* (Nichols and possibly others) retain their virulence for humans.

Laboratory Hazards: The agent may be present in materials collected from primary and secondary cutaneous and mucosal lesions and in blood. Accidental parenteral inoculation, contact of mucous membranes or broken skin with infectious clinical materials, and possibly infectious aerosols, are the primary hazards to laboratory personnel.

Recommended Precautions: Biosafety Level 2 practices, containment equipment and facilities are recommended for all activities involving the use or manipulation of blood or lesion materials from humans or infected rabbits. Gloves should be worn when there is a likelihood of direct skin contact with lesion materials. Periodic serological monitoring should be considered in personnel regularly working with infectious materials. Vaccines are not available for use in humans.

Agent: Vibrionic enteritis *(Vibrio cholerae, V. parahaemolyticus)*

Vibrionic enteritis due to *Vibrio cholerae* or *Vibrio parahaemolyticus* is a documented but rare cause of laboratory-associated illness.[153] Naturally and experimentally infected animals are a potential source of infection.

Laboratory Hazards: All pathogenic vibrios may occur in feces. Ingestion of *V. cholerae,* and ingestion or parenteral inoculation of other vibrios constitute the primary laboratory hazard. The human oral infecting dose of *V. cholerae* in healthy non-achlorhydric individuals is approximately 10^6 organisms.[111] The importance of aerosol exposure is not known. The risk of infection following oral exposure may be increased in achlorhydric individuals.

Recommended Precautions: Biosafety Level 2 practices, containment equipment and facilities are recommended for activities with cultures or potentially infectious clinical materials. Animal Biosafety Level 2 practices, containment equipment and facilities are recommended for activities with naturally or experimentally infected animals. Although vaccines have been shown to provide partial protection of short duration (3–6 months) to nonimmune individuals in highly endemic areas,[13] the routine use of cholera vaccine in laboratory staff is not recommended.

Agent: *Yersinia pestis*

Plague is a proven but rare laboratory hazard. Four cases have been reported in the United States.[17,151]

Laboratory Hazards: The agent may be present in bubo fluid, blood, sputum, cerebrospinal fluid (CSF), feces, and urine from humans, depending on the clinical form and stage of the disease. Primary hazards to laboratory personnel include: direct contact with cultures and infectious materials from humans or rodents; infectious aerosols or droplets generated during the manipulation of cultures, infected tissues, and in the necropsy of rodents; accidental autoinoculation; ingestion; and bites by infected fleas collected from rodents.

Recommended Precautions: Biosafety Level 2 practices, containment equipment and facilities are recommended for all activities involving the handling of potentially infectious clinical materials and cultures. Special care should be taken to avoid the generation of aerosols from infectious materials, and during the necropsy of naturally or experimentally infected rodents. Gloves should be worn when handling field-collected or infected laboratory rodents, and when there is the likelihood of direct skin contact with infectious materials. Necropsy of rodents is ideally conducted in a biological safety cabinet. Although field trials have not been conducted to determine the efficacy of a licensed inactivated vaccine, experience with this product has been favorable.[22] Immunization is recommended for personnel working regularly with cultures of *Y. pestis* or infected rodents.[34]

Additional primary containment and personnel precautions, such as those described for Biosafety Level 3, are recommended for activities with high potential for droplet or aerosol production, for work with antibiotic-resistant strains and for activities involving production quantities or concentrations of infectious materials.

Rickettsial Agents

Agent: *Coxiella burnetii*

Of the rickettsial agents, *Coxiella burnetii* probably presents the greatest risk of laboratory infection. The organism is highly infectious and remarkably resistant to drying and environmental conditions.[190] The infectious dose of virulent, Phase I organisms in laboratory animals has been calculated to be as small as a single organism. The estimated human ID_{25-50} (inhalation) for Q fever is 10 organisms.[191] Pike's summary indicates that Q fever is the second most commonly reported laboratory-associated infection, with outbreaks involving 15 or more persons recorded in several institutions.[141,151] A broad range of domestic and wild mammals are natural hosts for Q fever, and may serve as potential sources of infection for laboratory and animal care personnel. Exposure to naturally infected, often asymptomatic sheep, and to their birth products, is a documented hazard to personnel.[28,175] Although rare, *C. burnetii* is known to cause chronic infections such as endocarditis or granulomatous hepatitis. Genetic analyses[165] as well as structural analysis of the lipopolysaccharides[82] of endocarditis-associated isolates of *C. burnetii* suggest that specific strains may be associated with endocarditis.

Laboratory Hazards: The necessity of using embryonated eggs or cell culture techniques for the propagation of *C. burnetii* leads to extensive purification procedures. Exposure to infectious aerosols or parenteral inoculation are the most likely sources of infection to laboratory and animal care personnel.[141] The agent may be present in infected arthropods, and in the blood, urine, feces, milk, and tissues of infected animal or human hosts. The placenta of infected sheep may contain as many as 10^9 organisms per gram of tissue[195] and milk may contain 10^5 organisms per gram.

Recommended Precautions: Biosafety Level 2 practices and facilities are recommended for nonpropagative laboratory procedures including serological examinations and staining of impression smears. Biosafety Level 3 practices and facilities are recommended for activities involving the inoculation, incubation, and harvesting of embryonated eggs or cell cultures, the necropsy of infected animals and the manipulation of infected tissues. Since infected guinea pigs and other rodents may shed the organisms in urine or feces,[151] experimentally infected rodents should be maintained under Animal Biosafety Level 3.

Recommended precautions for facilities using sheep as experimental animals are described by

104

Spinelli[175] and by Bernard.[11] An investigational new Phase I Q fever vaccine (IND) is available from the Special Immunizations Program, U.S. Army Medical Research Institute for Infectious Diseases (USAMRIID), Fort Detrick, Maryland. The use of this vaccine should be limited to those at high risk of exposure and who have no demonstrated sensitivity to Q fever antigen. Individuals with valvular heart disease should not work with *C. burnetii.*

Agent: *Rickettsia prowazekii, Rickettsia typhi (R. mooseri), Rickettsia tsutsugamushi, Rickettsia canada,* and Spotted Fever Group agents of human disease; *Rickettsia rickettsii, Rickettsia conorii, Rickettsia akari, Rickettsia australis, Rickettsia siberica*

Pike reported 57 cases of laboratory-associated typhus (type not specified), 56 cases of epidemic typhus with 3 deaths, and cases of murine typhus.[151] More recently 3 cases of murine typhus were reported from a research facility.[26] Two of these 3 cases were associated with handling of infectious materials on the open bench; the third case resulted from an accidental parenteral inoculation. These 3 cases represented an attack rate of 20% in personnel working with infectious materials.

Rocky Mountain spotted fever is a documented hazard to laboratory personnel. Pike[151] reported 63 laboratory-associated cases, 11 of which were fatal. Oster[144] reported 9 cases occurring over a 6-year period in one laboratory, which were believed to have been acquired as a result of exposure to infectious aerosols.

Laboratory Hazards: Accidental parenteral inoculation and exposure to infectious aerosols are the most likely sources of laboratory-associated infection.[89] Successful aerosol transmission of *R. rickettsii* has been experimentally documented in nonhuman primates.[166] Five cases of rickettsial pox recorded by Pike[151] were associated with exposure to bites of infected mites.

105

Naturally and experimentally infected mammals, their ectoparasites, and their infected tissues are potential sources of human infection. The organisms are relatively unstable under ambient environmental conditions.

Recommended Precautions: Biosafety Level 2 practices and facilities are recommended for nonpropagative laboratory procedures, including serological and fluorescent antibody procedures, and for the staining of impression smears. Biosafety Level 3 practices and facilities are recommended for all other manipulations of known or potentially infectious materials including necropsy of experimentally infected animals and trituration of their tissues, and inoculation, incubation, and harvesting of embryonated eggs or tissue cultures. Animal Biosafety Level 2 practices and facilities are recommended for the holding of experimentally infected mammals other than arthropods.

Level 3 practices and facilities are recommended for animal studies with arthropods naturally or experimentally infected with rickettsial agents of human disease.

Because of the proven value of antibiotic therapy in the early stages of infection, it is essential that laboratories working with rickettsiae have an effective system for reporting febrile illnesses in laboratory personnel, medical evaluation of potential cases and, when indicated, institution of appropriate antibiotic therapy. Vaccines are not currently available for use in humans (see Appendix C).

Viral Agents (Other than arboviruses)

Agent: Hepatitis A Virus, Hepatitis E Virus

Laboratory-associated infections with hepatitis A or E viruses do not appear to be an important occupational risk among laboratory personnel. However, the disease is a documented hazard in animal handlers and others working with chimpanzees and other nonhuman primates which are naturally or experimentally infected.[153] Hepatitis E virus appears to be less of a risk to personnel than hepatitis A virus, except during pregnancy, when infection can result in severe or fatal disease. Workers handling other recently captured, susceptible primates (owl monkeys, marmosets) may also be at risk.

106

Laboratory Hazards: The agents may be present in feces, saliva, and blood of infected humans and nonhuman primates. Ingestion of feces, stool suspensions, and other contaminated materials is the primary hazard to laboratory personnel. The importance of aerosol exposure has not been demonstrated. Attenuated or avirulent strains of hepatitis A viruses have been described resulting from serial passage in cell culture.

Recommended Precautions: Biosafety Level 2 practices, safety equipment, and facilities are recommended for activities with known or potentially infected feces from humans or nonhuman primates. Animal Biosafety Level 2 practices and facilities are recommended for activities using naturally or experimentally infected nonhuman primates. Animal care personnel should wear gloves and take other appropriate precautions to avoid possible fecal-oral exposure. A licensed inactivated vaccine against hepatitis A is available in Europe; it is available as an investigational vaccine in the U.S., and is recommended for laboratory personnel. Vaccines against hepatitis E are not available for use in humans.

Agent: Hepatitis B Virus, Hepatitis C Virus (formerly known as nonA nonB Virus), Hepatitis D Virus

Hepatitis B has been one of the most frequently occurring laboratory-associated infections,[153] and laboratory workers are recognized as a high risk group for acquiring such infections.[170] Individuals who are infected with hepatitis B virus are at risk of infection with hepatitis D (delta) virus, which is defective and requires the presence of hepatitis B virus for replication.

107

Hepatitis C infection can occur in the laboratory situation. The prevalence of antibody to hepatitis C is slightly higher in medical care workers than in the general population. Epidemiologic evidence indicates that hepatitis C is spread predominantly by the parenteral route.[94,128]

Laboratory Hazards: Hepatitis B virus may be present in blood and blood products of human origin, in urine, semen, cerebrospinal fluid, and saliva. Parenteral inoculation, droplet exposure of mucous membranes, and contact exposure of broken skin are the primary laboratory hazards. The virus may be stable in dried blood or blood components for several days. Attenuated or avirulent strains have not been identified.

Hepatitis C virus has been detected primarily in blood and serum, less frequently in saliva and rarely or not at all in urine or semen. It appears to be relatively unstable to storage at room temperature, repeated freezing and thawing, etc.

Recommended Precautions: Biosafety Level 2 practices, containment equipment and facilities are recommended for all activities utilizing known or potentially infectious body fluids and tissues. Additional primary containment and personnel precautions, such as those described for Biosafety Level 3, may be indicated for activities with potential for droplet or aerosol production and for activities involving production quantities or concentrations of infectious materials. Animal Biosafety Level 2 practices, containment equipment and facilities are recommended for activities utilizing naturally or experimentally infected chimpanzees or other nonhuman primates. Gloves should be worn when working with infected animals and when there is the likelihood of skin contact with infectious materials. Licensed recombinant vaccines against hepatitis B are available and are highly recommended for laboratory personnel.[46] Vaccines against hepatitis C and D are not yet available for use in humans.

108

Agent: *Herpesvirus simiae* (B-virus)

B-virus is a naturally occurring alphaherpesvirus infecting free-living or captive *Macaca mulatta, M. fasicularis,* and other members of the genus. It is associated with acute vesicular oral lesions, as well as latent and often recrudescent infection.[193] Human infection has been documented in 25 instances, usually with a lethal outcome or serious sequelae from encephalitis.[7,92,145]

Although B-virus presents a potential hazard to laboratory personnel working with the agent, laboratory-associated human infections with B-virus have, with rare exceptions, been limited to those having direct contact with macaques. Primary macaque cell cultures, including commercially-prepared rhesus monkey kidney cells, occasionally are inapparently infected with B-virus and have been implicated in one human case.[92]

Sixteen fatal cases of human infections with B-virus have been reported.[193]

Laboratory Hazards: The highest risk of acquiring B-virus from macaques is through the bite of an infected monkey with active lesions. Contamination of broken skin or mucous membranes with oral, ocular, or genital secretions from animals with lesions, or experiencing clinically silent virus shedding, is also dangerous. Stability of viral infectivity on cages and other surfaces is not known, but the potential hazard must be recognized. The importance of aerosol exposure is thought to be minimal. Attenuated or avirulent strains have not been identified.

The agent also may be present in thoracic and abdominal viscera and nervous tissues of naturally infected macaques. These tissues, and the cultures prepared from them, are potential hazards.[196]

Recommended Precautions: Biosafety Level 2 practices and facilities are recommended for all activities involving the use or manipulation of tissues, body fluids, and primary tissue culture materials from macaques. Additional practices and personnel precautions, such as those detailed for Biosafety Level 3, are recommended for activities involving the use or manipulation of any material known to contain *Herpesvirus simiae. In vitro* propagation of the virus for diagnosis may be conducted under the same guidelines, but it would be prudent to confine manipulations of positive cultures which would contain high-titered virus to a Class 3 biosafety cabinet or BSL-4 facility, depending on the judgement of the laboratory director.

Biosafety Level 4 practices and facilities are recommended for activities involving the propagation and manipulation of production quantities or concentrates of *H. simiae.*

The wearing of gloves, masks, and laboratory coats is recommended for all personnel working with non-human primates—especially macaques and other Old World species—and for all persons entering animal rooms where non-human primates are housed. Any macaque colony not known to be free of B-virus infection should be presumed to be naturally infected. Guidelines are available for safely working with macaques and should be consulted.[39,150] Animals with oral lesions suggestive of active B-virus infection should be isolated and handled with extreme caution. Studies with animals experimentally infected with *H. simiae* should be conducted at ABSL-3.

Vaccines are not available for use in humans. Human to human transmission has only occurred in one case, suggesting that precautions should be taken with vesicle fluids, oral secretions, and conjunctival secretions of infected persons.[92]

Antiviral drugs have shown promise in the therapy of rabbits infected with *H. simiae,* and anecdotal observations suggest this may extend to man.[7,92] Because of the seriousness of infection with this virus, experienced medical personnel should be available for consultation to manage incidents involving exposure to the agent or suspected infections.

Agent: Human Herpesviruses

The herpesviruses are ubiquitous human pathogens and are commonly present in a variety of clinical materials submitted for virus isolation. While few of these viruses are demonstrated causes of clinical laboratory-associated infections, they are primary as well as opportunistic pathogens, especially in immunocompromised hosts. Herpes simplex viruses 1 and 2 and varicella virus pose some risk via direct contact and/or aerosols; cytomegalovirus and Epstein-Barr virus pose relatively low infection risks to laboratory personnel. The risk of laboratory infection from herpesviruses 6 and 7 is not known. Although this diverse group of indigenous viral agents does not meet the criteria for inclusion in agent-specific summary statements (i.e., demonstrated or high potential hazard for laboratory-associated infection; grave consequences should infection occur), the frequency of their presence in clinical materials and their common use in research warrants their inclusion in this publication.

Laboratory Hazards: Clinical materials and isolates of herpesviruses may pose a risk of infection following ingestion, accidental parenteral inoculation, droplet exposure of the mucous membranes of the eyes, nose, or mouth, or inhalation of concentrated aerosolized materials. Clinical specimens containing the more virulent *Herpesvirus simiae* (B-virus) may be inadvertently submitted for diagnosis of suspected herpes simplex infection. This virus has also been found in cultures of primary rhesus monkey kidney cells. Cytomegalovirus may pose a special risk during pregnancy because of potential infection of the fetus.

Recommended Precautions: Biosafety Level 2 practices, containment equipment, and facilities are recommended for activities utilizing known or potentially infectious clinical materials or cultures of indigenous viral agents which are associated or identified as a primary pathogen of human disease. Although there is little evidence that infectious aerosols are a significant source of laboratory-associated infections, it is prudent to avoid the generation of aerosols during the handling of clinical materials or isolates, or during the necropsy of animals. Primary containment devices (e.g., biological safety cabinets) constitute the basic barrier protecting personnel from exposure to infectious aerosols.

Agent: Influenza

Laboratory-associated infections with influenza are not normally documented in the literature, but are known to occur by informal accounts and published reports, particularly when new strains showing antigenic drift or shift are introduced into a laboratory for diagnostic/research purposes.[64]

Laboratory animal-associated infections are not reported; however, there is a high possibility of human infection from infected ferrets and vice-versa.

Laboratory Hazards: The agent may be present in respiratory tissues or secretions of humans or most infected animals, and in the cloaca of many infected avian species. The virus may be disseminated in multiple organs in some infected animal species.

The primary laboratory hazard is inhalation of virus from aerosols generated by infected animals, or by aspirating, dispensing, or mixing virus-infected samples. Genetic manipulation has the potential for altering the host range, pathogenicity, and antigenic composition of influenza viruses. There is unknown potential for introducing into man transmissible viruses with novel antigenic composition.

Recommended Precautions: Biosafety Level 2 practices and facilities are recommended when receiving and inoculating routine laboratory diagnostic specimens. Autopsy material should be handled in a biological safety cabinet using Biosafety Level 2 procedures.

Activities Utilizing Noncontemporary Virus Strains: Biosafety considerations should take into account the available information about infectiousness and virulence of the strains being used, and the potential for harm to the individual or society in the event that laboratory-acquired infection and subsequent transmission occurs. Research or production activities utilizing contemporary strains may be safely performed using Biosafety Level 2 containment practices. Susceptibility to infection with older noncontemporary human strains, with recombinants, or with animal isolates warrant the use of Biosafety Level 2 containment procedures. However, there is no evidence for laboratory-acquired infection with reference strains A/PR/8/34 and A/WS/33, or its commonly used neurotropic variants.

Agent: Lymphocytic Choriomeningitis Virus

Laboratory-associated infections with LCM virus are well documented in facilities where infections occur in laboratory rodents—especially mice, hamsters and guinea pigs.[14,98,151] Nude and SCID mice may pose a special risk of harboring silent chronic infections. Cell cultures that inadvertently have become infected represent a potential source of infection and dissemination of the agent. Natural infections are found in nonhuman primates, including macaques and marmosets (*Callitrichid* hepatitis virus is a lymphocytic choriomeningitis virus) and may be fatal to marmoset monkeys. Swine and dogs are less important vectors.

Laboratory Hazards: The agent may be present in blood, cerebrospinal fluid, urine, secretions of the nasopharynx, feces and tissues of infected animal hosts and possibly man. Parenteral inoculation, inhalation, contamination of mucous membranes or broken skin with infectious tissues or fluids from infected animals, are common hazards. Aerosol transmission is well documented.[14] The virus may pose a special risk during pregnancy because of potential infection of the fetus.

Recommended Precautions: Biosafety Level 2 practices and facilities are suitable for activities utilizing known or potentially infectious body fluids, and for tissue culture passage of laboratory-adapted, mouse brain-passaged strains. Animal Biosafety Level 2 practices and facilities are suitable for studies in adult mice with mouse brain-passaged strains. However, additional primary containment and personnel precautions, such as those described for Biosafety Level 3, are indicated for activities with high potential for aerosol production, or involving production quantities or concentrations of infectious materials; and for manipulation of infected transplantable tumors, field isolates and clinical materials from human cases. Animal Biosafety Level 3 practices and facilities are recommended for work with infected hamsters. Vaccines are not available for use in humans.[98]

Agent: Poliovirus

Laboratory-associated infections with polioviruses are uncommon and have been limited to unvaccinated laboratory personnel working directly with the agent.[151] Laboratory animal-associated infections have not been reported;[23] however, naturally or experimentally infected nonhuman primates could provide a source of infection to exposed unvaccinated persons. Transgenic mice expressing the human receptor for polioviruses can be infected with virulent polioviruses and are a potential source of human infection.

Laboratory Hazards: The agent is present in the feces and in throat secretions of infected persons. Ingestion or parenteral inoculation of infectious tissues or fluids by non-immunized personnel are the primary hazards in the laboratory. The importance of aerosol exposure is not known; it has not been reported as a hazard. Laboratory exposures pose negligible risk to appropriately immunized persons.

Recommended Precautions: Biosafety Level 2 practices and facilities are recommended for all activities utilizing known or potentially infectious culture fluids and clinical materials involving known or suspected wild-type strains. All laboratory personnel working directly with the agent must have documented polio vaccination or demonstrated serologic evidence of immunity to all three poliovirus types.[33] Animal Biosafety Level 2 practices and facilities are recommended for studies of virulent viruses in animals. Unless there are strong scientific reasons for working with virulent polioviruses (which have been eradicated from the United States), laboratories should use the attenuated Sabin oral poliovirus vaccine strains. These pose no significant risk to immunized laboratory personnel.

Agent: Poxviruses

Sporadic cases of laboratory-associated infections with pox viruses (smallpox, vaccinia, yaba, tanapox) have been reported.[151] Epidemiological evidence suggests that transmission to humans of monkeypox virus from nonhuman primates or rodents to humans may have occurred in nature, but not in the laboratory setting. Naturally or experimentally infected laboratory animals are a potential source of infection to exposed unvaccinated laboratory personnel. Genetically engineered recombinant vaccinia viruses pose an additional potential risk to laboratory personnel, through direct contact or contact with clinical materials from infected volunteers or animals.

Laboratory Hazards: The agents may be present in lesion fluids or crusts, respiratory secretions, or tissues of infected hosts. Ingestion, parenteral inoculation, and droplet or aerosol exposure of mucous membranes or broken skin with infectious fluids or tissues, are the primary hazards to laboratory and animal care personnel. Some poxviruses are stable at ambient temperature when dried and may be transmitted by fomites.

Recommended Precautions: The possession and use of variola viruses is restricted to the World Health Organization Collaborating Center for Smallpox and Other Poxvirus Infections, located at the Centers for Disease Control and Prevention, Atlanta, Georgia. Biosafety Level 2 practices and facilities are recommended for all activities involving the use or manipulation of poxviruses, other than variola, that pose an infection hazard to humans. All persons working in or entering laboratory or animal care areas where activities with vaccinia, monkey pox, or cow pox viruses are being conducted should have documented evidence of satisfactory vaccination within the preceding ten years.[31,52] Activities with vaccinia, cow pox, or monkey pox viruses, in quantities or concentrations greater than those present in diagnostic cultures, may also be conducted at Biosafety Level 2 by immunized personnel, provided that all manipulations of viable materials are conducted in Class I or II biological safety cabinets or other primary containment equipment. Immunosuppressed individuals are at greater risk of severe disease if infected with a poxvirus.[52]

Agent: Rabies Virus

Laboratory-associated infections are extremely rare. Two have been documented. Both resulted from presumed exposure to high titered infectious aerosols generated in a vaccine production facility[201] and a research facility,[25] respectively. Naturally or experimentally infected animals, their tissues, and their excretions are a potential source of exposure for laboratory and animal care personnel.

Laboratory Hazards: The agent may be present in all tissues of infected animals. Highest titers are present in CNS tissue, salivary glands, and saliva. Accidental parenteral inoculation, cuts, or sticks with contaminated laboratory equipment, bites by infected animals, and exposure of mucous membranes or broken skin to infectious tissue or fluids, are the most likely sources for exposure of laboratory and animal care personnel. Infectious aerosols have not been a demonstrated hazard to personnel working with clinical materials and conducting diagnostic examinations. Fixed and attenuated strains of virus are presumed to be less hazardous, but the only two recorded cases of laboratory associated rabies resulted from exposure to a fixed Challenge Virus Standard (CVS) and an attenuated strain derived from SAD (Street Alabama Dufferin) strain, respectively.[25,201]

Recommended Precautions: Biosafety Level 2 practices and facilities are recommended for all activities utilizing known or potentially infectious materials. Immunization is recommended for all individuals prior to working with rabies virus or infected animals, or engaging in diagnostic, production, or research activities with rabies virus. Immunization is also recommended for all individuals entering or working in the same room where rabies virus or infected animals are used. While it is not always feasible to open the skull or remove the brain of an infected animal within a biological safety cabinet, it is pertinent to wear heavy protective gloves to avoid cuts or sticks from cutting instruments or bone fragments, and to wear a face shield to protect the mucous membranes of the eyes, nose, and mouth from exposure to infectious droplets or tissue fragments. If a Stryker saw is used to open the skull, avoid contacting the brain with the blade of the saw. Additional primary containment and personnel precautions, such as those described for Biosafety Level 3, may be indicated for activities with a high potential for droplet or aerosol production, and for activities involving production quantities or concentrations of infectious materials.

Agent: Retroviruses, including Human and Simian Immunodeficiency Viruses (HIV and SIV)

Data on occupational HIV transmission in laboratory workers are collected through two CDC-supported national surveillance systems: surveillance for 1) AIDS and 2) HIV-infected persons who may have acquired their infection through occupational exposures. For surveillance purposes, laboratory workers are defined as those persons, including students and trainees, who have worked in a clinical or HIV laboratory setting at anytime since 1978. Persons reported from these two systems are classified as cases of either documented or possible occupational transmission. Those classified as documented occupational transmission had evidence of HIV seroconversion (a negative HIV-antibody test at the time of the exposure which converted to positive) following a discrete percutaneous or mucocutaneous occupational exposure to blood, body fluids, or other clinical or laboratory specimens. Those persons classified as possible occupational transmission do not have behavioral or transfusion risks for HIV infection which could be identified during follow-up investigation; each reported past percutaneous or mucocutaneous occupational exposures to blood, body fluids, or laboratory specimens, but seroconversion to HIV was not documented.

As of September 30, 1992, CDC had reports of 12 (11 clinical and 1 nonclinical) laboratory workers in the United States with documented occupational transmission and 13 (12 clinical and 1 nonclinical) classified as possible occupational transmission.[55]

Among those with documented occupational transmission, 8 had percutaneous exposure, 3 had mucocutaneous exposure, and 1 had both percutaneous and mucocutaneous exposures. Eleven were exposed to HIV-infected blood and one to concentrated live HIV. The procedure most often associated with transmission was phlebotomy. Three of these workers have developed AIDS.

Of the 13 laboratory workers classified as possible occupational transmission, 9 have developed AIDS.

In 1992, two workers were reported to have developed antibodies to simian immunodeficiency virus (SIV) following exposures in different laboratories. One was associated with a needle stick which occurred while the worker was manipulating a blood-contaminated needle after bleeding an SIV-infected macaque monkey.[103] The other involved a laboratory worker who handled macaque SIV-infected blood specimens without gloves. Though no specific incident was recalled, this worker had dermatitis on the forearms and hands while working with the infected blood specimens.[53] As of October 1992 neither of the two workers had developed any illness.

Laboratory Hazards: HIV has been isolated from blood, semen, saliva, tears, urine, cerebrospinal fluid, amniotic fluid, breast milk, cervical secretion, and tissue of infected persons and experimentally infected nonhuman primates.[167] CDC has recommended that blood and body fluid precautions be used consistently when handling any blood-contaminated specimens.[43,45] This approach, referred to as "universal precautions," precludes the need to identify clinical specimens obtained from HIV$^+$ patients or to speculate as to the HIV status of a specimen.

Although the risk of occupationally acquired HIV is primarily through exposure to infected blood, it is also prudent to wear gloves when manipulating other body fluids such as feces, saliva, urine, tears, sweat, vomitus, and human breast milk. This reduces the potential for exposure to other microorganisms that may cause other types of infections.

In the laboratory, virus should be presumed to be present in all blood or clinical specimens contaminated with blood, in any unfixed tissue or organ (other than intact skin) from a human (living or dead), in HIV cultures, in all materials derived from HIV cultures, and in/on all equipment and devices coming into direct contact with any of these materials.

SIV has been isolated from blood, cerebrospinal fluid, and a variety of tissues of infected nonhuman primates. Limited data exist on the concentration of virus in semen, saliva, cervical secretions, urine, breast milk, and amniotic fluid. In the laboratory, virus should be presumed to be present in all SIV cultures, in animals experimentally infected or inoculated with SIV, in all materials derived from HIV or SIV cultures, and in/on all equipment and devices coming into direct contact with any of these materials.[44]

In the laboratory, the skin (especially when scratches, cuts, abrasions, dermatitis, or other lesions are present) and mucous membranes of the eye, nose, and mouth should be considered as potential pathways for entry of these retroviruses. Whether infection can occur via the respiratory tract is unknown. Needles, sharp instruments, broken glass, and other sharp objects must be carefully handled and properly discarded. Care must be taken to avoid spilling and splashing infected cell-culture liquid and other virus-containing or potentially-infected materials.

Recommended Precautions:

In addition to these recommended precautions, persons working with HIV, SIV, or other bloodborne pathogens should consult the OSHA Bloodborne Pathogen Standard.[187] *Questions related to interpretation of this Standard should be directed to Federal, regional or state OSHA offices.*

1. BSL-2 standard and special practices, containment equipment and facilities are recommended for activities involving *all* blood-contaminated clinical specimens, body fluids and tissues from *all* humans or from HIV- or SIV-infected or inoculated laboratory animals.

2. Activities such as producing research-laboratory-scale quantities of HIV or SIV, manipulating concentrated virus preparations, and conducting procedures that may produce droplets or aerosols, are performed in a BSL-2 facility, but using the additional practices and containment equipment recommended for BSL-3.

3. Activities involving industrial-scale volumes or preparation of concentrated HIV or SIV are conducted in a BSL-3 facility, using BSL-3 practices and containment equipment.

4. Nonhuman primates or other animals infected with HIV or SIV are housed in ABSL-2 facilities using ABSL-2 special practices and containment equipment.

Additional Comments:

1. There is no evidence that laboratory clothing poses a risk for retrovirus transmission; however, clothing that becomes contaminated with HIV or SIV preparations should be decontaminated before being laundered or discarded. Laboratory personnel must remove laboratory clothing before going to non-laboratory areas.

2. Work surfaces are decontaminated with an appropriate chemical germicide after procedures are completed, when surfaces are overtly contaminated, and at the end of each work day. Many commercially available chemical disinfectants[70,116,158,164,188] can be used for decontaminating laboratory work surfaces and some laboratory instruments, for spot cleaning of contaminated laboratory clothing, and for spills of infectious materials. Prompt decontamination of spills should be standard practice.

3. Human serum from any source that is used as a control or reagent in a test procedure should be handled at BSL-2.

4. It is recommended that all institutions establish written policies regarding the management of laboratory exposure to HIV and SIV in conjunction with applicable federal, state, and local laws. Such policies should consider confidentiality, consent for testing, administration of appropriate prophylactic drug therapy,[48] counseling, and other related issues. If a laboratory worker has a parenteral or mucous-membrane exposure to blood, body fluid, or viral-culture material, the source material should be identified and, if possible, tested for the presence of virus. If the source material is positive for HIV antibody, virus, or antigen, or is not available for examination, the worker should be counseled regarding the risk of infection and should be evaluated clinically and serologically for evidence of HIV infection. The worker should be advised to report and seek medical evaluation of any acute febrile illness that occurs within 12 weeks after the exposure.[40] Such an illness—particularly one characterized by fever, rash, or lymphadenopathy—may indicate recent HIV infection. If seronegative, the worker should be retested 6 weeks after the exposure and periodically thereafter (i.e., at 12 weeks and 6, 9, and 12 months after exposure). During this follow-up period exposed workers should be counseled to follow Public Health Service recommendations for preventing transmission of HIV.[37,40,41,48,120]

5. Other primary and opportunistic pathogenic agents may be present in the body fluids and tissues of persons infected with HIV. Laboratory workers should follow accepted biosafety practices to ensure maximum protection against inadvertent laboratory exposure to agents that may also be present in clinical specimens or in specimens obtained from nonhuman primates.[36,38,42]

Research involving other human (i.e., human T-lymphotrophic virus types I and II) and simian retroviruses occurs in many laboratories. Transmission of such viruses has not been reported in the laboratory setting. The precautions outlined above are sufficient while working with these agents.

Between 1989 and 1992, 69 persons with $CD4^+$ T-lymphocyte depletion, but without evident HIV infection, were identified in the United States. This condition has been provisionally termed "Idiopathic $CD4^+$ T-lymphocytopenia" (ICL). To date, investigations of persons with idiopathic $CD4^+$ T-cell depletion indicate that ICL is rare, that these findings may represent various disorders, and in some cases, normal or transient variations in $CD4^+$ T-lymphocyte counts. No epidemiologic or laboratory evidence of a transmissible agent of immunodeficiency has been found as of October 1992.[172a]

Agent: Transmissible Spongiform Encephalopathies (Creutzfeldt-Jakob, kuru and related agents)

Laboratory-associated infections with the transmissible spongiform encephalopathies (prion diseases) have not been documented. However, there is evidence that Creutzfeldt-Jakob disease (CJD) has been transmitted iatrogenically to patients by corneal transplants, dura mater grafts and growth hormone extracted from human pituitary glands, and by exposure to contaminated electroencephalographic electrodes.[99] Infection is always fatal. There is no known nonhuman reservoir for CJD or kuru. Nonhuman primates and other laboratory animals have been infected by inoculation, but there is no evidence of secondary transmission. Scrapie of sheep and goats, bovine spongiform encephalopathy and mink encephalopathy are transmissible spongiform encephalopathies of animals that are similar to the human transmissible diseases. However, there is no evidence that the animal diseases can be transmitted to man.

Laboratory Hazards: High titers of a transmissible agent have been demonstrated in the brain and spinal cord of persons with kuru. In persons with Creutzfeldt-Jakob disease and its Gerstmann-Sträussler-Schenker Syndrome variants, a similar transmissible agent has been demonstrated in the brain, spleen, liver, lymph nodes, lungs, spinal cord, kidneys, cornea and lens, and in spinal fluid and blood. Accidental parenteral inoculation, especially of nerve tissues, including formalin-fixed specimens, is extremely hazardous. Although non-nerve tissues are less often infectious, all tissues of humans and animals infected with these agents should be considered potentially hazardous. The risk of infection from aerosols, droplets, and exposure to intact skin, gastric and mucous membranes is not known; however, there is no evidence of contact or aerosol transmission. These agents are characterized by extreme resistance to conventional inactivation procedures including irradiation, boiling, dry heat and chemicals (formalin, betapropiolactone, alcohols); however, they are inactivated by 1 N NaOH, sodium hypochlorite (≥2% free chlorine concentration) and steam autoclaving at 134°C for 1 hour.

Recommended Precautions: Biosafety Level 2 practices and facilities are recommended for all activities utilizing known or potentially infectious tissues and fluids from naturally-infected humans and from experimentally infected animals. Extreme care must be taken to avoid accidental autoinoculation, or other traumatic parenteral inoculations of infectious tissues and fluids.[76] Although there is no evidence to suggest that aerosol transmission occurs in the natural disease, it is prudent to avoid the generation of aerosols or droplets during the manipulation of tissues or fluids, and during the necropsy of experimental animals. It is further strongly recommended that gloves be worn for activities that provide the opportunity for skin contact with infectious tissues and fluids. Formaldehyde-fixed and paraffin-embedded tissues, especially of the brain, remain infectious. It is recommended that formalin-fixed tissues from suspected cases of transmissible encephalopathy be immersed in 96% formic acid for 30 minutes before histopathologic processing.[15] Vaccines are not available for use in humans.[51]

Agent: Vesicular Stomatitis Virus

A number of laboratory-associated infections with indigenous strains of VSV have been reported.[178] Laboratory activities with such strains present two different levels of risk to laboratory personnel and are related, at least in part, to the passage history of the strains utilized. Activities utilizing infected livestock, their infected tissues, and virulent isolates from these sources are a demonstrated hazard to laboratory and animal care personnel.[85,148] Rates of seroconversion and clinical illness in personnel working with these materials are high.[148] Similar risks may be associated with exotic strains such as Piry.[178]

In contrast, anecdotal information indicates that activities with less virulent laboratory-adapted strains (e.g., Indiana, San Juan, and Glascow) are rarely associated with seroconversion or illness. Such strains are commonly used by molecular biologists, often in large volumes and high concentrations, under conditions of minimal or no primary containment. Some strains of VSV are considered restricted organisms by USDA regulations (9CFR 122.2). Experimentally infected mice have not been a documented source of human infection.

Laboratory Hazards: The agent may be present in vesicular fluid, tissues, and blood of infected animals and in blood and throat secretions of infected humans. Exposure to infectious aerosols, infected droplets, direct skin and mucous membrane contact with infectious tissues and fluids, and accidental autoinoculation, are the primary laboratory hazards associated with virulent isolates. Accidental parenteral inoculation and exposure to infectious aerosols represent potential risks to personnel working with less virulent laboratory-adapted strains.

Recommended Precautions: Biosafety Level 3 practices and facilities are recommended for activities involving the use or manipulation of infected tissues and virulent isolates from naturally or experimentally infected livestock. Gloves and respiratory protection are recommended for the necropsy and handling of 157 infected animals. Biosafety Level 2 práctices and facilities are recommended for activities utilizing laboratory-adapted strains of demonstrated low virulence. Vaccines are not available for use in humans.

124

Agent: Arboviruses

Arboviruses Assigned to Biosafety Level 2. The American Committee on Arthropod-Borne Viruses (ACAV) registered 535 arboviruses as of December 1991. In 1979, the ACAV's Subcommittee on Arbovirus Laboratory Safety (SALS) categorized each of the 424 viruses then registered in the *Catalogue of Arboviruses and Certain other Viruses of Vertebrates*[96] into 1 of 4 recommended practices, safety equipment, and facilities described in this publication as Biosafety Levels 1–4.[178] Since 1980, SALS has periodically updated the 1980 publication by providing a supplemental listing and recommended levels of practice and containment for arboviruses registered since 1979. SALS categorizations were based on risk assessments from information provided by a worldwide survey of 585 laboratories working with arboviruses. SALS recommended that work with the majority of these agents should be conducted at the equivalent of Biosafety Level 2, Table A. SALS also recognizes five commonly used vaccine strains, for which attenuation is firmly established, which may be handled safely at BSL 2, provided that personnel working with these vaccine strains are immunized, Table B. SALS has classified all registered viruses for which insufficient laboratory experience exists as BSL 3, Table C, and reevaluates the classification whenever additional experience is reported.

The viruses classified as BSL 2 are listed alphabetically in Table A, and include the following agents which are reported to cause laboratory-associated infections.[86,151,178]

Virus	Cases
Vesicular stomatitis	46
Colorado tick fever	16
Dengue	11
Pichinde	17
Western equine encephalomyelitis	7 (2deaths)
Rio Bravo	7
Kunjin	6
Catu	5
Caraparu	5
Ross River	5
Bunyamwera	4
Eastern equine encephalomyelitis	4
Zika	4
Apeu	2
Marituba	2
Tacaribe	2
Muructucu	1
O'nyong nyong	1
Modoc	1
Oriboca	1
Ossa	1
Keystone	1
Bebaru	1
Bluetongue	1

125

The result of the SALS survey clearly indicate that the suspected source of the laboratory-associated infections listed above was other than exposure to infectious aerosols. Recommendations that work with the 341 arboviruses, listed in Table A, should be conducted at Biosafety Level 2 was based on the existence of adequate historical laboratory experience to assess risks for the virus which indicate that (a) no overt laboratory-associated infections are reported or (b) infections resulted from exposures other than to infectious aerosols or (c) if disease from aerosol exposure is documented, it is uncommon.

Laboratory Hazards: Agents listed in this group may be present in blood, CSF, central nervous system and other tissues, and infected arthropods, depending on the agent and the stage of infection. While the primary laboratory hazards are accidental parenteral inoculation, contact of the virus with broken skin or mucous membranes, and bites of infected laboratory rodents or arthropods, infectious aerosols may also be a potential source of infection.

126

Recommended Precautions: Biosafety Level 2 practices, safety equipment, and facilities are recommended for activities with potential infectious clinical materials and arthropods and for manipulations of infected tissue cultures, embryonated eggs, and rodents. Infection of newly hatched chickens with eastern and western equine encephalomyelitis viruses is especially hazardous and should be undertaken under Biosafety Level 3 conditions by immunized personnel. Investigational vaccines (IND) against eastern equine encephalomyelitis and western equine encephalomyelitis viruses are available through the Centers for Disease Control and Prevention and the U.S. Army Medical Research Institute for Infectious Diseases (USAMRIID), Fort Detrick, Maryland. The use of these vaccines is recom-

TABLE A Arboviruses and arenaviruses assigned to Biosafety Level 2

Acado
Acara
Aguacate
Alfuy
Almpiwar
Amapari
Ananindeua
Anhanga
Anhembi
Anopheles A
Anopheles B
Apeu
Apoi
Aride
Arkonam
Aroa
Aruac
Arumowot
Aura
Avalon
Abras
Abu Hammad
Aabahoyo
Bagaza
Bahig
Bakau
Baku
Bandia
Bangoran
Bangui
Banzi
Barmah Forest
Barur
Batai
Batama
Bauline
Bebaru
Belmont
Benevides
Benfica
Bertioga
Bimiti
Birao
Bluetongue
Boraceia
Botambi
Boteke
Bouboui
Bujaru
Bunyamwera
Bunyip
Burg E Arab
Bushbush
Bussuquara
Buttonwillow
Bwamba
Cacao
Cache Valley
Caimito
California enc.
Calovo
Candiru
Cape Wrath
Capim
Caraparu
Carey Island
Catu
Chaco
Chagres
Chandipura
Changuinola
Charleville
Chenuda
Chilibre
Chobar gorge
Clo Mor
Colorado tick fever
Corriparta
Cotia
Cowbone Ridge
Csiro Village
Cuiaba-D'aguilar
Dakar Bat
Dengue-1
Dengue-2
Dengue-3
Dengue-4
Dera Ghazi Khan
East. equine enc.[d]
Edge Hill
Entebbe Bat
Ep. Hem. Disease
Erve
Eubenangee
Eyach
Flanders
Fort Morgan
Frijoles
Gamboa
Gan Gan
Gomoka
Gossas
Grand Arbaud
Great Island
Guajara
Guama
Guaratuba
Guaroa
Gumbo Limbo
Hart Park
Hazara
Highlands J
Huacho
Hughes
Icoaraci
Ieri
Ilesha
Ilheus
Ingwavuma
Inkoo
Ippy
Irituia
Isfahan
Itaporanga
Itaqui
Jamestown Canyon
Japanaut
Jerry Slough
Johnston Atoll
Joinjakaka
Juan Diaz
Jugra
Jurona
Jutiapa
Kadam
Kaeng Khoi
Kaikalur
Kaisodi
Kamese
Kammavan pettai
Kannaman galam
Kao Shuan
Karimabad
Karshi
Kasba
Kemerovo
Kern Canyon
Ketapang
Keterah
Keuraliba
Keystone
Kismayo
Klamath
Kokobera
Kolongo
Koongol
Kotonkan
Kowanyama
Kunjin
Kununurra
Kwatta
La Crosse
La Joya
Lagos Bat
Landjia
Langat
Lanjan
Las Maloyas
Latino
Le Dantec
Lebombo
Lednice
Lipovnik
Lokern
Lone Star
Lukuni
M'poko
Madrid
Maguari
Mahogany Hammock
Main Drain
Malakal
Manawa
Manzanilla
Mapputta
Maprik
Marco
Marituba
Marrakai
Matariya
Matruh
Matucare
Melao
Mermet
Minatitlan
Minnal
Mirim
Mitchell River
Modoc
Moju
Mono Lake
Mont. myotis leuk.
Moriche
Mosqueiro
Mossuril
Mount Elgon BatMurutucu
Mykines
Navarro
Nepuyo
Ngaingan
Nique
Nkolbisson
Nola
Ntaya
Nugget
Nyamanini
Nyando
O'nyong-nyong'
Okhotskiy
Okola
Olifantsvlei
Oriboca
Ossa
Pacora
Pacui
Pahayokee
Palyam
Parana
Pata
Pathum Thani
Patois
Phnom-Penh Bat
Pichinde
Pixuna
Pongola
Ponteves
Precarious Point
Pretoria
Prospect Hill
Puchong
Punta Salinas
Punta Toro
Qalyub
Quaranfil
Restan
Rio Bravo
Rio Grande
Ross River
Royal Farm
Sabo
Saboya
Saint Floris
Sakhalin
Salehabad
San angelo
Sandfly f. (Naples)
Sandfly f. (Sicilian)
Sandjimba
Sango
Sathuperi
Sawgrass
Sebokele
Seletar
Sembalam
Serra do Navio
Shamonda
Shark River
Shuni
Silverwater
Simbu
Simian hem. fever
Sindbis
Sixgun City
Snowshoe Hare
Sokuluk
Soldado
Sororoca
Stratford
Sunday Canyon
Tacaiuma
Tacaribe
Taggert
Tahyna
Tamiami
Tanga
Tanjong Rabo
Tataguine
Tehran
Tembe
Tembusu
Tensaw
Tete
Tettnang
Thimiri
Thottapal
ayam
Tibrogargan
Timbo
Timboteua
Tindholmur
Toscana
Toure
Tribec
Triniti
Trivittatus
Trubanaman
Tsuruse
Turlock
Tyuleniy
Uganda S
Umatilla
Umbre
Una
Upolu
Urucuri
Usutu
Uukuniemi
Vellore
Venkatapuram
Vinces
Virgin River
VS-Indiana
VS-New Jersey
Wad Medani
Wallal
Wanowrie
Warrego
West. equine enc.[d]
Whataroa
Witwatersrand
Wonga
Wongorr
Wyeomyia
Yaquinea Head
Yata
Yogue
Zaliv
Terpeniya
Zegla
Zika
Zingilamo
Zirqa

[d]A vaccine is available and is recommended for all persons working with this agent.

TABLE B Vaccine strains of BSL 3/4 viruses which may be handled at BSL 2

Virus	Vaccine strain
Chikungunya	131/25
Junin	Candid #1
Rift Valley fever[20]	MP-12
Venezuelan equine	TC-83
Yellow fever	17-D

mended for personnel who work directly and regularly with these two agents in the laboratory. Western equine encephalomyelitis immune globulin (human) is also available from the Centers for Disease Control and Prevention. The efficacy of this product has not been established.

Prior to 1988, 12 laboratory-acquired dengue infections had been reported. However, from 1988 through 1991, four additional cases have been documented. In all four cases, proper protective gear (long sleeve lab gowns tying in back, gloves, masks, safety glasses) were not worn; and, in three instances, containment of potential aerosols in a laminar flow biosafety cabinet was ignored. These aerosols or infected fluids most likely produced contamination of broken, unprotected skin. An additional factor in these cases was work with highly concentrated amounts of virus. Safe manipulation of dengue viruses in the laboratory (particularly in concentrated preparations) requires strict adherence to Biosafety Level 2 recommendations.

Arboviruses and Arenaviruses Assigned to Biosafety Level 3. SALS has recommended that work with the 171 arboviruses included in the two alphabetical listings on page 343 should be conducted at the equivalent of Biosafety Level 3 practices, safety equipment and facilities. These recommendations are based on the following criteria: for Table C, SALS considered the laboratory experience inadequate to assess risk, regardless of the available information regarding disease severity. For Table D, SALS recorded overt laboratory-associated infections with these agents which occurred by the aerosol route if protective vaccines were not used or were unavailable, and that the natural disease in humans is potentially severe, life threatening, or causes residual damage. Arboviruses were also classified BSL 3 if they cause diseases in domestic animals in countries outside the USA. Laboratory or laboratory animal-associated infections have been reported with the following BSL agents:[86,151,178]

Virus	Cases (SALS)
Venezuelan equine encephalomyelitis	150 deaths
Rift Valley fever	47 (1 death)
Chikungunya	39
Yellow fever	38 (8 deaths)
Japanese encephalitis	22
Louping ill	22
West Nile	18
Lymphocytic choriomeningitis	15
Orungo	13
Pery	13
Wesselsbron	13
Mucambo	10
Oropouche	7
Germiston	6
Bhanja	6
Hantaan	6
Mayaro	5
Spondweni	4
Murray Valley encephalitis	3
Semliki Forest	3 (1 death)
Powassan	2
Dugbe	2
Issyk-kul	1
Koutango	1

Large quantities and high concentrations of Semliki Forest virus are commonly used or manipulated by molecular biologists under conditions of moderate or low containment. Although antibodies have been demonstrated in individuals working with this virus the first overt (and fatal) laboratory-associated infection with this virus was reported in 1979.[198] Because the outcome of this infection may have been influenced by a compromised host, an unusual route of exposure or high dosage, or a mutated strain of the virus, this case and its outcome are not typical. More recently, SFV was associated with an outbreak of febrile illness among European soldiers stationed in Bangui.[118] The route of exposure was not determined in the fatal laboratory infection; for the natural infections, mosquitoes were the probable vector. SALS continues to classify SFV as a BL 3 virus, with the caveat that most activities with this virus can be safety conducted at Biosafety Level 2.

Some viruses (e.g., Akabane, Israel turkey meningoencephalitis) are listed in Level 3, not because they pose a threat to human health, but because they are exotic diseases of domestic livestock or poultry.

Laboratory Hazards: The agents listed in this group may be present in blood, cerebrospinal fluid, urine and exudates depending on the specific agent and stage of disease. The primary laboratory hazards are exposure to aerosols of infectious solutions and animal bedding, accidental parenteral inoculation, and broken skin contact. Some of these agents (e.g., VEE) may be relatively stable in dried blood or exudates. Attenuated vaccine strains for a number of these agents are listed in Table B.

Recommended Precautions: Biosafety Level 3 practices, safety equipment, and facilities are recom-

TABLE C Arboviruses and certain other viruses assigned to Biosafety Level 3 (on the basis of insufficient experience)

Adelaide River	Iaco	Para
Agua Preta	Ibaraki	Paramushir
Alenquer	Ife	Paroo River
Almeirim	Ingangapi	Perinet
Altamira	Inini	Petevo
Andasibe	Issyk-Kul	Picola
Antequera	Itaituba	Playas
Araguari	Itimirim	Pueblo Viejo
Aransas Bay	Itupiranga	Purus
Arbia	Jacareacanga	Radi
Arboledas	Jamanxi	Razdan
Babanki	Jari	Resistencia
Batken	Kedougou	Rochambeau
Belem	Khasan	Salanga
Berrimah	Kindia	San Juan
Bimbo	Kyzylagach	Santa Rosa
Bobaya	Lake Clarendon	Santarem
Bobia	Llano Seco	Saraca
Bozo	Macaua	Saumarez Reef
Buenaventura	Mapuera	Sedlec
Cabassou[c,d]	Mboke	Sena Madureira
Cacipacore	Meaban	Sepik
Calchaqui	Mojui Dos Compos	Shokwe
Cananeia	Monte Dourado	Slovakia
Caninde	Munguba	Somone
Chim	Naranjal	Spipur
Coastal Plains	Nariva	Tai
Connecticut	Nasoule	Tamdy
Corfou	Ndelle	Telok Forest
Dabakala	New Minto	Termeil
Douglas	Ngari	Thiafora
Enseada	Ngoupe	Tilligerry
Estero Real	Nodamura	Tinaroo
Fomede	Northway	Tlacotalpan
Forecariah	Odrenisrou	Tonate[c,d]
Fort Sherman	Omo	Ttinga
Gabek Forest	Oriximina	Xiburema
Gadgets Gully	Ouango	Yacaaba
Garba	Oubangui	Yaounde
Gordil	Oubi	Yoka
Gray Lodge	Ourem	Yug Bogkanovac
Gurupi	Palestina	

[c]SALS recommends that work with this agent should be conducted only in Biosafety Level 3 facilities which provide for HEPA filtration of all exhaust air prior to discharge from the laboratory.

[d]A vaccine is available and is recommended for persons working with this agent.

TABLE D Arboviruses and certain other viruses assigned to Biosafety Level 3

Aino	Louping Ill[a,c]	Sagiyama
Akabane	Mayaro	Sal Vieja
Bhanja	Middelburg	San Perlita
Chikungunya[c,d]	Mobala	Semliki Forest
Cocal	Mopeia[e]	Seoul
Dhori	Mucambo[c,d]	Spondweni
Dugbe	Murray Valley enc.	St. Louis enc.
Everglades[c,d]	Nairobi sheep disease[a]	Thogoto
Flexal	Ndumu	Tocio[c]
Germiston[c]	Negishi	Turuna
Getah	Oropouche[c]	Venezuelan equine[c,d] encephalitis
Hantaan	Orungo	Vesicular stomatitis (alagoas)
Israel Turkey mening.	Peaton	Wesselsbron[a,c]
Japanese enc.	Piry	West Nile
Junin[c,d]	Powassan	Yellow fever[c,d]
Kairi	Puumala	Zinga[b]
Kimberley	Rift Valley fever[a,b,c,d]	
Koutango		

[a]The importation, possession, or use of this agent is restricted by USDA regulation or administrative policy. See Appendix E.

[b]Zinga virus is now recognized as being identical to Rift Valley Fever virus.

[c]See Table C.

[d]See Table C.

[e]This virus is presently being registered in the *Catalogue of Arboviruses*.[96]

mended for activities using potentially infectious clinical materials and infected tissue cultures, animals, or arthropods.

A licensed attenuated live virus is available for immunization against yellow fever and is recommended for all personnel who work with this agent or with infected animals, and those who enter rooms where the agents or infected animals are present. An investigational vaccine (IND) available for immunization against Venezuelan equine encephalomyelitis is recommended for all personnel working with VEE (and the related Everglades, Mucambo, Tonate, and Cabassou viruses), infected animals, or entering rooms where these agents or infected animals are present. Likewise, investigational vaccines for Rift Valley fever and Junin viruses are available from USAMRIID. Work with Hantaan (Korean hemorrhagic fever) virus and related viruses (Puumala and Seoul) in rats, voles, and other laboratory rodents should be conducted with special caution, because of the extreme hazard of aerosol infection.

Arboviruses, Arenaviruses, or Filoviruses Assigned to Biosafety Level 4. SALS has recommended that work with the 15 arboviruses, arenaviruses, or filoviruses[104] included in the alphabetical listing that follows should be conducted at the equivalent of Biosafety Level 4 practices, safety equipment, and facilities. These recommendations are based on documented cases of severe and frequently fatal naturally occurring human infections and aerosol-transmitted laboratory infections. SALS recommended that certain agents with a close antigenic relationship to the Biosafety Level 4 agents (e.g., Absettarov and Kumlinge viruses) also be handled at this level provisionally until sufficient laboratory experience was obtained to retain these agents at this level or to work with them at a lower level. Laboratory or laboratory animal-associated infections have been reported with the following agents:[65,86,90,109,151,178,194]

Virus	Cases (SALS)
Junin	21 (1 death)
Marburg	25 (5 deaths)
Russian Spring-Summer encephalitis	8
Congo-Crimean hemorrhagic fever	8 (1 death
Omsk hemorrhagic fever	5
Lassa	2 (1 death)
Machupo	1 (1 death)
Ebola	1

Rodents are natural reservoirs of Lassa fever virus (*Mastomys natalensis*), Junin and Machupo vi-

137

Arboviruses, arenaviruses and filoviruses assigned to Biosafety Level 4

Congo-Crimean hemorrhagic fever	Marburg
Tick-borne encephalitis virus complex	Ebola
(Absettarov, Hanzalova, Hypr, Kumlinge, Kyasanur Forest disease, Omsk hemorrhagic fever, and Russian Spring-Summer encephalitis)	Lassa Junin Machupo Guanarito

ruses (*Calomys* spp.) and perhaps other members of this group. Nonhuman primates were associated with the initial outbreaks of Kyasanur Forest disease (*Presbytis* spp.) and Marburg disease (*Cercopithecus* spp.); more recently, filoviruses related to Ebola were associated with *Macaca* spp. Arthropods are the natural vectors of the tick-borne encephalitis complex agents. Work with or exposure to rodents, nonhuman primates, or vectors naturally or experimentally infected with these agents represents a potential source of human infection.

Laboratory Hazards: The infectious agents may be present in blood, urine, respiratory and throat secretions, semen and tissues from human or animal hosts, and in arthropods, rodents, and nonhuman primates. Respiratory exposure to infectious aerosols, mucous membrane exposure to infectious droplets, and accidental parenteral inoculation are the primary hazards to laboratory or animal care personnel.[109,194]

Recommended Precautions: Biosafety Level 4 practices and facilities are recommended for all activities utilizing known or potentially infectious materials of human, animal, or arthropod origin. A new, live attenuated investigational (IND) Junin virus vaccine (Candid #1) is available from the U.S. Army Medical Research Institute for Infectious Diseases (USAMRIID) and is recommended for all laboratory and animal care personnel working with the agent or infected animals and for all personnel entering laboratories or animal rooms when the agent is in use. SALS has lowered the biohazard classification of Junin virus to BSL 3, provided all at-risk personnel are immunized and the laboratory is equipped with HEPA filtered exhaust. Clinical specimens from persons suspected of being infected with one of the agents listed in this summary should be submitted to a laboratory with a Biosafety Level 4 maximum containment facility.[23,141]

138

APPENDIX A: BIOLOGICAL SAFETY CABINETS

Biological Safety Cabinets (BSCs) are among the most effective, as well as the most commonly used primary containment devices in laboratories working with infectious agents. The three general types available (Class I, II, III) have performance characteristics and applications which are described in this appendix.

Properly maintained Class I and II BSCs, when used in conjunction with good microbiological techniques, provide an effective containment system for safe manipulation of moderate and high-risk microorganisms (Biosafety Level 2 and 3 agents). Both Class I and II BSCs have inward face velocities (75–100 linear feet per minute) that provide comparable levels of containment for laboratory workers and the immediate environment from infectious aerosols generated within the cabinet. Class II BSCs have the additional advantage of providing protection to the research material by high-efficiency particulate air (HEPA)-filtration of the air flow down across the work surface (vertical laminar flow). Class III cabinets offer the maximum protection to laboratory personnel, the community, and the environment because all hazardous materials are contained in a totally enclosed, ventilated cabinet.

Class I

(Note: Class I BSCs are no longer being manufactured on a regular basis; many have been replaced by Class II BSCs.)

The *Class I Biological Safety Cabinet* (Fig. 1) is a negative-pressure, ventilated cabinet usually operated with an open front and a minimum face velocity at the work opening of at least 75 linear feet per

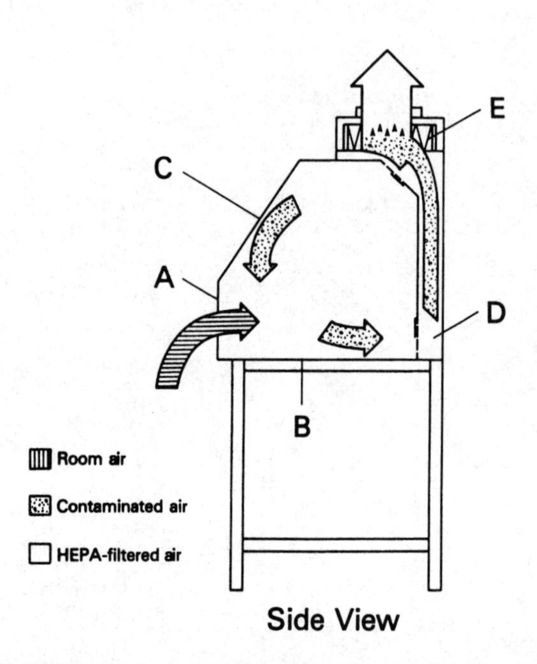

FIGURE 1 Class I Biological Safety Cabinet. A. front opening, B. work surface, C. window, D. exhaust plenum, E. HEPA filter.

142

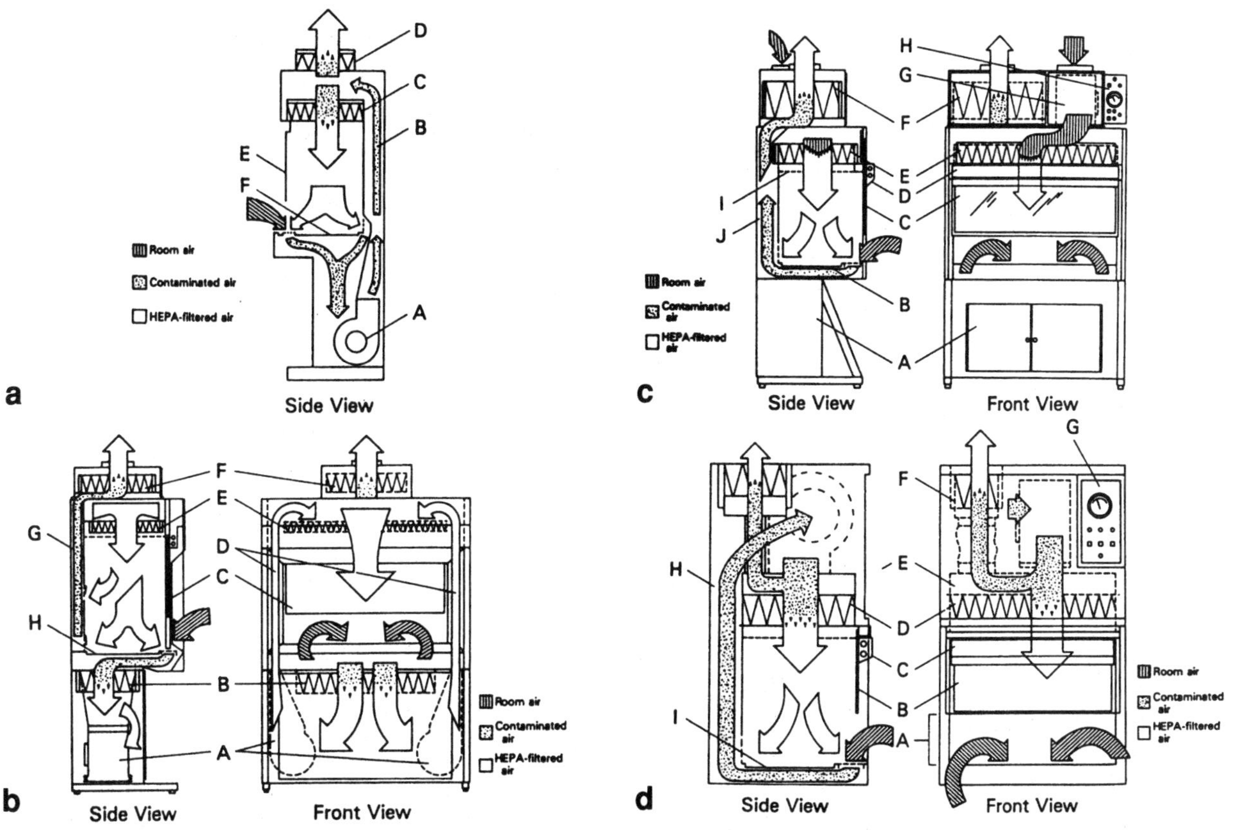

FIGURE 2 (a) Class II, Type A BSC. A. blower, B. rear plenum, C. supply HEPA filter, D. exhaust, E. sash, F. work surface. (b) Class II, Type B1 BSC. A. blowers, B. supply HEPA filters, C. sliding sash, D. positive pressure plenums, E. additional supply HEPA filter or back-pressure plate, F. exhaust HEPA filter, G. negative pressure exhaust plenum, H. work surface. (c) Class II, Type B2 BSC. A. storage cabinet, B. work surface, C. sliding sash, D. lights, E. supply HEPA filter, F. exhaust HEPA filter, G. supply blower, H. control panel, I. filter screen, J. negative pressure plenum. (d) Table-top model of a Class II, Type B3 BSC. A. front opening, B. sliding sash, C. light, D. supply HEPA filter, E. positive pressure plenum, F. exhaust HEPA filter, G. control panel, H. negative pressure plenum, I. work surface.

minute (lfpm). All of the air from the cabinet is exhausted through a HEPA filter either into the laboratory, or to the outside. The Class I BSC is designed for general microbiological research with low and moderate risk agents, and is useful for containment of mixers, blenders, and other equipment. These cabinets are *not* appropriate for handling research materials that are vulnerable to airborne contamination, since the inward flow of unfiltered air from the laboratory can carry microbial contaminants into the cabinet.

The Class I BSC can also be used with an installed front closure panel without gloves that will increase the inward flow velocity to approximately 150 lfpm. If such equipped cabinets are ducted to the outside exhaust, they may be used for toxic or radiolabelled materials used as an adjunct to microbiological research. Additionally, arm-length rubber gloves may be attached to the front panel with an inlet air pressure release for further protection.

Class II

The *Class II Biological Safety Cabinet* (Fig. 2) is designed with inward air flow at a velocity to protect personnel (75–100 lfpm), HEPA-filtered vertical laminar airflow for product protection, and HEPA-filtered exhaust air for environmental protection. Design, construction, and performance standards for Class II BSCs, as well as a list of products that meet these standards, have been developed by and are available from the National Sanitation Foundation International,[136] Ann Arbor, Michigan. Utilization of this standard and list should be the first step in selection and procurement of a Class II BSC.

Class II BSCs are classified into two types (A and B) based on construction, air flow velocities and patterns, and exhaust systems. Basically, Type A cabinets are suitable for work with microbiological research *in the absence of* volatile or toxic chemicals and radionuclides, since air is recirculated within the work area. Type A cabinets may be exhausted through HEPA filters into the laboratory, or to the outside via a "thimble" connection to the exhaust ductwork.

Type B cabinets are further sub-typed into types B1, B2, and B3. A comparison of the design features and applications is found in Fig. 2 of this appendix and Table 3. Type B cabinets are hard-ducted to the exhaust system, and contain negative pressure plena. These features, plus an increased face velocity of 100 lfpm, allow work to be done with toxic chemicals or radionuclides.

It is imperative that Class I and II biological safety cabinets are tested and certified *in situ* at the time of installation within the laboratory, at any time the BSC is moved, and at least annually thereafter. Certification at locations other than the final site may attest to the performance capability of the individual cabinet or model but does not supersede the critical certification prior to use in the laboratory.

As with any other piece of laboratory equipment, personnel must be trained in the proper use of the biological safety cabinets. Of particular note are those activities which may disrupt the inward directional airflow through the work opening of Class I and II cabinets. Repeated insertion and withdrawal of the workers' arms in and from the work chamber, opening and closing doors to the laboratory or isolation cubicle, improper placement or operation of materials or equipment within the work chamber, or brisk walking past the BSC while it is in use are demonstrated causes of the escape of aerosolized particles from within the cabinet. Class I and II cabinets should be located away from traffic patterns and doors. Fans, heating and air conditioning registers, and other air handling devices can also disrupt airflow patterns if located adjacent to the BSC. Strict adherence to recommended practices for the use of BSCs and proper placement in the laboratory are important in attaining the maximum containment capability of the equipment as is the mechanical performance of the equipment itself.

Class III

The *Class III Biological Safety Cabinet* (Fig. 3) is a totally enclosed, ventilated cabinet of gas-tight construction and offers the highest degree of personnel and environmental protection from infectious aerosols, as well as protection of research materials from microbiological contaminants. Class III cabinets are most suitable for work with hazardous agents that require Biosafety Level 3 or 4 containment.

All operations in the work area of the cabinet are performed through attached rubber gloves. The

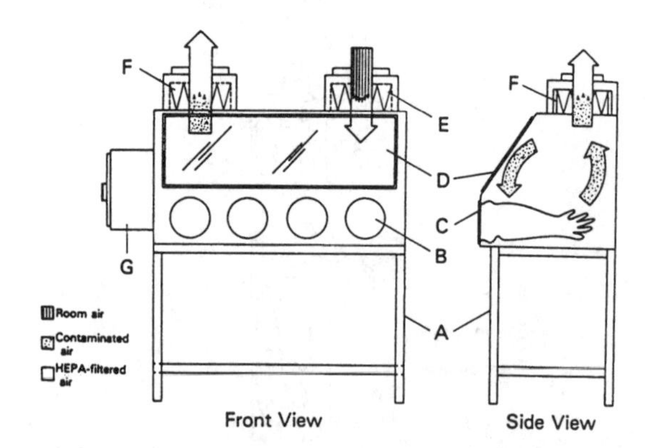

FIGURE 3 Class III BSC. A. stand, B. glove ports, C. O-ring for attaching arm-length gloves to cabinet, D. slopped glass viewing window, E. supply HEPA filter, F. exhaust HEPA filter (Note that the second exhaust HEPA filter required for Class III cabinets is not depicted in this diagram), G. double-ended autoclave.

Class III cabinet is operated under negative pressure. Supply air is HEPA-filtered, and the cabinet exhaust air is filtered by two HEPA filters in series, or HEPA filtration followed by incineration, before discharge outside of the facility.

All equipment required by the laboratory activity, such as incubators, refrigerators, and centrifuges, must be an integral part of the cabinet system. The Class III cabinet must be connected to double-doored autoclaves and chemical dunk tanks to sterilize or disinfect all materials exiting the cabinet, and to allow supplies to enter the cabinet. Several Class III cabinets are therefore typically set up as an interconnected system.

Positive-Pressure Personnel Suit

Personnel protection equivalent to that provided by Class III cabinets can also be obtained with the use of a one-piece, ventilated suit for the laboratory worker when working with Biosafety Level 3 or 4 agents in a "suit area" and using Class I or II BSCs. The personnel suit is maintained under positive pressure with a life support system to prevent leakage into the suit. In this containment system, the worker is isolated from the work materials.

The personnel suit area must be essentially equivalent to a large Class III cabinet. The area is entered through an air-lock fitted with airtight doors. A chemical shower is provided as a "dunk tank" to decontaminate the surfaces of the suit as the worker leaves the area. The exhaust air from the suit area is filtered by two HEPA filter units installed in series. The entire area must be under negative pressure.

As in the case with Class III BSCs, the gloves of the personnel suit are the most vulnerable component of the system, as they are subject to punctures by sharps or animal bites.

(Caution: Horizontal laminar flow "clean benches" are used in clinical, pharmaceutical, and laboratory facilities strictly for *product* protection. Since the worker sits in the immediate downstream exhaust from the "clean bench", this equipment must *never* be used for handling toxic, infectious, radioactive, or sensitizing materials.)

146

APPENDIX B: IMMUNOPROPHYLAXIS

An additional level of protection for at-risk personnel may be achieved with appropriate prophylactic immunizations. A written organizational policy which defines at-risk personnel, which specifies risks as well as benefits of specific vaccines, and which distinguishes between required and recommended vaccines is essential. In developing such an organizational policy, these recommendations and requirements should be specifically targeted at infectious diseases known or likely to be encountered in a particular facility.

Vaccines for which the benefits (levels of antibody considered to be protective) clearly exceed the risks (local or systemic reactions) should be required for all clearly identified at-risk personnel. Examples of such preparations include vaccines against hepatitis B, yellow fever, rabies, and poliomyelitis. Recommendations for giving less efficacious vaccines, those associated with high rates of local or systemic reactions, or those that produce increasingly severe reactions with repeated use should be carefully considered. Products with these characteristics (e.g., cholera, tularemia, and typhoid vaccines) may be recommended but should not ordinarily be required for employment. A complete record of vaccines received on the basis of occupational requirements or recommendations should be maintained in the employee's permanent medical file.

Recommendations for the use of vaccines, adapted from those of Public Health Service Advisory Committee on Immunization Practices, are included in the agent summary statements in Section VII. Particular attention must be given to individuals who are or may become immunocompromised, as recommendations for vaccine administration may be different than for immunologically competent adults.

147

APPENDIX C: SURVEILLANCE OF PERSONNEL FOR LABORATORY-ASSOCIATED RICKETTSIAL INFECTIONS

Under natural circumstances, the severity of disease caused by rickettsial agents varies considerably. In the laboratory, very large inocula which might produce unusual and perhaps very serious responses are possible. Surveillance of personnel for laboratory-associated infections with rickettsial agents can dramatically reduce the risk of serious consequences of disease.

Experience indicates that infections treated adequately with specific anti-rickettsial chemotherapy on the first day of disease do not generally present serious problems. Delay in instituting appropriate chemotherapy, however, may result in debilitating or severe acute disease ranging from increased periods of convalescence in typhus and scrub typhus to death in *R. rickettsii* infections. The key to reducing the severity of disease from laboratory-associated infections is a reliable surveillance system which includes (1) round-the-clock availability of an experienced medical officer, (2) indoctrination of all personnel into the potential hazards of working with rickettsial agents and advantages of early therapy, (3) a reporting system for all recognized overt exposures and accidents, (4) the reporting of all febrile illnesses, especially those associated with headache, malaise, prostration, when no other certain cause exists and, (5) a non-punitive atmosphere that encourages reporting of any febrile illness.

Rickettsial agents can be handled in the laboratory with minimal real danger to life when an adequate surveillance system complements a staff who are knowledgeable about the hazards of rickettsial infections and who put to use the safeguards recommended in the agent summary statements.

148

APPENDIX D: IMPORTATION AND INTERSTATE SHIPMENT OF HUMAN PATHOGENS AND RELATED MATERIALS

The importation or subsequent receipt of etiologic agents and vectors of human disease is subject to the Public Health Service Foreign Quarantine Regulations (42 CFR, Section 71.156). Permits authorizing the importation or receipt of regulated materials and specifying conditions under which the agent or vector is shipped, handled, and used are issued by the Centers for Disease Control and Prevention.

The interstate shipment of indigenous etiologic agents, diagnostic specimens, and biological products is subject to applicable packaging, labeling, and shipping requirements of the Interstate Shipment of Etiologic Agents (42 CFR Part 72). Packaging and labeling requirements for interstate shipment of etiologic agents are summarized and illustrated in Fig. 4.

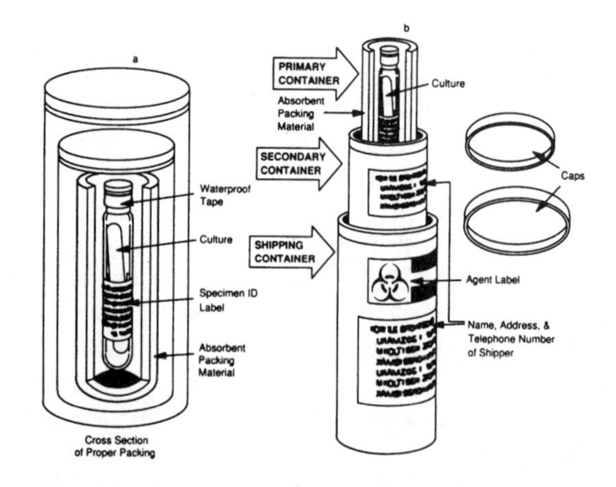

149

FIGURE 4 Packing and labeling of etiologic agents.

Figures 4a and 4b diagram the packaging and labeling of etiologic agents in volumes of less than 50 ml. in accordance with the provisions of subparagraph 72.3(a) of the regulation on Interstate Shipment of Etiologic Agents (42 CFR, Part 72). A revision has been proposed that may result in additional package labeling requirements, but this has not been issued as of the publication of this third edition of BMBL.

For further information on any provision of this regulation contact:

Centers for Disease Control and Prevention
Attn: Biosafety Branch Chief
Mail Stop F-05
1600 Clifton Road N.E.
Atlanta, GA 30333
Telephone: (404) 639-3883
Fax: (404) 639-2294

Note that the shipper's name, address, and telephone number must be on the outer and inner containers. The reader is also advised to refer to additional provisions of the Department of Transportation (49 CFR, Parts 171–180) Hazardous Materials Regulations.

Persons needing to report leaking, damaged packages of etiologic agents may call 1-800-232-0124.

Additional information on the importation and interstate shipment of etiologic agents of human disease, diagnostic specimens, and other related materials may be obtained by contacting:

Centers for Disease Control and Prevention
Attention: Biosafety Branch
Office of Health and Safety, Mail Stop F-05
1600 Clifton Road N.E.
Atlanta, Georgia 30333
Telephone: (404) 639-3883
Fax: (404) 639-2294

APPENDIX E: RESTRICTED ANIMAL PATHOGENS

150

Nonindigenous pathogens of domestic livestock and poultry may require special laboratory design, operation, and containment features not generally addressed in this publication. The importation, possession, or use of the following agents is prohibited or restricted by law or by U.S. Department of Agriculture regulations or administrative policies:

African horse sickness
African Swine fever virus
Akabane virus
Besnoitia besnoiti
Borna disease virus
Bovine spongiform encephalopathy
Bovine infectious petechial fever agent
Brucellosis melitensis
Camelpox virus
Cochliomyia hominivorax (screwworm)
Ephemeral fever virus
Foot and mouth disease virus
Fowl plague virus (lethal avian influenza)
Hog cholera virus
Histoplasma (Zymonema) farciminosum
Louping ill virus
Lumpy skin disease virus
Mycoplasma agalactiae
Mycoplasma mycoides
Nairobi sheep disease virus (Ganjam virus)
Newcastle disease virus (velogenic strains)
Peste des petits ruminants (pest of small ruminants)
Pseudomonas ruminantium (heartwater)
Rift Valley fever virus
Rinderpest virus
Sheep and goat pox
Swine vesicular disease virus
Teschen disease virus
Theileria annulata
Theileria lawrencei
Theileria bovis
Theileria hirci
Trypanosoma evansi
Trypanosoma vivax
Vesicular exanthema virus
Viral hemorrhagic disease of rabbits
Wesselsbron disease virus

The importation, possession, use, or interstate shipment of animal pathogens other than those listed above may also be subject to regulations of the U.S. Department of Agriculture.

Additional information may be obtained by writing to:

151

U.S. Department of Agriculture
Animal and Plant Health Inspection Service
Veterinary Services, Import-Export Products Staff
Room 756, Federal Building
6505 Belcrest Road
Hyattsville, Maryland 20782
Telephone: (301) 436-7830 or
(301) 436-8499
Fax: (301) 436-8226

APPENDIX F: RESOURCES FOR INFORMATION

152

Resources for information, consultation, and advice on biohazard control, decontamination procedures, and other aspects of laboratory safety management include:

Centers for Disease Control and Prevention
Attention: Biosafety Branch
Atlanta, Georgia 30333
Telephone: (404) 329-3883

National Institutes of Health
Attention: Division of Safety
Bethesda, Maryland 20205
Telephone: (301) 496-1357

National Animal Disease Center
U.S. Department of Agriculture
Ames, Iowa 50010
Telephone: (515) 862-8258

References

1. Allen, B.W. 1981. Survival of tubercle bacilli in heat-fixed sputum smears. J Clin Pathol **34:**719–722.
2. American Society of Heating, Refrigerating, and Air-Conditioning Engineers, Inc. 1991. "Laboratories". In: ASHRAE Handbook, *Heating, Ventilation, and Air-Conditioning Applications.* p. 14.1–14.17.
3. American Thoracic Society. 1983a. Levels of laboratory services for mycobacterial diseases. Am Rev Respir Dis **128:**213.
4. American Thoracic Society. 1983b. The levels of service concept in mycobacteriology. Am Thorac Soc News, Summer 1983. p. 19–25.
5. Anonymous. 1992. Drugs for Parasitic Infections. The Medical Letter on Drugs and Therapeutics. **34:**17–26.
6. Anonymous. 1980. Tuberculosis infection associated with tissue processing. Cal Morbid **30.**
7. Artenstein, A.W., Hicks, C.B., Goodwin, B.S., Hilliard, J.K. 1991. Human Infection with B Virus Following a Needlestick Injury. Rev Infect Dis **13:**288–91.
8. Bacteriologist dies of meningitis. 1936. JAMA **106:**129.
9. Baum, G.L., Lerner, P.I. 1971. Primary pulmonary blastomycosis: a laboratory acquired infection. Ann Intern Med **73:**263–265.
10. Beeman, E.A. 1950. Q fever—An epidemiological note. Pub Hlth Rep **65**(2):88–92.
11. Bernard, K.W., Parham, G.L., Winkler, W.G., Helmick, C.G. 1982. Q fever control measures: Recommendations for research of facilities using sheep. Inf Control **3:**461–465.
12. *Biosafety in the Laboratory:* Prudent Practices for the Handling and Disposal of Infectious Materials. 1989. National Research Council. National Academy Press, Washington, D.C.
13. Blaser, M.J., Hickman, F.W., Farmer, J.J., III, Brenner, D.J., Balows, A. and Feldman, R.A. 1980. *Salmonella typhi:* the laboratory as a reservoir of infection. J Infect Dis **142:**934–938.
14. Bowen, G.S., Calisher, C.H., Winkler, W.G., Kraus, A.L., Fowler, E.H., Garman, R.H., Fraser, D.W., Hinman, A.R. 1975. Laboratory studies of a lymphocytic choriomeningitis virus outbreak in man and laboratory animals. Am J Epidemiol **102:**233–40.
15. Brown, P., Wolff, A., Gadusek, D.C. 1990. A simple and effective method for inactivating virus infectivity in formalin-fixed tissue samples from patients with Cruetzfeldt-Jakob disease. Neurology **40:**887–890.
16. Burke, D.S. 1977. Immunization against tularemia: analysis of the effectiveness of live *Francisella tularensis* vaccine in prevention of laboratory-acquired tularemia. J Infect Dis **135:**55–60.
17. Burmeister, R.W., Tigertt, W.D., Overholt, E.L. 1962. Laboratory-acquired pneumonic plague. Ann Intern Med **56:**789–800.
18. Bush, J.D. 1943. Coccidioidomycosis. J Med Assoc Alabama **13:**159–166.
19. Burstyn, D.G., Baraff, L.J., Peppler M.S., Leake, R.D., et al. 1983. Serological response to filamentous hemagglutinin and lymphocytosis-promoting toxin of *Bordetella pertussis.* Infection and Immunity **41**(3):1150–6.
20. Caplen, H., Peters, C.J., and Bishop, D.H.L. 1985. Mutagen-directed Attenuation of Rift Valley Fever Virus as a Method for Vaccine Development. J Gen Virol **66:**2271–2277.
21. Carougeau, M. 1909. Premier cas Africain de sporotrichose de deBeurmann: Transmission de la sporotrichose du mulet a l'homme. Bull Mem Soc Med Hop (Paris) **28:**507–510.
22. Cavenaugh, D.C., Elisberg, B.L., Llewellyn, C.H., Marshall, J.D., Jr., Rust, J.H., Williams, J.E., and Meyer, K.F. 1974. Plague Immunization IV. Indirect evidence for the efficacy of plague vaccine. J Infect Dis **129:** (Supplement) S37-S40.
23. Center for Disease Control, Office of Biosafety. 1974. *Classification of Etiologic Agents on the Basis of Hazard,* 4th Edition. U.S. Department of Health, Education and Welfare, Public Health Service.
24. Center for Disease Control. 1976. Unpublished Data. Center for Infectious Diseases. U.S. Department of Health, Education and Welfare, Public Health Service.
25. Center for Disease Control. 1977. Rabies in a laboratory worker New York. MMWR **26**(22):183–184.
26. Center for Disease Control. 1978. Laboratory-acquired endemic typhus. MMWR **27**(26):215–216.
27. Center for Disease Control. 1978. Meningococcal Polysaccharide vaccines. Recommendations of the Immunization Practices Advisory Committee (ACIP) and Mortality Weekly Report **27**(35):327–328.
28. Centers for Disease Control. 1979. Q fever at a university research center—California. MMWR **28**(28):333–334.
29. Centers for Disease Control. 1980. Chagas' disease, Kalamazoo, Michigan. MMWR **20**(13):147–8.
30. Centers for Disease Control. 1980. Recommendations for initial management of suspected or confirmed cases of Lassa fever. MMWR **28**(52)**:** Suppl:3S-12S.
31. Centers for Disease Control. 1980. Smallpox Vaccines. Recommendation of the Immunization Practices Advisory Committee (ACIP). MMWR **29:** 417–420.
32. Centers for Disease Control. 1981. Recommendations of the Immunization Practices Advisory Committee (ACIP) Diphtheria, Tetanus, and Pertussis. MMWR **30**(32):392–396.
33. Centers for Disease Control. 1982. Recommendations of the Immunization Practices Advisory Committee

(ACIP). Poliomyelitis prevention. MMWR **29**(3):22–26 and 31–34.

34. Centers for Disease Control. 1982. Plague vaccine. Selected recommendations of the Public Health Service Advisory Committee on Immunization Practices (ACIP). MMWR **40** (RR-12): 41–42.
35. Centers for Disease Control. 1985. Pertussis—Washington, 1984. MMWR **34**(26):90–400.

156

36. Centers for Disease Control. 1985. Revision of the case definition of acquired immunodeficiency syndrome for national reporting—United States. MMWR **34:**373–5.
37. Centers for Disease Control. 1986. Additional recommendations to reduce sexual and drug abuse-related transmission of human T-lymphotrophic virus type III/lymphadenopathy-associated virus. MMWR **35:**152–5.
38. Centers for Disease Control. 1986. Diagnosis and management of mycobacterial infection and disease in persons with human T-lymphotrophic virus type III/lymphadenopathy-associated virus infection. MMWR **35:**448–52.
39. Centers for Disease Control. 1987. Guidelines for prevention of *Herpesvirus simiae* (B virus) infection in monkey handlers. MMWR, **36:**680–2,687–9.
40. Centers for Disease Control. 1987. Recommendations for prevention of HIV transmission in health-care settings. MMWR **36**(suppl 2):3S-18S.
41. Centers for Disease Control. 1987. Public Health Service guidelines for counseling and antibody testing to prevent HIV infections and AIDS. MMWR **36:**509–15.
42. Centers for Disease Control. 1987. Revision of the Centers for Disease Control surveillance case definition for acquired immunodeficiency syndrome. MMWR **36**(suppl 1):1S-15S.
43. Centers for Disease Control. 1988. Update: Universal Precautions for Prevention of Transmission of Human Immunodeficiency Virus, Hepatitis B Virus and Other Bloodborne Pathogens in Healthcare Settings. MMWR, **37:**377–382, 387, 388.
44. Centers for Disease Control. 1988. Guidelines to prevent simian immunodeficiency virus infection in laboratory workers. MMWR **37:**693–704.
45. Centers for Disease Control. 1987. Recommendations for Prevention of HIV Transmission in Health-Care Settings. MMWR, **36**(25):1–7.

157

46. Centers for Disease Control. 1990. Recommendations of the Immunizations Practices Advisory Committee (ACIP)—Inactivated hepatitis B virus vaccine. MMWR **39,** No. RR-2.
47. Centers for Disease Control. 1990. Guidelines for Preventing the Transmission of Tuberculosis in Health-Care Settings, with Special Focus on HIV-Related Issues. MMWR **39,** No. RR-17.
48. Centers for Disease Control. 1990. Public Health Service statement on management of occupational exposure to human immunodeficiency virus, including considerations regarding Zidovudine postexposure use. 1990. MMWR **39,** No. RR-1.
49. Centers for Disease Control. 1991. Laboratory-acquired meningococcemia—California and Massachusetts. MMWR **40**(3):46–47,55.
50. Centers for Disease Control. 1991. Recommendations of the Immunization Practices Advisory Committee (ACIP)—Diphtheria, Tetanus and Pertussis: Recommendations for Vaccine Use and Other Preventive Measures. MMWR **40,** No. RR-10.
51. Centers for Disease Control. 1991. Recommendations of the Immunization Practices Advisory Committee (ACIP) -MMWR **40,** No. RR-12.
52. Centers for Disease Control. 1991. Vaccinia (Smallpox) Vaccine, Recommendations of the Immunization Practices Advisory Committee (ACIP). MMWR **40,** No. RR-14.
53. Centers for Disease Control. 1992. Seroconversion to simian immunodeficiency virus in two laboratory workers. MMWR **41:**678–681.
54. Centers for Disease Control. 1992. National Action Plan to Combat Multi-drug-Resistant Tuberculosis. Meeting the Challenge of Multi-drug-Resistant Tuberculosis: Summary of a Conference. Management of Persons Exposed to Multi-drug-Resistant Tuberculosis. MMWR **41,** No. RR-11.
55. Centers for Disease Control. *HIV/AIDS Surveillance Report,* October 1992:14.

158

56. Conant, N.F. 1955. Development of a method for immunizing man against coccidioidomycosis, Third Quarterly Progress Report. Contract DA-18-064-CML-2563, Duke University, Durham, NC. Available from Defense Documents Center, AD 121–600.
57. Cooper, C.R., Dixon, D.M., and Salkin, I.F. 1992. Laboratory acquired sporotrichosis. J Med Vet Mycol **30:**169–171.
58. Denton, J.F., DiSalvo, A.F., Hirsch, M.L. 1967. Laboratory-acquired North American blastomycosis. JAMA **199:**935–936.
59. Dickson, E.C. 1937. Coccidioides infection: Part I. Arch Intern Med **59:**1029–1044.
60. Dickson, E.C. 1937. "Valley fever" of the San Joaquin Valley and fungus coccidioides. Calif Western Med **47:**151–155.
61. Dickson, E.C., Gifford, M.A. 1938. Coccidioides infection (coccidioidomycosis): II. The primary type of infection. Arch Intern Med **62:**853–871.
62. Diena, B.B., Wallace, R., Ashton, F.E., Johnson, W. and Patenaude, B. 1976. Gonococcal conjunctivitis: accidental infection. Can Med Assoc J **115:**609,612.
63. Donham, K.J. and Leininger, J.R. 1977. Spontaneous leprosy-like disease in a chimpanzee. J Infect Dis **136:**132–136.
64. Dowdle, W.R. and Hattwick, M.A.W. 1977. Swine influenza virus infections in humans. J Infect Dis **136:** Suppl: S386–389.
65. Edmond, R.T.D., Evans, B., Bowen, E.T.W. and Lloyd, G. 1977. A Case of Ebola virus infection. Br Med J **2:**541–544.
66. Ellingson, H.V., Kadull, P.J., Bookwalter, H.L., Howe, C. 1946. Cutaneous anthrax: report of twenty-five cases. JAMA **131:**1105–8.
67. Evans, N. 1903 . A clinical report of a case of blastomycosis of the skin from accidental inoculation. JAMA **40:**1172–1175.
68. Eyles, D.E., Coatney, G.R., and Getz, M.E. 1960. Vivax-type malaria parasite of macaques transmissible to man. Science **131:**1812–1813.

159

69. Fava, A. 1909. Un cas de sporotrichose conjonctivale et palpebrale primitives. Ann Ocul (Paris) **141:**338–343.
70. Favero M.S., Bond, W.W. 1991. Sterilization, Disinfection and Antisepsis in the Hospital. In: Lennette E.H., Balows A., Hausler W.J., Shadomy H.J., eds. *Manual*

of Clinical Microbiology, 4th ed. Washington, DC: American Society for Microbiology: 183–200.

71. *Federal Register.* 1976. Recombinant DNA Research Guidelines. **41:**27902–27943
72. *Federal Register.* 1986. Guidelines for Research Involving Recombinant DNA Molecules. **51:**16958–16968.
73. Fielitz, H. 1910. Ueber eine Laboratoriumsinfektion mit dem *Sporotrichum* de Beurmanni. Centralbl Bakteriol Parasitenk Abt I Orig **55:**361–370.
74. Fitzgerald, J.J., Johnson, R.C., Smith, M. 1976. Accidental laboratory infection with *Treponema pallidum.* Nichols strain. J Am Vener Dis Assoc **3:**76–78.
75. Furcolow, M.L. 1961. Airborne Histoplasmosis. Bact Rev **25:**301–309.
76. Gajdusek, D.C., Gibbs, C.J., Asher, D.M., Brown, P., Diwan, A., Hoffman, P., Nemo, G., Rohwer, R. and White, L. 1977. Precautions in the medical care and in handling materials from patients with transmissible virus dementia (Creutzfeldt-Jakob Disease). N Engl J Med **297:**1253–1258.
77. Good, R.C., Snider, D.E., Jr. 1982. Isolation of non-tuberculosis mycobacteria in the U.S., 1980. J Infect Dis **146:**829–833.
78. Green, R.N., Tuffnell, P.G. 1968. Laboratory acquired melioidosis. Am J Med **44:**599–605.
79. Grist, N.R., Emslie, J.A.N. 1985. Infections in British clinical laboratories, 1982–3. J Clin Pathol **38:**721–725.
80. Grist, N.R., Emslie, J.A.N. 1987. Infections in British clinical laboratories, 1984–5. J Clin Pathol **40:**826–829.

160

81. Gutteridge, W.E., Cover, B., Cooke, A.J.D. 1974. Safety precautions for working with *Trypanosoma cruzi.* Trans R Soc Trop Med Hyg **68:**161.
82. Hackstadt, T. 1986. Antigenic variation in the phase I lipopolysaccharide of Coxiella burnetii isolates. Infect Immun **52:**377–340.
83. Halde, C. 1964. Percutaneous *Cryptococcus neoformans* inoculation without infection. Arch Dermatol. **89:**545.
84. Hanel, E., Jr., and Kruse, R.H. 1967. Laboratory-acquired mycoses. Department of the Army, Miscellaneous Publication 28.
85. Hanson, R.P., et al. 1950. Human infections with the virus of vesicular stomatitis. J Lab Clin Med **36:**754–758.
86. Hanson, R.P., Sulkin, S.E., Buescher, E.L., Hammond, W.McD., McKinney, R.W., and Work, T.E. 1967. Arbovirus infections of laboratory workers. Science **158:**1283–1286.
87. Harrell, E.R. 1964. The known and the unknown of the occupational mycoses, p. 176–178. In: *Occupational Diseases Acquired From Animals.* Continued Education Series No. 124, Univ Mich Sch Pub Hlth, Ann Arbor, MI.
88. Harrington, J.M., and Shannon, H.S. 1976. Incidence of tuberculosis, hepatitis, brucellosis and shigellosis in British medical laboratory workers. Br Med J **1:**759–762.
89. Hattwick, M.A.W., O'Brien, R.J., Hanson, B.F. 1976. Rocky Mountain Spotted Fever: epidemiology of an increasing problem. Ann Intern Med **84:**732–739.
90. Hennessen, W. 1971. Epidemiology of "Marburg Virus" disease. In: Martini, G.A., Siegert, R. eds., *Marburg Virus Disease.* New York: Springer-Verlag 161–165.
91. Herwaldt, B.L., Juranek, D.D. 1993. Laboratory-acquired malaria, leishmaniasis, trypanosomiasis, and toxoplasmosis. Am J Trop Med Hyg **48:**313–323.

161

92. Holmes, G.P., Hilliard, J.K., Klontz, K.C., Rupert, A.H., Schindler, C.M., Parrish, E., Griffin, D.G., Ward, G.S., Bernstein, N.D., Bean, T.W., Ball, M.R., Brady, J.A., Wilder, M.H., Kaplan, J.E. 1990. B Virus *(Herpesvirus simiae)* Infection in Humans: Epidemiologic Investigation of a Cluster. Ann of Int Med **112:**833–839.
93. Holzer, E. 1962. Botulism caused by inhalation. Med Klin **41:**1735–1740.
94. Houghton, M., Weiner, A., Han, J., Kuo, G., Choo, Q-L. 1991. Molecular biology of the hepatitis C viruses: implications for diagnosis, development and control of viral disease. Hepatology **14:**381–388.
95. Huddleson, I.F. and Munger, M. 1940. A study of an epidemic of brucellosis due to *Brucella melitensis.* Am J Public Health **30:**944–954.
96. *International Catalog of Arboviruses Including Certain Other Viruses of Vertebrates.* 1985. The Subcommittee on Information Exchange of the American Committee on Arthropod-borne Viruses. Third edition. N. Karabatsos, Editor. American Society for Tropical Medicine and Hygiene. San Antonio, TX.
97. Jacobson, J.T., Orlob, R.B., Clayton, J.L. 1985. Infections acquired in clinical laboratories in Utah. J Clin Microbiol **21:**486–489.
98. Jahrling, P.B., Peters, C.J. 1992. Lymphocytic choriomeningitis virus, a neglected pathogen of man. Arch Pathol Lab Med **116:**486–8.
99. Jarvis, W.R. 1982. Precautions for Creutzfeldt-Jakob Disease. Infect Control **3:**238–239.
100. Jeanselme, E., Chevallier, P. 1910. Chancres sporotrichosiques des doigts produits par la morsure d'un rat inocule de sporotrichose. Bull Mem Soc Med Hop (Paris) **30:**176–178.
101. Jeanselme, E., Chevallier, P. 1911. Transmission de la sporotrichose a l'homme par les morsures d'un rat blanc inocule avec une nouvelle variete de *Sporotrichum:* Lymphangite gommeuse ascendante. Bull Mem Soc Med Hop (Paris) **31:**287–301.
102. Kaufmann, A.F., Anderson, D.C. 1978. Tuberculosis control in nonhuman primates. In: Montali, R.J. (ed.). *Mycobacterial Infections of Zoo Animals.* Washington, D.C.: Smithsonian Institution Press, 227–234.

162

103. Khabbaz, R.F, Rowe, T., Murphey-Corb, M., et al. 1992. Simian immunodeficiency virus needlestick accident in a laboratory worker. Lancet **340:**271–273.
104. Kiley, M.P., Bowel, E.T.W., Eddy, G.A., Isaacson, M., Johnson, K.M., McCormick, J.B., Murphy, F.A., Pattyn, S.R., Peters, D., Prozesky, W., Regnery, R., Simpson, D.I.H., Slenczka, W., Sureau, P., van der Groen, G., Webb, P.A., and Sulff, H. 1982. Filoviridae: a taxonomic home for Marburg and Ebola viruses? Intervirology **18:**24–32.
105. Klutsch, K., Hummer, N., Braun, H., Heidland, A. 1965. Zur Klinik der Coccidioidomykose. Deut Med Wochensch **90:**1498–1501.
106. Kurt, T.L., Yeager, A.S., Guenette S., et al. 1972. Spread of pertussis by hospital staff. JAMA **221:**264.
107. Larsh, H.W., Schwartz, J. 1977. Accidental inoculation-blastomycosis. Cutis **19:**334–336.
108. Larson, D.M., Eckman, M.R., Alber, C.L., Goldschmidt, V.G. 1983. Primary cutaneous (inoculation)

blastomycosis: an occupational hazard to pathologists. Amer J Clin Pathol **79**:253–25.

109. Leifer, E., Gocke, D.J., Bourne, H. 1970. Lassa fever, a new virus disease of man from West Africa. II. Report of a laboratory-acquired infection treated with plasma from a person recently recovered from the disease. Am J Trop Med Hyg **19**:677–9.
110. Lettau, L.A. 1991. Nosocomial transmission and infection control aspects of parasites and ectoparasitic diseases: Part I. Introduction/enteric parasites. Part II. Blood and tissue parasites. Infect Control Hosp Epidemiol **12**:59–65;111–121.
111. Levine, M.M., Kaper, J.B., Black, R.E., Clements, M.C. 1983. New knowledge on pathogenesis of bacterial enteric infections as applied to vaccine development. Microbiol Reviews **47**:510–550.
112. Linneman, C.C., Ramundo, N., Perlstein, P.H., et al. 1975. Use of pertussis vaccine in an epidemic involving hospital staff. Lancet **2**:540.
113. Looney, J.M., Stein, T. 1950. Coccidioidomycosis. N Engl J Med **242**:77–82.

163

114. Magneson, H.J., Thomas, E.W., Olansky, S., Kaplan, B.L., De Mello, L., Cutler, J.C. 1956. Inoculation syphilis in human volunteers. Medicine **35**:33–82.
115. Marchoux, P.E. 1934. Un cas d'inoculation accidentelle du bacille de Hanson en pays non lepreux. Int J Lepr **2**:1–7.
116. Martin L.S., McDougal, J.S., Loskoski, S.L. 1985. Disinfection and inactivation of the human T lymphotrophic virus type III/lymphadenopathy-associated virus. J Infect Dis **152**:400–3.
117. Martini, G.A., Schmidt, H.A. 1968. Spermatogenic transmission of Marburg virus. Klin Wschr **46**:398–400.
118. Mathiot. C.C., Grimaud,G., Garrey, P., Bouguety, J.C., Mada, A., Daguisy, A.M., Georges, A.J. 1990. An Outbreak of Human Semliki Forest Virus Infections in Central African Republic. Am J Trop Med Hyg **42**:386–393.
119. McAleer, R. 1980. An epizootic in laboratory guinea pigs due to *Trichophyton mentagrophytes*. Aust Vet J **56**:234–236.
120. McCray, E., Cooperative Needlestick Study Group. 1986. Occupational risk of the acquired immunodeficiency syndrome among health-care workers. N Engl J Med **314**:1127–32.
121. McDade, J.E., Shepard, C.C. 1979. Virulent to avirulent conversion of Legionnaire's disease bacterium (*Legionella pneumophila*)—its effect on isolation techniques. J Infect Dis **139**:707–711.
122. McKinney, R.W., Wasserman, B., Carpenter, M., Kent, M.D., Manning, C., Spillane, M., Richmond, J.Y. 1985. XXVII Biological Safety Conference, Salk Institute for Biological Studies, La Jolla, CA.
123. Melnick, J.L., Wenner, H.A., Phillips, C.A. 1979. Enteroviruses. In: *Diagnostic Procedures for Viral, Rickettsial and Chlamydial Infections*, 5th edition. Lennette, E.H., and Schmidt, N.J., Eds., Washington, D.C., American Public Health Association, pp. 471–534.
124. Meyer, K.F. 1915. The relationship of animal to human sporotrichosis: Studies on American sporotrichosis III. JAMA **65**:579–585.

164

125. Meyer, K.F., Eddie, B.: 1941. Laboratory infections due to *Brucella*. J Infect Dis **68**:24–32.
126. Meyers, W.M., Walsh, G.P., Brown, H.L., Fukunishi, Y., Binford, C.H., Gerone, P.J., Wolf, R.H. 1980. Naturally acquired leprosy in a mangabey monkey (*Cercocebus* sp.) Int J Lepr **48**:495–496.
127. Miller, C.D., Songer, J.R., Sullivan, J.F. 1987. A twenty-five year review of laboratory acquired human infections at the National Animal Disease Center. Am Ind Hyg Assoc J **48**:271–275.
128. Miyamura, T., Saito, I., Katayama, T., Kikuchi, S., Tateda, A., Houghton, M., Choo, Q-L., Kuo, G. 1990. Detection of antibody against antigen expressed by molecularly cloned hepatitis C virus cDNA: application to diagnosis and blood screening for post-transfusion hepatitis. Proc Natl Acad Sci USA **87**:983–987.
129. Morisset, R., Spink, W.W. 1969. Epidemic canine brucellosis due to a new species, *Brucella canis*. Lancet **2**:1000–2.
130. Morse, S.I. 1968. Pertussis in Adults (editorial). Ann Intern Med. **68**:953.
131. Müller, H.E. 1988. Laboratory-acquired mycobacterial infection. Lancet **2**:331.
132. Murray, J.F., Howard, D.H. 1964. Laboratory-acquired histoplasmosis. Am Rev Respir Dis **89**:631–640.
133. Nabarro, J.D.N. 1948. Primary pulmonary coccidioidomycosis: Case of laboratory infection in England. Lancet **1**:982–984.
134. National Committee for Clinical Laboratory Standards (NCCLS). 1991. *Protection of laboratory workers from infectious disease transmitted by blood, body fluids, and tissue*. Tentative Guideline. M29-T2, Vol. 11, No. 14.
135. Nelson, J.D. 1978. The changing epidemiology of pertussis in young infants. The role of adults as reservoirs of infection. Am J Dis Child **132**:371.
136. National Sanitation Foundation Standard 49. 1983. Class II (Laminar Flow) Biohazard Cabinetry. Ann Arbor, Michigan.

165

137. Norden, A. 1951. Sporotrichosis: Clinical and laboratory features and a serologic study in experimental animals and humans. Acta Pathol Microbiol Scand. Suppl. **89**:3–119.
138. Oates, J.D., Hodgin, U.G., Jr. 1981. Laboratory-acquired *Campylobacter* enteritis. South Med J **74**:83.
139. Office of Research Safety, National Cancer Institute, and the Special Committee of Safety and Health Experts. 1978. *Laboratory Safety Monograph:* A Supplement to the NIH Guidelines for Recombinant DNA Research. Bethesda, MD, National Institutes of Health.
140. Oliphant, J.W., Parker, R.R. 1948. Q fever: Three cases of laboratory infection. Public Health Rep **63(42)**:1364–1370.
141. Oliphant, J.W., Gordon, D.A., Meis, A., Parker, R.R. 1949. Q fever in laundry workers, presumably transmitted from contaminated clothing. Am J Hyg **49**(1):76–82.
142. Oliver, J., Reusser, T.R. 1942. Rapid method for the concentration of tubercle bacilli. Am Rev Tuberc **45**:450–452.
143. Olle-Goig, J. and Canela-Soler, J.C. 1987. An outbreak of *Brucella melitensis* infection by airborne transmission among laboratory workers. Am J Publ Hlth **77**:335–338.
144. Oster, C.N., et al. 1977. Laboratory-acquired Rocky Mountain spotted fever. The hazard of aerosol transmission. N Engl J Med **297**:859–862.

145. Palmer, A.E. 1987. B Virus, *Herpesvirus simiae:* historical perspective. J. Med. Primatol **16:**99–130.
146. Parker, C. 1992. Dept. of Microbiology, University of Missouri, Columbia, Missouri (personal communication).
147. Parritt, R.J., Olsen, R.E. 1947. Two simultaneous cases of leprosy developing in tattoos. Am J Pathol **23:**805–817.
148. Patterson, W.C., Mott, L.O., Jenney, E.W. 1958. A study of vesicular stomatitis in man. J Am Vet Med Assoc **133**(1)**:**57–62.

166

149. Penner, J.L., Hennessy, J.N., Mills, S.D., Bradbury, W.C. 1983. Application of serotyping and chromosomal restriction endonuclease digest analysis in investigating a laboratory-acquired case of *Campylobacter jejuni* enteritis. J Clin Microbiol **18:**1427–1428.
150. Perkins, F.T., Hartley, E.G. 1966. Precautions against B virus infection. Brit Med J **1:**899–901.
151. Pike, R.M. 1976. Laboratory-associated infections: Summary and analysis of 3,921 cases. Hlth Lab Sci **13:**105–114.
152. Pike, R.M. 1978. Past and present hazards of working with infectious agents. Arch Path Lab Med **102:**333–336.
153. Pike, R.M. 1979. Laboratory-associated infections: incidence, fatalities, causes and prevention. Ann Rev Microbiol **33:**41–66.
154. Pike, R.M., Sulkin, S.E., Schulze, M.L. 1965. Continuing importance of laboratory-acquired infections. Am J Public Health **55:**190–199.
155. Prescott, J.F., Karmali, M.A. 1978. Attempts to transmit *Campylobacter enteritis* to dogs and cats (letter). Can Med Assoc J **119:**1001–1002.
156. Reid, D.D. 1957. Incidence of tuberculosis among workers in medical laboratories. Brit Med J **2:**10–14.
157. Report of the Committee of Inquiry into the Smallpox Outbreak in London in March and April 1973. 1974. Her Majesty's Stationery Office, London.
158. Resnick L, Veren K, Salahuddin, SZ, Tondreau S, Markham PD. 1986. Stability and inactivation of HTLV-III/LAV under clinical and laboratory environments. JAMA **255:**1887–91.
159. Richardson, J.H. 1973. Provisional summary of 109 laboratory-associated infections at the Center for Disease Control, 1947–1973. Presented at the 16th Annual Biosafety Conference, Ames, Iowa.
160. Riley, R.L. 1957. Aerial dissemination of pulmonary tuberculosis. Am. Rev Tuberc **76:**931–941.
161. Riley, R.L. 1961. Airborne pulmonary tuberculosis. Bacteriol Rev **25:**243–248.

167

162. Robertson, D.H.H., Pickens, S., Lawson, J.H., Lennex, B. 1980. An accidental laboratory infection with African trypanosomes of a defined stock. I and II. J Inf **2:**105–112, 113–124.
163. Robinson, D.A. 1981. Infective dose of *Campylobacter jejuni* in milk. Brit Med J **282:**1584.
164. Rutala, WA. 1990. APIC guideline for selection and use of disinfectants. Am. J. Infection Control. **18:**99–117.
165. Samuel, J.E., Frazier, M.E., Mallavia, L.P. 1985. Correlation of plasmid type and disease caused by Coxiella burnetii. Infect Immun **49:**775–779.
166. Sastaw, S., Carlisle, H.N. 1966. Aerosol infection of monkeys with *Rickettsia rickettsii.* Bact Rev **30:**636–645.
166a. Schlech, W.F., Turchik, J.B., Westlake, R.E., Jr., Klein, G.C., Band, J.D., Weaver, R.E. 1981. Laboratory-acquired infection with *Pseudomona pseudomallie* (meliodosis). N Eng J Med **305:**1133–1135.
167. Schochetman, G., George, J.R. 1991. AIDS Testing: Methodology and Management Issues. Springer-Verlag, New York.
168. Schwarz, J., Baum, G.L. 1951. Blastomycosis. Amer J Clin Pathol **21:**999–1029.
169. Segretain, G. 1959. *Penicillium marnefii,* n. sp., agent d'une mycose du systeme reticulo-endothelial. Mycopathol Mycol Appl **11:**327–353.
170. Skinholj, P. 1974. Occupational risks in Danish clinical chemical laboratories. II Infections. Scand J Clin Lab Invest **33:**27–29.
171. Smith, C.E. 1950. The hazard of acquiring mycotic infections in the laboratory. Presented at 78th Ann. Meeting Am Pub Hlth Assoc, St. Louis, MO.
172. Smith, C.E., Pappagianis, D., Levine, H.B., Saito, 1961. Human coccidioidomycosis. Bacteriol Rev **25:**310–320.

168

172a.Smith, D.K., Neal, J.J., Holmberg, S.D., Centers for Disease Control Idiopathic CD4-T-lymphocytopenia Task Force. 1992. Unexplained opportunistic infections in CD4-T-lymphocytopenia in persons without HIV infection, U.S. New England Journal of Medicine (submitted by Journal request).
173. Smith, D.T., Harrell, E.R., Jr. 1948. Fatal Coccidioidomycosis: A case of laboratory infection. Am Rev Tuberc **57:**368–374.
174. Smithwick, R.W., Stratigos, C.B. 1978. Preparation of acid-fast microscopy smears for proficiency testing and quality control. J Clin Microbiol **8:**110–111.
175. Spinelli, J.S., et al. 1981. Q fever crisis in San Francisco: Controlling a sheep zoonosis in a lab animal facility. Lab Anim **10**(3)**:**24–27.
176. Spink, W.W. 1956. *The Nature of Brucellosis.* Minneapolis, The University of Minnesota Press, pp. 106–108.
177. Sterne, M., Wertzel, L.M. 1950. A new method of large-scale production of high-titer botulinum formol-toxoid types C and D. J Immunol **65:**175–183.
178. Subcommittee on Arbovirus Laboratory Safety for Arboviruses and Certain Other Viruses of Vertebrates. 1980. Am J Trop Med Hyg **29**(6)**:**1359–1381.
179. Sulkin, S.E., Pike, R.M. 1949. Viral Infections Contracted in the Laboratory. New Engl J Med **241**(5)**:**205–213.
180. Sulkin, S.E., Pike, R.M. 1951. Survey of laboratory-acquired infections. Am J Public Health **41**(7)**:**769–781.
181. Sullivan, J.F., Songer, J.R., Estrem, I.E. 1978. Laboratory-acquired infections at the National Animal Disease Center, 1960–1976. Health Lab Sci **15**(1)**:**58–64.
182. Tesh, R.B., and Schneidau, J.D., Jr. 1966. Primary cutaneous histoplasmosis. New Engl J Med **275:**597–599.
183. Thompson, D.W., Kaplan, W. 1977. Laboratory-acquired sporotrichosis. Sabouraudia **15:**167–170.

169

184. Tomlinson, C.C., Bancroft, P. 1928. Granuloma coccidioides: Report of a case responding favorably to antimony and potassium tartrate. JAMA **91:**947–951.
185. Tosh, F.E., Balhuizen, J., Yates, J.L., Brasher, C.A. 1964. Primary cutaneous histoplasmosis: Report of a case. Arch Intern Med **114:**118–119.
186. U.S. Congress, 1988. *Medical Waste Tracking Act of 1988*. H.R. 3515, 42 U.S.C. 6992–6992k.

187. U.S. Department of Labor, Occupational Safety and Health Administration. 1991. *Occupational Exposure to Bloodborne Pathogens, Final Rule.* Fed. Register **56:**64175–64182.
188. U.S. Environmental Protection Agency. 1986. EPA guide for infectious waste management. Washington DC: U.S. Environmental Protection Agency; Publication no. EPA/530-5W-86-014.
189. Walsh, G.P., Storrs, E.E., Burchfield, H.P. Cottrel, E.H., Vidrine, M.F., Binford, C.H. 1975. Leprosy-like disease occurring naturally in armadillos. J Reticuloendothel Soc **18:**347–351.
190. Wedum, A.G., Kruse, R.H. 1969. *Assessment of risk of human infection in the microbiology laboratory.* Misc Pub 30, Industrial Health and Safety Directorate, Fort Detrick, Frederick, Md.
191. Wedum, A.G., Barkley, W.E., Hellman, A. 1972. Handling of infectious agents. J Am Vet Med Assoc **161:**1557–1567.
192. Wedum, A.G. History of Microbiological Safety. 1975. 18th Biological Safety Conference. Lexington, Kentucky.
193. Weigler, B.J. 1992. Biology of B Virus in Macaque and Human Hosts: A Review. Clin Infect Dis **14:**555–67.
194. Weissenbacher, M.C., Grela, M.E., Sabattini, M.S., Maiztegui, J.I., Coto, C.E., Frigerio, M.J., Cossio, P.M., Rabinovich, A.S., Oro, J.G.B. 1978. Inapparent infections with Junin virus among laboratory workers. J Infect Dis **137:**309–313.
195. Welsh, H.H., Lennette, E.H., Abinanti, F.R., Winn, J.F. 1951. Q fever in California IV. Occurrence of *Coxiella burnetii* in the placenta of naturally infected sheep. Public Health Rep. **66:**1473–1477.
196. Wells, D.L., Lipper, S.L., Hilliard, J.K., Stewart, J.A., Holmes, G.P., Herrmann, K.L., Kiley, M.P., Schonberger, L.B. 1989. *Herpesvirus simiae* Contamination of Primary Rhesus Monkey Kidney Cell Cultures. Centers for Disease Control Recommendations to Minimize Risks to Laboratory Personnel. Diagn Microbiol Infect Dis **12:**333–336.
197. Wilder, W. H., McCullough, C.P. 1914. Sporotrichosis of the eye. JAMA **62:**1156–1160.
198. Willems, W.R., Kaluza, G., Boschek, C.B., Bauer, H. 1979. Semliki Forest Virus: Cause of a Fatal Case of Human Encephalitis. Science **203:**1127–1129.
199. Wilson, J.W., Cawley, E.P., Weidman, F.D., Gilmer, W.S. 1955. Primary cutaneous North American blastomycosis. Arch Dermatol **71:**39–45.
200. Wilson, J.W., Smith, C.E., Plunkett, O.A. 1953. Primary cutaneous coccidioidomycosis; the criteria for diagnosis and a report of a case. Calif Med **79:**233–239.
201. Winkler, W.G. 1973. Airborne rabies transmission in a laboratory worker. JAMA **226**(10):1219–1221.
202. World Health Organization. 1978. Smallpox surveillance. Weekly Epidemiological Record **53**(35):265–266.

APPENDIX

Occupational Exposure to Bloodborne Pathogens; Final Rule (29 CFR Part 1910.1030)†

U.S. DEPARTMENT OF LABOR, OCCUPATIONAL SAFETY AND HEALTH ADMINISTRATION

XI. THE STANDARD

General Industry

Part 1910 of title 29 of the Code of Federal Regulations is amended as follows:

PART 1910—[AMENDED]

Subpart Z—[Amended]

1. The general authority citation for subpart Z of 29 CFR part 1910 continues to read as follows and a new citation for § 1910.1030 is added:

Authority: Secs. 6 and 8, Occupational Safety and Health Act, 29 U.S.C. 655, 657, Secretary of Labor's Orders Nos. 12-71 (36 FR 8754), 8-76 (41 FR 25059), or 9-83 (48 FR 35736), as applicable; and 29 CFR part 1911.

* * * * * *

Section 1910.1030 also issued under 29 U.S.C. 653.

* * * * * *

2. Section 1910.1030 is added to read as follows:

§ 1910.1030 Bloodborne Pathogens.

(a) Scope and Application. This section applies to all occupational exposure to blood or other potentially infectious materials as defined by paragraph (b) of this section.

(b) Definitions. For purposes of this section, the following shall apply:

Assistant Secretary means the Assistant Secretary of Labor for Occupational Safety and Health, or designated representative.

Blood means human blood, human blood components, and products made from human blood.

Bloodborne Pathogens means pathogenic microorganisms that are present in human blood and can cause disease in humans. These pathogens include, but are not limited to, hepatitis B virus (HBV) and human immunodeficiency virus (HIV).

Clinical Laboratory means a workplace where diagnostic or other screening procedures are performed on blood or other potentially infectious materials.

Contaminated means the presence or the reasonably anticipated presence of blood or other potentially infectious materials on an item or surface.

Contaminated Laundry means laundry which has been soiled with blood or other potentially infectious materials or may contain sharps.

Contaminated Sharps means any contaminated object that can penetrate the skin including, but not limited to, needles, scalpels, broken glass, broken capillary tubes, and exposed ends of dental wires.

†Reprinted from *Fed. Regist.* **56**(235):64175–64182, Rules and Regulations, 6 December 1991.

Decontamination means the use of physical or chemical means to remove, inactivate, or destroy bloodborne pathogens on a surface or item to the point where they are no longer capable of transmitting infectious particles and the surface or item is rendered safe for handling, use, or disposal.

Director means the Director of the National Institute for Occupational Safety and Health, U.S. Department of Health and Human Services, or designated representative.

Engineering Controls means controls (e.g., sharps disposal containers, self-sheathing needles) that isolate or remove the bloodborne pathogens hazard from the workplace.

Exposure Incident means a specific eye, mouth, other mucous membrane, non-intact skin, or parenteral contact with blood or other potentially infectious materials that results from the performance of an employee's duties.

Handwashing Facilities means a facility providing an adequate supply of running potable water, soap and single use towels or hot air drying machines.

Licensed Healthcare Professional is a person whose legally permitted scope of practice allows him or her to independently perform the activities required by paragraph (f) Hepatitis B Vaccination and Post-exposure Evaluation and Follow-up.

HBV means hepatitis B virus.

HIV means human immunodeficiency virus.

Occupational Exposure means reasonably anticipated skin, eye, mucous membrane, or parenteral contact with blood or other potentially infectious materials that may result from the performance of an employee's duties.

Other Potentially Infectious Materials means

(1) The following human body fluids: semen, vaginal secretions, cerebrospinal fluid, synovial fluid, pleural fluid, pericardial fluid, peritoneal fluid, amniotic fluid, saliva in dental procedures, any body fluid that is visibly contaminated with blood, and all body fluids in situations where it is difficult or impossible to differentiate between body fluids;

(2) Any unfixed tissue or organ (other than intact skin) from a human (living or dead); and

(3) HIV-containing cell or tissue cultures, organ cultures, and HIV- or HBV-containing culture medium or other solutions; and blood, organs, or other tissues from experimental animals infected with HIV or HBV.

Parenteral means piercing mucous membranes or the skin barrier through such events as needlesticks, human bites, cuts, and abrasions.

Personal Protective Equipment is specialized clothing or equipment worn by an employee for protection against a hazard. General work clothes (e.g., uniforms, pants, shirts or blouses) not intended to function as protection against a hazard are not considered to be personal protective equipment.

Production Facility means a facility engaged in industrial-scale, large-volume or high concentration production of HIV or HBV.

Regulated Waste means liquid or semi-liquid blood or other potentially infectious materials; contaminated items that would release blood or other potentially infectious materials in a liquid or semi-liquid state if compressed; items that are caked with dried blood or other potentially infectious materials and are capable of releasing these materials during handling; contaminated sharps; and pathological and microbiological wastes containing blood or other potentially infectious materials.

Research Laboratory means a laboratory producing or using research-laboratory-scale amounts of HIV or HBV. Research laboratories may produce high concentrations of HIV or HBV but not in the volume found in production facilities.

Source Individual means any individual, living or dead, whose blood or other potentially infectious materials may be a source of occupational exposure to the employee. Examples include, but are not limited to, hospital and clinic patients; clients in institutions for the developmentally disabled; trauma victims; clients of drug and alcohol treatment facilities; residents of hospices and nursing homes; human remains; and individuals who donate or sell blood or blood components.

Sterilize means the use of a physical or chemical procedure to destroy all microbial life including highly resistant bacterial endospores.

Universal Precautions is an approach to infection control. According to the concept of Universal Precautions, all human blood and certain human body fluids are treated as if known to be infectious for HIV, HBV, and other bloodborne pathogens.

Work Practice Controls means controls that reduce the likelihood of exposure by altering the manner in which a task is performed (e.g., prohibiting recapping of needles by a two-handed technique).

(c) Exposure control—(1) Exposure Control Plan. (i) Each employer having an employee(s) with occupational exposure as defined by paragraph (b) of this section shall establish a written Exposure Control Plan designed to eliminate or minimize employee exposure.

(ii) The Exposure Control Plan shall contain at least the following elements:

(A) The exposure determination required by paragraph (c)(2),

(B) The schedule and method of implementation for paragraphs (d) Methods of Compliance, (e) HIV and HBV Research Laboratories and Production Facilities, (f) Hepatitis B Vaccination and Post-Exposure Evaluation and Follow-up, (g) Communication of Hazards to Employees, and (h) Recordkeeping, of this standard, and

(C) The procedure for the evaluation of circumstances surrounding exposure incidents as required by paragraph (f)(3)(i) of this standard.

(iii) Each employer shall ensure that a copy of the Exposure Control Plan is accessible to employees in accordance with 29 CFR 1910.20(e).

(iv) The Exposure Control Plan shall be reviewed and updated at least annually and whenever necessary to reflect new or modified tasks and procedures which affect occupational exposure and to reflect new or revised employee positions with occupational exposure.

(v) The Exposure Control Plan shall be made available to the Assistant Secretary and the Director upon request for examination and copying.

(2) Exposure determination. (i) Each employer who has an employee(s) with occupational exposure as defined by paragraph (b) of this section shall prepare an exposure determination. This exposure determination shall contain the following:

(A) A list of all job classifications in which all employees in those job classifications have occupational exposure;

(B) A list of job classifications in which some employees have occupational exposure, and

(C) A list of all tasks and procedures or groups of closely related task and procedures in which occupational exposure occurs and that are performed by employees in job classifications listed in accordance with the provisions of paragraph (c)(2)(i)(B) of this standard.

(ii) This exposure determination shall be made without regard to the use of personal protective equipment.

(d) Methods of compliance—(1) General—Universal precautions shall be observed to prevent contact with blood or other potentially infectious materials. Under circumstances in which differentiation between body fluid types is difficult or impossible, all body fluids shall be considered potentially infectious materials.

(2) Engineering and work practice controls. (i) Engineering and work practice controls shall be used to eliminate or minimize employee exposure. Where occupational exposure remains after institution of these controls, personal protective equipment shall also be used.

(ii) Engineering controls shall be examined and maintained or replaced on a regular schedule to ensure their effectiveness.

(iii) Employers shall provide handwashing facilities which are readily accessible to employees.

(iv) When provision of handwashing facilities is not feasible, the employer shall provide either an appropriate antiseptic hand cleanser in conjunction with clean cloth/paper towels or antiseptic towelettes. When antiseptic hand cleansers or towelettes are used, hands shall be washed with soap and running water as soon as feasible.

(v) Employers shall ensure that employees wash their hands immediately or as soon as feasible after removal of gloves or other personal protective equipment.

(vi) Employers shall ensure that employees wash hands and any other skin with soap and water, or flush mucous membranes with water immediately or as soon as feasible following contact of such body areas with blood or other potentially infectious materials.

(vii) Contaminated needles and other contaminated sharps shall not be bent, recapped, or removed except as noted in paragraphs (d)(2)(vii)(A) and (d)(2)(vii)(B) below. Shearing or breaking of contaminated needles is prohibited.

(A) Contaminated needles and other contaminated sharps shall not be recapped or removed unless the employer can demonstrate that no alternative is feasible or that such action is required by a specific medical procedure.

(B) Such recapping or needle removal must be accomplished through the use of a mechanical device or a one-handed technique.

(viii) Immediately or as soon as possible after use, contaminated reusable sharps shall be placed in appropriate containers until properly reprocessed.

These containers shall be:

(A) Puncture resistant;

(B) Labeled or color-coded in accordance with this standard;

(C) Leakproof on the sides and bottom; and

(D) In accordance with the requirements set forth in paragraph (d)(4)(ii)(E) for reusable sharps.

(ix) Eating, drinking, smoking, applying cosmetics or lip balm, and handling contact lenses are prohibited in work areas where there is a reasonable likelihood of occupational exposure.

(x) Food and drink shall not be kept in refrigerators, freezers, shelves, cabinets or on countertops or benchtops where blood or other potentially infectious materials are present.

(xi) All procedures involving blood or other potentially infectious materials shall be performed in such a manner as to minimize splashing, spraying, spattering, and generation of droplets of these substances.

(xii) Mouth pipetting/suctioning of blood or other potentially infectious materials is prohibited.

(xiii) Specimens of blood or other potentially infectious materials shall be placed in a container which prevents leakage during collection, handling, processing, storage, transport, or shipping.

(A) The container for storage, transport, or shipping shall be labeled or color-coded according to paragraph (g)(1)(i) and closed prior to being stored, transported, or shipped. When a facility utilizes Universal Precautions in the handling of all specimens, the labeling/color-coding of specimens is not necessary provided containers are recognizable as containing specimens. This exemption only applies while such specimens/containers remain within the facility. Labeling or color-coding in accordance with paragraph (g)(1)(i) is required when such specimens/containers leave the facility.

(B) If outside contamination of the primary container occurs, the primary container shall be placed within a second container which prevents leakage during handling, processing, storage, transport, or shipping and is labeled or color-coded according to the requirements of this standard.

(C) If the specimen could puncture the primary container, the primary container shall be placed within a secondary container which is puncture-resistant in addition to the above characteristics.

(xiv) Equipment which may become contaminated with blood or other potentially infectious materials shall be examined prior to servicing or shipping and shall be decontaminated as necessary, unless the employer can demonstrate that decontamination of such equipment or portions of such equipment is not feasible.

(A) A readily observable label in accordance with paragraph (g)(1)(i)(H) shall be attached to the equipment stating which portions remain contaminated.

(B) The employer shall ensure that this information is conveyed to all affected employees, the servicing representative, and/or the manufacturer, as appropriate, prior to handling, servicing, or shipping so that appropriate precautions will be taken.

(3) Personal protective equipment—(i) Provision. When there is occupational exposure, the employer shall provide, at no cost to the employee, appropriate personal protective equipment such as, but not limited to, gloves, gowns, laboratory coats, face shields or masks and eye protection, and mouthpieces, resuscitation bags, pocket masks, or other ventilation devices. Personal protective equipment will be considered "appropriate" only if it does not permit blood or other potentially infectious materials to pass through to or reach the employee's work clothes, street clothes, undergarments, skin, eyes, mouth, or other mucous membranes under normal conditions of use and for the duration of time which the protective equipment will be used.

(ii) Use. The employer shall ensure that the employee uses appropriate personal protective equipment unless the employer shows that the employee temporarily and briefly declined to use personal protective equipment when, under rare and extraordinary circumstances, it was the employee's professional judgment that in the specific instance its use would have prevented the delivery of health care or public safety services or would have posed an increased hazard to the safety of the worker or coworker. When the employee makes this judgment, the circumstances shall be investigated and documented in order to determine whether changes can be instituted to prevent such occurrences in the future.

(iii) Accessibility. The employer shall ensure that appropriate personal protective equipment in the appropriate sizes is readily accessible at the worksite or is issued to employees. Hypoallergenic gloves, glove liners, powderless gloves, or other similar alternatives shall be readily accessible to those employees who are allergic to the gloves normally provided.

(iv) Cleaning, Laundering, and Disposal. The employer shall clean, launder, and dispose of personal protective equipment required by paragraphs (d) and (e) of this standard, at no cost to the employee.

(v) Repair and Replacement. The employer shall repair or replace personal protective equipment as needed to maintain its effectiveness, at no cost to the employee.

(vi) If a garment(s) is penetrated by blood or other potentially infectious materials, the garment(s) shall be removed immediately or as soon as feasible.

(vii) All personal protective equipment shall be removed prior to leaving the work area.

(viii) When personal protective equipment is removed it shall be placed in an appropriately designated area or container for storage, washing, decontamination or disposal.

(ix) Gloves. Gloves shall be worn when it can be reasonably anticipated that the employee may have hand contact with blood, other potentially infectious materials, mucous membranes, and non-intact skin; when performing vascular access procedures except as specified in paragraph (d)(3)(ix)(D); and when handling or touching contaminated items or surfaces.

(A) Disposable (single use) gloves such as surgical or examination gloves, shall be replaced as soon as practical when contaminated or as soon as feasible if they are torn, punctured, or when their ability to function as a barrier is compromised.

(B) Disposable (single use) gloves shall not be washed or decontaminated for re-use.

(C) Utility gloves may be decontaminated for re-use if the integrity of the glove is not compromised. However, they must be discarded if they are cracked, peeling, torn, punctured, or exhibit other signs of deterioration or when their ability to function as a barrier is compromised.

(D) If an employer in a volunteer blood donation center judges that routine gloving for all phlebotomies is not necessary then the employer shall:

(1) Periodically reevaluate this policy;

(2) Make gloves available to all employees who wish to use them for phlebotomy;

(3) Not discourage the use of gloves for phlebotomy; and

(4) Require that gloves be used for phlebotomy in the following circumstances:

(i) When the employee has cuts, scratches, or other breaks in his or her skin;

(ii) When the employee judges that hand contamination with blood may occur, for example, when performing phlebotomy on an uncooperative source individual; and

(iii) When the employee is receiving training in phlebotomy.

(x) Masks, Eye Protection, and Face Shields. Masks in combination with eye protection devices, such as goggles or glasses with solid side shields, or chin-length face shields, shall be worn whenever splashes, spray, spatter, or droplets of blood or other potentially infectious materials may be generated and eye, nose, or mouth contamination can be reasonably anticipated.

(xi) Gowns, Aprons, and Other Protective Body Clothing. Appropriate protective clothing such as, but not limited to, gowns, aprons, lab coats, clinic jackets, or similar outer garments shall be worn in occupational exposure situations. The type and characteristics will depend upon the task and degree of exposure anticipated.

(xii) Surgical caps or hoods and/or shoe covers or boots shall be worn in instances when gross contamination can reasonably be anticipated (e.g., autopsies, orthopaedic surgery).

(4) Housekeeping. (i) General. Employers shall ensure that the worksite is maintained in a clean and sanitary condition. The employer shall determine and implement an appropriate written schedule for cleaning and method of decontamination based upon the location within the facility, type of surface to be cleaned, type of soil present, and tasks or procedures being performed in the area.

(ii) All equipment and environmental and working surfaces shall be cleaned and decontaminated after contact with blood or other potentially infectious materials.

(A) Contaminated work surfaces shall be decontaminated with an appropriate disinfectant after completion of procedures; immediately or as soon as feasible when surfaces are overtly contaminated or after any spill of blood or other potentially infectious materials; and at the end of the work shift if the surface may have become contaminated since the last cleaning.

(B) Protective coverings, such as plastic wrap, aluminum foil, or imperviously-backed absorbent paper used to cover equipment and environmental surfaces, shall be removed and replaced as soon as feasible when they become overtly contaminated or at the end of the workshift if they may have become contaminated during the shift.

(C) All bins, pails, cans, and similar receptacles intended for reuse which have a reasonable likelihood for becoming contaminated with blood or other potentially infectious materials shall be inspected and decontaminated on a regularly scheduled basis and cleaned and decontaminated immediately or as soon as feasible upon visible contamination.

(D) Broken glassware which may be contaminated shall not be picked up directly with the hands. It shall be cleaned up using mechanical means, such as a brush and dust pan, tongs, or forceps.

(E) Reusable sharps that are contaminated with blood or other potentially infectious materials shall not be stored or processed in a manner that requires employees to reach by hand into the containers where these sharps have been placed.

(iii) Regulated Waste.

(A) Contaminated Sharps Discarding and Containment.

(1) Contaminated sharps shall be discarded immediately or as soon as feasible in containers that are:

(i) Closable;

(ii) Puncture resistant;

(iii) Leakproof on sides and bottom; and

(iv) Labeled or color-coded in accordance with paragraph (g)(1)(i) of this standard.

(2) During use, containers for contaminated sharps shall be:

(i) Easily accessible to personnel and located as close as is feasible to the immediate area where sharps are used or can be reasonably anticipated to be found (e.g., laundries);

(ii) Maintained upright throughout use; and

(iii) Replaced routinely and not be allowed to overfill.

(3) When moving containers of contaminated sharps from the area of use, the containers shall be:

(i) Closed immediately prior to removal or replacement to prevent spillage or protrusion of contents during handling, storage, transport, or shipping;

(ii) Placed in a secondary container if leakage is possible. The second container shall be:

(A) Closable;

(B) Constructed to contain all contents and prevent leakage during handling, storage, transport, or shipping; and

(C) Labeled or color-coded according to paragraph (g)(1)(i) of this standard.

(4) Reusable containers shall not be opened, emptied, or cleaned manually or in any other manner which would expose employees to the risk of percutaneous injury.

(B) Other Regulated Waste Containment. (1) Regulated waste shall be placed in containers which are:

(i) Closable;

(ii) Constructed to contain all contents and prevent leakage of fluids during handling, storage, transport or shipping;

(iii) Labeled or color-coded in accordance with paragraph (g)(1)(i) of this standard; and

(iv) Closed prior to removal to prevent spillage or protrusion of contents during handling, storage, transport, or shipping.

(2) If outside contamination of the regulated waste container occurs, it shall be placed in a second container. The second container shall be:

(i) Closable;

(ii) Constructed to contain all contents and prevent leakage of fluids during handling, storage, transport, or shipping;

(iii) Labeled or color-coded in accordance with paragraph (g)(1)(i) of this standard; and

(iv) Closed prior to removal to prevent spillage or protrusion of contents during handling, storage, transport, or shipping.

(C) Disposal of all regulated waste shall be in accordance with applicable regulations of the United States, States and Territories, and political subdivisions of States and Territories.

(iv) Laundry.

(A) Contaminated laundry shall be handled as little as possible with a minimum of agitation.

(1) Contaminated laundry shall be bagged or containerized at the location where it was used and shall not be sorted or rinsed in the location of use.

(2) Contaminated laundry shall be placed and transported in bags or containers labeled or color-coded in accordance with paragraph (g)(1)(i) of this standard. When a facility utilizes Universal Precautions in the handling of all soiled laundry, alternative labeling or color-coding is sufficient if it permits all employees to recognize the containers as requiring compliance with Universal Precautions.

(3) Whenever contaminated laundry is wet and presents a reasonable likelihood of soak-through of or leakage from the bag or container, the laundry shall be placed and transported in bags or containers which prevent soak-through and/or leakage of fluids to the exterior.

(B) The employer shall ensure that employees who have contact with contaminated laundry wear protective gloves and other appropriate personal protective equipment.

(C) When a facility ships contaminated laundry off-site to a second facility which does not utilize Universal Precautions in the handling of all laundry, the facility generating the contaminated laundry must place such laundry in bags or containers which are labeled or color-coded in accordance with paragraph (g)(1)(i).

(e) HIV and HBV Research Laboratories and Production Facilities.

(1) This paragraph applies to research laboratories and production facilities engaged in the culture, production, concentration, experimentation, and manipulation of HIV and HBV. It does not apply to clinical or diagnostic laboratories engaged solely in the analysis of blood, tissues, or organs. These requirements apply in addition to the other requirements of the standard.

(2) Research laboratories and production facilities shall meet the following criteria:

(i) Standard microbiological practices. All regulated waste shall either be incinerated or decontaminated by a method such as autoclaving known to effectively destroy bloodborne pathogens.

(ii) Special practices.

(A) Laboratory doors shall be kept closed when work involving HIV or HBV is in progress.

(B) Contaminated materials that are to be decontaminated at a site away from the work area shall be placed in a durable, leakproof, labeled or color-coded container that is closed before being removed from the work area.

(C) Access to the work area shall be limited to authorized persons. Written policies and procedures shall be established whereby only persons who have been advised of the potential biohazard, who meet any specific entry requirements, and who comply with all entry and exit procedures shall be allowed to enter the work areas and animal rooms.

(D) When other potentially infectious materials or infected animals are present in the work area or containment module, a hazard warning sign incorporating the universal biohazard symbol shall be posted on all access doors. The hazard warning sign shall comply with paragraph (g)(1)(ii) of this standard.

(E) All activities involving other potentially infectious materials shall be conducted in biological safety cabinets or other physical-containment devices within the containment module. No work with these other potentially infectious materials shall be conducted on the open bench.

(F) Laboratory coats, gowns, smocks, uniforms, or other appropriate protective clothing shall be used in the work area and animal rooms. Protective clothing shall not be worn outside of the work area and shall be decontaminated before being laundered.

(G) Special care shall be taken to avoid skin contact with other potentially infectious materials. Gloves shall be worn when handling infected animals and when making hand contact with other potentially infectious materials is unavoidable.

(H) Before disposal all waste from work areas and from animal rooms shall either be incinerated or decontaminated by a method such as autoclaving known to effectively destroy bloodborne pathogens.

(I) Vacuum lines shall be protected with liquid disinfectant traps and high-efficiency particulate air (HEPA) filters or filters of equivalent or superior efficiency and which are checked routinely and maintained or replaced as necessary.

(J) Hypodermic needles and syringes shall be used only for parenteral injection and aspiration of fluids from laboratory animals and diaphragm bottles. Only needle-locking syringes or disposable syringe-needle units (i.e., the needle is integral to the syringe) shall be used for the injection or aspiration of other potentially infectious materials. Extreme caution shall be used when handling needles and syringes. A needle shall not be bent, sheared, replaced in the sheath or guard, or removed from the syringe following use. The needle and syringe shall be promptly placed in a puncture-resistant container and autoclaved or decontaminated before reuse or disposal.

(K) All spills shall be immediately contained and cleaned up by appropriate professional staff or others properly trained and equipped to work with potentially concentrated infectious materials.

(L) A spill or accident that results in an exposure incident shall be immediately reported to the laboratory director or other responsible person.

(M) A biosafety manual shall be prepared or adopted and periodically reviewed and updated at least annually or more often if necessary. Personnel shall be advised of potential hazards, shall be required to read instructions on practices and procedures, and shall be required to follow them.

(iii) Containment equipment.

(A) Certified biological safety cabinets (Class I, II, or III) or other appropriate combinations of personal protection or physical containment devices, such as special protective clothing, respirators, centrifuge safety cups, sealed centrifuge rotors, and containment caging for animals, shall be used for all activities with other potentially infectious materials that pose a threat of exposure to droplets, splashes, spills, or aerosols.

(B) Biological safety cabinets shall be certified when installed, whenever they are moved and at least annually.

(3) HIV and HBV research laboratories shall meet the following criteria:

(i) Each laboratory shall contain a facility for hand washing and an eye wash facility which is readily available within the work area.

(ii) An autoclave for decontamination of regulated waste shall be available.

(4) HIV and HBV production facilities shall meet the following criteria:

(i) The work areas shall be separated from areas that are open to unrestricted traffic flow within the building. Passage through two sets of doors shall be the basic requirement for entry into the work area from access corridors or other contiguous areas. Physical separation of the high-containment work area from access corridors or other areas or activities may also be provided by a double-doored clothes-change room (showers may be included), airlock, or other access facility that requires passing through two sets of doors before entering the work area.

(ii) The surfaces of doors, walls, floors and ceilings in the work area shall be water resistant so that they can be easily cleaned. Penetrations in these surfaces shall be sealed or capable of being sealed to facilitate decontamination.

(iii) Each work area shall contain a sink for washing hands and a readily available eye wash facility. The sink shall be foot, elbow, or automatically operated and shall be located near the exit door of the work area.

(iv) Access doors to the work area or containment module shall be self-closing.

(v) An autoclave for decontamination of regulated waste shall be available within or as near as possible to the work area.

(vi) A ducted exhaust-air ventilation system shall be provided. This system shall create directional airflow that draws air into the work area through the entry area. The exhaust air shall not be recirculated to any other area of the building, shall be discharged to the outside, and shall be dispersed away from occupied areas and air intakes. The proper direction of the airflow shall be verified (i.e., into the work area).

(5) Training Requirements. Additional training requirements for employees in HIV and HBV research laboratories and HIV and HBV production facilities are specified in paragraph (g)(2)(ix).

(f) Hepatitis B vaccination and post-exposure evaluation and follow-up—(1) General. (i) The employer shall make available the hepatitis B vaccine and vaccination series to all employees who have occupational exposure, and post-exposure evaluation and follow-up to all employees who have had an exposure incident.

(ii) The employer shall ensure that all medical evaluations and procedures including the hepatitis B vaccine and vaccination series and post-exposure evaluation and follow-up, including prophylaxis, are:

(A) Made available at no cost to the employee;

(B) Made available to the employee at a reasonable time and place;

(C) Performed by or under the supervision of a licensed physician or by or under the supervision of another licensed healthcare professional; and

(D) Provided according to recommendations of the U.S. Public Health Service current at the time these evaluations and procedures take place, except as specified by this paragraph (f).

(iii) The employer shall ensure that all laboratory tests are conducted by an accredited laboratory at no cost to the employee.

(2) Hepatitis B Vaccination. (i) Hepatitis B vaccination shall be made available after the employee has received the training required in paragraph (g)(2)(vii)(I) and within 10 working days of initial assignment to all employees who have occupational exposure unless the employee has previously received the complete hepatitis B vaccination series, antibody testing has revealed that the employee is immune, or the vaccine is contraindicated for medical reasons.

(ii) The employer shall not make participation in a prescreening program a prerequisite for receiving hepatitis B vaccination.

(iii) If the employee initially declines hepatitis B vaccination but at a later date while still covered under the standard decides to accept the vaccination, the employer shall make available hepatitis B vaccination at that time.

(iv) The employer shall assure that employees who decline to accept hepatitis B vaccination offered by the employer sign the statement in appendix A.

(v) If a routine booster dose(s) of hepatitis B vaccine is recommended by the U.S. Public Health Service at a future date, such booster dose(s) shall be made available in accordance with section (f)(1)(ii).

(3) Post-exposure Evaluation and Follow-up. Following a report of an exposure incident, the employer shall make immediately available to the exposed employee a confidential medical evaluation and follow-up, including at least the following elements:

(i) Documentation of the route(s) of exposure, and the circumstances under which the exposure incident occurred;

(ii) Identification and documentation of the source individual, unless the employer can establish that identification is infeasible or prohibited by state or local law;

(A) The source individual's blood shall be tested as soon as feasible and after consent is obtained in order to determine HBV and HIV infectivity. If consent is not obtained, the employer shall establish that legally required consent cannot be obtained. When the source individual's consent is not required by law, the source individual's blood, if available, shall be tested and the results documented.

(B) When the source individual is already known to be infected with HBV or HIV, testing for the source individual's known HBV or HIV status need not be repeated.

(C) Results of the source individual's testing shall be made available to the exposed employee, and the employee shall be informed of applicable laws and regulations concerning disclosure of the identity and infectious status of the source individual.

(iii) Collection and testing of blood for HBV and HIV serological status;

(A) The exposed employee's blood shall be collected as soon as feasible and tested after consent is obtained.

(B) If the employee consents to baseline blood collection, but does not give consent at that time for

HIV serologic testing, the sample shall be preserved for at least 90 days. If, within 90 days of the exposure incident, the employee elects to have the baseline sample tested, such testing shall be done as soon as feasible.

(iv) Post-exposure prophylaxis, when medically indicated, as recommended by the U.S. Public Health Service;

(v) Counseling; and

(vi) Evaluation of reported illnesses.

(4) Information Provided to the Healthcare Professional. (i) The employer shall ensure that the healthcare professional responsible for the employee's Hepatitis B vaccination is provided a copy of this regulation.

(ii) The employer shall ensure that the healthcare professional evaluating an employee after an exposure incident is provided the following information:

(A) A copy of this regulation;

(B) A description of the exposed employee's duties as they relate to the exposure incident;

(C) Documentation of the route(s) of exposure and circumstances under which exposure occurred;

(D) Results of the source individual's blood testing, if available; and

(E) All medical records relevant to the appropriate treatment of the employee including vaccination status which are the employer's responsibility to maintain.

(5) Healthcare Professional's Written Opinion. The employer shall obtain and provide the employee with a copy of the evaluating healthcare professional's written opinion within 15 days of the completion of the evaluation.

(i) The healthcare professional's written opinion for Hepatitis B vaccination shall be limited to whether Hepatitis B vaccination is indicated for an employee, and if the employee has received such vaccination.

(ii) The healthcare professional's written opinion for post-exposure evaluation and follow-up shall be limited to the following information:

(A) That the employee has been informed of the results of the evaluation; and

(B) That the employee has been told about any medical conditions resulting from exposure to blood or other potentially infectious materials which require further evaluation or treatment.

(iii) All other findings or diagnoses shall remain confidential and shall not be included in the written report.

(6) Medical recordkeeping. Medical records required by this standard shall be maintained in accordance with paragraph (h)(1) of this section.

(g) Communication of hazards to employees—(1) Labels and signs. (i) Labels.

(A) Warning labels shall be affixed to containers of regulated waste, refrigerators and freezers containing blood or other potentially infectious material; and other containers used to store, transport or ship blood or other potentially infectious materials, except as provided in paragraph (g)(1)(i)(E), (F) and (G).

(B) Labels required by this section shall include the following legend:

(C) These labels shall be fluorescent orange or orange-red or predominantly so, with lettering or symbols in a contrasting color.

(D) Labels [are required to be] affixed as close as feasible to the container by string, wire, adhesive, or other method that prevents their loss or unintentional removal.

(E) Red bags or red containers may be substituted for labels.

(F) Containers of blood, blood components, or blood products that are labeled as to their contents and have been released for transfusion or other clinical use are exempted from the labeling requirements of paragraph (g).

(G) Individual containers of blood or other potentially infectious materials that are placed in a labeled container during storage, transport, shipment or disposal are exempted from the labeling requirement.

(H) Labels required for contaminated equipment shall be in accordance with this paragraph and shall also state which portions of the equipment remain contaminated.

(I) Regulated waste that has been decontaminated need not be labeled or color-coded.

(ii) Signs.

(A) The employer shall post signs at the entrance to work areas specified in paragraph (e), HIV and HBV Research Laboratory and Production Facilities, which shall bear the following legend:

(Name of the Infectious Agent)

(Special requirements for entering the area)

(Name, telephone number of the laboratory director or other responsible person.)

(B) These signs shall be fluorescent orange-red or predominantly so, with lettering or symbols in a contrasting color.

(2) Information and Training. (i) Employers shall ensure that all employees with occupational exposure participate in a training program which must be provided at no cost to the employee and during working hours.

(ii) Training shall be provided as follows:

(A) At the time of initial assignment to tasks where occupational exposure may take place;

(B) Within 90 days after the effective date of the standard; and

(C) At least annually thereafter.

(iii) For employees who have received training on bloodborne pathogens in the year preceding the effective date of the standard, only training with respect to the provisions of the standard which were not included need be provided.

(iv) Annual training for all employees shall be provided within one year of their previous training.

(v) Employers shall provide additional training when changes such as modification of tasks or procedures or institution of new tasks or procedures affect the employee's occupational exposure. The additional training may be limited to addressing the new exposures created.

(vi) Material appropriate in content and vocabulary to educational level, literacy, and language of employees shall be used.

(vii) The training program shall contain at a minimum the following elements:

(A) An accessible copy of the regulatory text of this standard and an explanation of its contents;

(B) A general explanation of the epidemiology and symptoms of bloodborne diseases;

(C) An explanation of the modes of transmission of bloodborne pathogens;

(D) An explanation of the employer's exposure control plan and the means by which the employee can obtain a copy of the written plan;

(E) An explanation of the appropriate methods for recognizing tasks and other activities that may involve exposure to blood and other potentially infectious materials;

(F) An explanation of the use and limitations of methods that will prevent or reduce exposure including appropriate engineering controls, work practices, and personal protective equipment;

(G) Information on the types, proper use, location, removal, handling, decontamination and disposal of personal protective equipment;

(H) An explanation of the basis for selection of personal protective equipment;

(I) Information on the hepatitis B vaccine, including information on its efficacy, safety, method of administration, the benefits of being vaccinated, and that the vaccine and vaccination will be offered free of charge;

(J) Information on the appropriate actions to take and persons to contact in an emergency involving blood or other potentially infectious materials;

(K) An explanation of the procedure to follow if an exposure incident occurs, including the method of reporting the incident and the medical follow-up that will be made available;

(L) Information on the post-exposure evaluation and follow-up that the employer is required to provide for the employee following an exposure incident;

(M) An explanation of the signs and labels and/or color coding required by paragraph (g)(1); and

(N) An opportunity for interactive questions and answers with the person conducting the training session.

(viii) The person conducting the training shall be knowledgeable in the subject matter covered by the elements contained in the training program as it relates to the workplace that the training will address.

(ix) Additional Initial Training for Employees in HIV and HBV Laboratories and Production Facilities. Employees in HIV or HBV research laboratories and HIV or HBV production facilities shall receive the following initial training in addition to the above training requirements.

(A) The employer shall assure that employees demonstrate proficiency in standard microbiological practices and techniques and in the practices and operations specific to the facility before being allowed to work with HIV or HBV.

(B) The employer shall assure that employees have prior experience in the handling of human pathogens or tissue cultures before working with HIV or HBV.

(C) The employer shall provide a training program to employees who have no prior experience in handling human pathogens. Initial work activities shall not include the handling of infectious agents. A progression of work activities shall be assigned as techniques are learned and proficiency is developed. The employer shall assure that employees participate in work activities involving infectious agents only after proficiency has been demonstrated.

(h) Recordkeeping—(1) Medical Records. (i) The employer shall establish and maintain an accurate record for each employee with occupational exposure, in accordance with 29 CFR 1910.20.

(ii) This record shall include:

(A) The name and social security number of the employee;

(B) A copy of the employee's hepatitis B vaccination status including the dates of all the hepatitis B vaccinations and any medical records relative to the employee's ability to receive vaccination as required by paragraph (f)(2);

(C) A copy of all results of examinations, medical testing, and follow-up procedures as required by paragraph (f)(3);

(D) The employer's copy of the healthcare professional's written opinion as required by paragraph (f)(5); and

(E) A copy of the information provided to the healthcare professional as required by paragraphs (f)(4)(ii)(B)(C) and (D).

(iii) Confidentiality. The employer shall ensure that employee medical records required by paragraph (h)(1) are:

(A) Kept confidential; and

(B) Are not disclosed or reported without the employee's express written consent to any person within or outside the workplace except as required by this section or as may be required by law.

(iv) The employer shall maintain the records required by paragraph (h) for at least the duration of employment plus 30 years in accordance with 29 CFR 1910.20.

(2) Training Records.

(i) Training records shall include the following information:

(A) The dates of the training sessions;

(B) The contents or a summary of the training sessions;

(C) The names and qualifications of persons conducting the training; and

(D) The names and job titles of all persons attending the training sessions.

(ii) Training records shall be maintained for 3 years from the date on which the training occurred.

(3) Availability. (i) The employer shall ensure that all records required to be maintained by this section shall be made available upon request to the Assistant Secretary and the Director for examination and copying.

(ii) Employee training records required by this paragraph shall be provided upon request for examination and copying to employees, to employee representatives, to the Director, and to the Assistant Secretary in accordance with 29 CFR 1910.20.

(iii) Employee medical records required by this paragraph shall be provided upon request for examination and copying to the subject employee, to anyone having written consent of the subject employee, to the Director, and to the Assistant Secretary in accordance with 29 CFR 1910.20.

(4) Transfer of Records. (i) The employer shall comply with the requirements involving transfer of records set forth in 29 CFR 1910.20(h).

(ii) If the employer ceases to do business and there is no successor employer to receive and retain the records for the prescribed period, the employer shall notify the Director, at least three months prior to their disposal and transmit them to the Director, if required by the Director to do so, within that three month period.

(i) Dates—(1) Effective Date. The standard shall become effective on March 6, 1992.

(2) The Exposure Control Plan required by paragraph (c)(2) of this section shall be completed on or before May 5, 1992.

(3) Paragraph (g)(2) Information and Training and (h) Recordkeeping shall take effect on or before June 4, 1992.

(4) Paragraphs (d)(2) Engineering and Work Practice Controls, (d)(3) Personal Protective Equipment, (d)(4) Housekeeping, (e) HIV and HBV Research Laboratories and Production Facilities, (f) Hepatitis B Vaccination and Post-Exposure Evaluation and Follow-up, and (g) (1) Labels and Signs, shall take effect July 6, 1992.

Appendix A to Section 1910.1030—Hepatitis B Vaccine Declination (Mandatory)

I understand that due to my occupational exposure to blood or other potentially infectious materials I may be at risk of acquiring hepatitis B virus (HBV) infection. I have been given the opportunity to be vaccinated with hepatitis B vaccine, at no charge to myself. However, I decline hepatitis B vaccination at this time. I understand that by declining this vaccine, I continue to be at risk of acquiring hepatitis B, a serious disease. If in the future I continue to have occupational exposure to blood or other potentially infectious materials and I want to be vaccinated with hepatitis B vaccine, I can receive the vaccination series at no charge to me.

APPENDIX

Recommendations Concerning Chemical Hygiene in Laboratories (Nonmandatory)†

NATIONAL RESEARCH COUNCIL

FOREWORD

As guidance for each employer's development of an appropriate laboratory Chemical Hygiene Plan, the following non-mandatory recommendations are provided. They were extracted from "Prudent Practices for Handling Hazardous Chemicals in Laboratories" (referred to below as "Prudent Practices"), which was published in 1981 by the National Research Council and is available from the National Academy Press, 2101 Constitution Ave., NW, Washington, DC 20418. "Prudent Practices" is cited because of its wide distribution and acceptance and because of its preparation by members of the laboratory community through the sponsorship of the National Research Council.

However, none of the recommendations given here will modify any requirements of the laboratory standard. This appendix merely presents pertinent recommendations from "Prudent Practices," organized into a form convenient for quick reference during operation of a laboratory facility and during development and application of a Chemical Hygiene Plan. Users of this appendix should consult "Prudent Practices" for a more extended presentation and justification for each recommendation.

†This is Appendix A (pp. 3331–3335) to OSHA Standard 1910.1450: *Occupational Exposures to Hazardous Chemicals in Laboratories.* Reprinted from *Fed. Regist.* **55:**3300–3335, 31 January 1990.

"Prudent Practices" deals with both safety and chemical hazards while the laboratory standard is concerned primarily with chemical hazards.

Therefore, only those recommendations directed primarily toward control of toxic exposures are cited in this appendix, with the term *chemical hygiene* being substituted for the word *safety.* However, since conditions producing or threatening physical injury often pose toxic risks as well, page references concerning major categories of safety hazards in the laboratory are given in section F. The recommendations from "Prudent Practices" have been paraphrased, combined, or otherwise reorganized, and headings have been added. However, their sense has not been changed.

CORRESPONDING SECTIONS OF THE STANDARD AND THIS APPENDIX

The following table is given for the convenience of those who are developing a Chemical Hygiene Plan which will satisfy the requirements of paragraph (e) of the standard. It indicates those sections of this appendix which are most pertinent to each of the sections of paragraph (e) and related paragraphs.

In this appendix, those recommendations directed primarily at administrators and supervisors are given in sections A–D. Those recommendations of primary concern to employees who are actually handling laboratory chemicals are given in section E. (References to page numbers in "Prudent Practices" are given in parentheses.)

Paragraph and topic in laboratory standard	Relevant section
(e)(3)(i) Standard operating procedures for handling toxic chemicals	C, D, E
(e)(3)(ii) Criteria to be used for implementation of measures to reduce exposures	D
(e)(3)(iii) Fume hood performance	C4b
(e)(3)(iv) Employee information and training (including emergency procedures)	D9, D10
(e)(3)(v) Requirements for prior approval of laboratory activities	E2b, E4b
(e)(3)(vi) Medical consultation and medical examinations	D5, E4f
(e)(3)(vii) Chemical hygiene responsibilities	B
(e)(3)(viii) Special precautions for work with particularly hazardous substances	E2, E3, E4

A. GENERAL PRINCIPLES FOR WORK WITH LABORATORY CHEMICALS

In addition to the more detailed recommendations listed below in sections B–E, "Prudent Practices" expresses certain general principles, including the following:

1. It is prudent to minimize all chemical exposures. Because few laboratory chemicals are without hazards, general precautions for handling all laboratory chemicals should be adopted, rather than specific guidelines for particular chemicals (2, 10). Skin contact with chemicals should be avoided as a cardinal rule (198).

2. Avoid underestimation of risk. Even for substances of no known significant hazard, exposure should be minimized; for work with substances which present special hazards, special precautions should be taken (10, 37, 38). One should assume that any mixture will be more toxic than its most toxic component (30, 103) and that all substances of unknown toxicity are toxic (3, 34).

3. Provide adequate ventilation. The best way to prevent exposure to airborne substances is to prevent their escape into the working atmosphere by use of hoods and other ventilation devices (32, 198).

4. Institute a chemical hygiene program. A mandatory chemical hygiene program designed to minimize exposures is needed; it should be a regular, continuing effort, not merely a standby or short-term activity (6, 11).

Its recommendations should be followed in academic teaching laboratories as well as by full-time laboratory workers (13).

5. Observe the PELs, TLVs. The Permissible Exposure Limits of OSHA and the Threshold Limit Values of the American Conference of Governmental Industrial Hygienists should not be exceeded (13).

B. CHEMICAL HYGIENE RESPONSIBILITIES

Responsibility for chemical hygiene rests at all levels (6, 11, 21) including the:

1. Chief executive officer, who has ultimate responsibility for chemical hygiene within the institution and must, with other administrators, provide continuing support for institutional chemical hygiene (7, 11).

2. Supervisor of the department or other administrative unit, who is responsible for chemical hygiene in that unit (7).

3. Chemical hygiene officer(s), whose appointment is essential (7) and who must:

(a) Work with administrators and other employees to develop and implement appropriate chemical hygiene policies and practices (7);

(b) Monitor procurement, use, and disposal of chemicals used in the lab (8);

(c) See that appropriate audits are maintained (8);

(d) Help project directors develop precautions and adequate facilities (10);

(e) Know the current legal requirements concerning regulated substances (50); and

(f) Seek ways to improve the chemical hygiene program (8, 11).

4. Laboratory supervisor, who has overall responsibility for chemical hygiene in the laboratory (21) including responsibility to:

(a) Ensure that workers know and follow the chemical hygiene rules, that protective equipment is available and in working order, and that appropriate training has been provided (21, 22);

(b) Provide regular, formal chemical hygiene and housekeeping inspections including routine inspections of emergency equipment (21, 171);

(c) Know the current legal requirements concerning regulated substances (50, 231);

(d) Determine the required levels of protective apparel and equipment (156, 160, 162); and

(e) Ensure that facilities and training for use of any material being ordered are adequate (215).

5. Project director or director of other specific operation, who has primary responsibility for chemical hygiene procedures for that operation (7).

6. Laboratory worker, who is responsible for:

(a) Planning and conducting each operation in accordance with the institutional chemical hygiene procedures (7, 21, 22, 230); and (b) Developing good personal chemical hygiene habits (22).

C. THE LABORATORY FACILITY

1. Design. The laboratory facility should have:

(a) An appropriate general ventilation system (see C4 below) with air intakes and exhausts located so as to avoid intake of contaminated air (194);

(b) Adequate, well-ventilated stockrooms/storerooms (218, 219);

(c) Laboratory hoods and sinks (12, 162);

(d) Other safety equipment including eyewash fountains and drench showers (162, 169); and

(e) Arrangements for waste disposal (12, 240).

2. Maintenance. Chemical-hygiene-related equipment (hoods, incinerator, etc.) should undergo continual appraisal and be modified if inadequate (11, 12).

3. Usage. The work conducted (10) and its scale (12) must be appropriate to the physical facilities available and, especially, to the quality of ventilation (13).

4. Ventilation—(a) General laboratory ventilation. This system should: Provide a source of air for breathing and for input to local ventilation devices (199); it should not be relied on for protection from toxic substances released into the laboratory (198); ensure that laboratory air is continually replaced, preventing increase of air concentrations of toxic substances during the working day (194); direct air flow into the laboratory from non-laboratory areas and out to the exterior of the building (194).

(b) Hoods. A laboratory hood with 2.5 linear feet of hood space per person should be provided for every 2 workers if they spend most of their time working with chemicals (199); each hood should have a continuous monitoring device to allow convenient confirmation of adequate hood performance before use (200, 209). If this is not possible, work with substances of unknown toxicity should be avoided (13) or other types of local ventilation devices should be provided (199). See pp. 201–206 for a discussion of hood design, construction, and evaluation.

(c) Other local ventilation devices. Ventilated storage cabinets, canopy hoods, snorkels, etc. should be provided as needed (199). Each canopy hood and snorkel should have a separate exhaust duct (207).

(d) Special ventilation areas. Exhaust air from glove boxes and isolation rooms should be passed through scrubbers or other treatment before release into the regular exhaust system (208). Cold rooms and warm rooms should have provisions for rapid escape and for escape in the event of electrical failure (209).

(e) Modifications. Any alteration of the ventilation system should be made only if thorough testing indicates that worker protection from airborne toxic substances will continue to be adequate (12, 193, 204).

(f) Performance. Rate: 4–12 room air changes/hour is normally adequate general ventilation if local exhaust systems such as hoods are used as the primary method of control (194).

(g) Quality. General air flow should not be turbulent and should be relatively uniform throughout the laboratory, with no high velocity or static areas (194, 195); airflow into and within the hood should not be excessively turbulent (200); hood face velocity should be adequate (typically 60–100 lfm) (200, 204).

(h) Evaluation. Quality and quantity of ventilation should be evaluated on installation (202), regularly monitored (at least every 3 months) (6, 12, 14, 195), and reevaluated whenever a change in local ventilation devices is made (12, 195, 207). See pp. 195–196 for methods of evaluation and for calculation of estimated airborne contaminant concentrations.

D. COMPONENTS OF THE CHEMICAL HYGIENE PLAN

1. Basic Rules and Procedures (Recommendations for these are given in section E, below)

2. Chemical Procurement, Distribution, and Storage

(a) Procurement. Before a substance is received, information on proper handling, storage, and disposal should be known to those who will be involved (215, 216). No container should be accepted without an adequate identifying label (216). Preferably, all substances should be received in a central location (216).

(b) Stockrooms/storerooms. Toxic substances should be segregated in a well-identified area with local exhaust ventilation (221). Chemicals which are highly toxic (227) or other chemicals whose containers have been opened should be in unbreakable secondary containers (219). Stored chemicals should be examined periodically (at least annually) for replacement, deterioration, and container integrity (218–219). Stockrooms/storerooms should not be used as preparation or repackaging areas, should be open during normal working hours, and should be controlled by one person (219).

(c) Distribution. When chemicals are hand carried, the container should be placed in an outside container or bucket. Freight-only elevators should be used if possible (223).

(d) Laboratory storage. Amounts permitted should be as small as practical. Storage on bench tops and in hoods is inadvisable. Exposure to heat or direct sunlight should be avoided. Periodic inventories should be conducted, with unneeded items being discarded or returned to the storeroom/stockroom (225–226, 229).

3. Environmental Monitoring

Regular instrumental monitoring of airborne concentrations is not usually justified or practical in laboratories but may be appropriate when testing or redesigning hoods or other ventilation devices (12) or when a highly toxic substance is stored or used regularly (e.g., 3 times/week) (13).

4. Housekeeping, Maintenance, and Inspections

(a) Cleaning. Floors should be cleaned regularly (24).

(b) Inspections. Formal housekeeping and chemical hygiene inspections should be held at least quarterly (6, 21) for units which have frequent personnel changes and semiannually for others; informal inspections should be continual (21).

(c) Maintenance. Eyewash fountains should be inspected at intervals of not less than 3 months (6). Respirators for routine use should be inspected periodically by the laboratory supervisor (169). Other safety equipment should be inspected regularly (e.g., every 3–6 months) (6, 24, 171). Procedures to prevent restarting of out-of-service equipment should be established (25).

(d) Passageways. Stairways and hallways should not be used as storage areas (24). Access to exits, emergency equipment, and utility controls should never be blocked (24).

5. Medical Program

(a) Compliance with regulations. Regular medical surveillance should be established to the extent required by regulations (12).

(b) Routine surveillance. Anyone whose work involves regular and frequent handling of toxicologically significant quantities of a chemical should consult a qualified physician to determine on an individual basis whether a regular schedule of medical surveillance is desirable (11, 50).

(c) First aid. Personnel trained in first aid should be available during working hours and an emergency room with medical personnel should be nearby (173). See pp. 176–178 for description of some emergency first aid procedures.

6. Protective Apparel and Equipment

These should include for each laboratory:

(a) Protective apparel compatible with the required degree of protection for substances being handled (158–161);

(b) An easily accessible drench-type safety shower (162, 169); (c) An eyewash fountain (162);

(d) A fire extinguisher (162–164);

(e) Respiratory protection (164–169), fire alarm and telephone for emergency use (162) should be available nearby; and

(f) Other items designated by the laboratory supervisor (156, 160).

7. Records

(a) Accident records should be written and retained (174).

(b) Chemical Hygiene Plan records should document that the facilities and precautions were compatible with current knowledge and regulations (7).

(c) Inventory and usage records for high-risk substances should be kept as specified in sections E3e below.

(d) Medical records should be retained by the institution in accordance with the requirements of state and federal regulations (12).

8. Signs and Labels

Prominent signs and labels of the following types should be posted:

(a) Emergency telephone numbers of emergency personnel/facilities, supervisors, and laboratory workers (28);

(b) Identity labels, showing contents of containers (including waste receptacles) and associated hazards (27, 48);

(c) Location signs for safety showers, eyewash stations, other safety and first aid equipment, exits (27) and areas where food and beverage consumption and storage are permitted (24); and

(d) Warnings at areas or equipment where special or unusual hazards exist (27).

9. Spills and Accidents

(a) A written emergency plan should be established and communicated to all personnel; it should include procedures for ventilation failure (200), evacuation, medical care, reporting, and drills (172).

(b) There should be an alarm system to alert people in all parts of the facility including isolation areas such as cold rooms (172).

(c) A spill control policy should be developed and should include consideration of prevention, containment, cleanup, and reporting (175).

(d) All accidents or near accidents should be carefully analyzed with the results distributed to all who might benefit (8, 28).

10. Information and Training Program

(a) Aim: To assure that all individuals at risk are adequately informed about the work in the laboratory, its risks, and what to do if an accident occurs (5, 15).

(b) Emergency and Personal Protection Training: Every laboratory worker should know the location and proper use of available protective apparel and equipment (154, 169). Some of the full-time personnel of the laboratory should be trained in the proper use of emergency equipment and procedures (6). Such training as well as first aid instruction should be available to (154) and encouraged for (176) everyone who might need it.

(c) Receiving and stockroom/storeroom personnel should know about hazards, handling equipment, protective apparel, and relevant regulations (217).

(d) Frequency of Training: The training and education program should be a regular, continuing activity—not simply an annual presentation (15).

(e) Literature/Consultation: Literature and consulting advice concerning chemical hygiene should be readily available to laboratory personnel, who should be encouraged to use these information resources (14).

11. Waste Disposal Program.

(a) Aim: To assure that minimal harm to people, other organisms, and the environment will result from the disposal of waste laboratory chemicals (5).

(b) Content (14, 232, 233, 240): The waste disposal program should specify how waste is to be collected, segregated, stored, and transported and include consideration of what materials can be incinerated. Transport from the institution must be in accordance with DOT regulations (244).

(c) Discarding Chemical Stocks: Unlabeled containers of chemicals and solutions should undergo prompt disposal; if partially used, they should not be opened (24, 27). Before a worker's employment in the laboratory ends, chemicals for which that person was responsible should be discarded or returned to storage (226).

(d) Frequency of Disposal: Waste should be removed from laboratories to a central waste storage area at least once per week and from the central waste storage area at regular intervals (14).

(e) Method of Disposal: Incineration in an environmentally acceptable manner is the most practical disposal method for combustible laboratory waste (14, 238, 241). Indiscriminate disposal by pouring waste chemicals down the drain (14, 231, 242) or adding them to mixed refuse for landfill burial is unacceptable (14). Hoods should not be used as a means of disposal for volatile chemicals (40, 200). Disposal by recycling (233, 243) or chemical decontamination (40, 230) should be used when possible.

E. BASIC RULES AND PROCEDURES FOR WORKING WITH CHEMICALS

The Chemical Hygiene Plan should require that laboratory workers know and follow its rules and procedures. In addition to the procedures of the subprograms mentioned above, these should include the rules listed below.

1. General Rules

The following should be used for essentially all laboratory work with chemicals:

(a) Accidents and spills—Eye Contact: Promptly flush eyes with water for a prolonged period (15 minutes) and seek medical attention (33, 172). Ingestion: Encourage the victim to drink large amounts of water (178). Skin Contact: Promptly flush the affected area with water (33, 172, 178) and remove any contaminated clothing (172, 178). If symptoms persist after washing, seek medical attention (33). Clean-up: Promptly clean up spills, using appropriate protective apparel and equipment and proper disposal (24, 33). See pp. 233–237 for specific clean-up recommendations.

(b) Avoidance of "routine" exposure: Develop and encourage safe habits (23); avoid unnecessary exposure to chemicals by any route (23); Do not smell or taste chemicals (32). Vent apparatus which may discharge toxic chemicals (vacuum pumps, distillation columns, etc.) into local exhaust devices (199). Inspect gloves (157) and test glove boxes (208) before use. Do not allow release of toxic substances in cold rooms and warm rooms, since these have contained recirculated atmospheres (209).

(c) Choice of chemicals: Use only those chemicals for which the quality of the available ventilation system is appropriate (13).

(d) Eating, smoking, etc.: Avoid eating, drinking, smoking, gum chewing, or application of cosmetics in areas where laboratory chemicals are present (22, 24, 32, 40); wash hands before conducting these activities (23, 24). Avoid storage, handling, or consumption of food or beverages in storage areas, refrigerators, glassware or utensils which are also used for laboratory operations (23, 24, 226).

(e) Equipment and glassware: Handle and store laboratory glassware with care to avoid damage; do not use damaged glassware (25). Use extra care with Dewar flasks and other evacuated glass apparatus; shield or wrap them to contain chemicals and fragments should implosion occur (25). Use equipment only for its designed purpose (23, 26).

(f) Exiting: Wash areas of exposed skin well before leaving the laboratory (23).

(g) Horseplay: Avoid practical jokes or other behavior which might confuse, startle or distract another worker (23).

(h) Mouth suction: Do not use mouth suction for pipeting or starting a siphon (23, 32).

(i) Personal apparel: Confine long hair and loose clothing (23, 158). Wear shoes at all times in the laboratory but do not wear sandals, perforated shoes, or sneakers (158).

(j) Personal housekeeping: Keep the work area clean and uncluttered, with chemicals and equipment being properly labeled and stored; clean up the work area on completion of an operation or at the end of each day (24).

(k) Personal protection: Assure that appropriate eye protection (154–156) is worn by all persons, including visitors, where chemicals are stored or handled (22, 23, 33, 154). Wear appropriate gloves when the potential for contact with toxic materials exists (157); inspect the gloves before each use, wash them before removal, and replace them periodically (157). (A table of resistance to chemicals of common glove materials is given on p. 159). Use appropriate (164–168) respiratory equipment when air contaminant concentrations are not sufficiently restricted by engineering controls (164–165), inspecting the respirator before use (169). Use any other protective and emergency apparel and equipment as appropriate (22, 157–162). Avoid use of contact lenses in the laboratory unless necessary; if they are used, inform supervisor so special precautions can be taken (155). Remove laboratory coats immediately on significant contamination (161).

(l) Planning: Seek information and advice about hazards (7), plan appropriate protective procedures, and plan positioning of equipment before beginning any new operation (22, 23).

(m) Unattended operations: Leave lights on, place an appropriate sign on the door, and provide for containment of toxic substances in the event of failure of a utility service (such as cooling water) to an unattended operation (27, 128).

(n) Use of hood: Use the hood for operations which might result in release of toxic chemical vapors or dust (198–199). As a rule of thumb, use a hood or other local ventilation device when working with any appreciably volatile substance with a TLV of less than 50 ppm (13). Confirm adequate hood performance before use; keep hood closed at all times except when adjustments within the hood are being made (200); keep materials stored in hoods to a minimum and do not allow them to block vents or airflow (200). Leave the hood "on" when it is not in active use if toxic substances are stored in it or if it is uncertain whether adequate general laboratory ventilation will be maintained when it is "off" (200).

(o) Vigilance: Be alert to unsafe conditions and see that they are corrected when detected (22).

(p) Waste disposal: Assure that the plan for each laboratory operation includes plans and training for waste disposal (230). Deposit chemical waste in appropriately labeled receptacles and follow all other waste disposal procedures of the Chemical Hygiene Plan (22, 24). Do not discharge to the sewer concentrated acids or bases (231); highly toxic, malodorous, or lachrymatory substances (231); or any substances which might interfere with the biological activity of waste water treatment plants, create fire or explosion hazards, cause structural damage or obstruct flow (242).

(q) Working alone: Avoid working alone in a building; do not work alone in a laboratory if the procedures being conducted are hazardous (28).

2. Working with Allergens and Embryotoxins

(a) Allergens (examples: diazomethane, isocyanates, bichromates): Wear suitable gloves to prevent hand contact with allergens or substances of unknown allergenic activity (35).

(b) Embryotoxins (34–35) (examples: organomercurials, lead compounds, formamide): If you are a woman of childbearing age, handle these substances only in a hood whose satisfactory performance has been confirmed, using appropriate protective apparel (especially gloves) to prevent skin contact. Review each use of these materials with the research supervisor and review continuing uses annually or whenever a procedural change is made.

Store these substances, properly labeled, in an adequately ventilated area in an unbreakable secondary container.

Notify supervisors of all incidents of exposure or spills; consult a qualified physician when appropriate.

3. Work with Chemicals of Moderate Chronic or High Acute Toxicity

Examples: diisopropylfluorophosphate (41), hydrofluoric acid (43), hydrogen cyanide (45).

Supplemental rules to be followed in addition to those mentioned above (Procedure B of "Prudent Practices," pp. 39–41):

(a) Aim: To minimize exposure to these toxic substances by any route using all reasonable precautions (39).

(b) Applicability: These precautions are appropriate for substances with moderate chronic or high acute toxicity used in significant quantities (39).

(c) Location: Use and store these substances only in areas of restricted access with special warning signs (40, 229).

Always use a hood (previously evaluated to confirm adequate performance with a face velocity of at least 60 linear feet per minute) (40) or other containment device for procedures which may result in the generation of aerosols or vapors containing the substance (39); trap released vapors to revent their discharge with the hood exhaust (40).

(d) Personal protection: Always avoid skin contact by use of gloves and long sleeves (and other protective apparel as appropriate) (39). Always wash hands and arms immediately after working with these materials (40).

(e) Records: Maintain records of the amounts of these materials on hand, amounts used, and the names of the workers involved (40, 229).

(f) Prevention of spills and accidents: Be prepared for accidents and spills (41). Assure that at least 2 people are present at all times if a compound in use is highly toxic or of unknown toxicity (39). Store breakable containers of these substances in chemically resistant trays; also work and mount apparatus above such trays or cover work and storage surfaces with removable, absorbent, plastic backed paper (40). If a major spill occurs outside the hood, evacuate the area; assure that cleanup personnel wear suitable protective apparel and equipment (41).

(g) Waste: Thoroughly decontaminate or incinerate contaminated clothing or shoes (41). If possible, chemically decontaminate by chemical conversion (40). Store contaminated waste in closed, suitably labeled, impervious containers (for liquids, in glass or plastic bottles half-filled with vermiculite) (40).

4. Work with Chemicals of High Chronic Toxicity

(Examples: dimethylmercury and nickel carbonyl (48), benzo-a-pyrene (51), N-nitroso-diethylamine (54), other human carcinogens or substances with high carcinogenic potency in animals (38).) Further supplemental rules to be followed, in addition to all these mentioned above, for work with substances of known high chronic toxicity (in quantities above a few milligrams to a few grams, depending on the substance) (47). (Procedure A of "Prudent Practices" pp. 47–50).

(a) Access: Conduct all transfers and work with these substances in a "controlled area": a restricted access hood, glove box, or portion of a lab, designated for use of highly toxic substances, for which all people with access are aware of the substances being used and necessary precautions (48).

(b) Approvals: Prepare a plan for use and disposal of these materials and obtain the approval of the laboratory supervisor (48).

(c) Non-contamination/Decontamination: Protect vacuum pumps against contamination by scrubbers or HEPA filters and vent them into the hood (49). Decontaminate vacuum pumps or other contaminated equipment, including glassware, in the hood before removing them from the controlled area (49, 50). Decontaminate the controlled area before normal work is resumed there (50).

(d) Exiting: On leaving a controlled area, remove any protective apparel (placing it in an appropriate, labeled container) and thoroughly wash hands, forearms, face, and neck (49).

(e) Housekeeping: Use a wet mop or a vacuum cleaner equipped with a HEPA filter instead of dry sweeping if the toxic substance was a dry powder (50).

(f) Medical surveillance: If using toxicologically significant quantities of such a substance on a regular basis (e.g., 3 times per week), consult a qualified physician concerning desirability of regular medical surveillance (50).

(g) Records: Keep accurate records of the amounts of these substances stored (229) and used, the dates of use, and names of users (48).

(h) Signs and labels: Assure that the controlled area is conspicuously marked with warning and restricted access signs (49) and that all containers of these substances are appropriately labeled with identity and warning labels (48).

(i) Spills: Assure that contingency plans, equipment, and materials to minimize exposures of people and property in case of accident are available (233–234).

(j) Storage: Store containers of these chemicals only in a ventilated, limited access (48, 227, 229) area in appropriately labeled, unbreakable, chemically resistant, secondary containers (48, 229).

(k) Glove boxes: For a negative pressure glove box, ventilation rate must be at least 2 volume changes/hour and pressure at least 0.5 inches of water (48). For a positive pressure glove box, thoroughly check for leaks before each use (49). In either case, trap the exit gases or filter them through a HEPA filter and then release them into the hood (49).

(l) Waste: Use chemical decontamination whenever possible; ensure that containers of contaminated waste (including washings from contaminated flasks) are transferred from the controlled area in a secondary container under the supervision of authorized personnel (49, 50, 233).

5. Animal Work with Chemicals of High Chronic Toxicity

(a) Access: For large scale studies, special facilities with restricted access are preferable (56).

(b) Administration of the toxic substance: When possible, administer the substance by injection or gavage instead of in the diet. If administration is in the diet, use a caging system under negative pressure or under laminar airflow directed toward HEPA filters (56).

(c) Aerosol suppression: Devise procedures which minimize formation and dispersal of contaminated aerosols, including those from food, urine, and feces (e.g., use HEPA filtered vacuum equipment for cleaning, moisten contaminated bedding before removal from the cage, mix diets in closed containers in a hood) (55, 56).

(d) Personal protection: When working in the animal room, wear plastic or rubber gloves, fully buttoned laboratory coat or jumpsuit and, if needed because of incomplete suppression of aerosols, other apparel and equipment (shoe and head coverings, respirator) (56).

(e) Waste disposal: Dispose of contaminated animal tissues and excreta by incineration if the available incinerator can convert the contaminant to non-toxic products (238); otherwise, package the waste appropriately for burial in an EPA-approved site (239).

F. SAFETY RECOMMENDATIONS

The above recommendations from "Prudent Practices" do not include those which are directed primarily toward prevention of physical injury rather than toxic exposure. However, failure of precautions against injury will often have the secondary effect of causing toxic exposures. Therefore, we list below page references for recommendations concerning some of the major categories of safety hazards which also have implications for chemical hygiene:

1. Corrosive agents: (35–36)
2. Electrically powered laboratory apparatus: (179–192)
3. Fires, explosions: (26, 57–74, 162–164, 174–175, 219–220, 226–227)
4. Low temperature procedures: (26, 88)
5. Pressurized and vacuum operations (including use of compressed gas cylinders): (27, 75–101)

G. MATERIAL SAFETY DATA SHEETS

Material safety data sheets are presented in "Prudent Practices" for the chemicals listed below. (Asterisks denote that comprehensive material safety data sheets are provided.)

*Acetyl peroxide (105)
*Acrolein (106)
*Acrylonitrile
Ammonia (anhydrous)(91)
*Aniline (109)
*Benzene (110)
*Benzo[α]pyrene (112)
*Bis(chloromethyl) ether (113)
Boron trichloride (91)
Boron trifluoride (92)
Bromine (114)
*Tert-butyl hydroperoxide (148)
*Carbon disulfide (116)
Carbon monoxide (92)
*Carbon tetrachloride (118)
*Chlorine (119)
Chlorine trifluoride (94)
*Chloroform (121)
Chloromethane (93)
*Diethyl ether (122)
Diisopropyl fluorophosphate (41)
*Dimethylformamide (123)
*Dimethyl sulfate (125)
*Dioxane (126)
*Ethylene dibromide (128)
*Fluorine (95)
*Formaldehyde (130)
*Hydrazine and salts (132)
Hydrofluoric acid (43)
Hydrogen bromide (98)
Hydrogen chloride (98)
*Hydrogen cyanide (133)
*Hydrogen sulfide (135)
Mercury and compounds (52)
*Methanol (137)
*Morpholine (138)
*Nickel carbonyl (99)
*Nitrobenzene (139)
Nitrogen dioxide (100)
N-nitrosodiethylamine (54)
*Peracetic acid (141)
*Phenol (142)
*Phosgene (143)
*Pyridine (144)
*Sodium azide (145)
*Sodium cyanide (147)
Sulfur dioxide (101)
*Trichloroethylene (149)
*Vinyl chloride (150)

APPENDIX

Emergency First-Aid Guide†

IV

In an emergency, proper treatment should be institute by you as soon as possible. If care of the victim who is not breathing is delayed more than 4 min, death may occur. Patients with profuse bleeding that persists longer than 15 min may die.

CALL FOR HELP

1. If an injured person needs help and is breathing, telephone for help immediately.
2. If the victim is not breathing, start life-saving treatment and then ask someone to telephone your local rescue squad.
3. When you are talking to the emergency personnel:

    (i) Give the telephone number from which you are calling.
    (ii) Indicate the address of the victim's location, relating any special directions that would be helpful.
    (iii) Describe the victim's condition as clearly as possible—unconscious, burned, bleeding, and so on. If poison is involved, name the substance if possible.
    (iv) Give your name.
    (v) Do not hang up until the emergency personnel end the conversation. Additional information may be needed, or they may wish to give you instructions on what to do until help arrives.

†From the office of Richard Edlich.

BLEEDING

External

1. Immediately apply continuous, firm pressure over the wound, using the cleanest material that is available to you. Even your hand can be used, if necessary.
2. Elevate the bleeding site above the victim's heart.
3. If bleeding continues, apply more forceful, direct pressure on the wound.
4. If extensive bleeding has occurred, have the victim lie down and elevate his or her legs 12 in.
5. Do not give the victim anything to eat or drink. If part of the person's body has been cut off, wrap the part in the cleanest material you have and give it to the rescue squad when they arrive.

Internal

Even though you may not be able to see the bleeding, internal bleeding is a serious problem. Coughing or vomiting blood and/or passing blood in the urine or stool (a black, tarlike stool) are signs of internal bleeding. Victims of accidental injury may have bleeding that you cannot see.

1. Have the victim lie down and elevate his or her legs 12 in.

2. DO NOT GIVE THE VICTIM ANYTHING TO EAT OR DRINK WHILE WAITING FOR THE RESCUE SQUAD.

BURNS

Flame

1. Stop the burning by having the victim drop to the floor or ground and roll.
2. Smother any residual flames with a blanket.
3. If the victim is unconscious:
   (i) Open the airway by tilting the head back.
   (ii) Check for breathing.
   (iii) If the victim is not breathing, start CPR (see Unconscious Person, below).
4. Put a cool, moist compress on the burned area, or immerse it in cold water (1 to 5°C; 34 to 41°F) until the rescue squad arrives.
5. Cover the breathing victim with a clean sheet.

Chemical

Liquid Chemical

1. Pour large volumes of tap water over the burned area.
2. Remove clothes from burned area.
3. Continue flushing the area with water until the rescue squad arrives.

Dry Chemical

1. Brush chemical off clothing before removal of clothing.
2. After removal of clothing, visible solid particles should be brushed from the skin.
3. Pour large volumes of water, from any source, over the site contacted by the chemical until the rescue squad arrives.

DO NOT USE ANTIDOTES.

Electrical

1. Turn off the electrical circuit.
2. If the circuit cannot be disconnected, disengage the victim by using a long piece of wood (e.g., broom) or another nonconductive object. During the rescue, avoid contact with the electrical source.
3. If the victim is unconscious:
   (i) Open the airway by tilting the head back.

FIGURE 1 Choking, step 2.

   (ii) Check for breathing.
   (iii) If the victim is not breathing, start mouth-to-mouth ventilation (see Unconscious Person, below). If there is no pulse, start CPR.

CHOKING

If the person is choking and can speak or is coughing, do not interfere.

If the choking person cannot speak or cough, he or she may die within a few minutes unless the object stuck in his or her throat is dislodged.

1. Support the victim's chest with one hand. With the heel of your other hand, administer four sharp blows between the victim's shoulder blades.

2. Stand behind the victim with your arms directly under the victim's armpits and encircling his or her chest. Place the thumb side of one fist on the middle of the victim's breastbone. You should then grasp your fist with your other hand and exert four backward thrusts with the intent of relieving the obstruction (Fig. 1).

If the person still cannot speak or cough:

3. Alternate between the above two maneuvers until the victim either is no longer choking or loses consciousness.

4. If the person loses consciousness, help him or her to the floor, tilt his or her head back, and attempt to perform mouth-to-mouth ventilation (see Unconscious Person, below). Check for a pulse; if there is none, perform CPR.

5. If the chest does not rise when you perform mouth-to-mouth ventilation, roll the person onto his or her side and give four sharp blows between the shoulder blades.

6. Roll the victim onto his or her back and kneel close to the side of the body. Place the heel of one hand against the lower half of the breastbone. Cover

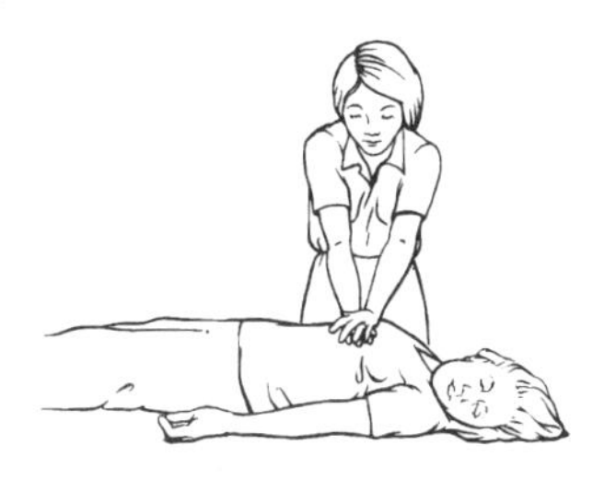

FIGURE 2 Choking, step 6.

this hand with your other hand. Quickly press down on the breastbone four times (Fig. 2).

7. You should attempt to remove the foreign body by using the tongue-jaw lift maneuver (Fig. 3). Open the victim's mouth by grasping both the tongue and lower jaw between your thumb and index finger of one hand and lift the tongue and jaw forward. This maneuver draws the tongue and lower jaw from the back of the throat and allows the index finger of your other hand to remove and dislodge the foreign body.

8. Attempt to give mouth-to-mouth ventilation again. If the victim's chest still does not rise, repeat steps 5 through 8 until the rescue squad arrives.

FROSTBITE

1. Gently remove all clothing from the affected area.
2. The skin of the affected area will appear gray white, or waxy.
3. Immerse the affected part in warm water (102 to 108°F; 39 to 42°C) until the skin becomes flushed or until the rescue squad arrives.
4. Cover the rewarmed part with dry, sterile dressings.
5. Elevate and immobilize the rewarmed part.

HEART ATTACK

Heart attacks are a leading cause of death. Signs and symptoms of a heart attack include one or more of the following:

FIGURE 3 Choking, step 7.

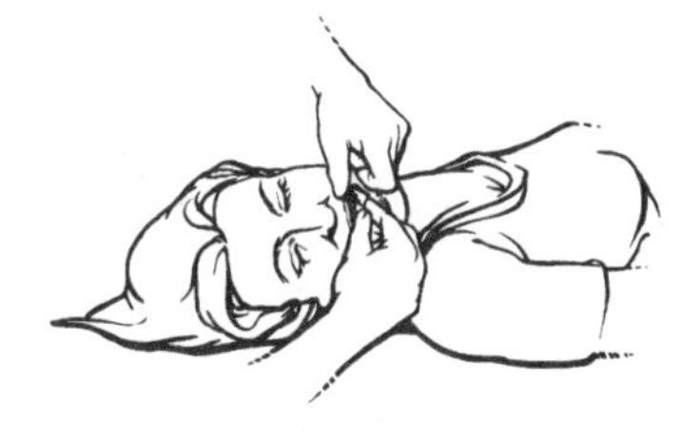

1. Squeezing chest pain that may radiate to the arms, stomach, jaw, or shoulders
2. Shortness of breath
3. Profuse sweating
4. Nausea or vomiting
5. Dizziness, faint-feeling, or weakness
6. Anxiety

If any of these signs and symptoms are detected, call the rescue squad immediately. While waiting for the rescue squad to arrive:

1. Reassure the patient that help is on the way. If possible, do not leave the patient unattended.
2. Loosen tight clothing (e.g., collars, belts).
3. Place the patient in a semisitting position to allow him or her to breathe more easily.
4. If the patient becomes unconscious:
   (i) Open the airway by tilting the head back.
   (ii) Check for breathing and for pulse.
   (iii) If the victim's heart and breathing have stopped, CPR is needed to avoid brain damage (see Unconscious Person, below).

HEAT STROKE

Signs and symptoms: hot or dry skin, altered state of consciousness, body temperature greater than 105°F (40.6°C).

Heat stroke occurs most frequently in the inactive aged and the overly active young and healthy.

1. Remove the victim's outer clothing.
2. If possible, wrap the patient in a sheet soaked with cool water and fan vigorously until the rescue squad arrives or until his or her temperature is less than 103°F (39.4°C).

INJURIES

If possible, an injured person should remain still until the rescue squad arrives.

Broken Bones

If bone ends are sticking out of the skin, do not push them back in. Cover the exposed ends with the cleanest material that is available to you.

Head Injury

A person who has an injury to his or her head may also have a neck injury and must not move or be moved.

1. Stabilize the victim's head with your hands.

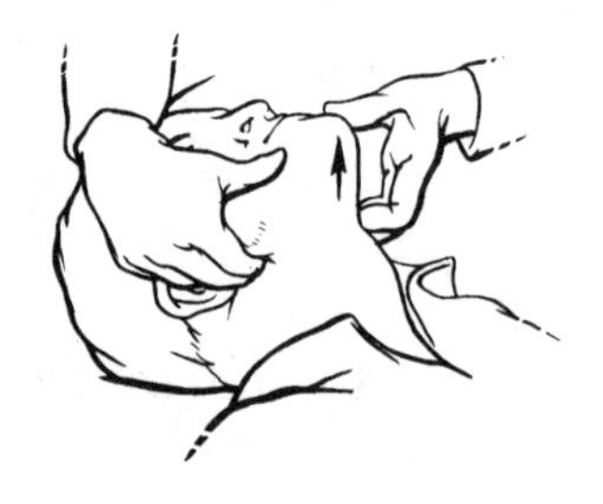

FIGURE 4 Modified jaw thrust maneuver.

2. If the victim is unconscious, open the airway by using the modified jaw thrust maneuver (Fig. 4) (see Unconscious Person, below).
   (i) Kneel beside the victim's head.
   (ii) Place the index and long fingers of your hands at the angles of the jaw (below each ear lobe). After lifting upward on the jaw, open the victim's mouth by pushing with your thumbs on his or her chin.
3. Look, listen, and feel for breathing.
4. If the victim is not breathing, start mouth-to-mouth ventilation; if there is no pulse, start CPR.

Impaled Object

1. Do not remove any object that is impaled or imbedded in a person.
2. Keep the victim still.
3. If there is external bleeding resulting from the impaled object, control the bleeding by applying direct pressure on the person's skin *around* the object. NEVER PRESS ON THE IMPALED OBJECT.

POISONING

If you suspect a person has swallowed a potentially dangerous substance:

1. Call the nearest Poison Control Center whose staff will give you instructions on what to do.
2. Always have syrup of ipecac on hand because the Poison Control Center may advise you to give it to the victim. Syrup of ipecac can be purchased at any drugstore.
3. If the person is unconscious and starts to vomit, turn him or her onto the side and clear the vomitus out of the mouth and throat.
4. If the victim becomes unconscious:
   (i) Open the airway by tilting the head back.
   (ii) Check for breathing.
   (iii) If the victim is not breathing, start mouth-to-mouth ventilation (see Unconscious Person, below).
5. Do not throw away anything that might be associated with the poisonous substance (e.g., bottles, pills, vomitus, leaves). Give them to the rescue squad when they arrive.

SEIZURES

A seizure can result from epilepsy, head injuries, high fever, poisoning, or many other problems. During a seizure, the victim must be protected from injury:

1. If the victim's mouth can be opened easily, place something (e.g., a spoon wrapped with a handkerchief) between his or her teeth. Do not force the mouth open.
2. Loosen any tight clothing.
3. Move surrounding objects (e.g., furniture) away from the victim or place yourself between the objects and the victim.

The victim may stop breathing during the seizure but will usually resume breathing when the seizure activity stops. If the victim does not resume breathing after the seizure:

1. Open the airway by tilting the head back.
2. Check for breathing.
3. If the victim is not breathing, start mouth-to-mouth ventilation (see Unconscious Person, below).

UNCONSCIOUS PERSON

Unconsciousness can be the result of many different illnesses or injuries. Regardless of the cause, you should initiate the following specific treatments:

1. Establish unresponsiveness by shaking the patient's shoulder and shouting at him or her. If the person does not respond, have someone call the rescue squad. Check for breathing and pulse.
2. Kneel beside the victim. Before moving him or her, check for possible spinal injury (see Injuries, above).
3. Position the victim for CPR. Place one of your hands beneath the victim's neck and the other on the forehead. Open the victim's airway by tilting the head back (Fig. 5). This maneuver will lift the

FIGURE 5 Open the victim's airway.

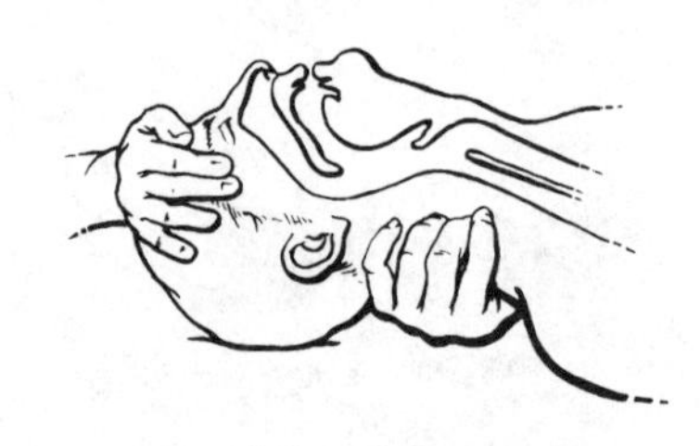

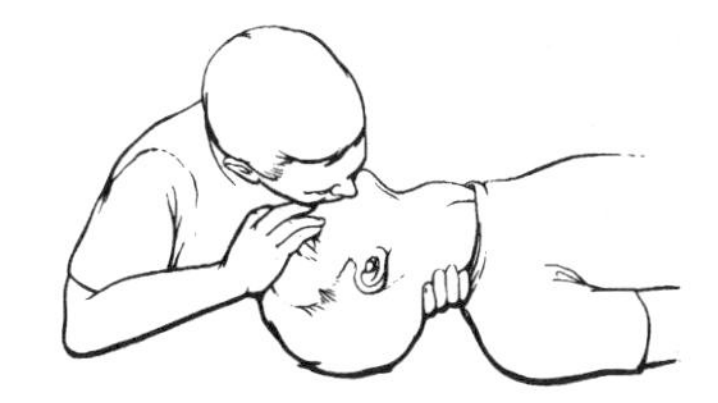

FIGURE 6 Mouth-to-mouth ventilation.

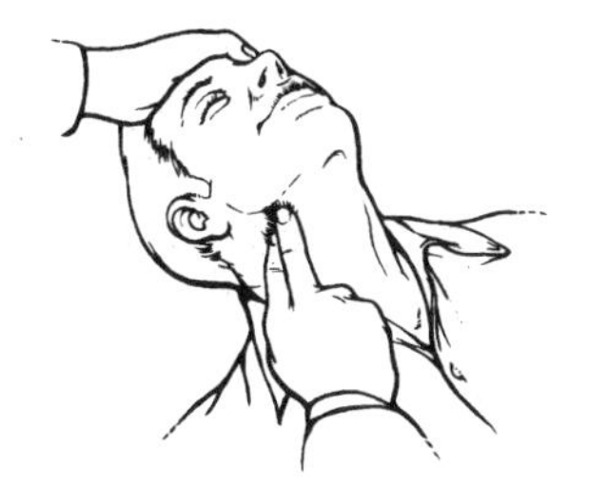

FIGURE 7 Check for pulse.

victim's tongue (the most common cause of airway obstruction in the unconscious person) off the back of the throat. If this does not open the airway, use the modified jaw thrust maneuver (see Injuries, above).

4. Check for breathing by placing your ear as close as possible to the victim's mouth. Look for movement of the victim's chest. Feel for movement of air against your cheek. Listen for breath sounds. If breathing is detected, keep the person's head tilted back until the rescue squad arrives.

5. If the victim is not breathing, start mouth-to-mouth ventilation (Fig. 6). While pinching the nostrils, cover the mouth completely with your mouth. Exhale completely into the victim's mouth until the chest moves. This should be repeated four times. After each ventilation, take your mouth off the victim's mouth to allow exhalation.

6. Check for pulse by placing two or three fingers (not your thumb) gently on the person's neck, to one side of the Adam's apple (Fig. 7).

7. If the pulse is present, continue mouth-to-mouth ventilation at the rate of one breath every 5 s until the rescue squad arrives.

8. If the pulse is absent, external cardiac compression should be started by a person properly trained in the technique. Mouth-to-mouth ventilation and external cardiac compression (known as CPR) should be continued until the rescue squad arrives.

If an infant or small child is unconscious, do not tilt his or her head as far back as you would for an adult. If you need to administer mouth-to-mouth ventilation, small puffs of air blown into both the mouth and nose should be sufficient to make the chest rise.

Index